Advances in Sport, Leisure and Ergonomics

This important volume brings together research by leading international ergonomists and sport and exercise scientists, as presented at the 4th International Conference on Sport, Leisure and Ergonomics. The book presents a wide range of studies in occupational ergonomics, each utilizing techniques that are also employed by sport and exercise science research groups. It therefore breaks new ground in the interface between sport and industry.

Arranged into sections examining environment, special populations, human factors interface, sports technology and occupational health, this book will be an essential purchase for all those involved in sports science or ergonomics research.

Thomas Reilly is Director of the Research Institute for Sport and Exercise Sciences at Liverpool John Moores University, UK.

Julie Greeves is Senior Scientist at QinetiQ, Farnborough, UK.

Advances in Sport, Leisure and Ergonomics

Edited by Thomas Reilly and Julie Greeves

London and New York

First published 2002
by Routledge
11 New Fetter Lane, London EC4P 4EE

Simultaneously published in the USA and Canada
by Routledge
29 West 35th Street, New York, NY 10001

Routledge is an imprint of the Taylor & Francis Group

All chapters except chapter 5 © 2002 Taylor and Francis;
chapter 5 © US Government

Typeset in Times by
Elite Typesetting Techniques
Printed and bound in Great Britain by
MPG Books Ltd, Bodmin

All rights reserved. No part of this book may be reprinted or
reproduced or utilised in any form or by any electronic,
mechanical, or other means, now known or hereafter
invented, including photocopying and recording, or in any
information storage or retrieval system, without permission in
writing from the publishers.

British Library Cataloguing in Publication Data
A catalogue record for this book is available
from the British Library

Library of Congress Cataloging in Publication Data
Advances in sport, leisure and ergonomics/
edited by Thomas Reilly and Julie Greeves.
p. cm.
"Proceedings of the 4th International Conference on Sport, Leisure
and Ergonomics, held at Burton Manor College, Wirral in November
1999"–Pref.
Includes bibliographical references and index.
1. Sports–Physiological aspects–Congresses. 2. Human factors–
Congresses. I. Reilly, Thomas, 1941 - II. Greeves, Julie. III.
International Conference on Sport, Leisure and Ergonomics
(4th : 1999 : Wirral, England).

RC1235.A384 2002
612'.044–dc21
2001048951

ISBN 0-415-27125-8

Contents

Preface ix
Acknowledgements xi

PART I
Introduction to Ergonomics: Sport and Leisure

1 Sports ergonomics: An introduction 3
  T. REILLY

PART II
Exercise and the Environment

2 Exercise and the cold 13
  T. D. NOAKES

3 Pacing strategies during a cycling time trial with simulated head winds and tailwinds 33
  G. ATKINSON AND A. BRUNSKILL

4 The effects of phase control materials on hand skin temperature within gloves of soccer goalkeepers 45
  A. J. PURVIS AND N. T. CABLE

5 Aerobic performance of Special Operations Forces personnel after a prolonged submarine deployment 55
  D. M. FOTHERGILL AND J. R. SIMS

6 Use of melatonin in recovery from jet-lag following an eastward flight across 10 time-zones 67
  B. J. EDWARDS, G. ATKINSON, J. WATERHOUSE, T. REILLY,
  R. GODFREY AND R. BUDGETT

| 7 | Do subjective symptoms predict our perception of jet-lag?
J. WATERHOUSE, B. EDWARDS, A. NEVILL, G. ATKINSON, T. REILLY, P. DAVIES AND R. GODFREY | 81 |

PART III
Ergonomics in Competitive Sports

8	Effect of carbohydrate supplementation on simulated exercise of rugby league referees D. P. M. MACLAREN AND G. CLOSE	97
9	Determining the protective function of sports footwear M. J. LAKE	107
10	How the free limbs are used by elite high jumpers in generating vertical velocity A. LEES, J. ROJAS, M. CEPERO, V. SOTO AND M. GUTIERREZ	119
11	Kinematic adjustments in the basketball jump shot against an opponent F. J. ROJAS, M. CEPRO, A OÑA AND M. GUTIERREZ	135

PART IV
Ergonomics in Special Populations

12	Decreased submaximal oxygen uptake during short duration oral contraceptive use: a randomized cross-over trial in premenopausal women M. GIACOMONI AND G. FALGAIRETTE	147
13	Promoting children's physical activity in primary school: an intervention study using playground markings G. STRATTON	159
14	Exercise testing in children: an alternative approach D. B. CLAXTON, J. H. CHAPMAN, N. V. CHALLIS AND M. L. FYSH	169
15	Walking in visually handicapped children and its cost V. BUNC, J. SEGETOVA AND L. SAFARIKOVA	179
16	Touch-down and take-off characteristics of the long-jump performance of world level above- and below-knee amputee athletes L. NOLAN AND A. LEES	187

17 Dynamic control and conventional strength ratios of the
    quadriceps and hamstrings in subjects with anterior cruciate
    ligament deficiency   201
    C. D. HOLE, G. H. SMITH, J. HAMMOND, A. KUMAR, J. SAXTON AND
    T. COCHRANE

PART V
Human Factors and Psychology

18 Performance and human factors: considerations about cognition
    and attention for self-paced and externally-paced events   211
    R. N. SINGER

19 Effects of cricket-ball colour and illuminance levels on catching
    behaviour in professional cricketers   231
    K. SCOTT, D. KINGSBURY, S. BENNETT, K. DAVIDS AND M. LANGLEY

20 Individual differences, exercise and leisure activity in predicting
    affective well-being in young adults   239
    C. SALE, A. GUPPY AND M. EL-SAYED

21 Transfer and motor skill learning in association football   249
    C. WEIGELT, A. M. WILLIAMS, T. WINGROVE AND M. A. SCOTT

PART VI
Methodological Studies in Sport and Ergonomics

22 Electromyography in sports and occupational settings: an update
    of its limits and possibilities   261
    J. P. CLARYS

23 The relationship between heart rate and oxygen uptake during
    non-steady state exercise   275
    S. D. M. BOT AND A. P. HOLLANDER

24 Modelling handgrip strength in the presence of confounding
    variables: results from the Allied Dunbar National Fitness Survey 291
    A. M. NEVILL AND R. L. HOLDER

25 Prediction and validation of fat-free mass in the lower limbs of
    young adult male Rugby Union players using dual-energy X-ray
    absorptiometry as the criterion measure   303
    W. BELL, D. M. COBNER AND W. D. EVANS

26 Variations of anatomical elements contributing to subtalar joint stability: intrinsic risk factors for post-traumatic lateral instability of the ankle? 313
E. BARBAIX, P. VAN ROY AND J. P. CLARYS

27 Inter-articular kinematics of the normal glenohumeral joint in the late preparatory phase of throwing: Kaltenborn's rule revisited 321
J. P. BAEYENS, P. VAN ROY AND J. P. CLARYS

PART VII
Ergonomics and Health in the Workplace

28 Developing a holistic understanding of workplace health: the case of bank workers 335
L. DUGDILL

29 Effects of activity-rest schedules on physiological strain and spinal load in hospital-based porters 347
C. BEYNON, J. BURKE, D. DORAN AND A. NEVILL

30 Implications of an adjustable bed height during standard nursing tasks on spinal motion, perceived exertion and muscular activity 355
D. E. CABOOR, M. O. VERLINDEN, E. ZINZEN, P. VAN ROY, M. P. VAN RIEL AND J. P. CLARYS

31 Work-based musculoskeletal problems: initiative to improve health 365
D. LEIGHTON

32 Will the use of different prevalence rates influence the development of a primary prevention programme for low-back problems? 373
E. ZINZEN, D. CABOOR, M. VERLINDEN, E. CATRYSSE, W. DUQUET, P. VAN ROY AND J. P. CLARYS

Preface

The content of this volume represents the refereed proceedings of the 4th International Conference on Sport, Leisure and Ergonomics, held at Burton Manor, Wirral in November 1999. The event is held every 4 years in affiliation with the Ergonomics Society: co-sponsors on this occasion included the World Commission of Science and Sports, the International Society for Advancement of Kinanthropometry, and the British Association of Sport and Exercise Sciences. The involvement of these bodies is reflected in the range of topics addressed within this volume.

The material published here was peer reviewed prior to its initial publication in a special issue of Ergonomics (October, 2000). For this book, the content has been reorganised into six main parts, each forming a coherent group of related research reports. These have been preceded by a new introductory chapter which sets the scene for building bridges between conventional ergonomics on one hand and the sport and exercise sciences on the other.

The environment in which sports training and competitions are held may represent both a physiological stress and a training stimulus. Following the introductory chapter, the next part of this book is focussed on environmental factors, especially inclement weather and time-zone transitions. In some instances these concerns may be examined under simulated conditions; in others, field circumstances may be the more appropriate setting for investigation. The studies of jet-lag in athletes reflect the global nature of contemporary sports, rapid travel across multiple time-zones being a common feature of the itinerary of athletes.

Competitive sports participants, especially at the professional level, are now more than ever backed up by sports science support systems. The major input of such personnel is typically a detailed analysis of the individual and team's performance, presented as feedback to the athlete or to squad management. The interpretation of performance, especially when disrupted, is reliant on a reference database and a theoretical framework. The improvement of performance may necessitate research into interventions (e.g. nutritional ergogenic aids), simulations (such as biomechanical modelling of performance) or protective equipment. Examples of ergogenic interventions are included in Part 3.

Some of the difficulties in ergonomics and sports science relate to the inferences that can be made from discrete studies and the influence of individual differences on the results. In sport especially, the child is not simply an adult in miniature but requires specific study and understanding. Similarly disabled athletes, whether chronically or temporarily affected, represent a heterogenous population exemplified by both the multiple events organised for them and the growing status of the Paralympic Games. The need to study effects specific to age, gender and sport is acknowledged increasingly within the sports sciences. Part 4 therefore is composed of reports addressing ergonomic issues in special populations.

A hallmark of ergonomics is its interdisciplinary nature in which the human sciences—notably applied anatomy, biomechanics, physiology and psychology—complement each other. The current content incorporates a range of research methodologies from laboratory-based experimental work to qualitative methods for use in occupational contexts. In some instances special insights are obtained by employing a specific disciplinary approach, examples of which are biomechanics and anthropometric studies in this volume. Discrete chapters are devoted to topics that fall predominantly within the field of psychology (Part 5) and the studies that address methodological questions (Part 6).

At the previous conferences, the topics researched invariably constituted problems within sport and leisure but ergonomics approaches were employed in reaching solutions. For the fourth conference the organisers encouraged discrete studies in occupational ergonomics which utilise techniques that were regularly employed also by sports and exercise science research groups. Besides, research policy outlines from occupational perspectives are included, for example as viewed by the Health and Safety Executive, in order to inform those human scientists who are interested in cross-fertilisation of research between sport and industry. For these reasons the present volume should be of relevance to research workers and practitioners not only in human factors and ergonomics, the sport and exercise sciences but also to many others across the human sciences.

Thomas Reilly, Liverpool John Moores University, UK
Julie Greeves, QinetiQ, Farnborough, UK.

Acknowledgements

All papers in this volume, with the exception of the Preface and Introduction, were first published in *Ergonomics*, Volume 43, Number 10 (2000), ISSN 0014-0139 print/ISSN 1366-5847 online, http://www.tandf.co.uk/journals. Reproduced with permission.

Part I
Introduction to Ergonomics: Sport and Leisure

1 Sports Ergonomics: An Introduction

T. Reilly

Background

Ergonomics refers to the application of human sciences to the individual whether the context is activity at work, at home or at play. The subject area is interdisciplinary, drawing on a knowledge base and methods from specialisms that include applied anatomy, biomechanics, engineering, physiology and psychology. The focus is on the individual or 'human operator' and how he or she harmonises with the task, the equipment being used, and the environment in which the activity takes place.

Sport and leisure were viewed traditionally as entirely separate from occupational 'work' in the domain of people's lives. The advent of mass participation in sport, exercise and recreational activities over the last couple of decades has necessitated a reappraisal of human factors in these areas. Over the period also there has been a growth in professional sport which is now readily accepted as a form of mass entertainment. Parallel with these developments has been an increasingly systematic approach towards analysing the stresses of sport on its participants and the preparation of athletes for competitive engagements. Furthermore, innovative engineering technologies have led to changes in design of sports equipment with the aim of enhancing performance.

Conventional ergonomics interventions were framed around relieving stress, improving safety, increasing comfort, avoiding fatigue, enhancing efficiency and so on. These concepts are relevant both to competitive sport and to occupational work, although with different emphasis. The industrial ergonomist, for example, seeks to reduce energy requirements and the task may be redesigned to achieve this end without loss of external output, whereas in sport the notion that energy expenditure per unit task output should be minimised is accepted but the main aim is to maximise output even if there are energetic consequences. Besides, stress is acknowledged as a corollary of competing in sport so the objective is to control rather than reduce stress. These differences are likely to be due to the competitiveness intrinsic in sport.

There is therefore a constant drive to gain a competitive edge over sports opponents in order to secure victory in major competitions. At the lower levels of participation, the objectives may be participation for personal enjoyment, social reasons, or health-related benefits. Irrespective of the aspirations of participants, safety is a fundamental criterion. Risk of injury varies among sports as it does in occupational contexts but in both domains the employment of preventative strategies is a feature of an ergonomics perspective.

Ergonomics and the sport and exercise sciences emerged from different backgrounds, now having their own professional bodies and scientific periodicals. They share many common concepts, concerns and research methods, a fact which has led to the forging of some links between the disciplines. The scope of sports science and its overlap with the field of ergonomics has been outlined some years ago (Reilly, 1984). The overlapping methodologies included physiological and psychological techniques for monitoring work stress, motion analysis, computer modelling and simulations whilst areas of mutual concerns included equipment design, footwear and clothing, and systems performance. In both domains particular strengths are the insights and flexibility offered by adopting an interdisciplinary approach to tackling the particular problem which is presented.

Fitting the task to the person
A fundamental principle in ergonomics is that human characteristics are taken into account when designing the task and the work-place. In practice, tasks often have to be redesigned in order for individual workers to cope (feature known as regenerative ergonomics). The designer has to take into account the range of human characteristics (e.g. anthropometry) and capabilities (e.g. information processing, power output and so on) in specifying the layout. The principle of 'fitting the task to the human' has been a hallmark of classical ergonomics (Grandjean, 1969). If the individual is not properly fitted for the task, the performance is likely to be sub-optimal and the ensuing strain may cause breakdown or injury.

The tenet of 'fitting the task to the person' is an acknowledgement that there are limits to human capabilities. The athlete continually endeavours to raise the ceiling of his or her physiological capacity by means of hard training in order to cope best with a fixed task. Monitoring of individual capabilities is achieved by means of regular physical fitness testing. The purpose may be to identify not only individual strengths but also weaknesses which can then be remedied in training; the effectiveness of the training intervention can be monitored at the next occasion on which testing is conducted.

The desired match between task demands and individual capabilities has implications for choice of activity, training and selection as well as fitness assessment (see Figure 1). The harmony must be extended from individual and activity to the equipment being used whether this is a machine such as bicycle, rowing shell or Formula 1 racing car, or equipment such as ski boots and skis. For this reason the traditional measures of physiological limits (maximal oxygen uptake, maximal voluntary contraction and so on) need to be re-examined and replaced where appropriate by performance measures on sports-specific ergometers.

There is a range of laboratory-based protocols available for assessing the fitness of athletes. Protocols are outlined in various manuals for purposes of standardisation and comparison between laboratories (Eston and Reilly, 1996). As far as possible the ergometry employed in testing is compatible with the mode of exercise in the sport concerned. Consequently ergometers have been designed not just for cycling and running but for kayak specialists, rowers, swim-benches for swimmers, skiing simulators for skiers and sailboard simulation for 'wind-surfers' (Reilly et al., 1993). Where possible, field tests are employed for convenience as much as for sports specificity, in hockey (Reilly and Borrie, 1992) and football (Reilly and Doran, 2001) for example.

The prediction of performance from physiological responses to laboratory-based assessments is more straightforward in individual sports such as running, cycling or rowing than in team games where performance is much more complex. Success in games is dependent on a host of factors both personal and circumstantial, not least

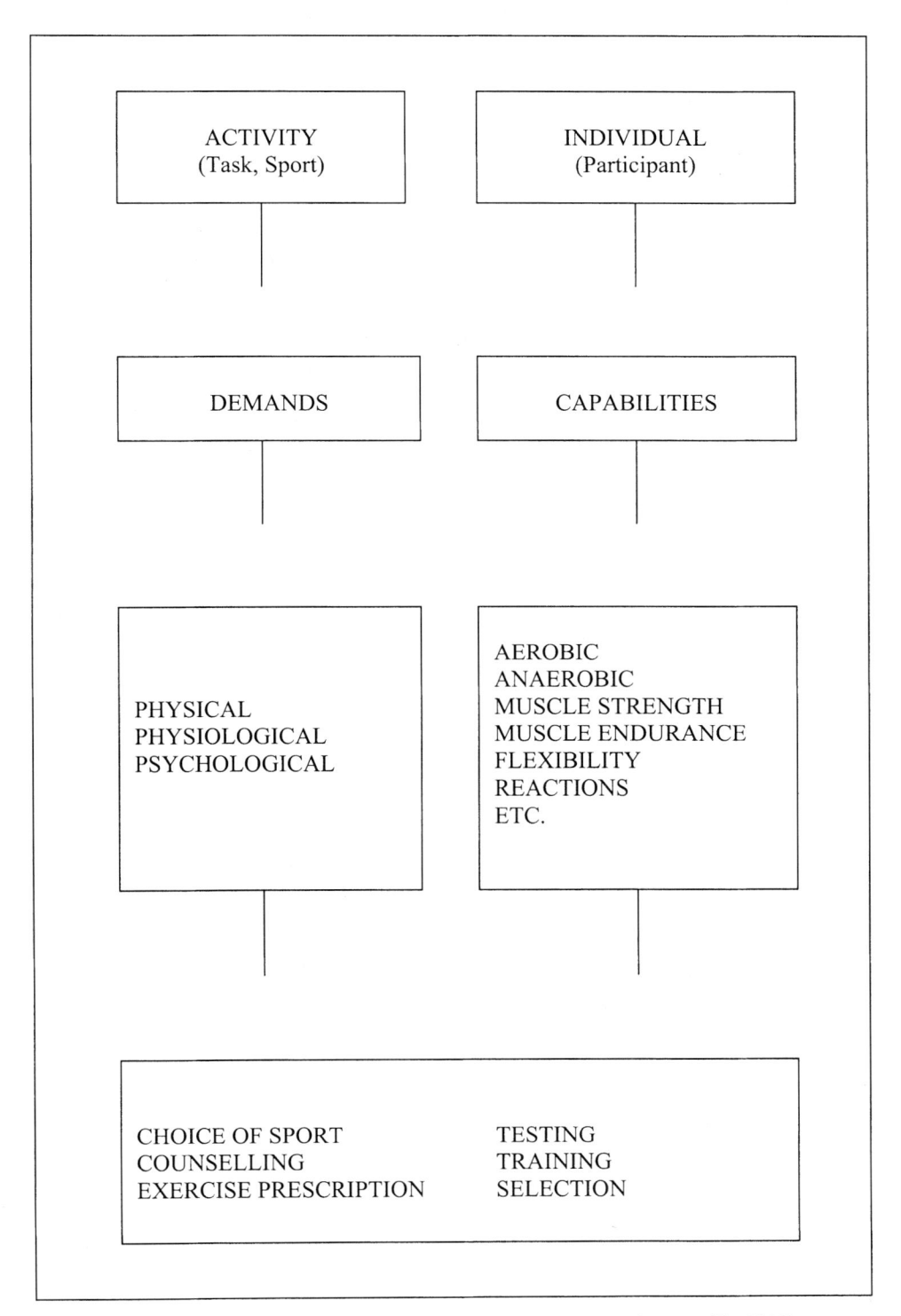

```
┌─────────────────────────┬─────────────────────────┐
│       ACTIVITY          │      INDIVIDUAL         │
│     (Task, Sport)       │     (Participant)       │
└───────────┬─────────────┴────────────┬────────────┘
            │                          │
       ┌────┴────┐              ┌──────┴──────┐
       │ DEMANDS │              │ CAPABILITIES│
       └────┬────┘              └──────┬──────┘
            │                          │
   ┌────────┴────────┐        ┌────────┴─────────┐
   │                 │        │ AEROBIC          │
   │                 │        │ ANAEROBIC        │
   │  PHYSICAL       │        │ MUSCLE STRENGTH  │
   │  PHYSIOLOGICAL  │        │ MUSCLE ENDURANCE │
   │  PSYCHOLOGICAL  │        │ FLEXIBILITY      │
   │                 │        │ REACTIONS        │
   │                 │        │ ETC.             │
   └────────┬────────┘        └────────┬─────────┘
            │                          │
   ┌────────┴──────────────────────────┴─────────┐
   │  CHOICE OF SPORT        TESTING             │
   │  COUNSELLING            TRAINING            │
   │  EXERCISE PRESCRIPTION  SELECTION           │
   └─────────────────────────────────────────────┘
```

Figure 1. An ergonomics model of sports participation (from Reilly 1991).

of which is the coherence of the team as a whole and the availability of good coaching expertise. In such cases team selection may be based on how the configuration of players best fit the team's overall strategy, balanced with decisions about what playing style best suits the team as a collective unit. In such instances the overall framework is a combination of 'fitting the task to the individual' and 'fitting the individual to the task' by virtue of training, selection and tactical choices.

The challenge of identifying talented players for specialist development programmes is one that is of contemporary concern in football games in particular. The process of talent detection and identification has until recently been based on a relatively flimsy scientific foundation. Williams and Reilly (2000) focusing on the world's most popular game, presented a comprehensive account of the array of factors impinging on performance in association football. The potential predictors of performance for the various sports science disciplines are demonstrated in Figure 2.

Demands of activity

There is a variety of ways in which the physiological strain associated with occupational work and physical activity can be assessed. In classical ergonomics the indices of work severity included body temperature, oxygen consumption, energy expenditure and heart rate. Some attention was given to the nature of the activity, whether intermittent or continuous, the duration of the activity, muscle groups engaged as indicated by electromyography and the rest periods interspersed between the bouts of activity.

The heart rate has been used as an indication of the overall physiological strain of activity. Its use as a predictor of energy expenditure has been criticised due to the influence of factors which influence heart rate but not metabolism. These include emotional factors, heat stress and muscle groups involved. Nevertheless the averaging out of heart rates, monitored nowadays by short-range radio telemetry,

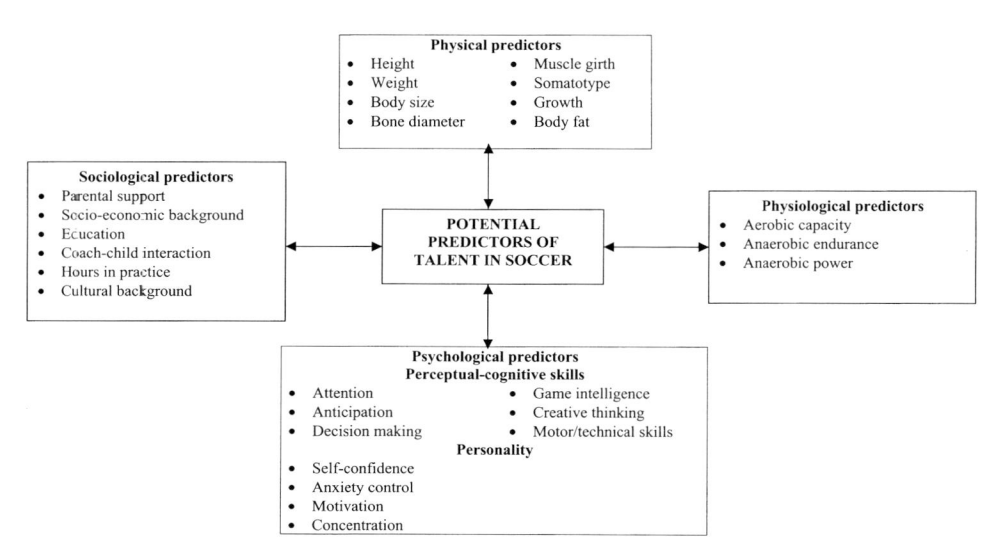

Figure 2. Potential predictors of talent in soccer from each sports science discipline (adapted from Williams and Franks, 1998).

provides a reasonable indication of the metabolic load, even in many non-steady state conditions.

Technological developments have helped to increase the range of methods available for monitoring energy expenditure. The use of doubly-labelled water (Westerterp and Saris, 1991) has improved the accuracy of measuring energy expended in sports event such as the Tour de France. It has not been utilised in occupational contexts where there is a habitually high outlay of physical exertion. In industrial circumstances a concern has been the periodic high efforts such as lifting and manual handling. In these cases the posture of the worker is of particular concern. Three-dimensional film analysis combined with electromyography has been a useful investigative tool for ergonomists in assessing the real demands of such activities.

In team games the analysis of overall performance is complex and lends itself towards computer-aided study. Hughes (1988) described how notation analysis can be used to identify patterns of play. Such systems have now been extended to incorporate observations on individual players' movements – which are quantified in terms of acceleration, speed and distance – and parameters of possession. The information gathered may be vast and require filtering before being used as feedback for coaching and playing staff.

Training
Training programmes embrace the concept of fitting the human to the task. Ideally training regimes are designed with specific goals in mind and can be evaluated in terms of reaching specific physiological targets. In practice, training goals are multifaceted and programmes vary in a progressive fashion throughout the annual cycle of competition. Furthermore, performance itself is mostly multivariate and requires an interplay of different physiological systems for optimal fine-tuning.

Circuit-weight training, for example, can be employed for aerobic purposes as well as for effects on the musculoskeletal system. Plyometric training focuses on stretch-shortening cycles of muscle action as employed in bounding, hopping or drop-jumping regimes. A corollary of this form of training is referred to as 'delayed onset muscle soreness' in which microdamage is caused to intramuscular structures. A prerequisite to plyometric training is a foundation of strength training. During the days in which 'delayed onset muscle soreness' is experienced, exposure to deep-water running can help maintain fitness and alleviate muscle soreness (Dowzer and Reilly, 1998).

Sports events such as triathlon call upon a combination of skills and specific fitness measures. In such cases special attention is needed with respect to how the training components complement each other rather than cause interference (Godfrey, 1998). Cross-training programmes may have particular benefits for health-related exercise, helping by virtue of their variety to maintain motivation and reduce the drop-out rate normally associated with a population of new exercisers.

The down-side of physical training is when adaptation does not accrue and breakdown occurs. Over-training refers to the condition in which under-performance is experienced despite continued or even increased training. A vicious cycle of more training produces lower performance and chronic fatigue. The phenomenon represents the classical ergonomics model of task demands outstripping human capability to cope with the training load.

Environment

Sports contests are often held in hostile climatic conditions, whether in heat, cold or at altitude. Furthermore there are many recreational activities in which participants are exposed to risk of accidents due to the environment, whether these activities are hill-walking, caving or pot-holing, breath-hold or scuba-diving and so on. Crossing multiple time-zones thereby disrupting the circadian body clock is a common experience, sometimes to locations that are environmentally polluted.

The environmental temperature in which athletes have to compete or train is rarely conducive to optimal performance. Often the participant is endangered in extremes of heat by a risk of hyperthermia. Climatic factors to consider are ambient temperature, humidity, radiant heat, air velocity and precipitation (rain, sleet and so on). These factors are incorporated into an index to indicate overall environmental heat stress as follows:

$$WBGT = 0.7 \, WBT + 0.2 \, GBT + 0.1 \, DBT$$

Where WB represents wet bulb

G indicates globe

DB represents dry bulb

T indicates temperature

Athletes due to compete in hot conditions are helped by undergoing a period of acclimatisation. The main physiological mechanisms that adjust to exercise in the heat are summarised in Table 1. Acclimatisation may be achieved by spending time at a warm-weather training camp or by utilising an environmental chamber in which combinations of heat and humidity are simulated. Such facilities have been used to good effect in athletes training for the Olympic Games (Reilly et al., 1997). The physiological adaptations may also need to be accompanied by behavioural strategies to cope with the heat and choice of appropriate clothing.

Clothing and behaviour are especially important when individuals are exposed to cold conditions. The environmental risk is reflected in the wind-chill index (Reilly, 2000) from which survival time under different conditions can be calculated. Body heat is lost much more quickly in water than in air and many water sports entail a risk of hypothermia or drowning. In underwater recreations such as diving (breath-hold or with self-contained underwater breathing apparatus) the hazards are well known, yet participants readily accept the risks for the transient thrills of these activities.

Hypoxia is a feature of high altitude sports, this stress being extreme in the high Himalayan peaks. The unpredictable weather and difficult climbs accentuate the

Table 1. Main effects associated with acclimatisation to heat

- Blood volume increases
- Sweat rate increases
    Onset at a lower core temperature
    Greater distribution over body surface
- Sodium and chloride content of sweat and urine decrease
- Perception of effort is reduced
- Glycogen utilization is decreased

respiratory stress. Nevertheless the high mountains continue to attract elite climbers keen to pit their skills against the environmental elements. At moderate altitudes aerobic performance is known to be adversely affected, yet world championship events (e.g. the 1968 Oympic Games and the 1986 World Cup) have been held at altitudes over 2 km. Training at altitude in order to compete well at such events is essential. The physiological adjustments that occur have led to the use of training camps at altitude in order to prepare for performance at sea level. Various forms of 'altitude huts' have been designed in recent years for training and/or sleeping in order to derive physiological benefits. At their simplest they take the form of normobaric hypoxic tents or living rooms.

Travel is now regarded as a necessary part of the habitual activity of athletes. It may entail travelling overseas for purposes of competition or across one's own country for domestic contests. Travel fatigue associated with long journeys on rail or road is an entirely different phenomenon from the 'jet-lag' syndrome induced by crossing multiple time zones. The desynchronisation of the body's circadian rhythms caused by a mismatch between the body's internal signals and environmental cues is the root cause of jet-lag and is associated with adverse effects on performance (Waterhouse et al., 1997). Contemporary efforts have been devoted towards behavioural rather than pharmacological means of reducing the effects of jet-lag (Waterhouse et al., 2000). There is also a concern about susceptibility to deep vein thrombosis provoked by spending many hours inactive and in a cramped position on board the airplane.

The quality of environmental air is also a concern for the ergonomist. Much attention has been devoted towards reducing environmental pollutants that might affect both health and human performance adversely (Atkinson, 1997). Such considerations might apply to indoor as well as outdoor sports arena, and to passive inhalation of pollutants in social circumstances.

Enhancing health and performance
Exercise is a potent means both of contributing to well-being and supporting quality of life. Its effects are not limited to musculoskeletal or cardiovascular systems but extend to mental health and affective states. 'Sport for all' campaigns have emphasised the value of exercise for weight-control purposes, offsetting inactive lifestyles and contributing to skeletal health in ageing individuals. Exercise programmes also help to counteract the 'computer play-station' culture of youngsters and instil positive attitudes towards active lifestyles.

The level of exercise performance is influenced both by genetic and environmental factors. Nevertheless, the influence of systematic training and development programmes should not be underestimated (see Williams and Reilly, 2000). Sustainable improvements in performance may be secured by appropriate choice of footwear, sports equipment or implements to match the prevailing surface or environment. Furthermore, the utilisation of nutritional strategies can be of clear benefit to performers when the metabolic limitations to exercise are firmly established.

Neither exercise nor life-style factors such as diet can on their own provide an elixir for health and well-being. A balanced and holistic approach is advocated, whether the focus is on the individual worker or on the athlete. Yet for solutions to be effective they must be acceptable to the individual and harmonise with individual needs.

References

ATKINSON, G. 1997, Air pollution and exercise, *Sports Exercise and Injury*, **3**, 2–8.
DOWZER, C. N. and REILLY, T. 1998, Deep-water running, *Sports Exercise and Injury*, **4**, 56–61.
ESTON, R. G. and REILLY, T. 1996. *Kinanthropometry and Exercise Physiology Laboratory Manual: Test, procedures and data*. London: E. and F. N. Spon.
GODFREY, R. J. 1998, Cross-training, *Sports Exercise and Injury*, **4**, 50–55.
GRANDJEAN, E. 1969, *Fitting the Task to the Man*. London: Taylor and Francis.
HUGHES, M. 1988, Computerized notation analysis in field games, *Ergonomics*, **31**, 1585–1592.
REILLY, T. 1984, Ergonomics in sport: an overview, *Applied Ergonomics*, **15**, 243–244.
REILLY, T. 1991, Physical fitness – for whom and for what? In: P. Oja and R. Telama (eds.), *Sport for All*, pp. 81–88. Amsterdam: Elsevier.
REILLY, T. 2000, Temperature and performance: cold. In: M. Harries, G. McLatchie, C. Williams and J. King (eds.), *ABC of Sports Medicine*, pp. 72–75. London: BMJ Books.
REILLY, T. and BORRIE, A. 1992, Physiology applied to field hockey, *Sports Medicine*, **14**, 10–26.
REILLY, T. and DORAN, D. 2001, Science and Gaelic football: a review, *Journal of Sports Sciences*, **19**, 181–193.
REILLY, T., BRYMER, E and TOWNEND, M. S. 1993, An ergonomic evaluation of boardsailing harnesses. In: E. J. Lovesey (ed.), *Contemporary Ergonomics 1993*, pp. 445–450. London: Taylor and Francis.
REILLY, T., MAUGHAN, R. J., BUDGETT, R. and DAVIES, B. 1997, The acclimatisation of international athletes. In: S. A. Robertson (ed.), *Contemporary Ergonomics 1997*, pp. 136–140. London: Taylor and Francis.
WATERHOUSE, J., REILLY, T. and ATKINSON, G. 1997, Jet-lag. *The Lancet*, **350**, 1611–1616.
WATERHOUSE, J., EDWARDS, B., NEVILL, A., ATKINSON, G., REILLY, T., DAVIS, P. and GODFREY, R. 2000, Do subjective symptoms predict our perception of jet-lag? *Ergonomics*, **43**, 1514–1527.
WESTERTERP, K. R. and SARIS, W. H. M. 1991, Limits of energy turnover in relation to physical performance, achievement of energy balance on a daily basis, *Journal of Sports Sciences*, **9**, Special Issue, 1–15.
WILLIAMS, A. M. and FRANKS, A. 1998, Talent identification in soccer, *Sports Exercise and Injury*, **4**, 159–165.
WILLIAMS, A. M. and REILLY, T. 2000, Talent identification and development in soccer, *Journal of Sports Sciences*, **18**, 657–667.

Part II
Exercise and the Environment

2 Exercise and the cold

Timothy David Noakes*

Research Unit for Exercise Science and Sports Medicine of the Medical Research Council and the University of Cape Town, Sports Science Institute of South Africa, Boundary Road, Newlands, 7700, South Africa

Keywords: Hypothermia; Running; Swimming; Environment; Anthropometry.

The generation of heat by the human body has been likened to that of a furnace. In response to winter conditions or prolonged immersion in cold water, heat may be lost from the body more quickly than it is produced leading to hypothermia. Various factors, environmental and individual, predispose a person to hypothermia when walking on dry land or during cold water immersion. Retention of the insulating properties of the clothing worn is of crucial importance in protecting against cold injury both on land and in water. Anthropometric characteristics and behavioural and physiological responses also influence the probability of survival under these conditions. Practical recommendations for behaviour that will enhance survival during prolonged exposure to cold on land or to immersion in cold water are considered.

1. Introduction

The Norwegian explorer, Roald Amundsen, leader of the first team to reach the South Pole in 1911 wrote astutely that: 'the human body is a furnace' (Huntford 1985: 103). Accordingly, many of the thermoregulatory problems that occur during exercise, when the furnace burns even more intensely than normal, result from the inability to cool the furnace; that is, to lose the excess heat produced during exercise. Only when environmental conditions are particularly cold, for example (1) during winter conditions at latitudes above about 50° in either hemisphere, or (2) when cold is associated with windy and especially wet conditions, or (3) when the athlete exercises in cold water for prolonged periods, does the risk arise that the athlete will lose heat faster than he or she can produce it. Under these conditions the body may be unable to maintain its core body temperature, which may fall progressively leading to hypothermia (core temperature $<35°C$) and the risk of death from the exposure/exhaustion syndrome.

Accordingly, this review of exercise in the cold will consider (1) the physiological effects of exercising in dry cold conditions in which the risk of hypothermia is somewhat less than the risk of frost-bite and other medical conditions, and (2) the effects of exercising in cold and wet conditions as exemplified by (a) exercising on land in cold, wet and windy conditions in wet clothing, and (b) swimming for prolonged periods in cold ($<20°C$) water.

*e-mail: tdnoakes@sports.uct.ac.za

2. The physiological effects of exercising in cold, wet and windy conditions on land

2.1. *Classic reports of hypothermia during exercise on land*

Hypothermia has long been recognized as a most serious condition, often with fatal consequences, for mountain hikers, fell runners and high altitude mountain climbers as, for example, struck the Fischer and Hall expeditions on Mount Everest on 9 May 1996 (Krakauer 1997).

Sutton (1972: 951) was probably the first to consider hypothermia as the cause of death in two runners competing in a 'Go-As-You-Please' race to the summit of Mount Wellington in Tasmania in 1903. The race was held in a snowstorm with a strong wind blowing and the runners were dressed only in 'singlets and light knickers'. Soon the competitors in the race were 'lying over logs, on the ground, and under trees, too exhausted to continue'. Almost certainly the deceased runners froze to death before they could be rescued.

The typical sequence of events in these tragedies is exemplified by those which occurred in the 1964 Four Inns Walking Competition in Derbyshire as described in detail by Pugh (1966, 1967, 1971). The event was held over 2 days and started at 06:00 h on 14 March 1964 with 80, three-man teams leaving at 2-min intervals. The temperatures for that day were estimated at between 0 and 4°C with a minimum wind speed of 25 knots. By 13:15 h the first reports were received that some hikers were in trouble. By late afternoon five distressed hikers had been brought to safety; one subsequently died. The bodies of a further two hikers were found 2 days later only after a search was conducted by more than 800 persons.

Characteristics common to all three fatalities were the following.

> Their clothing was inadequate as none had waterproof outer garments such as oilskins. When wet their clothing would have provided essentially no insulation (Pugh 1966, 1971).
> Subjects were lean with little subcutaneous fat. When the insulating effect of clothing is lost, the subcutaneous fat and muscle becomes the sole remaining insulator.
> All subjects had collapsed 1.5 to 2 h after fatiguing and starting to walk more slowly. In contrast, those who finished the event were able to maintain high rates of energy expenditure (and hence high rates of heat production) for the duration of the event.

Hypothermia has also been encountered even in long-distance runners. This is surprising as the most attention is usually paid to the supposed dangers of dehydration and heat retention leading to heatstroke and hyperthermia during exercise, especially running, in dry conditions on land (Noakes 1995, 1998). Thus physiologists and physicians have been slow to appreciate that the opposite condition, hypothermia, can also develop even in runners. It was perhaps the growth in popularity of mass-participation marathons in the Northern hemisphere, especially Britain, that drew attention to the risk of hypothermia developing during long distance running events.

The first recently documented case of hypothermia in a marathon runner was that of Ledingham *et al.* (1982), who reported a rectal temperature of 34.4°C in a runner who collapsed in the Glasgow Marathon, which was run under dry but cold conditions (dry bulb temperature 12°C) with a strong wind of 16 to 40 km h^{-1}.

Subsequently, Maughan (1985) measured the rectal temperature of 59 runners completing the 1982 Aberdeen Marathon, run under more favourable weather conditions (dry bulb temperature 12°C; dry with humidity of 75% and a wind speed of about 26 km h^{-1}). Despite the relatively mild conditions of the race, including the absence of rain, four runners finished the race with rectal temperatures below the normal 37°C.

South African studies have shown that conditions even in the generally warmer Southern hemisphere can, on occasion, be sufficiently unfavourable to cause hypothermia in long-distance runners. A study of all the runners admitted to the medical tent at the end of the 1985 56 km Two Oceans Marathon, run under unusually cold conditions for the Southern Hemisphere (wet bulb temperature 19.8°C; rain and a wind of 30 km h^{-1}), showed that eight (28%) of the collapsed runners had rectal temperatures below 37°C (Sandell *et al.* 1988). Despite maintaining a high running speed in excess of 17 km h^{-1}, one very thin élite runner collapsed on the course and was taken to the medical tent, where his rectal temperature was found to be 35.0°C.

2.2. Factors predisposing to the development of hypothermia in persons exercising on land

Factors that either predispose to or alternatively protect humans from developing hypothermia during exercise on land include the following: the environmental conditions, in particular the presence or absence of cold, windy and most especially wet conditions; the type of clothing worn, in particular its ability to keep its wearer completely dry in wet, rainy and windy conditions; the person's body build; the intensity at which he or she exercises, as this determines the rate of body heat production; the nearness of buildings or geographical features that will allow the stranded hiker, runner or climber to shelter from both the wind and rain; the availability of dry clothing and food; and the ability to generate artificial heat, for example by using a gas stove or a warm ($\sim 40°C$) bath to re-heat persons who have developed more marked levels of hypothermia.

2.3. A physiological explanation for the development of hypothermia during exercise in cold, wet and windy conditions on land

The human body temperature, which is regulated in a healthy body at 36–37°C depending on the time of day, represents a balance between the rate of heat production by, and heat loss from, the body. Hence changes in the rates of either heat production or heat loss, or more commonly both, determine whether hypothermia is likely to develop and under what conditions.

When exercising, the body is an inefficient machine so that only 25% of the chemical energy that is used mainly in the exercising muscles is turned into mechanical work; the remaining 75% is lost as heat — the furnace to which Amundsen referred. The rate at which the body produces heat is a linear function of speed when running (at any velocity) and when walking at speeds of up to about 7 km h^{-1}. At higher walking speeds, energy production rises as an exponential function of walking speed.

As a result, continuing to exercise vigorously will protect against the development of hypothermia if it maintains a high rate of body heat production that equals or exceeds the rate of heat loss from the body. In contrast, once the hiker, runner or mountain climber fatigues and starts to walk or stops walking altogether, the rate of heat production falls dramatically. This alone predisposes to the development of hypothermia in a cold environment, as occurred tragically in the case of the Mount Wellington 'Go-As-You-Please' race (Sutton 1972) and the Four Inns Walking Competition (Pugh 1966, 1967, 1971).

Environmental conditions, in particular the environmental temperature, the wind speed and the water content of the air (humidity) determine the rate at which heat is lost from the body. Hence they determine the rate of body cooling at any given rate of heat production.

The normal average human skin temperature is 33°C. At any lower environmental temperature, heat will be lost from the skin to the environment as the body attempts to heat up the air in direct contact with the body. This process is known as conduction. The rate at which this heat will be lost by conduction from the body will, in turn, be determined by the magnitude of the temperature gradient — the steeper the gradient, the greater the heat loss — and the rapidity with which the cooler air in contact with the skin is replaced by colder air. Continual replacement of warmed air by cooler air causes loss of heat from the body by means of convection. Convective heat loss rises as an exponential function of the speed at which air courses across the body, in effect the prevailing wind speed.

Therefore wind of increasing speed dramatically increases the 'coldness' of any environmental (dry bulb) temperature, in effect reducing the effective temperature to which the body is exposed and thereby increasing the rate of heat loss from the body. This is known as the wind chill factor. Figure 1 is a plot of the effective temperature for different actual (still-air) dry bulb temperatures at different prevailing wind speeds. To calculate the effective air temperature it is important to remember that the athlete's speed of movement into the wind increases the effective wind speed by a

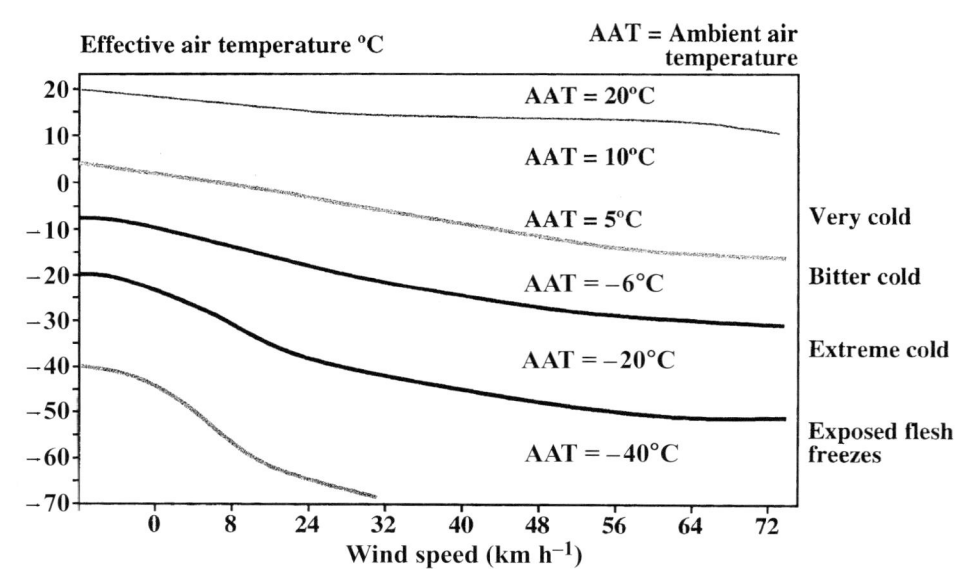

Figure 1. Effective air temperatures as a function of the (facing) wind speed at various ambient air temperatures (AATs). Note that frostbite occurs when naked skin is exposed to effective temperatures below −20°C. Redrawn from Noakes (1992: 106).

speed equal to the athlete's forward velocity, whereas running with the wind from behind reduces the effective wind speed by an equivalent amount.

Figure 1 also indicates the conditions in which frost-bite is likely to occur. Frost-bite occurs when the tissues, usually in the fingers and toes, freeze. As ice occupies a larger volume than water, the frozen tissues swell causing rupture and subsequent death of the cells in the tissues involved. Frost-bite becomes an increasing risk at effective air temperatures below $-20°C$. High mountains and the Polar regions notoriously combine very low temperatures and very high wind speeds, the combination that accounts for most deaths due to hypothermia in those locations. The warmer temperatures at lower altitudes cause the ice and snow to melt forming water. Water penetrates clothing more easily than does ice or snow, thereby reducing the usual insulation provided by clothing. It is this characteristic that increases the risk that hypothermia will develop even at the slightly warmer temperatures at lower altitudes.

The humidity of the air determines the extent to which heat can be lost to the environment in the form of evaporating sweat, the process known as evaporation. When body heat losses from routes other than sweating are great, as occurs during cold exposure, then the sweat rate is greatly reduced. As a result, the humidity of the air, which is likely to be very low in cold conditions, is not a factor influencing the risk that hypothermia will develop. Rather it is the presence of cold water, perhaps including sweat, directly in contact with the skin that becomes the most important factor predisposing to hypothermia during exercise on land when the environmental temperature is not as low as is found, for example, near the summits of high mountains or in the Polar regions.

Whereas air is a poor conductor of heat and hence a good *insulator*, water conducts heat approximately 25-fold faster than does air and hence is a very poor insulator (Tipton and Golden 1998). Thus the thin layer of air trapped next to the skin by clothing is rapidly heated to the skin temperature, thereby producing a layer of insulation, but the saturation of clothing with water removes this insulating layer and essentially exposes the skin to whatever is the external temperature. Under these conditions, the exposed human must either find dry, warm clothing and a warm shelter, or warm all the air in the entire planet to $33°C$, or die from hypothermia. There is no other alternative.

This loss of insulation caused by water explains why saturated, wet clothing or alternatively swimming in cold water predisposes to the development of hypothermia. This applies especially in windy, wet conditions such as experienced by those who died in the Four Inns Walking Competition disaster (Pugh 1966, 1967, 1971).

The ability of clothing to retain body heat, known as the insulating capacity of clothing, is expressed in clothing (CLO) units. One CLO unit is equivalent to the amount of insulation provided by ordinary business apparel, which provides comfort at temperatures of $21°C$ when both wind speed and humidity are low. The clothing of the Eskimo provides 10–12 CLO units and is essential for life in Polar conditions. The considerable heat production during exercise means that clothing that will provide 1 CLO unit of insulation is all that is required when running in temperatures as low as $-22°C$, provided that there is little or no wind or rain.

Wetness reduces the insulating properties of clothing as already described. For example, the insulating properties of the wet clothing found on those who died during the Four Inns Walking Competition was calculated to be only 0.2 CLO units

(Pugh 1966, 1967) or about one-tenth of the insulating capacity of that same clothing when dry.

Studies of long-distance swimmers, described in § 3, have shown that body build, especially the body fat and body muscle content, is a critical factor in determining the rate at which a swimmer will cool down during a long-distance swim. This is because both fat and inactive muscles receive a low blood flow and therefore act together as important insulators. Once the muscles become active, their blood flow increases turning this potential insulator into a very effective heat conductor (Tipton and Golden 1998).

Probably the same principle applies to exercisers exposed to cold, wet conditions on land; once their clothing becomes saturated and essentially loses all its insulating properties, those who have little body fat or skeletal muscle will probably be the most affected by the cold, and the most likely to become hypothermic. It was found, for example, that those who died in the Four Inns Walking Competition tragedy were both lean and relatively amuscular.

Once the climber or runner becomes too tired to continue exercising in cold, wet and windy conditions either because he or she is already developing hypothermia or because of the onset of exhaustion, the development of hypothermia will only be prevented if most of the following steps are undertaken. The hypothermic subject is taken expeditiously to buildings or geographical features that provide shelter from the wind and rain and where there is an external source of heating as well as a change of dry, water-repellent clothes, as well as a source of food. More clothing is required because the change from running or walking to the resting state has a marked effect on the rate of heat production and thus the amount of clothing needed to maintain

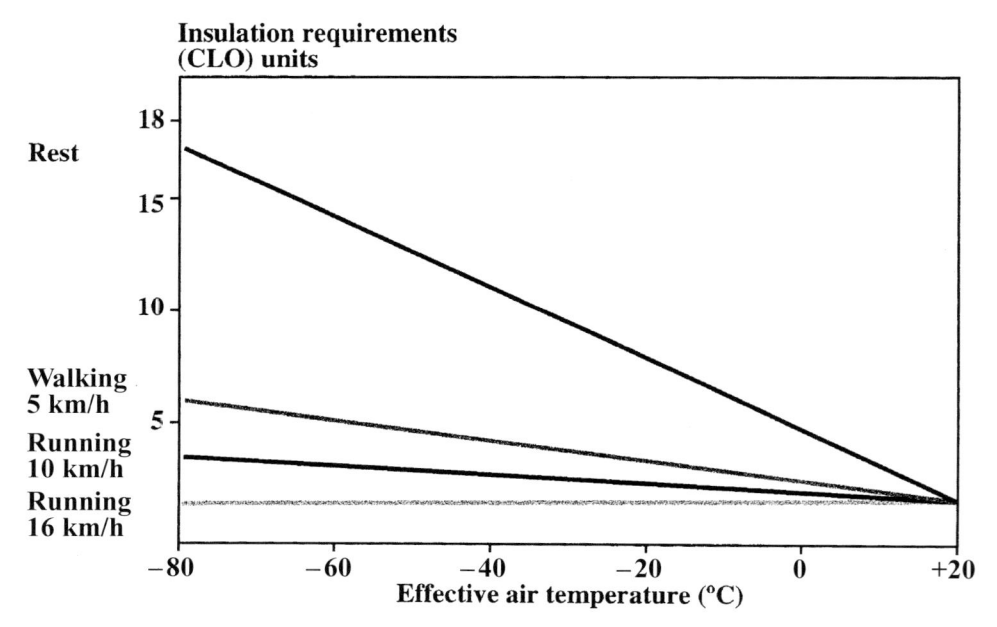

Figure 2. Effect of exercise on insulation requirements of clothing (measured in CLO units) at various effective air temperatures. Running at 16 km h^{-1} dramatically reduces the amount of clothing needed to maintain the normal body temperature. Redrawn from Noakes (1992: 106).

body temperature even at relatively mild effective temperatures. Thus clothing with at least four times as much insulation is required to maintain body temperature at rest at an effective air temperature of 0°C, as when running at 16 km h^{-1} at the same temperature (figure 2). For this reason it is essential that extra clothing including water-repellent outer clothing, both jacket and overtrousers, are always taken when exercising in cold and potentially wet and windy conditions. A problem is that clothing able to provide 4 CLO units of insulation may weigh as much as 10–15 kg (Pugh 1966).

Whilst it is usually assumed that a progressive whole-body hypothermia is the principal cause of this progressive inability to continue exercising in the cold, it also appears that cooling of the limb muscles may be an even more important factor. For example, as early as 1946, Adolph and Molnar (1946) reported that when working nude at 0°C, subjects became exhausted and confused within about 1 h even though they were exercising at a rate that they could easily sustain for 4 h in warmer conditions. Mean skin temperature fell to 15–17°C whereas rectal temperature did not fall below 36.9°C at exhaustion. Hence exhaustion was not due to whole-body hypothermia but resulted from the inability to continue exercising when the skin and presumably muscle temperature fell, perhaps causing associated pain and discomfort.

This progressive cooling of the active muscles may also make the exercise more difficult. During prolonged cold exposure, either on land (Pugh 1966, 1967) or in cold water (Tipton et al. 1999), there is a progressive increase in the energy cost of the activity; that is, the efficiency of movement is reduced. As a result, exercise of the same intensity becomes increasingly more difficult as it demands more energy, and this might also explain the earlier onset of fatigue when exercising in the cold.

3. Factors predisposing to the development of hypothermia in persons immersed in cold water

3.1. *Early studies of the physiological factors explaining long survival time in long-distance Channel swimmers*

The physiological challenges posed by exposure to cold water for prolonged periods have only been established relatively recently. For example as recently as 1960, it was not known: 'How are (English) Channel swimmers able to swim for 12–22 hours in water at a temperature of about 60°F (15.5°C) when after shipwreck persons survive in water at this temperature usually for only 4–6 hours?' (Pugh et al. 1960: 257). However, even Channel swimmers are not immune to the problems of swimming in cold water.

Pugh and Edholm (1955) reported the first study of the physiology of Channel swimmers. They noted that body temperatures were uniformly low in persons who completed the English Channel swim in 8 or more hours. Furthermore fewer than half the competitors completed the 1954 race studied by those workers; the assumption was that hypothermia explained these failures to complete the swim. They also reported that in 1953 one swimmer, J. Zirganos from Greece, had attempted to swim for 4 h in the Bosphorus at 8°C (46.4°F). After 3 h he was removed from the water, semi-conscious, in which condition he remained for a further 3 h. Zirganos and his trainer were 'convinced he had been poisoned' (Pugh and Edholm 1955: 767).

In 1955, Zirganos attempted to swim across the North Channel, dividing Northern Ireland from Scotland, a distance equivalent to that of the English

Channel. When he entered the water, the temperature was 8°C, substantially colder than the usual temperatures of 15–18°C experienced by himself and other swimmers in the English Channel, but as cold as the temperature in the Bosphorus where he had been 'poisoned' within 3 h. The tragic outcome of the 4-h swim in the icy water by Zirganos was described subsequently in chilling, if naive, detail.

> He went unconscious with a horrible gasp. The dusky blue that Zirganos had turned meant serious things to the doctor ... Quickly he asked for a pen-knife and started cutting Zirganos open. When the heart was exposed, he could see the rapid fluttering of ventricular fibrillation. Immediately the doctor began direct heart massaging in an attempt to re-establish the normal heart rhythm. Five minutes later the doctor realised that Zirganos had swum his last ... He was only three miles from the Scottish shore (Wennerberg 1974).

In fact, Zirganos's heart would have recovered and his life would have been saved had he been brought aboard the boat and treated with whole-body heating for the correct diagnosis of hypothermia. Perhaps the point of the story is to confirm the level of ignorance then present of the cause, diagnosis and management of hypothermia in Channel swimmers.

Zirganos had been a central figure in Pugh and Edholm's (1955) first study to establish why Channel swimmers could survive prolonged exposure in cold water. At the time it was assumed that the high rates of heat production achieved by the swimmers explained their superior cold tolerance. For example, the popular advice for shipwreck survivors at that time was the following:

> Fit men, who are in danger of immersion in cold water, might be advised to swim or struggle as hard as they can for as long as they can. If they try to preserve their strength by clinging to wreckage or floating on their lifebelts, they will die from cold. Perhaps more lives would have been saved in the past if this had been understood (Glaser 1950: 1068).

Accordingly Pugh and Edholm (1955) devised an experiment to evaluate this hypothesis. They compared changes in rectal temperature of an exercise physiologist, Griffiths Pugh, and of the fatter Channel swimmer, Zirganos, when they either lay in a bath or swam in cold water at the same temperature of 16°C.

The results depicted in figure 3, show that rectal temperatures fell in both subjects when they lay motionless in the bath, with the fall being somewhat greater in the thinner Pugh. When they swam in equally cold water, their responses were quite different. Whereas Zirganos's rectal temperature rose when he swam in cold water (figure 3), that of Pugh fell precipitously so that after 70 min he was unable to continue. His rectal temperature at that time was below 34°C.

Interestingly, Pugh was much more uncomfortable when he lay still in the water at 16°C for about 30 min. He experienced much more severe symptoms and greater pain than when swimming in cold water. This indicates that peripheral muscle cooling also contributes to discomfort and risk of fatality during cold water exposure. Despite a much higher rectal temperature (figure 3) after 33 min Pugh asked to be taken out of the water. He had to be lifted from the bath because muscular rigidity prevented any voluntary movements. Later Pugh and Edholm (1955: 764) wrote that: 'The experience was so unpleasant that he was unwilling to

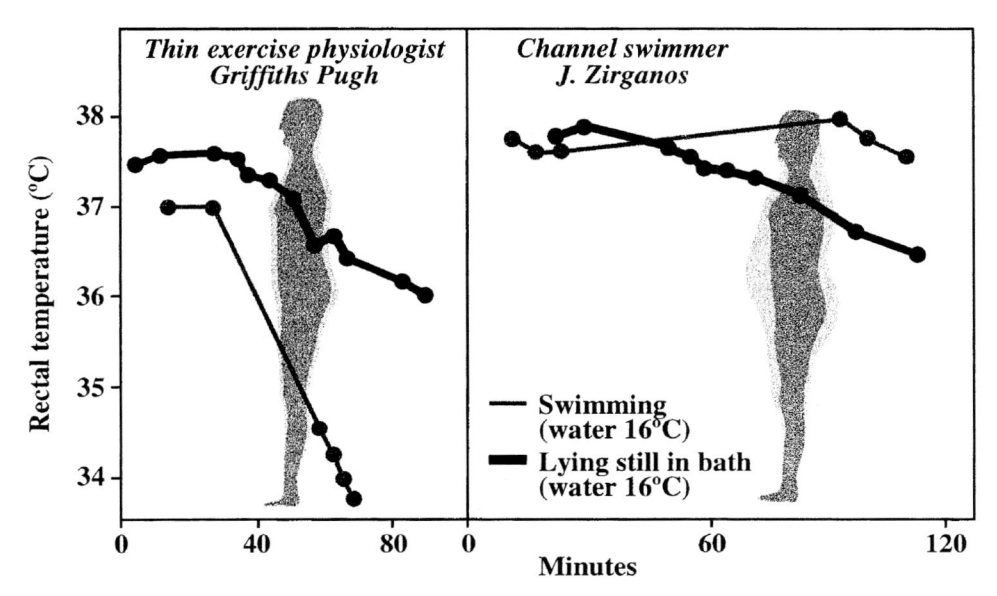

Figure 3. Changes in rectal temperatures in a thin exercise physiologist and a fatter experienced English Channel swimmer when lying or swimming in cold water at 16°C. Redrawn from Pugh and Edholm (1955).

repeat the experiment'. This finding is analogous to the original finding of Adolph and Molnar (1946) that work in extreme cold on land was terminated before the core body temperature had dropped to profoundly hypothermic levels.

Thus the conclusions from these studies and observations (Pugh and Edholm 1955, Pugh et al. 1960) were that:

> increased subcutaneous body fatness explained why Channel swimmers were better able to maintain heat balance when exposed to cold water;
> when immersed in cold water, fatter swimmers were better able to maintain their body temperatures when they increased their rates of heat production by swimming; and
> in contrast, when they swam in cold water, thinner non-swimmers were unable to maintain their body temperatures as their rates of cooling increased more than did their rates of body heat production. This effect is due to the loss of insulation induced by exercising the arms in persons with little insulation in their arms.

Accordingly Pugh and Edholm (1955: 767) recommended that, to survive in cold water, fatter individuals could either swim or stay motionless whereas a thin individual 'would almost certainly survive longer if he refrained from swimming or struggling'. They concluded that: 'The traditional naval advice to cling to wreckage and not to waste energy by swimming is probably correct' (Pugh and Edholm 1955: 767).

Their data do not necessarily support this conclusion. For Pugh himself became physically incapacitated sooner and at a higher rectal temperature when he lay still in the cold water than when he swam in water at the same temperature. Thus by staying still he risked drowning from physical incapacitation presumably due to cooling and paralysis of his limb muscles. In contrast, if he chose to swim he would ultimately lose consciousness and drown as a result of hypothermia. The key to survival would be to know which strategy would delay the onset of physical incapacitation by either mechanism, and hence increase the possibility of rescue.

This study raises the intriguing possibility that survival times in cold water could be determined by either of two physiological processes.

(1) The loss of consciousness due to a fall in the whole-body (rectal) temperature. This would be the ultimate fate of both thin and fat persons who chose to swim when exposed to cold water.
(2) Physical incapacitation due to a fall in the temperature of the limb muscles ('freezing' of the limbs). This would occur in persons who chose to lie still in the cold water and perhaps also in some swimmers in moderately-cold water (Tipton et al. 1999).

Ultimately the choice not to swim would cause the person to be unable to move the limbs sufficiently to maintain his or her head above water thereby also drowning. Presumably a thicker fat layer over the limbs would delay cooling and enhance survival under these conditions.

3.2. *Physiological responses during cold water immersion*

The physiological responses to immersion in cold water have been classified into four different categories: (1) initial responses to immersion (0–3 min); (2) short-term immersion (3–15 min); (3) long-term immersion (>30 min), and (4) post-immersion (Tipton 1989). The nature and extent of the physiological response and the duration of survival are determined principally by the water temperature.

3.2.1. *Physiological responses during acute immersion to freezing water—the cold-shock response*: Few will survive exposure to ice-cold water for more than a few minutes and may not be able to swim even 100 m to safety because of the unique physiological responses invoked by immersion in cold water. Factors determining survival under those conditions are different from those determining survival during exposure to water that is somewhat warmer. This is because of the cold-shock response that develops on acute exposure to freezing water (0°C) (Tipton 1989).

In brief, immersion in ice-cold water induces an immediate peripheral vasoconstriction, tachycardia and an increased cardiac output. As a result systolic blood pressure may rise to 180 mmHg and heart rate to 120 beats min^{-1}. Such changes do not pose a risk to healthy individuals. Rather it is the respiratory and musculoskeletal responses that pose the greatest risk to life and which explain why the majority (>60%) of fatalities in open water occur within 3 m of safety, even in persons considered to be good swimmers (Tipton 1989).

On immersion to ice-cold water, there is an immediate and uncontrollable hyperventilation. Ventilation volume may increase to 100 l min^{-1} and ventilation rate to 60 breaths min^{-1} (Keatinge et al. 1969). These changes induce four effects. First, there is a loss of the normal coupling of respiration and locomotor activity in which one or two breaths are taken for each cycle of muscular activity be it in swimming, running or rowing. Such high ventilation rates disturb this normal coupling, resulting in the likely inhalation of water and swim failure within a few minutes, even in competent swimmers.

Second, the duration of a voluntary breath-hold is reduced to a maximum of about 10 s (Tipton and Golden 1998). This increases the risk of sea water aspiration. This response also explains why surfing large waves in cold water is particularly treacherous as breath-holds of up to 30 s may be required to survive a wipeout under a large wave. Third, the hyperventilation produces a marked respiratory alkalosis which may cause a reduction in cerebral blood flow leading to disorientation and a clouding of consciousness. Fourth, the hyperventilation induces a profound feeling of dyspnoea and may initiate a panic response.

3.2.2. *Factors influencing the magnitude of the cold-shock response*: It appears that the magnitude of the cold-shock response is similar at all temperatures below 5–15°C. The magnitude of the response is also determined by the surface area exposed to the cold stimulus and appears to be maximum from the contact of cold water with the front of the trunk and the back of the chest. Hence any clothing reduces the cold-shock response. Exercise does not influence the cold-shock response but physical fitness and cold habituation reduce the magnitude of the cardiovascular response. Pre-heating the skin, for example by sitting in a sauna, reduces the ventilatory response on exposure to cold water.

Increased body fatness is of value only because the lower specific gravity allows fatter people to float higher in the water with less physical exertion, therefore reducing the risk of sea water inhalation (Keatinge *et al.* 1969). The value of a floatation device is that it lifts the face out of the water thereby reducing the risk of sea water aspiration.

3.2.3. *Factors influencing survival during prolonged cold water immersion at higher (10–20°C) water temperatures*: The cold-shock response explains why death from drowning can occur within 5–10 min of immersion in ice-cold water and well before whole-body hypothermia can develop. More prolonged survival is possible when the water is warmer (>10–15°C). Under these conditions the survival time in cold water is probably determined by the rate of either whole-body or limb muscle cooling, whichever occurs most rapidly. Until recently, the rate of whole-body cooling has been considered the more important so that most of the research has focused on factors determining the rate of whole-body cooling rather than limb muscle cooling.

There are relatively few studies that assess the influence of different factors on the rate of cooling of the skeletal muscles in the peripheral limbs, clearly an area of considerable interest.

3.2.4. *Water temperature*: As already emphasized, water is an excellent conductor of heat. As it is not possible for a lone swimmer to heat all the Oceans of the world to his or her skin temperature (33°C), it follows that anyone exposed to cold water must develop hypothermia in times that will depend on the water temperature (Molnar 1946, Tipton and Golden 1998). For example, figure 4 shows the changes in rectal temperature in the thin exercise physiologist, Griffiths Pugh, when he swam in water temperatures from 15 to 28°C. At no water temperature was Pugh able to maintain his rectal temperature when he swam. Furthermore, his rectal temperature fell steeply at all water temperatures below 24°C.

Figure 5 depicts likely survival times following immersion in cold water. It emphasizes that survival times in water temperatures below 5°C are very short, less than 30 min, mainly due to the deleterious physical effects of the cold-shock response.

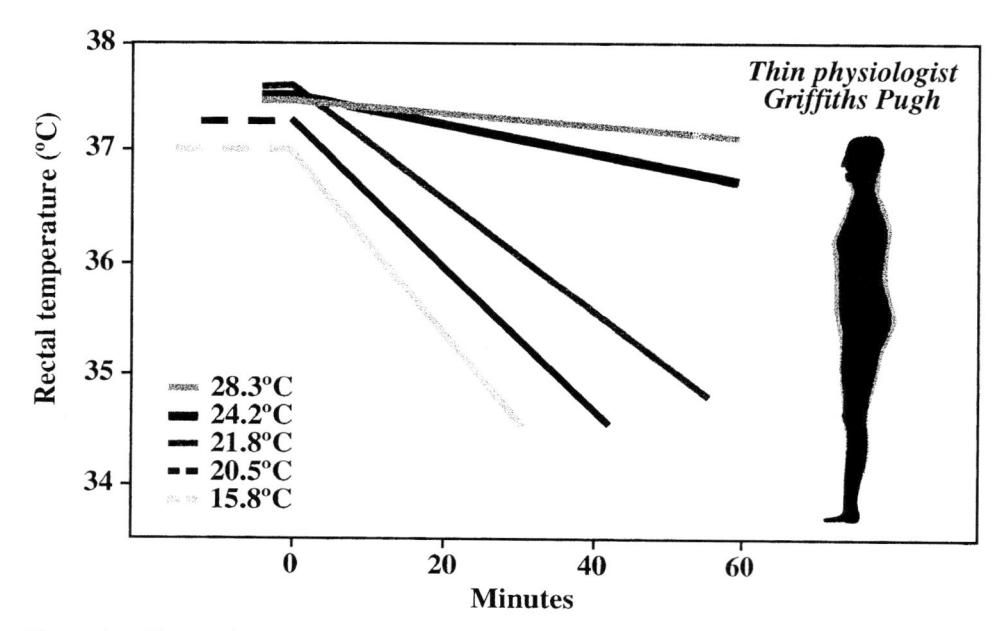

Figure 4. Changes in rectal temperature when the thin exercise physiologist swam in water at different temperatures. Note that the rate at which the rectal temperature fell increased sharply at all temperatures below 24°C. At no temperature below 28°C was the thin exercise physiologist able to maintain his body temperature when he swam. Redrawn from Pugh and Edholm (1955: 763).

In contrast, survival times in water temperatures above 15°C are relatively much more prolonged because (1) the subject is able to survive the cold-shock response, and (2) the rate of heat loss from the body and hence the rate of fall of the rectal temperature is inversely related to the water temperature (Pugh and Edholm 1955, figure 4).

3.2.5. *Clothing*: Clothing decreases the rate at which the body temperature falls during exposure to cold water (figure 5). It does this by maintaining higher skin and therefore higher body temperatures. For example in one experiment people working unclothed in water at 5°C for 20 min reduced their body temperature by 1.6°C, whereas when clothed the fall was only 0.6°C in the same time. In contrast if they remained inactive in the water when fully clothed, the fall in temperature was reduced to 0.3°C (Keatinge 1969).

The use of a dry- or wet-suit obviously reduces the rate of cooling the most effectively. Figure 5 shows that the use of a wet-suit increases survival time at 5°C about six-fold whereas a dry-suit in which air, not water, is in contact with the skin, increases survival time 10-fold at that water temperature.

3.2.6. *Body fat content*: There is a linear relationship between the extent of the fall in rectal temperature and the reciprocal of the mean skinfold thickness in subjects immersed in cold water at 15°C for 30 min (figure 6). Hence, as predicted from the study of Channel swimmers, increasing body fatness reduces the rate of body cooling

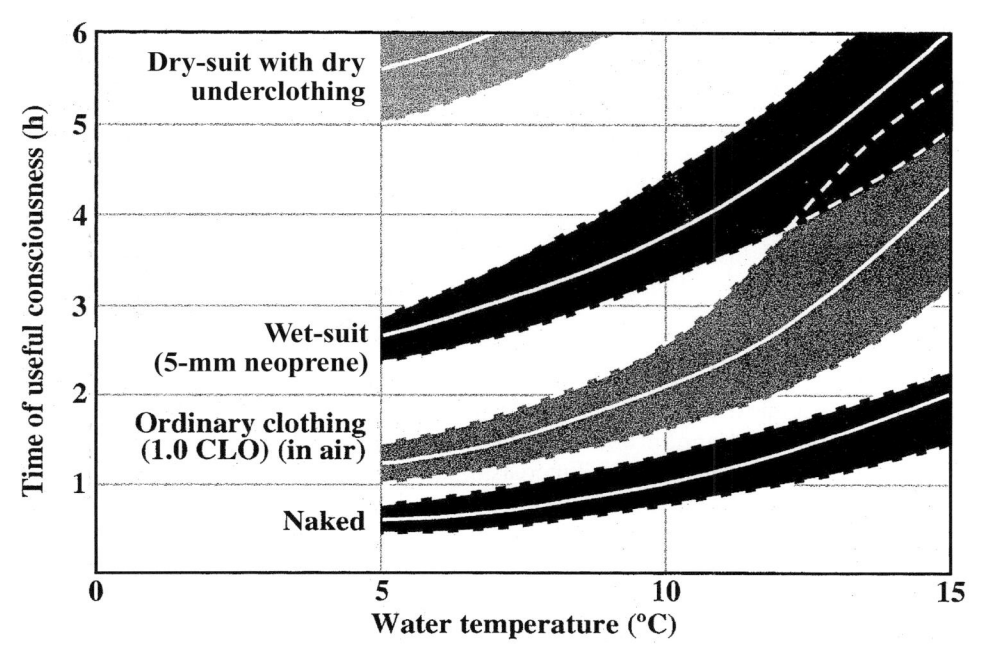

Figure 5. Survival times in water of different temperatures when wearing clothing of different insulation. Note that clothing influences survival times and that this effect becomes greater at warmer temperatures (>10°C). Survival times are greatly enhanced by the wearing of dry- or wet-suits. Redrawn from Tipton and Golden (1998: 249).

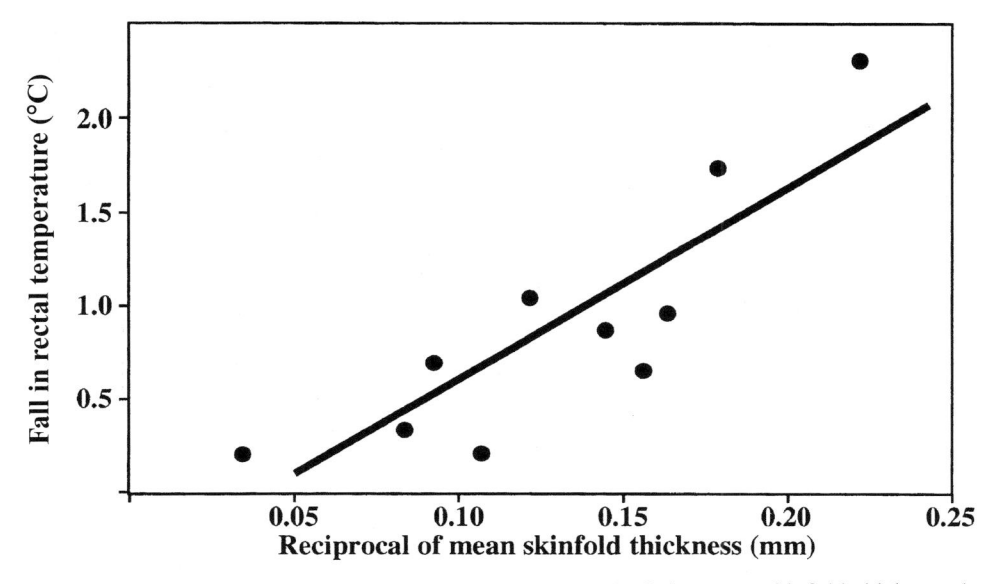

Figure 6. The linear relationship between the reciprocal of the mean skinfold thickness (a measure of body fatness) and the rate at which the rectal temperature falls on immersion in cold water. Redrawn from Keatinge (1969).

during immersion in cold water (Cannon and Keatinge 1960, Rennie *et al.* 1962, Holmer and Bergh 1974, Nadel *et al.* 1974, Dulac *et al.* 1987). Pugh *et al.* (1960) calculated that each 1-mm increase in the thickness of the subcutaneous fat layer effectively raises the water temperature by 1.5°C. They also noted that the Channel swimmers applied lanolin grease to a depth of 1 mm to their skins; this is equivalent to a 1 mm increase in the subcutaneous fat layer and an equivalent raising of the effective water temperature by 1.5°C.

3.2.7. *Metabolic rate*: Figure 7 shows that when exposed to cold water a thin person will begin to shiver, thereby increasing the metabolic rate at much higher water temperatures than does a fatter person. This gives rise to the concept of the critical water temperature for each individual, which is the water temperature below which, on prolonged immersion for up to 3 h, an individual is unable to maintain body temperature (Cannon and Keatinge 1960, Rennie *et al.* 1962, Holmer and Bergh 1974) without shivering and increasing the metabolic rate. Rennie *et al.* (1962) and Holmer and Bergh (1974) showed that the critical water temperature is, predictably, related to each individual's body fat content.

The benefit of shivering is that it increases the metabolic rate substantially. However, shivering also induces vasodilation in the arterioles supplying the shivering muscles. As a result muscle blood flow increases. The effect is that shivering reduces the insulating effect normally provided by the thickness and mass of the inactive muscles (figure 7). The extent of this effect is much larger than was originally recognized by those who assumed that the subcutaneous body fat layer was the sole insulator of the naked body.

Veicsteinas *et al.* (1982) and Rennie *et al.* (1980) have proposed that vasoconstricted (non-perfused) muscle acts in series with fat and skin to provide

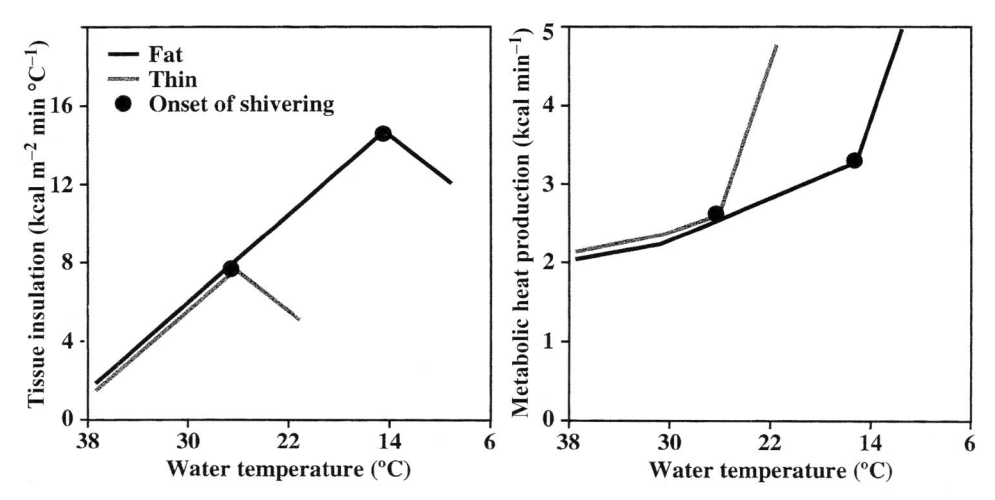

Figure 7. Comparison of the water temperature inducing shivering and a rise in metabolic rate in a thin and a fatter individual. Note that the thinner individual begins to shiver at a higher water temperature than does the fatter individual and that shivering reduces whole body insulation. As a result shivering in cold water simultaneously increases both the rates of heat loss and of heat production. Redrawn from Keatinge (1969).

insulation such that 75% of the body's insulation is provided by the vasoconstricted muscle and only 25% by the subcutaneous fat and skin. As a result, during exercise or shivering, the 'variable' resistance of muscle vasoconstriction is lost due to increased muscle perfusion leaving only the 'fixed' resistance of the subcutaneous fat and skin (Golden and Tipton 1987, Tipton and Golden 1998). As a result shivering increases both the rate of heat production and of heat loss and the effect on the body temperature will depend on which predominates. The insulative properties of muscle are lost at quite low exercise intensities; a whole-body oxygen consumption of 1 1 min^{-1} or about 4 METS (metabolic equivalents) of exercise (Tipton and Golden 1998).

Similar arguments apply to the effects of exercising as opposing to staying still during immersion in cold water. The potential benefit of exercising is that it increases the rate of heat production. However, it removes this 'variable' insulation of muscle and also prevents the formation of a thin layer of warmed water surrounding the body. Thus movement also increases the conduction of heat from the body by preventing the development of this thin layer of warmed water. As a result the rate of heat loss is increased during exercise in the water and the effect on the whole-body temperature will be determined by the water temperature, the person's body fatness, the intensity of the exercise and the nature of the activity undertaken.

3.2.8. *Nature of the activity undertaken*: An observation was made by the author when working with four thin triathletes preparing for the 1984 and 1985 London-to-Paris team triathlon competitions during which those thin athletes were required to swim the English Channel in relay. The original swim wet-suits under development at that time and which covered only the torso, leaving the arms and axilla exposed, were of little value in preventing the development of hypothermia in the leanest swimmers. In contrast, once the arms were completely encased by a new swimming wet-suit used in the 1985 race, even the thinnest triathletes could swim for extended periods ($\sim$30 min) in cold water (10–12°C) without distress and without clinical evidence for the development of either hypothermia or upper limb cooling. The physiological basis for this observation was already beginning to be understood at that time.

For example Toner *et al.* (1984) showed that the rate of cooling during immersion in cold water at 20, 26 or 33°C was always greatest for arm exercise compared to either leg exercise alone or arm and leg exercise at the identical metabolic rate. They concluded that heat loss was greater from the arms than from the legs when either muscle group was exercised at the same absolute metabolic rate.

Next, Golden and Tipton (1987) showed that when immersed in cold water at 18.5°C, subjects' body temperatures fell more rapidly when they lay still than when they did exclusively lower body (leg) exercise. This finding clearly conflicted with the original study of Pugh and Edholm (1955) (see figure 3) and others (Keatinge *et al.* 1969) in which whole-body exercise, that is, exercise which included the arms, was studied.

Hence Golden and Tipton (1987) concluded that the arms are a likely important source of heat loss during whole-body exercise in cold water, in keeping with the conclusions of Toner *et al.* (1984). They also noted that the majority of recognized swimming strokes require the greater use of the arms than the legs. The consequence is that the arms receive more heat, as a result of a greater blood flow, than the legs but are less able to retain it.

Accordingly they concluded that the type of exercise must be considered an important factor influencing heat balance during exercise in cold water. Exercising the legs while keeping the arms close to the body will conserve heat more effectively than will swimming at the same metabolic rate using mainly the arms. They concluded that: 'it is very difficult to give general advice concerning whether or not individuals should exercise following accidental immersion in cold water' (Golden and Tipton 1987: 404).

3.2.9. *Habituation*: Swimmers who train frequently in cold water develop a habituation in which they are able to maintain higher temperatures and swim for longer before they fatigue (Golden *et al*. 1980).

4. The role of limb cooling in determining survival during cold water immersion

The recent study of Tipton *et al*. (1999) has addressed the question of why some deaths following immersion in moderately cold water seem to occur too quickly for profound hypothermia to have developed. They wondered whether this resulted from either the uncontrollable hyperventilation that develops on exposure to cold water (Tipton 1989) which increased the probability that aspiration and drowning would occur, or whether it was due to a deterioration in physical capacity as described by Pugh himself when he lay in cold water and had to be removed, essentially paralysed, even though his whole body temperature was not particularly low (Pugh and Edholm 1955). This is analogous to the early finding of Adolph and Molnar (1946), already described, that subjects exercising in the nude at 0°C on land became confused and unable to exercise within 1 h when their rectal temperatures were approximately 37°C so that they were not hypothermic as traditionally defined.

Tipton *et al*. (1999) showed that when exposed to prolonged (90 min) swims in water at temperatures of 25, 18 and 10°C, heart rate, oxygen consumption and inspiratory volume were all increased significantly at the two colder water temperatures as also previously reported by Nadel *et al*. (1974) and Costill *et al*. (1967). As predicted, rectal temperatures were lower and rates of cooling were faster in the colder water. Tipton *et al*. (1999) argued that it was a progressive reduction in swimming ability rather than a critical whole-body hypothermia which predicted swimming failure. Hence in water at 10°C, there was a progressive reduction in swimming efficiency and in stroke length so that the tiring swimmer became progressively more vertical in the water, thereby increasing drag, impairing swimming efficiency and increasing the sinking force. Ultimately this would lead to swim failure.

Tipton *et al*. (1999) also noted that there was a relationship between the thickness of the subcutaneous fat layer over the skin and swimming efficiency in cold water supporting the hypothesis that (1) the arms are especially susceptible to cooling when swimming in cold water; (2) a greater subcutaneous fat layer reduces the rate of cooling of the arm muscles when swimming in cold water so that (3) arm muscle function is better maintained with the result that a greater stroke length and a more horizontal position in the water can be maintained for longer when swimming in cold water.

Accordingly they concluded that local (arm) muscle cooling rather than a progressive whole-body hypothermia is the main cause of drowning in cold water because the 'local muscle cooling impairs swimming performance and, consequently, the ability to keep the airway clear of the water' (Tipton *et al*. 1999: 629).

Furthermore they suggested that with the exception of children rescued following prolonged immersion in ice-cold water, the major focus for the treatment of survivors rescued following immersion in cold water should perhaps be the correction of symptoms due to near drowning rather than severe hypothermia.

For historical completeness it is interesting to record that as early as 1946 Molnar reported that in one shipwreck in water at 4°C, 'some of the survivors who held on to rope couldn't let go and rescuers had to cut frozen rope to release them ... Most of them who were rescued were in an unconscious state and, when they became conscious, complained of numbness of extremities and hands ... (Molnar 1946: 1047).

Thus local limb cooling has been described but only more recently recognized as a significant factor in cold water immersion.

5. Practical conclusions from these studies

In summary, hypothermia is less common during exercise on land than in water at the same temperature because still air is a good insulator whereas cold water is an excellent heat conductor. In addition, there is the opportunity to wear layers of dry clothing on land. Furthermore exercise reduces the clothing requirements and improves survival at any cold air temperature on land whereas the opposite applies in swimming as exercise reduces the insulating capacity, especially on the muscles of the upper limb, thereby increasing heat loss from the body. Thus the key for survival during exercise in extreme conditions on land (as in water) is to maintain a high metabolic rate and to stay dry. In water this is achieved by wearing either a wet-suit or a dry-suit, both of which should cover the arms.

When exposed accidentally to ice-cold water, the initial crisis is to survive the cold-shock response. It is also unlikely that most will be able to cover more than 50–100 m should they attempt to swim to land. Thus the prospect of survival is determined almost entirely by the speed with which dry land is reached. Any delay increases the probability that drowning by inhalation, secondary to physical incapacitation, will occur.

If the accidental exposure is to warmer water, the survival time may be substantially prolonged. The popular advice under these conditions remains that given in 1974; 'the choice between vigorous swimming and floating would depend upon the water temperature, the resistance to heat flow provided by clothing or fatty tissue and the length of time one could tolerate heavy exercise' (Nadel *et al.* 1974). Few victims of immersion will, however, be in a position to understand the subtle interaction of all these factors. Hence this advice is, at best, largely meaningless to all but a few experts in thermodynamics and exercise physiology.

Perhaps the only clear advice is that, when suddenly immersed in very cold water, relatively lean people wearing few clothes are better off if they restrict their body movements, most especially of their upper arms, whereas at temperatures above 25°C exercise will usually slow the rate of cooling in both thin and fat individuals (Tipton and Golden 1998) and thereby enhance the prospects of survival.

References

ADOLPH, E. F. and MOLNAR, G. W. 1946, Exchanges of heat and tolerances to cold in men exposed to outdoor weather, *American Journal of Physiology*, **146**, 507–537.

CANNON, P. and KEATINGE, W. R. 1960, The metabolic rate and heat loss of fat and thin men in heat balance in cold and warm water, *Journal of Physiology*, **154**, 329–344.

COSTILL, D. L., CAHILL, P. J. and EDDY, D. 1967, Metabolic responses to submaximal exercise in three water temperatures, *Journal of Applied Physiology*, **22**, 628–632.
DULAC, S., QUIRION, A., DECARUFEL, D., LEBLANC, J., JOBIN, M., CÔTE, J., BRISSON, G. R., LAVOIE, J. M. and DIAMOND, P. 1987, Metabolic and hormonal responses to long-distance swimming in cold water, *International Journal of Sports Medicine*, **8**, 352–356.
GLASER, E. M. 1950, Immersion and survival in cold water, *Nature* (London), **166**, 1068.
GOLDEN, F. S. T. C. and TIPTON, M. J. 1987, Human thermal responses during leg-only exercise in cold water, *Journal of Physiology*, **391**, 399–405.
GOLDEN, F. S. T. C., HAMPTON, I. F. G. and SMITH, D. J. 1980, Lean long distance swimmers, *Journal of the Royal Naval Medical Service*, **66**, 26–30.
HOLMÉR, I. and BERGH, U. L. F. 1974, Metabolic and thermal response to swimming in water at varying temperatures, *Journal of Applied Physiology*, **37**, 702–705.
HUNTFORD, R. 1985, *The Last Place on Earth* (London: Pan Books).
KEATINGE, W. R. 1969, *Survival in Cold Water* (Oxford: Blackwell Scientific).
KEATINGE, W. R., PRYS-ROBERTS, C., COOPER, K. E., HONOUR, A. J. and HAIGHT, J. 1969, Sudden failure of swimming in cold water, *British Medical Journal*, **1**, 480–483.
KRAKAUER, J. 1997, *Into Thin Air* (New York: Villard Books).
LEDINGHAM, I. M., MACVICAR, S., WATT, I. and WESTON, G. A. 1982, Early resuscitation after marathon collapse, *Lancet*, **2**, 1096–1097.
MAUGHAN, R. J. 1985, Thermoregulation in marathon competitors at low ambient temperature, *International Journal of Sports Medicine*, **6**, 15–19.
MOLNAR, G. W. 1946, Survival of hypothermia by men immersed in the ocean, *Journal of the American Medical Association*, **131**, 1046–1050.
NADEL, E. R., HOLMÉR, I., BERGH, U., ÅSTRAND, P.-O. and STOLWIJK, J. A. J. 1974, Energy exchanges of swimming man, *Journal of Applied Physiology*, **36**, 465–471.
NOAKES, T. D. 1992, *Lore of Running*, 3rd ed. (Cape Town: Oxford University Press).
NOAKES, T. D. 1995, Dehydration during exercise — what are the real dangers? *Clinical Journal of Sport Medicine*, **5**, 123–128.
NOAKES, T. D. 1998, Fluid and electrolyte disturbances in heat illness, *International Journal of Sports Medicine*, **19**, S146–S149.
PUGH, L. G. C. E. 1966, Accidental hypothermia in walkers, climbers, and campers: Report to the Medical Commission on Accident Prevention, *British Medical Journal*, **1**, 123–129.
PUGH, L. G. C. E. 1967, Cold stress and muscular exercise, with special reference to accidental hypothermia, *British Medical Journal*, **2**, 333–337.
PUGH, L. G. C. E. 1971, Deaths from exposure on Four Inns Walking Competition, March 14–15, 1964, in E. Jokl and J. T. McClellan (eds), *Exercise and Cardiac Death* (Baltimore, MD: University Park Press), 112–120.
PUGH, L. G. C. E. and EDHOLM, O. G. 1955, The physiology of Channel swimmers, *Lancet*, **2**, 761–767.
PUGH, L. G. C. E., EDHOLM, O. G., FOX, R. H., WOLFF, H. S., HERVEY, G. R., HAMMOND, W. H., TANNER, J. M. and WHITEHOUSE, R. H. 1960, A physiological study of channel swimming, *Clinical Science*, **19**, 257–273.
RENNIE, D. W., COVINO, B. G., HOWELL, B. J., SONG, S. H., KANG, B. S. and HONG, S. K. 1962, Physical insulation of Korean diving women, *Journal of Applied Physiology*, **17**, 961–966.
RENNIE, D. W., PARK, A., VEICSTEINAS, A. and PENDERGAST, D. 1980, Metabolic and circulatory adaptation to cold water stress, in P. Cerretelli and B. J. Whipp (eds), *Exercise Bioenergetics and Gas Exchange* (Amsterdam: Elsevier/North-Holland Biomedical Press), 315–322.
SANDELL, R. C., PASCOE, M. D. and NOAKES, T. D. 1988, Factors associated with collapse following ultramarathon footraces. A preliminary study, *Physician and Sportsmedicine*, **16**, 86–94.
SUTTON, J. R. 1972, Community jogging versus arduous racing, *New England Journal of Medicine*, **286**, 951.
TIPTON, M. J. 1989, The initial responses to cold-water immersion in man, *Clinical Science*, **77**, 581–588.

TIPTON, M. J. and GOLDEN, F. S. T. C. 1998, Immersion in cold water: effects on performance and safety, in M. Harries, C. Williams, W. D. Stanish and L. J. Micheli (eds), *Oxford Textbook of Sports Medicine*, 2nd ed. (Oxford: Oxford University Press), 241–254.

TIPTON, M., EGLIN, C., GENNSER, M. and GOLDEN, F. 1999, Immersion deaths and deterioration in swimming performance in cold water: a volunteer trial, *Lancet*, **354**, 626–629.

TONER, M. M., SAWKA, M. N. and PANDOLF, K. B. 1984, Thermal responses during arm and leg and combined arm-leg exercise in water, *Journal of Applied Physiology*, **56**, 1355–1360.

VEICSTEINAS, A., FERRETTI, G. and RENNIE, D. W. 1982, Superficial shell insulation in resting and exercising men in cold water, *Journal of Applied Physiology*, **52**, 1557–1564.

WENNERBERG, C. 1974, *Wind, Waves and Sunburn. The Book of Swimming and Swimmers* (Cranbury, NJ: A. S. Barnes).

3 Pacing strategies during a cycling time trial with simulated headwinds and tailwinds

GREG ATKINSON* and ADAM BRUNSKILL

Research Institute for Sport and Exercise Sciences, Liverpool John Moores University, Trueman Building, Webster Street, Liverpool L3 2ET, UK

Keywords: Cycling performance; Pacing strategy; Headwind; Tailwind; Time trial.

The aims of this study were to examine the effects of one self-selected and two enforced pacing strategies (constant and variable power output) on cycling performance during a time trial in which variable wind conditions were simulated. Seven male cyclists rode their own bicycles on a Computrainer cycle ergometer, which was programmed to simulate a 16.1 km time trial on a flat course with a 8.05 km h^{-1} headwind in the first half of the race and a 8.05 km h^{-1} tailwind in the second half of the race. Subjects rode an initial time trial (ITT) at a self-selected pace to the best of their ability. The mean power output from this trial was then used to calculate the pacing strategies in the subsequent two trials: Constant (C)—riders rode the whole time trial at this mean power output; and Variable (V)—riders rode the first headwind section at a power output 5% higher than the mean and then reduced the power output in the last 8.05 km so that the mean power output was the same as in the initial time trial and in trial C. Power output, heart rate and ratings of perceived exertion (RPE) were recorded every 1.61 km. Finish times, 8.05 km split times and blood lactate levels, pre- and post-exercise (to calculate Δ lactate), were also recorded in each trial. In the ITT, riders chose a mean ± SD power output of 267 ± 56 W in the first 1.61 km which was 14% higher than the overall race mean ± SD of 235 ± 41 W. Power outputs then dropped to below the race mean after the first few kilometres. Mean ± SD finish times in the C and V time trials were 1661 ± 130 and 1659 ± 135 s, respectively. These were significantly faster than the 1671 ± 131 s recorded in the initial time trial ($p = 0.009$), even though overall mean power outputs were similar (234 – 235 W) between all trials ($p = 0.26$). Overall mean RPE and Δ lactate were lowest in trial V ($p < 0.05$). Perceived exertion showed a pacing strategy by race split interaction ($p < 0.0001$), but it was not increased significantly during the first 8.05 km of the V condition when power outputs were 5% higher than in condition C. Heart rate showed no main effect of pacing strategy ($p = 0.80$) and the interaction between strategy and race split did not reach statistical significance ($p = 0.07$). These results suggest that in a 16.1 km time trial with equal 8.05 km headwind and tailwind sections, riders habitually set off too fast in the first few kilometres and will benefit (10 s improvement) from a constant pacing strategy and, to a slightly greater degree (12 s improvement), from a variable (5% ±mean) pacing strategy in line with the variations in wind direction during the race. Riders should choose a constant power when external conditions are constant, but when there are hilly or variable wind sections in the race, a variable power strategy should be planned. This strategy would be best monitored with 'power-measuring devices' rather than heart rate or subjective feelings as the sensitivity of these variables to small but meaningful changes in power during a race is low.

*Author for correspondence. e-mail: g.atkinson@livjm.ac.uk

1. Introduction

Pacing strategy can be defined as the within-race distribution of work-rate (power output). Despite the fact that pacing strategy can, to some degree, be predicted by 'relative split-times' in actual races (Atkinson and Edwards 1998), research on pacing strategy during cycling time trials is sparse (Foster *et al.* 1994). Foster *et al.* (1993) studied cyclists who each performed five, 2 km time trials in the laboratory with various imposed pacing strategies (very slow to very fast starts) over the first 1 km. It was found that the fastest overall times were produced when the most 'even' pace was adopted (i.e. first 1 km = 50.9% of total time). Although the results of Foster *et al.* (1994) show quite clearly that pacing strategy is important, it is difficult to extrapolate their findings to actual cycling events. The 'closest' event to that modelled by Foster *et al.* (1993) is the 1 km time trial held in a velodrome. Van Ingen-Schenau *et al.* (1994) maintained that, in reality, the time taken to move a stationary bicycle to maximum speed at the start of such an event means that an even pace would not be optimal. These authors showed that more time is spent in the first 200 m of the time trial, so this part of the race is relatively more important than the remaining 800 m. In agreement with this view, Wilberg and Pratt (1988) observed that the first lap (333.3 m) in elite 1 km races is completed up to 3 s slower than the other two laps. To examine pacing strategies over such short distances, researchers need to be sure that they have simulated the standing start accurately.

In the longer cycling time trial competitions that are held on highways, the contribution of the standing start to total time taken is much less important, but the examination of pacing strategy is still complicated by the variable environmental conditions within a race. It is known that in hilly races or in windy conditions, race times will always be worse than in races on flat courses with no wind for the same amount of work done and even with no net change in altitude or wind velocity (Kyle 1988, White 1994, Martin *et al.* 1998). Swain (1997) employed the equation of motion of a cyclist provided by Di Prampero *et al.* (1979) to show that, when a race is held on a hilly course or on a course with equal sections of headwind and tailwind, faster times would be recorded if the cyclist increased power on the uphill or headwind sections and compensated with reduced power on the downhill or tailwind sections (average power output for the race remaining the same). This pacing strategy is based on the fact that relatively more time is spent in the uphill/headwind sections of a race than in the downhill/tailwind sections, so a relatively faster velocity in the former situation is more important to overall time than in the latter situation.

In agreement with the predictions of Swain (1997), Atkinson and Edwards (1998) found that the best cyclists in British national time trial championships did spend less time (relative to their own finish times) in the headwind sections of races. Although this observation could have been due to the better riders increasing power in these sections, the influence of the elite riders having better aerodynamic positions could not be ruled out. Differences in aerodynamic qualities do not compromise the 'time saving' hypothesis of Swain (1997), but affect the interpretation of why a rider has spent relatively less time in the headwind section of the race, i.e. the relative increase in speed could be due to the rider being more aerodynamic, rather than being able to increase power. The influence of aerodynamics could be fully controlled by employing one of the recently developed laboratory-based cycle ergometers (e.g. the Computrainer) to simulate race courses (hills and/or winds) by varying external resistance above and below the calibrated value for cycling on flat ground with no wind.

The aim of this study was to employ the Computrainer ergometer (Computrainer Pro, Racer Mate Inc., Seattle, USA) to examine the effects of one self-selected and two enforced pacing strategies (constant and variable) on cycling performance in a time trial with simulated variable wind conditions. Every race is held in novel external conditions, therefore knowledge of speed, gear ratio or pedal cadence does not help in monitoring pace in a real race. Power output can be measured directly with bicycle-mounted strain gauges such as those employed in Powercranks (Shoberer Rad Meβtechnik, Königshamp, Germany) (Jeukendrup and Van Diemen 1998), but these devices are expensive. It was, therefore, deemed important to ascertain if heart rate or rating of perceived exertion (RPE) could be used by a rider to monitor subtle changes in power output within a race, since these variables can be easily measured and are more directly related to human power output than speed or cadence.

2. Methods

Seven male cyclists, ranging from recreational to professional status, volunteered (giving signed informed consent) to ride three time trials with their own bicycles mounted on an ergometer. The mean ± SD age and rider-plus-bicycle mass were 25.7 ± 4.6 years and 83.7 ± 4.2 kg, respectively. Subjects rode the time trials in 'skinshorts' and a t-shirt. The mean ± SD temperature and barometric pressure in the laboratory over the test period were 19.3 ± 1.6°C and 738 ± 7 mmHg, respectively.

After a prior familiarization session with the equipment and protocols, riders were asked to complete a 16.1 km initial time trial (ITT) on a Computrainer Pro ergometer in the fastest time possible. Such a time trial protocol has been found in a previous study to show low enough levels of measurement error for it to be used with a relatively low sample size, while retaining statistical power (Atkinson et al. 1999). The simulated course profile was flat, but included a simulated 8.05 km h^{-1} headwind for the first 8.05 km of the time trial, followed by an 8.05 km h^{-1} tailwind for the remaining 8.05 km. This protocol was designed to simulate a time trial on a typical 'out and home' course, with a turn at half-distance. An 8.05 km h^{-1} wind velocity is lower than the annual average for most monitoring stations in the UK (Meteorological Office 1999). The only feedback provided to the rider was distance elapsed, as would be the case in an actual time trial. As mentioned in Section 1, the usefulness of information on speed, gear ratio and pedal cadence would be low in any one time trial even if the course had been ridden before, since the environmental conditions are unique to every race.

The mean power output from this time trial was used to formulate the pacing strategies in the subsequent two trials, the order of which was randomized. In the constant (C) trial, subjects were instructed to ride the whole time trial at this mean power output. In the variable (V) trial, riders were told to ride the first headwind section at a power output 5% higher than the mean and then to reduce the power output in the last 8.05 km by about 5% so that the mean power output at the end of trial V was the same as in ITT and trial C. During these trials the riders followed the Computrainer pacer, which was programmed with the calculated pacing strategy. No verbal motivation was given to the riders. The ergometer was calibrated according to the manufacturer's instructions prior to each time trial. Tyres were inflated to 6.9 bar. The ergometer and subject were 'warmed up' for 5 min at 100 W. The rolling resistance was calibrated at 0.907 ± 0.023 kg. This value is higher than that found to be typical of actual cycling (Di Prampero et al. 1979, Olds et al. 1993) and is to allow the Computrainer to decrease as well as increase external resistance in order to simulate a range of tailwinds/downhills and headwinds/uphills. It is stressed that this value of rolling resistance was controlled both within- and between-subjects in the experiment.

During each experimental session, subjects were first weighed while holding their bicycle on a balance scale (Avery, Smethwick, UK). A finger-prick blood sample was then taken. The following variables were recorded every 1.61 km during the time trial: ratings of perceived exertion (RPE) from the 6–20 Borg scale (Borg 1970); heart rate from a photo-reflectance, earlobe clip interfaced with the Computrainer system (Burke *et al.* 1993); and power output averaged over the previous 1.61 km of the time trial. Subjects rested for 5 min after the time trial and another finger-prick blood sample was taken. Blood samples were analysed for lactate concentration using a GM7 Multi-assay Analyser (Analox, London, UK). The difference in the pre- and post-time trial lactates was calculated and referred to as Δ lactate. Blood lactate and oxygen consumption were not measured during the trials, since this would interfere with the riders' performance (Jeukendrup *et al.* 1997). It was especially important that the riders performed to their best ability in ITT, since this governed the power profiles in the subsequent trials. Blood lactate and oxygen consumption cannot be measured during a real time trial and therefore cannot be used as indicators of pace.

All analyses were performed using Statistica for Windows (StatSoft Inc., Tulsa, OK). Power output, heart rate and RPE data were analysed using two-factor (3 × 10) analysis of variance (ANOVA) models. The two within-subjects (repeated measures) factors were pacing strategy (ITT, C and V) and time trial split (1–10). The other dependent variables (finish time, time for the first 8.05 km, time for the second 8.05 km, Δ lactate) were analysed with a one-factor (pacing strategy) repeated measures ANOVA. The assumption of sphericity was tested according to the methods of Vincent (1995). All analyses met this assumption (Huynn-Feldt epsilon >0.75). A significant main effect of pacing strategy was followed up with Newman-Keuls multiple comparisons. Significant main effects of race split and interactions between pacing strategy and race split in the two-factor ANOVAs were examined by viewing the standard interaction plots.

3. Results

In the initial time trial (ITT) riders started the race with a mean $\pm$ SD power output of 267 ± 56 W in the first 1.61 km, which was 14% higher than the overall race mean $\pm$ SD of 235 ± 41 W (figure 1). Power outputs then dropped by up to 7% below the race mean until the last 2 km when the subjects increased power output as the finish approached. The profile of power output for the constant (C) condition is also shown in figure 1. As expected, power outputs were relatively constant about the mean, although riders did find it difficult to maintain this mean power output in the last 2 km. The increase in power output during the first 8.05 km of the variable time trial (V) is also shown in figure 1. When the tailwind section was reached riders should have initially reduced power output by 5%. It can be seen from figure 1 that power outputs were actually lowered by 2–3% and then by 8–10% in the last few kilometres in order to ensure that overall mean power output did not differ between the trials (table 1). Despite these slight departures from the desired pacing strategies, the differences between conditions in within-race changes in power output were significant, as indicated by the interaction term of the ANOVA ($F(18,108) = 5.20$, $p < 0.0001$).

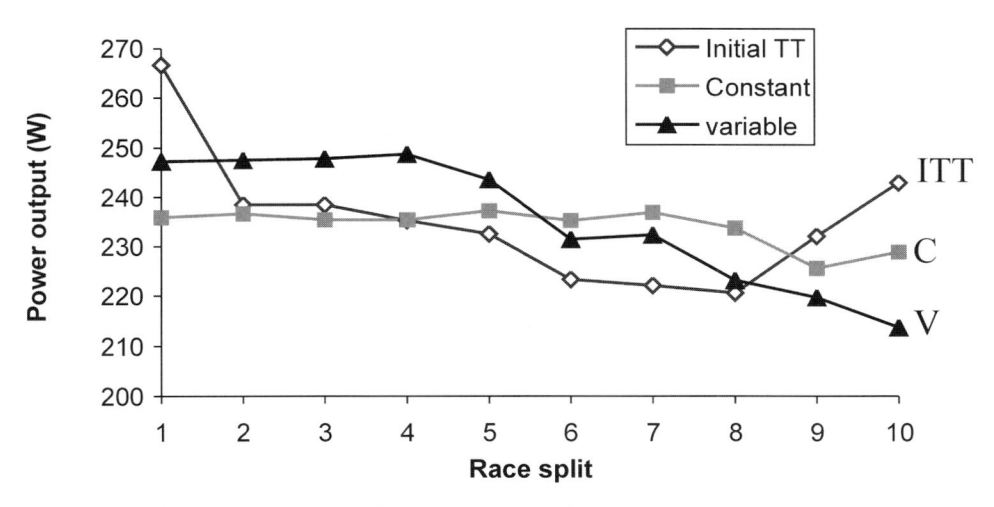

Figure 1. The within-race profile of power output for the three time trials (ITT, C and V). Data are mean values. Each race split represents 1.61 km. ◇, initial time trial (ITT); ■, constant power (C); ▲, variable power (V).

Table 1. Mean (SD) values of the dependent variables for the three time trials.

Variable	Initial time trial (ITT)	Constant power output (C)	Variable power output (V)	Significance of condition main effect	Multiple contrasts
Power output (W)	235.3 (40.5)	234.1 (38.6)	235.6 (40.5)	$F(2,12) = 1.5$ ($p = 0.26$)	n/a
Finish time (s)	1671.3 (131.4)	1661.1 (129.9)	1659.0 (134.8)	$F(2,12) = 7.2$ ($p = 0.009$)	ITT > C = V
Time for 1st 8.01 km (s)	930.4 (76.2)	942.4 (80.1)	922.0 (76.9)	$F(2,12) = 5.1$ ($p = 0.025$)	ITT = C > V
Time for 2nd 8.01 km (s)	740.9 (62.1)	718.7 (50.1)	737.0 (58.4)	$F(2,12) = 8.5$ ($p = 0.005$)	ITT = V > C
Heart rate (beats min^{-1})	173.4 (10.5)	172.1 (10.8)	171.0 (8.6)	$F(2,12) = 0.2$ ($p = 0.80$)	n/a
Δ lactate (mmol l^{-1})	3.16 (1.66)	2.81 (2.27)	1.54 (1.68)	$F(2,12) = 4.1$ ($p = 0.045$)	ITT = C > V
RPE	15.9 (1.1)	15.3 (1.5)	15.0 (1.1)	$F(2,12) = 4.0$ ($p = 0.046$)	ITT = C > V

n/a = not applicable.

Even though the overall mean power outputs were similar between all conditions, mean ± SD finish times in trials C and V were significantly faster than in ITT (table 1). Finish times in trials C and V were faster than in ITT by 10.2 (95% CI = 3.4–16.9) s and 12.3 (95% CI = 2.4–22.2) s, respectively. Table 1 also shows the mean ± SD times to complete each 8.05 km section of the race. As expected, the fastest time over the first 8.05 km was recorded in trial V, whilst the fastest time for the second wind-assisted 8.05 km was recorded in trial C. It can be observed that, in general, there was a difference of about 200 s (≈ 10 km h^{-1}) between the headwind and tailwind sections of the race.

For ITT, the mean RPE was 15.9 on the 6–20 Borg scale which represents 'hard'. Mean heart rate was 173.4 beats min^{-1}, which represents an exercise intensity of about 90% of predicted maximum heart rate (from 220-age). The mean post-race lactate level was 5 mmol l^{-1}. Overall mean RPE and Δ lactate were lowest in trial V (table 1). Perceived exertion showed a pacing strategy by race-split interaction ($F(18,108) = 3.34$, $p = 0.0001$), suggesting that it was sensitive to the different pacing strategies adopted. Nevertheless, inspection of figure 2 reveals that this interaction is not explained by differences in RPE in the first 8.05 km of the time trial which were small (<0.5 units), but by the relatively larger (>1 unit) differences in RPE during the last half of the race. Heart rate showed no main effect of pacing strategy (table 1) or an interaction between strategy and race split ($F(18,108) = 1.62$, $p = 0.07$). Although this interaction does approach significance with the relatively low sample size employed in the present study, inspection of figure 3 reveals that differences in the within-race profiles of heart rate are small. Moreover, after the first 1.61 km when there were obvious differences in power output between trials, heart rate showed no apparent differences. There was a slight increase in heart rate in ITT at splits 4 and 5, which coincides with the drop in self-paced power output following the very fast start in this time trial. There is also a trend for heart rate to be higher at the end of ITT, which coincides with the self-selected increase in power output towards the finish of the race.

4. Discussion

The main finding of the present study is that pacing strategy influences performance in a time trial in which external conditions vary and that this influence is independent of any differences in average power output over a race.

One aim of the sports scientist is to ensure that any predictive test of performance is analogous to actual performance, i.e. that it is externally valid (Hopkins et al. 1999). The degree to which sports scientists have realized this aim in the past can be

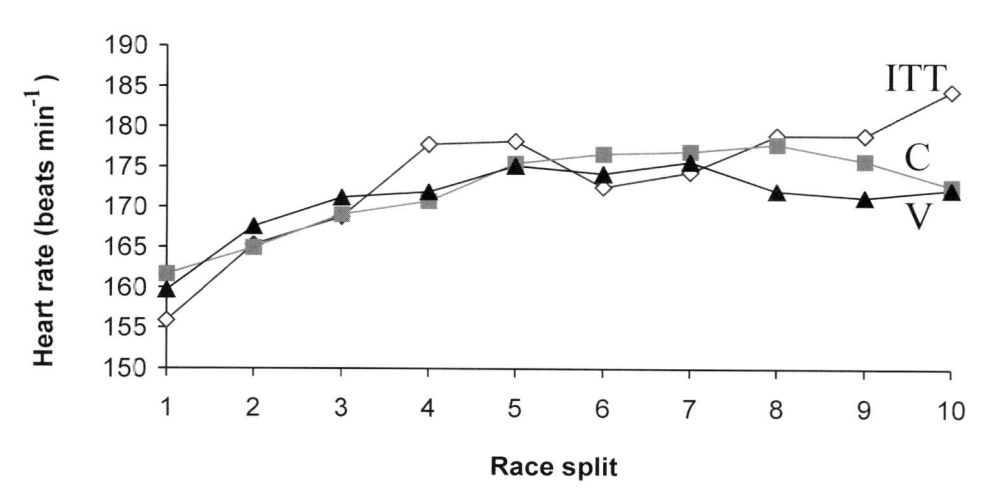

Figure 2. The within-race profile of RPE for the three time trials (ITT, C and V). Data are mean values. Each race split represents 1.61 km. ◇, initial time trial (ITT); ■, constant power (C); ▲, variable power (V).

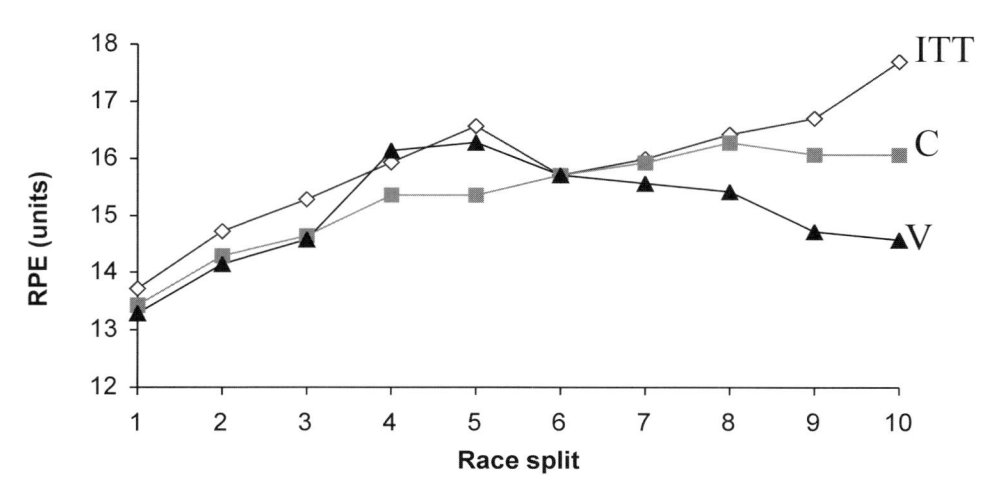

Figure 3. The within-race profile of heart rate for the three time trials (ITT, C and V). Data are mean values. Each race split represents 1.61 km. ◇, initial time trial (ITT); ■, constant power (C); ▲, variable power (V).

criticized. For example, only relatively recently have some performance tests been designed to reflect the actual intermittent nature of bunched road races in cycling (Coleman et al. 1998, Schabort et al. 1998). For cycling time trials, it has also taken a long time to arrive at externally valid tests. Commonly used incremental-power output or 'time to exhaustion' (at a set power output) tests may enable the measurement of any physiological limiting factors to performance in these tests, but they have little relevance to the actual characteristics of cycling or indeed *any* sport, and may in fact be very unreliable (Krebs and Powers 1989). Simulated cycling time trials in the laboratory have been found to be highly reliable (Hickey et al. 1992, Palmer et al. 1996, Atkinson et al. 1999), but are only valid to events with no variations in external conditions (wind and hills). The presence of hills and variable winds is very common in actual cycling time trials. For example, 'calm' days with no wind are observed for only 1–15% of the time when studied over a year in the UK (Meteorological Office 1999). To the authors' knowledge, the present study is the first to employ a laboratory-based time trial test in which variable wind conditions were modelled.

There is evidence that, although the Computrainer seems to simulate variations in wind velocity accurately, the way it does this may lead to slower times in general than in actual road time trials. The difference in speed of about 10 km h^{-1} between the headwind and tailwind sections of the race agree well with the data of Martin et al. (1998) who examined the influence of wind direction and velocity on actual road speeds at a similar cycling power of 255 W. They found that wind affects riding velocity by approximately 62% of a given wind velocity; 62% of the headwind and tailwind velocities examined in the present study is about +5 and −5 km h^{-1}, respectively, which gives a difference of 10 km h^{-1} between the race sections. These data confirm that the race conditions were simulated accurately compared to the best model yet for assessing external influences on cycling power (Martin et al. 1998). Nevertheless, the Computrainer does use a higher value of rolling resistance than has been measured for a real racing bicycle for the reasons outlined in Section 2. This may explain the relatively slow times during the time trial; even though time trialling was not his main discipline, there was one professional rider in the subject sample and his best time on the Computrainer was 24.2 min, compared to the 22–23 min that he usually recorded in local 16.1 km time trials. Future researchers could extend the work of Palmer et al. (1996) by comparing times obtained on the Computrainer with those in real time trials.

The first observation relevant to pacing strategy from the present data is that riders started the self-paced ITT with a power output that could not be sustained after the first 1.61 km. This observation agrees with the work of Firth (1998) who performed a series of case studies on rider pacing strategy, also over a 16.1 km time trial but with no variable wind conditions. For one subject, Firth (1998) reported that mean self-selected power output over the first 2 km was 23 W higher than the race mean (≈ 165 W). This difference translates to an increase in power output of 12% above the mean, which compares well with the mean increase in the present study of 14% above the mean for the first 1.61 km. It was apparent that the riders paid heavily for this fast start; power outputs dropped to below the mean after the first few kilometres in the self-paced ride leading to a slower time compared to condition C.

It was deemed important to include the normal acceleration phase of the time trial in the Computrainer simulation, i.e. the standing start. In general, the riders reached their eventual average speed after about 10 s, irrespective of experimental trial. This portion of the race is very small relative to average overall time (≈ 27 min). Besides, if the time spent in the acceleration phase was important to overall race time, it would be expected that the high power at the very start of ITT would lead to faster times than in condition C. Such a difference in times was not observed. Nevertheless, it is not known whether the subjects would have taken 10 s to reach their average speed from a standing start in an actual race.

Foster et al. (1994: 81) cited data on the influence of intramuscular pH on fatigue during exercise (Kostka and Cafarelli 1982) and the time-course in lactate accumulation during intense exercise, which they maintained suggested that 'athletes learn how to sense low values of muscle pH and adjust their pace accordingly so that they ideally reach critically low values of pH near the end of a race'. The present data suggest that this 'sense' is not particularly good in the first few kilometres of a time trial, even for a professional cyclist who volunteered for investigation (an increase of mean power output in the first 1.61 km of 16% above race mean). Unfortunately, the lactate analysis in the present study was delimited to pre- and post-exercise, since an invasive procedure during the time trial may have affected adherence to the pacing strategies and, subsequently, performance. Nevertheless, a comparison of lactate measurements between ITT and trial C during the first few kilometres of the time trial may have helped to examine the mechanisms behind the fast self-selected start and subsequent drop in power output in ITT.

Despite there being a physiological rationale for not adopting a fast start, the present results show that there is also a power-speed rationale for the time trial distance that was studied (16.1 km), since overall power outputs did not differ between the trials. As can be seen in figure 1, power outputs in ITT dropped to 2% below the race mean by the end of the headwind section and 8–10% below the race mean in the tailwind section. This reduction meant that more time was lost than was gained with the fast start. A slight reduction in power output in the tailwind section when speed was high would have been negligible, especially if power output was increased in the headwind section (see the discussion on variable power output below), but this reduction in ITT was obviously more than optimum.

The discouragement of an over-fast start in a cycling time trial is an important, but not the only, issue in pacing. A slightly faster time and lower overall levels of RPE and blood lactate were recorded when the cyclists completed the first half of the time trial into the headwind at a power 5% higher than the race mean (without an overly fast start). This was so even though there was an approximate 5% reduction in power output in the second half of the race when the riders were wind-assisted. These findings agree fully with the predictions of Swain (1997). It is stressed that both conditions C and V showed no difference in power output averaged over the whole race. As Swain (1997) discussed in full, the reason for the time saving and more favourable physiological and subjective responses to condition V is because, in general, relatively more time is spent in the headwind section of a race than the tailwind. It follows, therefore, that a rider who increases power output into the headwind and decreases power output with a tailwind will achieve a faster time than if he/she rides at a 'flatline' power output.

The mean time saving of about 2 s for condition V versus C is slightly less than the expected mean of 5 s if all the riders kept strictly to the administered pacing strategies. This time saving was estimated by running Computrainer programs with power outputs inputted manually, i.e. using the pacer control facility. One reason for this difference could have been the fact that mean power outputs were displayed by the Computrainer only in whole numbers of Watts. In the last few kilometres of condition V when the rider was receiving instructions to lower power output in order to finish with the same mean of, for example, 235 W, the actual race mean power output could have been as high as 235.4 W or as low as 234.6 W. This variability in power output would have added to the variability in time. One subject was unusually poor at adhering to the administered pacing strategy in condition C. A mean power output of 227 W was the aim but even after repeated warnings to lower power output, the subject finished with a mean power output of 229 W. This discrepancy could have also led to the lower than expected improvement in time for condition V versus C. It is stressed that together with the slightly improved time for condition V compared to C, RPE and lactate values *were* significantly lower in V suggesting that the 'joint concept' of performance and responses to performance was lower in condition V.

There may have been threats to the internal validity of the physiological and subjective response comparisons between ITT and the other two time trials, since the research design dictated that ITT was completed by all the subjects first. Therefore, reductions in RPE for trials C or V compared to ITT may have been due, in part, to learning or training effects. This criticism is difficult to defend as the whole point of the study was to ensure that the enforced pacing strategies did not differ in mean power output with an initial time trial. For this reason, it is stressed that any differences between conditions C and V are not influenced by this factor, since they were administered in a randomized order. Besides, the physiological and subjective data from ITT suggest that the well-motivated subjects were exercising to the best of their ability; mean heart rate was 90% of age-predicted maximum, mean RPE was 15.9. Oxygen consumption could have been measured to assess relative exercise intensity but it was thought that the protocol and equipment would have interfered with riders' performances. For the same reasons, this variable is too invasive to be used in actual races to monitor pacing strategy and so it was deemed not to be relevant to the present research questions.

Some time trials could be associated with more severe conditions than those modelled in the present study. The work of Swain (1997) can be summarized thus: that the longer the race, the greater the number of hilly and windy sections; and the greater the magnitude of hills and winds, the greater will be the time-savings if a rider adopts a variable power output distribution in parallel with these changes in external conditions. For example, Swain (1997) predicted that a rider who varies power output by 10% in variable conditions of a strong (24 km h^{-1}) wind could be more than 1 min faster over 40 km compared to a rider who chooses a flatline pacing strategy at the same mean power output. It is suggested that future research work examines the effects of variable power pacing strategies under different conditions of hills and wind.

Swain (1997) also predicted that the greater the magnitude of power output variation during a race in which external conditions vary, the greater the time savings. Nevertheless, there are obvious physiological limitations to how much a rider can vary power output during a race, as discussed above when the start that was too fast in ITT was considered. The degree of power output variability also depends on the time that is spent in each section of a race. In the present study, riders were able to increase power output by 5% for half the race distance (8.05 km) and overall RPE was still lower than in trial C and ITT. It could be speculated that riders could have increased power output in trial V slightly more (or decreased power output slightly less in the tailwind section) for RPE in all the time trials to be the same. This pacing strategy would have been the optimal increase for the race conditions that were simulated. An increase of power output larger than that administered in the present study might also be feasible if the course had only short distances of harder external conditions (hills and wind). Swain (1997) modelled courses with very short distances of hills and wind and he speculated that power output variations of up to 15% might be physiologically acceptable. Liedl et al. (1998) performed a study in which the mean power output (corresponding to 78% O2 max) from an initial 1-h time trial was used to examine the physiological and subjective strain of a constant and a 5% variable pace. No differences were found between the two pacing strategies in terms of mean O2, mean heart rate, mean blood lactate concentration and mean ratings of perceived exertion. Nevertheless, Liedl et al. (1998) varied power output according to time elapsed while holding external resistance constant, instead of varying power output and external conditions according to distance covered, which is the case in a real time trial and a Computrainer simulation.

One aim of the present study was to ascertain whether any variables are sensitive to small within-race changes in pacing strategy and can, therefore, be employed in actual races to monitor pace. It was clear from figure 1 and the statistical analyses that power output (as measured by the Computrainer) is sensitive to the changes in pace that were studied. Bicycle-mounted devices that measure power output, such as SRM Powercranks may, therefore, be useful in monitoring pacing strategy. Although the relationship between power output and oxygen consumption (O2) is well known, clearly it is impractical to attempt to measure O2 in a real time trial, even with an ambulatory device since this will affect performance. In the present study, the authors therefore concentrated on the most practical of the possible indicators of pace, which are heart rate and subjective responses. The within-race profile of heart rate was not significantly different between the trials, although the p-value of 0.07 and effect size of 0.2 for this analysis should not lead to outright acceptance of the null hypothesis. Nevertheless, it is clear from figure 3 that there were no differences in heart rate at the start of the time trials, a time when differences in power output between trials were greatest. In fact the lowest heart rates were found after the first few kilometres of ITT. For RPE there was evidence that the reduction in power output of about 5% in the tailwind section of condition V was accompanied by about a 1-unit drop in RPE compared to condition C. It is not known whether athletes could be trained to distinguish such a drop in RPE and hence employ this variable to monitor pace. Surprisingly, the 5% increase in power output into the headwind in condition V did not seem to mediate any change in RPE in this section of the race. To summarize, although RPE and heart rate might be very useful in monitoring general training 'zones' and despite the fact that there are obvious relationships between power output, RPE and heart rate when studied over a wide range of exercise intensities (Jeukendrup and Van Diemen 1998), it appears that the sensitivity of heart rate and RPE to detect small changes in power output that nevertheless affect performance is low, especially at the start of a simulated race.

There were four main dependent variables in the present study of finish time, heart rate, Δ lactate and RPE. Three of these variables showed significant differences between trials. Strictly, to allow for the possibility of making a type I error, α should have been corrected according to the Bonferroni method; $\alpha = 0.05/4 = 0.0125$. Therefore, it could be possible that the differences in Δ lactate and RPE could be a type I error. Nevertheless, the relatively high effect sizes of 0.3–0.5 are suggestive of a meaningful effect that demands further study. Unfortunately, a multivariate analysis could not be employed in the present study because of the co-relationships between, power, performance, heart rate and RPE.

5. Conclusions

These results suggest that in a 16.1 km time trial with equal 8.05 km headwind and tailwind sections, riders habitually vary pace in a sub-optimal manner, setting off too fast in the first few kilometres with a subsequent drop in power output below the race mean and a 'surge' in power output towards the finish. Swain (1997) predicted that the optimal pacing strategy is one that holds power output constant during constant external conditions, but varies power when external factors vary. The present results confirm this prediction in that the fastest times and lowest physiological and subjective responses to exercise were observed with the pacing strategy that varied power output in parallel with the simulated wind conditions. The sensitivity of heart rate and RPE to detect small ($\pm 5\%$) within-race changes in power output is low and, therefore, their usefulness for monitoring pacing strategies during races can be questioned.

References

ATKINSON, G. and EDWARDS, B. J. 1998, Pacing strategy and cycling performance: field data from the British women's 16 km time trial championships. In A. J. Sargeant and H. Siddons (eds), *Proceedings of Third Annual Congress of the European College of Sports Science* (Liverpool: Centre for Health Care Development), 211.

ATKINSON, G., NEVILL, A. M. and EDWARDS, B. 1999, What is an acceptable amount of measurement error? *Journal of Sports Sciences*, **17**, 18.

BORG, G. 1970, Perceived exertion as an indicator of somatic stress, *Scandinavian Journal of Rehabilitative Medicine*, **2**, 92–98.

BURKE, E. R., PORTER, M. and BARZUKAS, A. 1993, Precision and variability monitored by EKG, photo-reflectance and telemetered heart rate monitors, *Medicine and Science in Sports and Exercise*, **25**, S10.
COLEMAN, D. A., DAVISON, R. C. R., BALMER, J. and BIRD, S. R. 1998, The energy demands of competitive road race cycling, *Medicine and Science in Sports and Exercise*, **30**, S305.
DI PRAMPERO, P. E., CORTILI, G., MOGNONI, P. and SAIBENE, F. 1979, Equation of motion of a cyclist, *Journal of Applied Physiology*, **47**, 201–206.
FIRTH, M. 1998, From high-tech to low tech: another look at time trial pacing strategy, *Coaching News*, **3**, 7–10.
FOSTER, C., SCHRAGER, M., SNYDER, A. C. and THOMPSON, N. N. 1994, Pacing strategy and athletic performance, *Sports Medicine* **17**, 77–85.
FOSTER, C., SNYDER, A. C., THOMPSON, N. N., GREEN, M. A. and FOLEY, M. 1993, Effect of pacing strategy on cycle time trial performance, *Medicine and Science in Sports and Exercise*, **25**, 383–388.
HICKEY, M. S., COSTILL, D. L., MCCONELL, G. K., WIDRICK, J. J. and TANAKA, H. 1992, Day to day variation in time trial cycling performance, *International Journal of Sports Medicine*, **13**, 467–470.
HOPKINS, W. G., HAWLEY, J. A. and BURKE, L. M. 1999, Design and analysis of research on sport performance enhancement, *Medicine and Science in Sports and Exercise*, **31**, 472–485.
JEUKENDRUP, A. and VAN DIEMEN, A. 1998, Heart rate monitoring during training and competition in cyclists, *Journal of Sports Sciences*, **16**, S91–S99.
JEUKENDRUP, A., BROUNS, F., WAGENMAKERS, A. J. M. and SARIS, W. H. M. 1997, Carbohydrate-electrolyte feedings improve 1 h time trial cycling performance, *International Journal of Sports Medicine*, **18**, 125–129.
KOSTKA, C. E. and CAFARELLI, E. 1982, Effect of pH on sensation and vastus lateralis electromyogram during intense exercise, *Journal of Applied Physiology*, **52**, 1181–1185.
KREBS, P. S. and POWERS, S. K. 1989, Reliability of laboratory endurance tests, *Medicine and Science in Sports and Exercise*, **31**, S10.
KYLE, C. R. 1988, The mechanics and aerodynamics of cycling, in E. R. Burke and M. M. Newsome (eds), *Medical and Scientific Aspects of Cycling* (Champaign, Il: Human Kinetics), 235–251.
LIEDL, M. A., SWAIN, D. P., BRANCH, J. D., BRYANT, T. L., CORY, L. M. and LEETE, D. S. 1998, Physiological effects of constant vs. variable power during endurance cycling, *Medicine and Science in Sports and Exercise*, **30**, S112.
MARTIN, J. C., MILLIKEN, D. L., COBB, J. E., MCFADDEN, K. L. and COGGAN, A. R. 1998, Validation of a mathematical model for road cycling power, *Journal of Applied Biomechanics*, **14**, 276–291.
METEOROLOGICAL OFFICE 1999, Personal communication.
OLDS, T. S., NORTON, K. I. and CRAIG, N. P. 1993, Mathematical model of cycling performance, *Journal of Applied Physiology*, **75**, 730–737.
PALMER, G. S., DENNIS, S. C., NOAKES, T. D. and HAWLEY, J. A. 1996, Assessment of the reproducibility of performance testing on an air-braked cycle ergometer, *International Journal of Sports Medicine*, **17**, 293–298.
SCHABORT, E. J., HAWLEY, J. A., NOAKES, T. D., HOPKINS, W. G. and MUJIKA, I. 1998, A novel reliable laboratory test of endurance performance for road cyclists, *Medicine and Science in Sports and Exercise*, **30**, S305.
SWAIN, D. P. 1997, A model for optimizing cycling performance by varying power on hills and in wind, *Medicine and Science in Sports and Exercise*, **29**, 1104–1108.
VAN INGEN-SCHENAU, G. J., DE KONING, J. J. and DE GROOT, G. 1994, Optimisation of sprinting performance in running, cycling and speed skating, *Sports Medicine*, **17**, 259–275.
VINCENT, W. J. 1995, *Statistics in Kinesiology* (Champaign, IL: Human Kinetics).
WHITE, A. P. 1994, Factors affecting speed in human-powered vehicles, *Journal of Sports Sciences*, **12**, 419–424.
WILBERG, R. B. and PRATT, J. 1988, A survey of the race profiles of cyclists in the pursuit and kill track events, *Canadian Journal of Sports Sciences*, **13**, 208–213.

4 The effects of phase control materials on hand skin temperature within gloves of soccer goalkeepers

A. J. PURVIS* and N. T. CABLE

Research Institute for Sport and Exercise Sciences, Liverpool John Moores University, Henry Cotton Campus, Webster Street, Liverpool L3 2ET, UK

Keywords: Thermoregulatory responses; Clothing; Sweating; Simulated soccer activity.

In soccer, goalkeepers routinely wear gloves that may restrict heat loss from the hands and cause thermal discomfort. In order to alleviate this problem phase control materials (PCMs) have been incorporated into gloves to reduce heat load inside the glove, thereby maintaining a comfortable temperature. The aim of this study was to assess the efficacy of these materials during a simulation of goalkeeping activities. Seven subjects carried out two sessions of goalkeeper-specific exercise on a non-motorized treadmill, one session with a PCM glove and one session with a normal foam material glove (NFM). All sites of skin temperature measurement, except mean whole-body skin temperature, showed uniformly that the PCM glove caused a greater increase in skin temperature of the hand compared to the NFM glove. These results suggest that this particular specification of PCM promotes heat gain rather than heat loss and is therefore inappropriate to enhance thermal comfort in this setting.

1. Introduction

Soccer goalkeepers are different to outfield players in the skills that they utilize, the energy requirements of their game and the clothing that they wear. The most obvious clothing difference is the use of foam padded gloves which provide a functional and protective role. These gloves provide protection against injury to the hand and also assist in gripping the soccer ball.

When items such as goalkeeping gloves are worn, heat loss by convection and evaporation is impeded owing to the creation of a microenvironment close to the surface of the skin that is dissimilar to the ambient environment (Sullivan *et al.* 1987, Sullivan and Mekjavic 1992, Pascoe *et al.* 1994a, Pascoe *et al.* 1994b). The surface area of the hands is no longer exposed to the surrounding air and the convective exchange of heat from the body to air is virtually eliminated. Evaporative resistance of gloves at the surface of the hands is also increased, as air cannot remove the water vapour from the area of the hand. Reduced evaporation of sweat causes moisture to accumulate within the gloves and increases thermal discomfort. This can have physiological and performance ramifications (Smith *et al.* 1997).

---

*Author for correspondence.

Phase control materials (PCMs) have been designed to change their physical state from solid to liquid to gas over a certain range of temperatures. When fabrics are coated or impregnated with PCM the thermal properties of the material can be altered to encourage heat loss or heat gain. The aim of this study was to determine if the use of a PCM contained within the palmar area (extending from the wrist to the tips of the fingers) of a goalkeeper's glove increases heat loss from the hand and reduces the demands on evaporative heat dissipation.

2. Methods

2.1. Subjects

Experiments were conducted on 7 healthy, active subjects unacclimated to heat (5 males and 2 females). Their characteristics, mean and range, were 25.4 (20–29) years of age, body mass 82.0 (70.8–114.3) kg, hand length 21.1 (18.5–25.0) cm and hand width 12.2 (9.5–13.3) cm. The participants freely volunteered for the study, which was approved by the Human Ethics Committee at Liverpool John Moores University. Written informed consent was obtained from each of the participating individuals.

2.2. Exercise and environmental conditions

Each participant visited the laboratory twice and on each occasion performed 46 min and 10 s of intermittent exercise in neutral environmental conditions (approximately 21°C, 50% relative humidity). This time period was chosen to represent the average duration of the first half of a soccer game. The conditions were presented to subjects in a counter-balanced arrangement. The exercise protocol employed activity patterns specific to goalkeepers. One session was carried out with the subjects wearing gloves with PCM and in a second session the subjects wore gloves of normal foam material (NFM). The PCM was attached to the inner layer of the palmar region of the glove and was in contact with the skin. A normal foam material of a similar thickness was used as a control material. The PCM was set by the manufacturers to maintain temperature at 31°C. This temperature allows for accelerated heat loss in warm environments and diminished heat loss in cold environments.

The intermittent protocol was formulated using data published by Reilly and Thomas (1976) and was conducted on a non-motorized treadmill (Woodway, Seattle, WA). The protocol (figure 1) represents the actions of a goalkeeper during match-play conditions and utilized four self-paced actions—walk, jog, cruise and sprint. Both visits to the laboratory used self-paced actions.

2.3. Procedures

For each session subjects wore the same clothing ensemble which comprised underwear, polyester shorts, cotton T-shirt, trainers and goalkeeping gloves. Prior to testing, and upon completion, the subjects weighed themselves nude. The participant rested for 20 min before testing during which time thermistors were placed upon the skin with sterile surgical adhesive tape (Transpore, 3M, St Paul, Michigan, USA). An additional 10 min of supine rest allowed for the measurement of skin blood flow of the thumb and forearm. The subject then stood on the treadmill and put on the gloves immediately prior to the exercise period. Immediately after the completion of the treadmill exercise the subject removed the gloves and returned to the supine position and the measurement of skin blood flow was repeated.

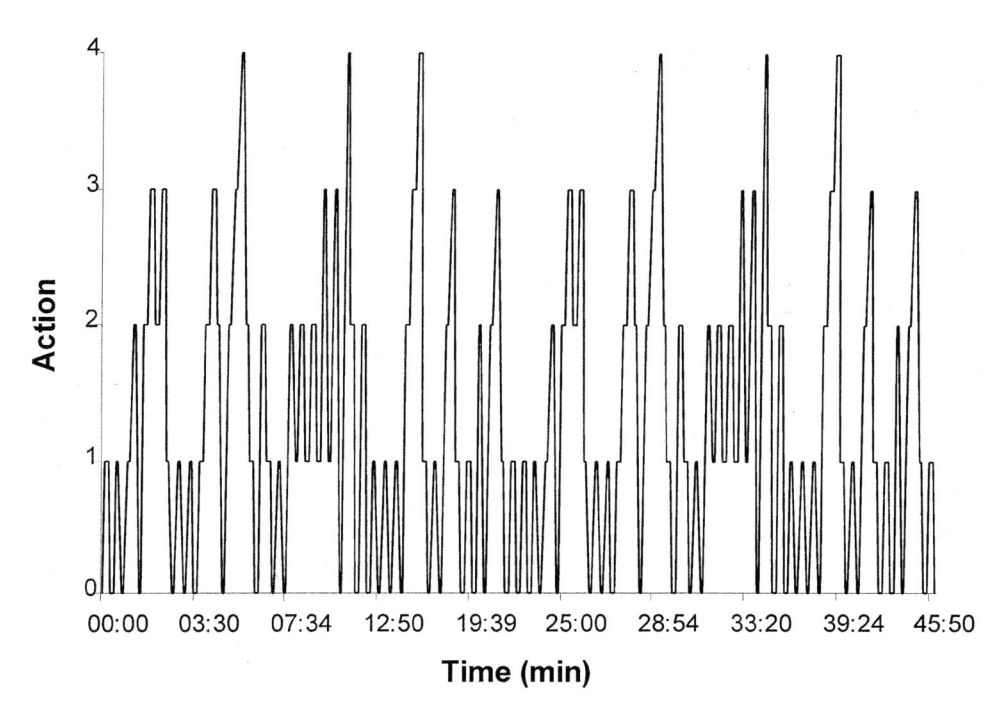

Figure 1. Activity changes over the 46 min 10 s goalkeeping protocol. The actions used are 0 = standing, 1 = walking, 2 = jogging, 3 = cruising, 4 = sprinting.

2.4. *Measurements and analysis*

Skin blood flow was measured using laser Doppler flowmetry (Periflux, Perimed, Järfälla, Sweden) with measurements recorded every 60 s for 10 min before exercise and for 10 min immediately following exercise. Prior to measurement of skin blood flow, the subject was fitted with a short-range radio telemetry system for the measurement of heart rate (Sports Tester, Polar Electro, Kempele, Finland). Heart rate and temperatures were recorded at 60-s intervals during the measurement of skin blood flow, during exercise and while measuring post-exercise skin blood flow.

Mean skin temperature was measured according to the methods of Ramanathan (1964) by the placement of thermistors (Grant thermistors, Grant Instruments, Cambridge, UK) on the chest, forearm, thigh and calf. In addition, hand skin temperature was measured at seven sites on the right hand and wrist (thumb, index finger, middle finger, hand palm, hand dorsal, wrist palm and wrist dorsal). Thermistors were connected to a data logger (Squirrel meter 1250, Grant Instruments, Cambridge, UK) that recorded temperatures every 60 s. Concurrently, participants were requested to rate perceived exertion on a scale between 6 to 20 (Borg 1970) at 5-min intervals.

2.5. *Statistical analysis*

The data were analysed using descriptive statistics and Student's t tests. Analysis of serial measurements was also employed (Mathews *et al.* 1990). This method uses a Student's t-test to examine the changes within a dependent variable from baseline to end-of-test. A p value of 0.05 or less was considered to be significant.

3. Results

Figure 2 displays the mean heart rate for each of the two conditions. The overall mean heart rate during exercise was 137 ± 12 beats min^{-1} and 137 ± 11 beats min^{-1} for subjects wearing PCM and NFM gloves, respectively ($p > 0.05$).

Exercise resulted in a mean absolute body mass loss of 0.7 ± 0.7 kg vs. 0.8 ± 0.5 kg for subjects wearing NFM and PCM gloves, respectively. This loss corresponds to a 0.8% decrease in body mass in both conditions ($p > 0.05$).

Ratings of perceived exertion were 11.7 ± 1.7 for the PCM condition and 12.3 ± 1.6 for the NFM condition. This RPE value equates to a perceived exercise intensity of approximately 'somewhat hard'. There was a trend for the RPE to increase with exercise duration and all initial ratings were lower than the ratings recorded at the end of exercise. There were no significant differences between conditions ($p > 0.05$).

Figure 3 displays absolute skin temperature of the hand presented as a mean of all seven sites measured. Over the duration of the exercise test the PCM glove caused a significant increase of temperature above baseline measurements in all areas of the hand ($p < 0.05$).

Figure 4 shows the change in temperature for each individual hand site. These differences were greater when wearing the PCM glove and most evident at the thumb, index finger and middle finger. Change in temperature at the thumb with PCM was $3.4 \pm 2.5°C$ compared to $1.9 \pm 2.5°C$ for NFM, at the index finger the change in temperatures were $4.2 \pm 2.4°C$ (PCM) and $2.8 \pm 1.8°C$ (NFM) and at the middle finger the change in temperatures were $4.3 \pm 2.6°C$ (PCM) and $2.5 \pm 1.8°C$ (NFM).

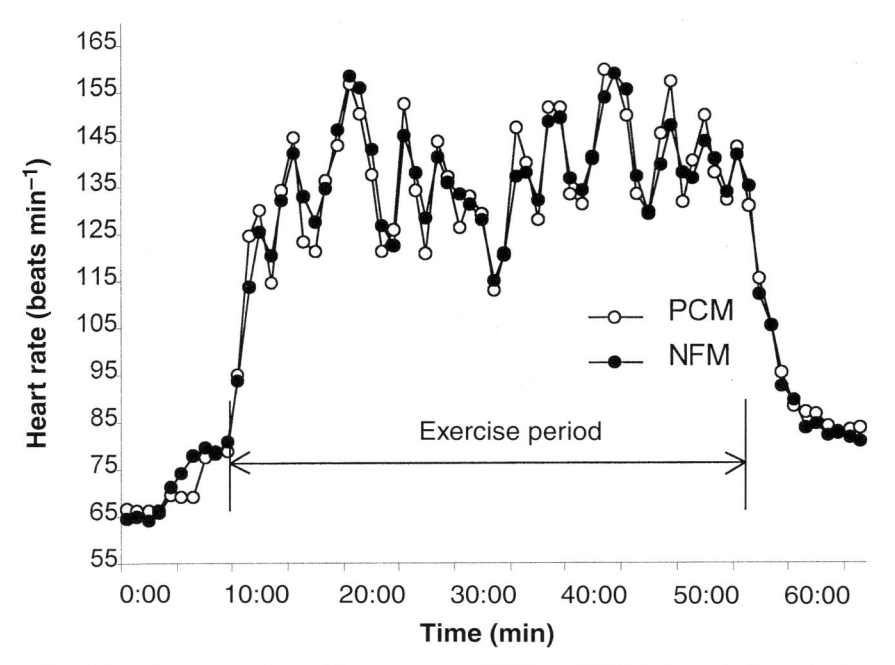

Figure 2. Mean heart rate for subjects wearing PCM and NFM gloves before, during and after exercise tests ($p > 0.05$).

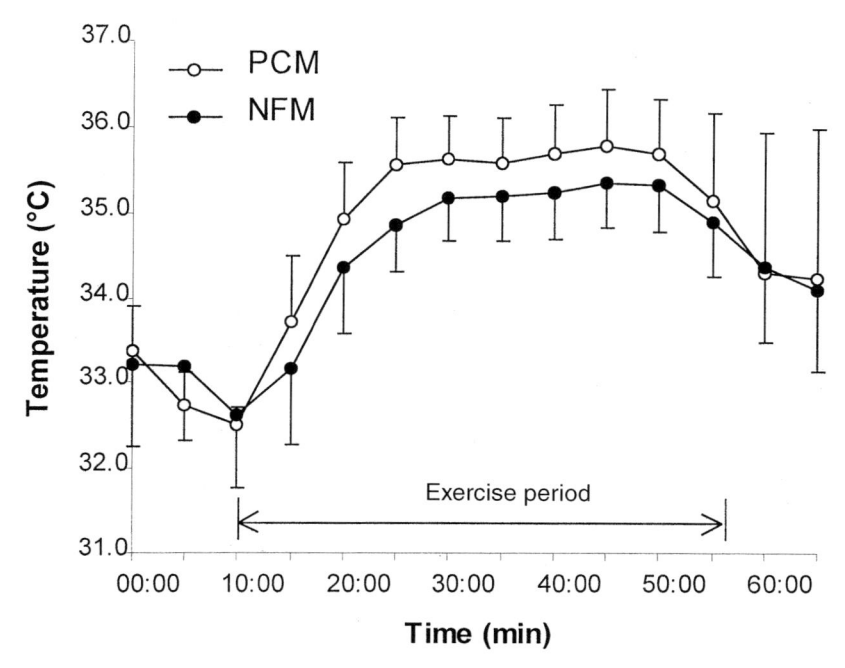

Figure 3. Mean (±SD) changes in mean hand temperature for PCM and NFM glove conditions ($p < 0.05$) before, during and after exercise.

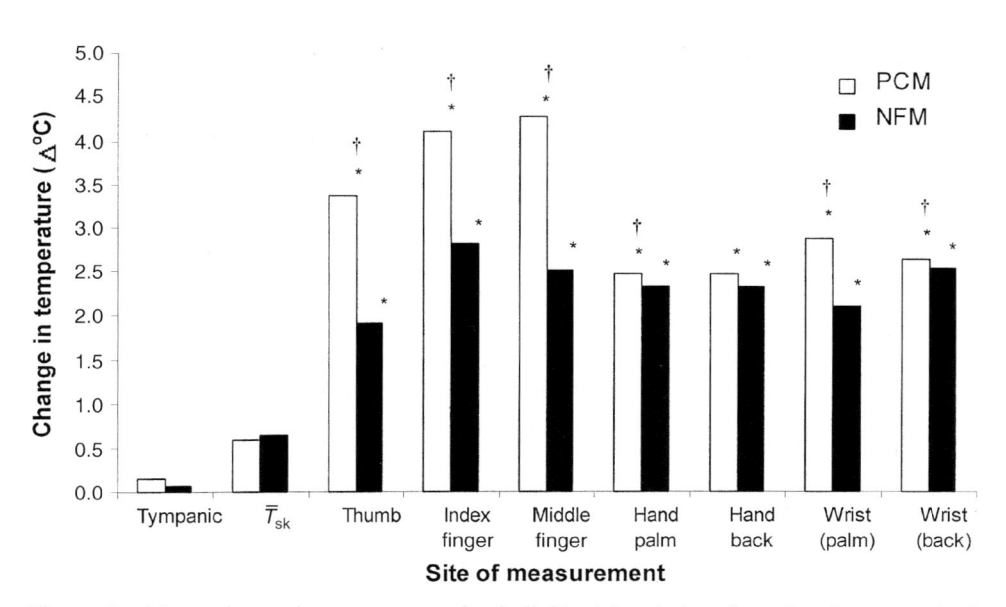

Figure 4. Mean change in temperature for individual hand sites from baseline to end-of-exercise. *, Significant increase from baseline ($p < 0.05$); †, significant difference between glove materials ($p < 0.05$).

Table 1. Mean absolute values ($\pm$SD) and significant differences of heart rate and skin temperatures for PCM and NFM conditions during exercise.

Measurement	PCM	NFM	p value
Heart rate (beats min^{-1})	137$\pm$12	137$\pm$11	0.99
Temperatures (°C)			
T_{sk}, mean skin temperature	32.1$\pm$0.2	32.1$\pm$0.3	0.86
Thumb	35.2$\pm$1.1	34.8$\pm$0.8	0.02
Index finger	35.4$\pm$1.3	34.7$\pm$1.2	0.00
Middle finger	35.4$\pm$1.2	34.9$\pm$1.1	0.00
Hand (palmar)	35.7$\pm$0.8	35.1$\pm$0.5	0.00
Hand (dorsal)	34.8$\pm$0.9	34.8$\pm$1.1	0.07
Wrist (palmar)	35.2$\pm$1.1	34.8$\pm$1.0	0.00
Wrist (dorsal)	34.0$\pm$1.0	33.4$\pm$1.1	0.00

The PCM condition exhibited higher absolute mean skin temperatures than the NFM condition for all hand skin sites with only the dorsal area of the hand not showing a significant increase above the NFM condition ($p = 0.07$) (table 1). There was no significant difference between conditions for mean skin temperature or heart rate. Overall mean absolute skin temperature of the hand was 35.1$\pm$1.0°C and 34.6$\pm$1.0°C for PCM and NFM conditions, respectively.

Gloves were weighed immediately before and after each exercise test session to determine moisture uptake. There was significantly greater ($p<0.05$) sweat retention in PCM gloves compared to NFM gloves, 11.0$\pm$2.3% vs. 8.0$\pm$2.9% for the left glove and 11.4$\pm$2.5% vs. 8.9$\pm$3.2% for the right glove.

Forearm skin blood flow increased post-exercise for both glove materials. Thumb skin blood flow exhibited opposing responses to the glove material. The PCM gloves caused post-exercise attenuation of cutaneous blood flow while NFM gloves caused a post-exercise increase. There was a significant difference ($p<0.05$) between skin blood flow of the thumb before and after exercise for both the PCM glove and NFM glove (figure 5). Post-exercise thumb skin blood flow was significantly higher for the NFM glove ($p<0.05$). Post-exercise forearm skin blood flow was significantly higher compared to pre-exercise for both conditions ($p<0.05$). Forearm skin blood flow did not show a significant difference between conditions.

4. Discussion

The similarity in mean heart rate between the two conditions indicated that the same intensity level was maintained for both trials. The close matching of the two heart rate recordings confirms that all subjects strictly adhered to the protocol of treadmill actions. The protocol incorporated many rapid changes in speed lasting a median of 5 s, therefore strict reproduction of the changes in speed was necessary. The parity of workload is also confirmed by the similarity of perceived exertion ratings ($p>0.05$). The mean RPE values illustrate the moderate overall intensity of the exercise protocol.

The protocol was intended to simulate the activity profile of a soccer goalkeeper as closely as is possible on a treadmill. Although the most realistic testing of soccer activity is during a competitive match, this situation does not allow for detailed physiological assessment, control of environment or the accurate repetition of experimental trials.

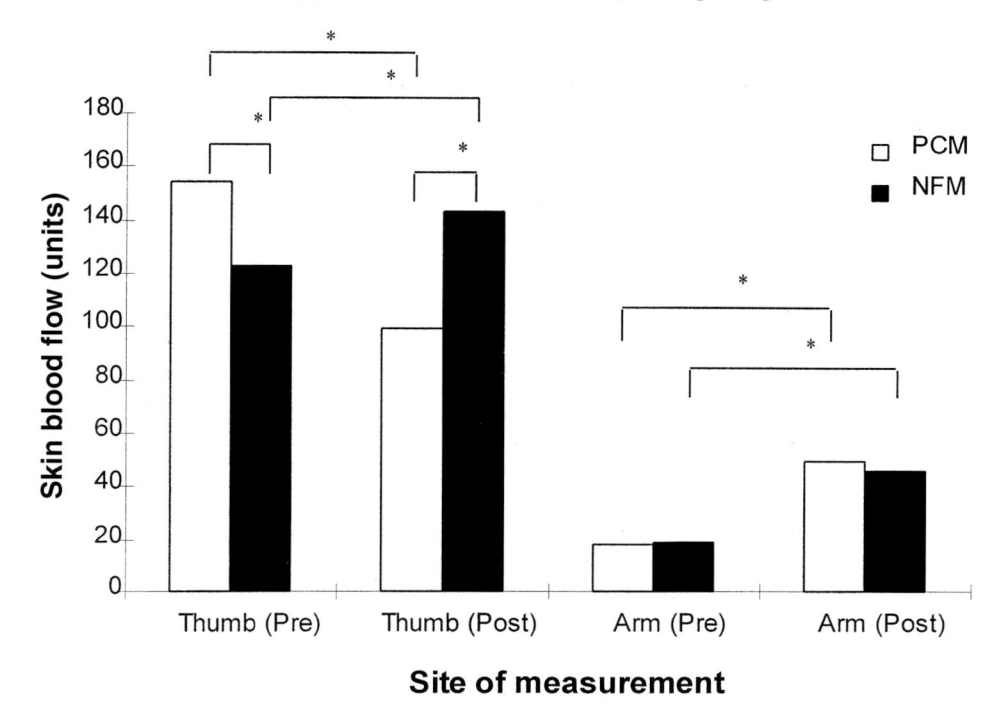

Figure 5. Changes in thumb and forearm skin blood flow for PCM and NFM glove conditions, pre-test and post-test (*$p<0.05$).

As the changes in nude body weight did not differ between conditions the overall sweat losses due to environmental and exercise thermal load were similar in the two trials. The mean loss of body mass of 0.8% due to sweat production reflects the lower work intensity of goalkeepers compared with outfield players (Reilly and Thomas 1976). Nevertheless, even mild dehydration can diminish performance (Maughan, 1993, MacLaren 1996).

Whole-body mean skin temperature did not differ between conditions. In contrast, analysis of serial measurements indicated that, apart from the dorsal hand ($p = 0.07$), all sites of measurement on the hand and wrist were significantly warmer when wearing the glove containing PCM. Immediately on wearing the gloves there was a sharp increase in skin temperature, which was more pronounced in the glove containing PCM. The PCM was designed to absorb heat from the skin in order to maintain a steady and comfortable temperature at approximately 31°C. The present data do not support these claims and, moreover, suggest that the material's performance is significantly inferior to that of a material used routinely in goalkeeping gloves.

The increase in mass due to sweat production and absorption was higher for the PCM glove than that seen in the glove containing NFM. The gain in mass was slightly higher in the right glove for both conditions and this may have been due to the presence of the thermistors (only in the right hand glove) causing a greater thermal load.

The properties of clothing that affect heat exchange are thermal insulation (affecting air velocity) and evaporative resistance (Holmér 1995). Skin temperature has been found to be reduced in conditions of high air movement compared to lower air movement (Adams et al. 1992, Bouskill et al. 1998). Elevated air velocity increases convective and evaporative heat loss and improves heat dissipation from the skin. If the PCM was predisposed to sticking to the surface of the skin when damp with sweat, the small amount of air movement within the enclosed microenvironment of the glove would be further reduced or eliminated. Sweat would be absorbed directly from the skin into the foam resulting in an increased sweat accumulation in the glove and increased skin temperature due to ineffective evaporative cooling. This results in production of extra sweat in an attempt to cool the skin (Nielsen and Endrusick 1992), further exacerbating the problem of heat loss within the gloves. These effects are seen in the results described above.

The build-up of condensation within the clothing layer is an important factor in heat loss (Lotens et al. 1995). During exercise there is significant production of sweat which accumulates in the gloves. If the trapped moisture in the gloves was allowed to evaporate, a contribution to heat transfer would be made. This was not evident in the gloves examined here. As sweat is trapped within the glove and does not evaporate, heat loss into the environment is restricted thereby causing an elevation in hand skin temperature.

Arm skin blood flow increased following exercise in both conditions. This response is as expected during recovery in order to dissipate the exercise-induced thermal load. However, the effects of glove material on skin blood flow in the thumb are more difficult to interpret. The measurements of post-exercise blood flow in the thumb showed a greater blood flow in the NFM condition ($p < 0.05$), suggesting that the PCM glove attenuates thumb skin blood flow.

Compared with rest, skin blood flow in the thumb increased significantly in the NFM condition, reflecting the local thermal drive to lose heat. The decrease in thumb skin blood flow in the PCM condition is somewhat surprising, given the more marked elevation in hand skin temperature during the PCM trial. Such a pattern was unexpected as there is usually a linear relationship between skin temperature and skin blood flow (Gleeson 1998). On closer inspection of these data, it is apparent that 5 out of 7 subjects responded with an increase in thumb skin blood flow during the PCM trial and the mean data are somewhat skewed by a marked decrease for two of the subjects. Nevertheless, it remains difficult to explain why these individuals have responded in this manner.

In conclusion, it is evident that further material development is necessary to combat the problems of moisture build-up and heat gain within the microenvironment of the glove. The particular specification of PCM evaluated in this study actually promotes heat gain rather than heat loss and is therefore inappropriate for enhancing thermal comfort in this setting.

References

ADAMS, W. C., MACK, G. W., LANGHANS, G. W. and NADEL, E. R. 1992, Effects of varied air velocity on sweating and evaporative rates during exercise, *Journal of Applied Physiology*, **73**, 2668–2674.

BORG, G. A. V. 1970, Perceived exertion as an indicator of somatic stress, *Scandinavian Journal of Rehabilitation Medicine*, **2**, 92–98.

BOUSKILL, L., SHELDON, N., PARSONS, K. and WITHEY, W. R. 1998, The effect of clothing fit on the clothing ventilation index, in M. A. Hanson (ed.), *Contemporary Ergonomics* (London: Taylor & Francis), 510–514.

GLEESON, M. 1998, Temperature regulation during exercise, *International Journal of Sports Medicine*, **19**, S96–S99.

HOLMÉR, I. 1995, Protective clothing and heat stress, *Ergonomics*, **38**, 166–182.
LOTENS, W. A., VAN DE LINDE, F. J. G. and HAVENITH, G. 1995, Effects of condensation in clothing on heat transfer, *Ergonomics*, **38**, 1114–1131.
MACLAREN, D. 1996, Nutrition, in T. Reilly (ed.), *Science and Soccer* (London: E. and F. N. Spon), 83–107.
MATHEWS, J. N. S., ALTMAN, D. G., CAMBELL, M. J. and ROYSTON, P. 1990, Analysis of serial measurements in medical research, *British Medical Journal*, **300**, 230–235.
MAUGHAN, R. 1993, Nutrition and diet in sport—an introduction, in D. A. D. Macleod, R. J. Maughan, C. Williams, C. R. Madeley, J. C. M. Sharp and R. W. Nutton (eds), *Intermittent High Intensity Exercise* (London: E. and F. N. Spon), 91–100.
NIELSEN, R. and ENDRUSICK, T. L. 1992, Localized temperatures and water vapour pressures within clothing during alternate exercise/rest in the cold, *Ergonomics*, **35**, 313–327.
PASCOE, D. D., SHANLEY, L. A. and SMITH, E. W. 1994a, Clothing and exercise. I: Biophysics of heat transfer between the individual, clothing and environment, *Sports Medicine*, **18**, 38–54.
PASCOE, D. D., BELLINGAR, T. A. and MCCLUSKEY, B. S. 1994b, Clothing and Exercise. II: Influence of clothing during exercise/work in environmental extremes, *Sports Medicine*, **18**, 94–108.
RAMANATHAN, L. N. 1964, A new weighting system for mean surface temperature of the human body, *Journal of Applied Physiology*, **19**, 531–534.
REILLY, T. and THOMAS, V. 1976, Motion analysis of work rate in different positional roles in professional football match-play, *Journal of Human Movement Studies*, **2**, 87–97.
SMITH, D. L., PETRUZZELLO, S. J., KRAMER, J. M. and MISNER, J. E. 1997, The effects of different thermal environments on the physiological and psychological responses of firefighters to a training drill, *Ergonomics*, **40**, 500–510.
SULLIVAN, P. J. and MEKJAVIC, I. B. 1992, Temperature and humidity within the clothing microenvironment, *Aviation, Space and Environmental Medicine*, **63**, 186–192.
SULLIVAN, P. J., MEKJAVIC, I. B. and KAKITSUBA, N. 1987, Determination of clothing microenvironment volume, *Ergonomics*, **30**, 1043–1052.

5 Aerobic performance of Special Operations Forces personnel after a prolonged submarine deployment

D. M. FOTHERGILL* and J. R. SIMS

Naval Submarine Medical Research Laboratory, Box 900, Groton, CT 06349-5900, USA

Keywords: Cooper test; Running; Deconditioning; Heart rate; Recovery.

The US Navy's Sea, Air and Land Special Operations Forces personnel (SEALs) perform a physically demanding job that requires them to maintain fitness levels equivalent to elite athletes. As some missions require SEALs to be deployed aboard submarines for extended periods of time, the prolonged confinement could lead to deconditioning and impaired mission-related performance. The objective of this field study was to quantify changes in aerobic performance of SEAL personnel following a 33-day submarine deployment.

Two age-matched groups of SEALs, a non-deployed SEAL team (NDST, $n=9$) and a deployed SEAL team (DST, $n=10$), performed two 12-min runs for distance (Cooper tests) 5 days apart pre-deployment and one Cooper test post-deployment. Subjects wore a Polar Vantage NV^{TM} heart rate (HR) monitor during the tests to record exercise and recovery HR. Variables calculated from the HR profiles included mean exercise heart rate (HR_{mean}), maximum exercise heart rate (HR_{max}), the initial slope of the HR recovery curve ($HR_{recslope}$) and HR recovery time ($HR_{rectime}$).

The second pre-deployment test (which was used in the comparison with the post-deployment test) showed a 2% mean increase in the distance achieved compared with the first ($n=18$, $p<0.05$) with no difference in HR_{mean}, HR_{max}, $HR_{recslope}$ and $HR_{rectime}$. The test-retest correlation coefficient and 95% limits of agreement for the Cooper tests were 0.79 ($p<0.001$) and -68.6 ± 267.5 m, respectively. For the NDST there were no changes in any of the HR measures or the distance run between the pre- and post-deployment tests. When individual running performances were expressed as a percentage change in the distance run between the pre- and post-deployment tests, the DST performed significantly worse than the NDST ($p<0.01$). The DST showed a 7% mean decrement in the distance run following deployment ($p<0.01$). The decrement in performance of the DST was not associated with any changes in HR_{mean} or HR_{max}; however, there was a 17% decrease in the $HR_{recslope}$ ($p<0.05$) and a 47% increase in $HR_{rectime}$ following the deployment ($p<0.05$).

In conclusion, prolonged confinement aboard a submarine compromises the aerobic performance of SEAL personnel. The resulting deconditioning could influence mission success.

1. Introduction

Over the past several years, the role of the submarine has seen increasing focus as a stealthy platform for insertion and extraction of Special Operations Forces personnel into and from littoral areas (Graves and Whitman 1999, Undersea Warfare 1999). Consequently, Special Operations Forces are often deployed aboard

*Author for correspondence. e-mail: fothergill@NSMRL.NAVY.MIL

submarines for extended periods of time. The US Navy's Sea, Air and Land Special Operations Forces personnel (SEALs) perform a physically demanding job that requires them to maintain fitness levels equivalent to those of elite athletes (Barnes and Strauss 1986, Prusaczyk et al. 1991, 1994, 1995). If the submarine deployment is prolonged, reduced physical activity resulting from the limited space and facilities aboard the submarine could lead to physical deconditioning. Significant decreases in fitness levels of the deployed SEALs may subsequently affect combat readiness.

Significant physiological changes occur when regular physical activity is reduced or stopped for periods of 2 weeks or more (Greenleaf and Kozlowski 1982, Coyle et al. 1984, American College of Sports Medicine 1998). Typical cardiorespiratory changes induced by physical inactivity include reductions in $\dot{V}O_2$max, anaerobic threshold, cardiac output and work capacity (Coyle et al. 1984, Bennett et al. 1985, Hickson et al. 1985, Raven et al. 1998). The magnitude of change in these parameters with detraining is dependent on the level of physical inactivity (Hickson et al. 1985) as well as the initial training status of the individual (Greenleaf and Kozlowski 1982, Raven et al. 1998). For combat personnel operating from the confines of a nuclear submarine, the degree of deconditioning during a deployment will largely depend on the duration of the mission. Limited space and reduced exercise facilities also affect the ability and motivation of the individuals to maintain regular physical training. Previous studies on submariners have documented an 8% reduction in exercise tolerance (Duett 1982), a 7–24% decrease in $\dot{V}O_2$max (Knight et al. 1981, Bennett et al. 1985), a 20% decrease in anaerobic threshold (Bennett et al. 1985) and elevated post-exercise recovery heart rates (Knight et al. 1981) following patrols of between 6 weeks and 6 months. Unfortunately, there is little information on the effects of a prolonged confinement aboard a submarine on readiness capabilities of Special Operations Forces. Therefore, the objective of the current field-based study was to determine whether there are significant changes in aerobic performance of SEALs following a 33-day submarine deployment.

2. Methods

2.1. Subjects

All of the 19 male subjects were active duty military volunteers from SEAL Delivery Vehicle (SDV) Team One stationed at Ford Island, Pearl Harbor, Hawaii (HI). Ten of the subjects participated in SDV operations aboard a converted ballistic missile class submarine (USS Kamehameha SSN 642), and were designated the deployed SEAL team (DST). The other nine subjects remained at their parent command on Ford Island for the duration of the study and were designated as the non-deployed (control) SEAL team (NDST). The deployed and control SEAL teams were not significantly different (mean ± SD) in age (DST = 29.2 ± 4.2 years; NDST = 29.7 ± 5.8 years), height (DST = 1.80 ± 0.047 m; NDST = 1.782 ± 0.082 m) and body mass (DST = 83.7 ± 6.3 kg; NDST = 84.7 ± 11.5 kg). Subjects were made aware of the nature of the experiment and signed informed consent forms. The experimental protocol was reviewed and passed by the Naval Submarine Medical Research Laboratory's Committee for the Protection of Human Subjects.

2.2. Protocol

Two weeks prior to deployment, the subjects performed a 12-min maximal run for distance (Cooper 1968) on a rectangular outdoor running course set up on the airfield at Ford Island, HI. The Cooper test was chosen as the measure of aerobic performance because: (1) previous research has shown good correlations ($r = 0.90$) between Cooper test performance and $\dot{V}O_2$max (Cooper 1968); (2) it can be performed at most places where the SEALs are deployed with minimal set-up time and equipment, thus making it an ideal field test of aerobic performance; (3) the performance measure (i.e. the distance run in 12 min) is more directly associated with SEAL mission-related performance than cardiovascular measures of performance; and (4) many subjects can be tested at the same time.

In order to foster competition and promote maximal performance, the SEALs performed the Cooper test in groups. Each subject wore a Polar® Vantage NV™ heart rate (HR) monitor (Model # 1901001, Polar Electro Oy, Kempele, Finland) during the test and for 5 min of recovery to monitor HR profiles during exercise and recovery. The watches were set to record and store HR data every 5 s. Data from the HR watches were downloaded to a Pentium II IBM ThinkPad™ using the Polar Advantage Interface™ and Polar Precision Performance Software™ (version 2.02.004) for later analysis. At the end of the test, the total distance run by each subject was marked and then measured with a surveyor's wheel (Road Runner™, Model RR-318, Keson®, Naperville, Il., USA).

Test-retest reliability of running performance was assessed by having the SEALs complete a second pre-deployment Cooper test 5 days later. Four days after their second pre-deployment Cooper test the deployed SEAL team boarded the USS Kamameha for a 33-day deployment. At the end of deployment they were housed in a tent camp for 7 days in Korea awaiting air transportation back to Pearl Harbor, HI. Post-deployment testing of the deployed SEAL team was conducted 1 day after their return to Ford Island, HI, and 9 days after leaving the submarine. The final Cooper test was conducted on the NDST 2 days prior to the return of the DST. All the tests were conducted at the same time of day (09:00–10:00 h) and under similar weather conditions (sunny, mean air temperature = 24.8–26.2°C, mean wind speed = 3.4–10.5 knots).

No specific intervention or instructions regarding maintenance of a regular exercise regimen were provided to the subjects by the research team. However, command doctrine encourages all SEALs to maintain a regular exercise regimen. In order to estimate the frequency and duration of physical activity during the study, the SEALs completed daily exercise logs for 4 days during the week prior to deployment, for 3 days during early deployment, and three days during late deployment. They recorded the total amount of time spent exercising per day to the nearest 30 min.

2.3. Statistical analysis

The dependent variables recorded from each SEAL's Cooper test included the distance run (D), mean exercise heart rate (HR_{mean}), maximum exercise heart rate (HR_{max}), the slope of the initial heart rate recovery curve ($HR_{recslope}$) and heart rate recovery time ($HR_{rectime}$). The $HR_{recslope}$ was derived from linear regression analysis of the first 30 s of HR recovery data; $HR_{rectime}$ was calculated as the time for the HR to drop to 70% of the final exercise HR value. Battery failure prevented analysis of the heart rate data from three of the control subjects. Also, one subject failed to perform a second pre-deployment Cooper test; however, data from this subject were still included in the preversus post-deployment analysis. Paired t-tests were conducted between the two pre-deployment Cooper tests to detect any systematic bias in the dependent variables that may occur as result of practice. Relative and absolute reliability of test-retest performance on the Cooper test was determined using the Pearson Product moment correlation and limits of agreement analysis, respectively, according to the methods outlined by Atkinson and Nevill (1998). Planned comparisons between the deployed and control groups were conducted using independent t-tests, while paired t-tests were used to compare changes in the dependent variables between pre- and post-deployment. Data from the second pre-deployment test were used in all comparisons with the post-deployment data.

A split plot repeated measures analysis of variance was used to compare differences in daily exercise time between the two SEAL groups across pre-deployment and deployment phases. Significance for all statistical tests was set at $p < 0.05$. Data are presented as means $\pm$ SD together with $\pm 95\%$ confidence limits for the percentage change in the dependent variable between the second pre-deployment test and the post-deployment test.

3. Results

During the second pre-deployment test, the SEALs ran on average 68.6 m further than during the first pre-deployment test (SD = ± 136.5 m, $n = 18$, $p < 0.05$) with no significant change in any of the HR variables. The test-retest correlation coefficient for the distance run was 0.79 ($p < 0.001$). Analysis of the differences in running distance between test and retest performance on the pre-deployment Cooper tests indicated the data to be homoscedastic (heteroscedastic correlation, $r = 0.01$, $p = 0.959$) and normally distributed (Shapiro-Wilks' W test). The 95% limits of agreement for the Cooper test were -68.6 ± 267.5 m.

Individual changes in running performance between the second pre-deployment Cooper test and the post-deployment Cooper test are shown in figure 1. For the NDST there were approximately an equal number of subjects who increased their performance as decreased their performance post-deployment. In contrast all but two individuals ran a shorter distance post-deployment than pre-deployment in the DST. A typical heart rate profile over the 12-min run and during recovery is shown in figure 2. Maximal heart rate was usually attained during the final minute of exercise and reached over 90% of the theoretical age-predicted maximum heart rate (i.e. predicted maximum HR = 220 − age) for all subjects during the pre- and post-deployment tests. Furthermore, the group average HR_{mean} maintained over the 12-min run was 90% of the predicted maximum heart rate on each of the three Cooper tests.

Table 1 summarizes the results of the comparison between the second pre-deployment test and the post-deployment test for both SEAL teams. For the NDST none of the variables showed any significant change between pre- and post-deployment testing. The only variable to show a significant difference between the NDST and DST was the pre-deployment HR_{mean}, which was 9 beats $\times$ min^{-1} lower for the NDST ($p < 0.05$). However, when running performance was expressed as a percentage change in the distance run between the pre- and post-deployment tests, there was a significant difference between the DST and NDST ($p < 0.01$). The NDST showed a non-significant $2 \pm 4.7\%$ (mean $\pm 95\%$ confidence limits) increase in running distance whereas the DST showed a $7 \pm 3.7\%$ decrement in the distance run following deployment ($p < 0.01$). The decrement in performance of the DST was not associated with any changes in HR_{mean} and HR_{max}, however $HR_{rectime}$ was $47 \pm 39.1\%$ (mean $\pm 95\%$ confidence limits) longer following the deployment ($p < 0.05$). In addition, $HR_{recslope}$ decreased by $17 \pm 16.7\%$ (mean $\pm 95\%$ confidence limits, $p < 0.05$) for the DST following deployment (table 1).

Analysis of the exercise logs revealed no differences in the amount of daily exercise activity between pre-deployment and deployment [$F_{(1,15)} = 1.74$, $p > 0.05$] for both the NDST (mean ± SD pre-deployment = 2.2 ± 2.3 h day^{-1}, during deployment = 2.7 ± 1.3 h day^{-1}, $n = 8$) and DST (mean ± SD pre-deployment = 0.7 ± 0.3 h day^{-1}, during deployment = 0.8 ± 0.6 h day^{-1}, $n = 9$). However, the DST subjects exercised approximately onethird the amount of time of the NDST over the study period [$F_{(1,15)} = 8.77$, $p < 0.01$].

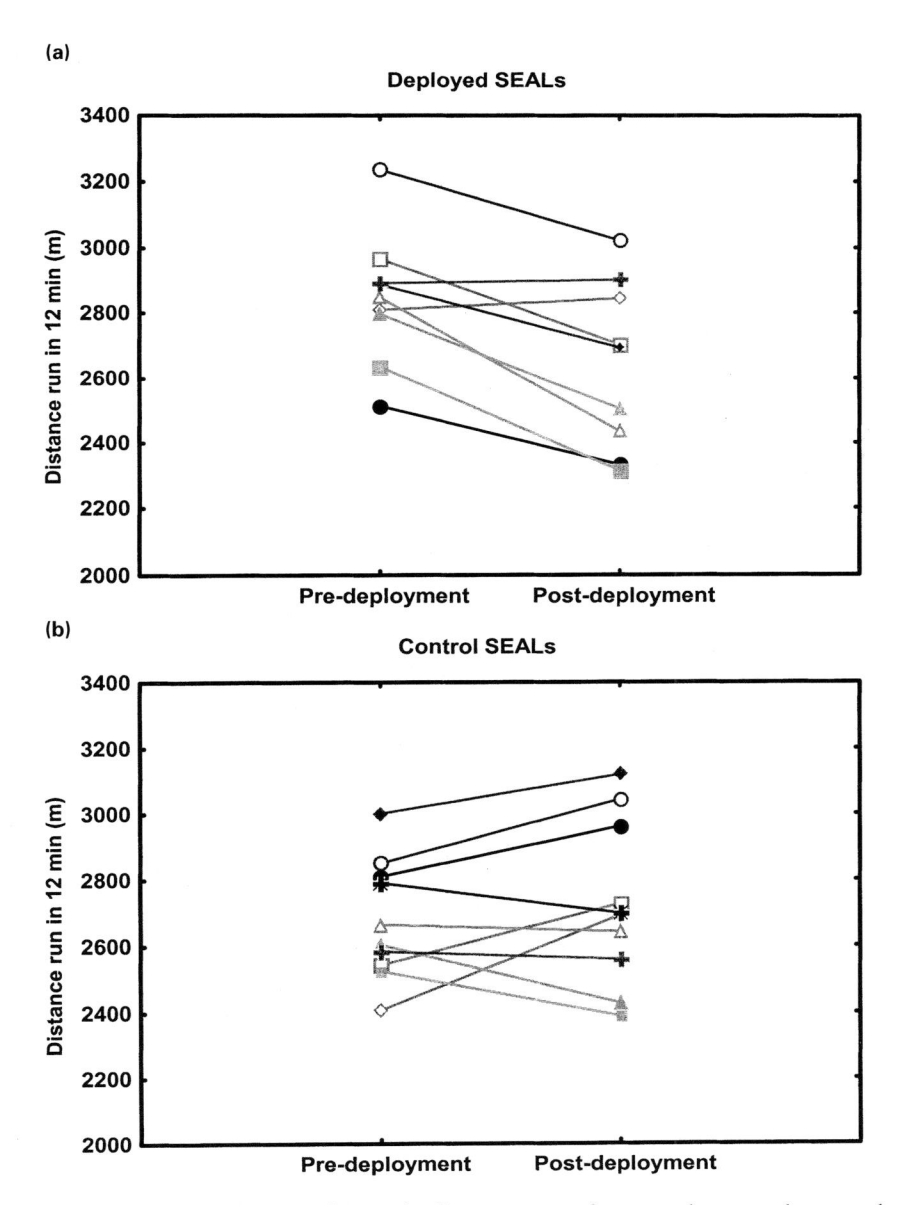

Figure 1. Case profiles showing changes in Cooper test performance between the second pre-deployment test and the post-deployment test for (a) the deployed SEALs and (b) the control SEALs.

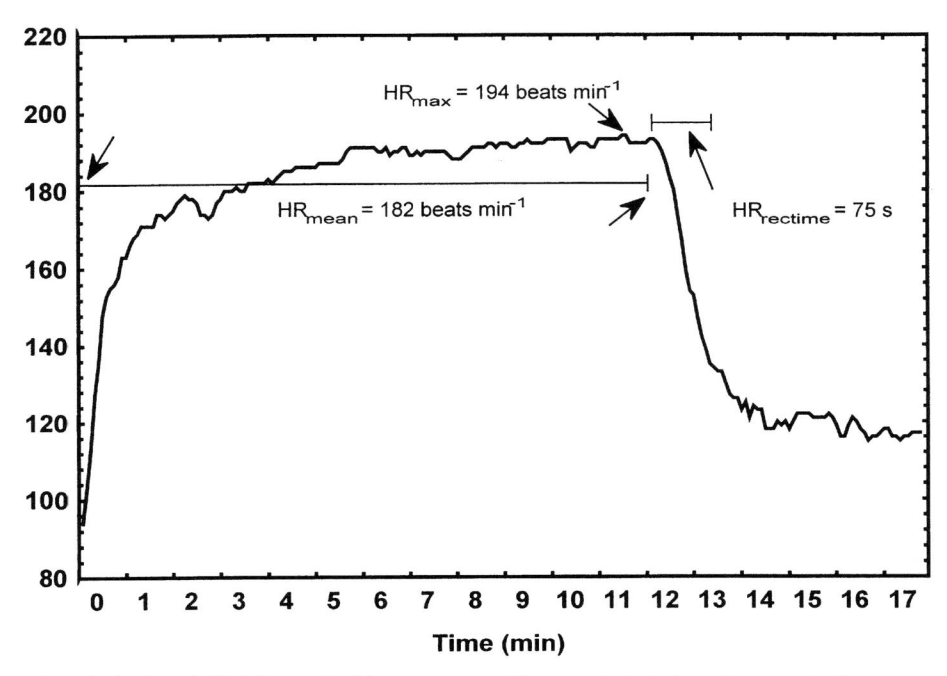

[Figure: heart rate profile graph with HR_max = 194 beats min⁻¹, HR_mean = 182 beats min⁻¹, HR_rectime = 75 s]

Figure 2. A deployed SEAL's typical heart rate profile recorded during the post-deployment Cooper test. The theoretical age predicted maximal heart rate (i.e. predicted maximal HR = 220−age) for this subject was 191 beats min^{-1}.

4. Discussion

4.1. *General discussion*

The results of the current study indicate that SEALs experience a significant decrease in running performance following a 33-day submarine deployment. This decrease in performance was not associated with changes in exercise heart rate since HR_{mean} and HR_{max} were similar during the pre- and post-deployment Cooper tests. However, it is well known that changes in running performance after training or detraining greatly exceed any associated changes in the cardiovascular system involving oxygen consumption or heart rate (Lambert *et al.* 1998). Consequently, the distance run on the Cooper test will give a better indication of an individual's state of aerobic conditioning than measures of exercise heart rate. Furthermore, from a practical perspective, running performance is likely to be more directly related to the physical demands of actual mission performance than physiological measures of cardiovascular function.

Misleading or inappropriate conclusions could result if running performance is used as the sole criterion for determining the state of aerobic conditioning in individuals whose motivation to perform maximally may differ between testing conditions. Although SEALs are known to be highly competitive and motivated individuals (Braun *et al.* 1994), the fact that they knew the objectives of the study could have influenced them to under-perform following deployment to exaggerate deconditioning effects. This strategy is motivated by the fact that large decrements in aerobic performance following the submarine deployment may prompt SEAL commanders to find alternative methods to deploy their troops for future missions. From the deployed SEALs perspective this would mean less time away from home

Table 1. Summary of results comparing running performance and heart rate measures on the Cooper test pre-† and post-deployment (means ± SD).

	Pre-deployment	Post-deployment	n	% Change	−95% Confidence limit (%)‡	+95% Confidence limit (%)‡
Distance run, D (m)						
DST	2839 ± 192	2646 ± 242**	10	−6.8**	−10.5	−3.2
NDST	2667 ± 187	2730 ± 262	9	+2.4	−2.4	+7.1
HR_{mean} (beats min^{-1})						
DST	174 ± 6.1	175 ± 9.6	10	+0.6	−2.0	+3.2
NDST	164 ± 5.6	166 ± 7.8	6	+1.5	−0.6	+3.6
HR_{max} (beats min^{-1})						
DST	185 ± 5.7	185 ± 10.5	10	+0.3	−2.3	+3.0
NDST	176 ± 7.2	179 ± 7.6	6	+1.3	−1.9	+4.4
$HR_{rectime}$ (s)						
DST	92 ± 21	136 ± 64	10	+46.8*	+7.8	+85.9
NDST	110 ± 31	140 ± 48	5	+31.5	−30.2	+93.1
$HR_{recslope}$ (beats min^{-2})						
DST	−0.748 ± 0.226	−0.605 ± 0.192	10	−17.1*	−31.3	+3.0
NDST	−0.566 ± 0.179	−0.497 ± 0.157	5	−6.9	−42.7	+28.8

†Data are from the second pre-deployment test.
‡The ±95% confidence limits shown are for the percentage change.
Significant differences between pre- and post-deployment tests: *$p < 0.05$, **$p < 0.01$.
HR_{mean} = Mean exercise heart rate; HR_{max} = Maximum exercise heart rate; $HR_{rectime}$ = Time taken for the heart rate to drop by 30% from that recorded during the final 5 s of exercise; $HR_{recslope}$ = Slope of the heart rate response during the first 30 s of recovery; n = number of subjects used in the dependent t-test analysis; DST = Deployed SEAL team; NDST = Non-deployed SEAL team.

and family, as well as less time within the cramped confinement of the submarine. Despite this possibility, the exercise heart rate data indicated that all the SEALs maintained a high intensity running pace throughout each of the Cooper tests. The fact that there was no difference in HR_{max} or HR_{mean} between pre- and post-deployment tests in both groups suggests that motivation factors were not a significant factor in the reduced post-deployment running performance of the DST.

In addition to the decrease in running performance, evidence of aerobic deconditioning in the DST is reflected in their change in the HR recovery profiles. Heart rate recovery is correspondingly faster in those individuals who have a higher aerobic capacity (Cardus and Spencer 1976, Hagberg et al. 1980, Kirby and Hartung 1980, Darr et al. 1988). Furthermore, endurance training results in a more rapid rate of recovery in HR following exercise (Hagberg et al. 1980). Consequently, in the present study, slower rates of heart rate recovery following exertion were attributed to a deconditioning effect. We provide two indices of HR recovery that may be useful to follow changes in the aerobic training status of an individual. The first index, $HR_{rectime}$, provides a quick field-based estimate of aerobic training status that requires few calculations or knowledge about the characteristics of the HR recovery profile. This index includes both the initial fast component and a portion of the subsequent slow component of the exponential decline in HR following exercise (Imai et al. 1994). The time constant of the slow component of the HR recovery curve depends on the exercise workload (Imai et al. 1994), hence $HR_{rectime}$ will be somewhat sensitive to differences in the running speed between tests. Results of the present study indicate that the DST had a substantial increase in recovery time following the submarine deployment as measured by the $HR_{rectime}$ index. This finding is tempered somewhat by the fact that the NDST also showed an increase in $HR_{rectime}$, albeit a non-significant one.

The second index of HR recovery, $HR_{recslope}$, focuses on the initial fast component of the HR recovery curve which is unaffected by differences in the workload of the preceding exercise (Millahn and Helke 1968, Nandi and Spodick 1977, Imai et al. 1994). Past research has shown that there are marked differences in the fast phase of recovery between aerobically trained and untrained individuals (Darr et al. 1988). In the present study, the lack of a significant difference in $HR_{recslope}$ between the deployed and control groups post-deployment most likely reflects the large variability of this measure within the groups as well as the small sample size of the groups. However, results from the more powerful within-group analysis show a significant 17% decrease in $HR_{recslope}$ of the deployed SEALs following deployment. This finding provides further evidence that the deployed SEALs experienced a significant deconditioning effect as a result of the deployment.

Currently, the mechanism by which HR recovery is altered with training status is poorly understood. This partly reflects the fact that the underlying mechanism for cardiac deceleration following exercise is still under debate. Based on experiments in which HR recovery was studied under conditions of sympathetic, parasympathetic and double blockade, Savin et al. (1982) concluded that intrinsic rather than neural or hormonal mechanisms regulate the initial post-exercise HR. However, Darr et al. (1988) questioned the conclusions of Savin et al. (1982) and suggested that neural factors are important in HR recovery and that the individual's training status can modulate the balance between sympathetic and parasympathetic influences on HR. More recently, several investigators have pointed to restoration of vagal tone as the primary mechanism for cardiac deceleration during the early phase of HR recovery (Perini et al. 1989, Imai et al. 1994). Based upon current knowledge it is therefore possible that the slower HR recovery of the deployed SEALs following deployment was the result of alterations in neural and intrinsic HR control subsequent to a detraining effect from their prolonged confinement.

The precise cause of the deconditioning of the deployed SEALs is less clear. Unlike submarine personnel, SEALs do not have to perform any watch standing or other submarine duties during their transit time to the theatre of operations. Thus, during this transit period SEALs have considerably more free time to exercise compared to when they are based at their home unit. Despite this factor the DST exercise logs showed no change in the amount of daily exercise performed between the pre-deployment and deployment study period. However, the exercise logs did show that the DST spent less time exercising than their non-deployed colleagues did. The large differences in daily exercise time between the NDST and DST during the week before deployment may reflect the lack of free time available for the DST to do additional physical exercise due to the demands of preparing for the mission. In contrast, the smaller amount of daily exercise performed by the DST relative to the NDST during deployment most likely reflects a combination of factors, including being away from the command physical training programme, reduced motivation, and limited availability of exercise equipment for the DST.

One of the limitations of the exercise logs is that they do not show whether or not there were any differences in the intensity or type of exercise performed while on land versus aboard the submarine. Owing to the limited space and availability of aerobic equipment on board the submarine, it is possible that the deployed SEALs switched from a predominantly aerobic training programme pre-deployment to a largely calisthenics and strength training programme during the deployment and/or that the intensity of their aerobic workouts while aboard the submarine were significantly reduced. Even if the duration and frequency of exercise is maintained at a constant level, reductions in training intensity can lead to a significant loss in aerobic power (Hickson et al. 1985). Furthermore, due to the specificity of training, the substitution of other non-running aerobic activities (e.g. bicycling) for running training have been shown to have limited cross-training benefits in highly trained athletes (Pierce et al. 1990, Tanaka 1994). Consequently, a prolonged layoff from training specific to running could lead to reduced running performance despite substituting running training with other forms of aerobic exercise.

4.2. Study limitations

As field studies typically show larger inter-subject variability than do laboratory-based studies it is desirable to employ large sample sizes to ensure sufficient statistical power to observe significant differences in the dependent variables of interest when they exist. In the present study operational requirements and the small subject population pool limited the number of subjects in each group. For example, for the DST the entire complement of SEALs on the mission volunteered as subjects. A further decrease in statistical power was evident in the present study owing to loss of some of the NDST recovery HR data. Consequently, there may be some between-group comparisons where even though large differences were shown, the large between-subject variability and small number of subjects resulted in insufficient statistical power to avoid committing a Type II error (i.e. that of falsely accepting the null hypothesis of no difference between two means).

A further limitation of the study was the fact that post-deployment testing on most of the DST members was conducted 9 days after they left the submarine (two SEALs remained on the submarine for a further 4 days before flying home). During the first 7 days after leaving the submarine the DST subjects were in Korea waiting for commercial air transportation to Pearl Harbor, HI. As no monitoring of activity was conducted during this period little is known of their activity levels during this time. Given the free time and opportunity it is likely that the SEALs continued some sort of aerobic training. If activities such as running were engaged in during this free time it is possible that some of the decrements in aerobic performance resulting from the submarine deployment may have been offset. However, the extent of any activity during this time period and its potential effects on post-deployment test performance are unknown.

An additional confounding factor that may have potentially affected post-deployment performance of the DST is the 5 h time difference between Korea and Hawaii. Although the DST performed the pre- and post-deployment tests at the same time of day, it is probable that their circadian rhythms will not have been fully synchronized with Hawaii Time during the post-deployment testing. Studies on the effects of circadian rhythms on performance have shown that factors such as flexibility, muscle strength and shortterm high power output vary with time of day in a sinusoidal manner and peak in the early evening close to the daily maximum in body temperature (Atkinson and Reilly 1996). Furthermore prolonged submaximal exercise carried out in hot conditions shows optimal performance in the morning

(Atkinson and Reilly 1996). In the present study any jet-lag or circadian asynchrony in the DST will probably have had only a small impact on their post-deployment performance tests for a number of reasons. First, the DST subjects were not tested until at least 24 h after arriving in Hawaii, thus permitting a certain amount of recovery from jet-lag before testing. Second, the considerable inter-individual variation in responses to sleep loss (Meney *et al.* 1998) as well as individual differences in performance rhythms (Atkinson and Reilly 1996) will probably have reduced the chances of observing a measurable circadian change in performance between the pre- and post-deployment tests. Nevertheless, evidence from the scientific literature suggests that small decreases in running performance could be expected immediately following an eastward flight across multiple time zones (Wright *et al.* 1983).

One final note that should be considered when interpreting the results of the current study is that this research was performed on a single submarine platform using SEALs from one command. Personal and group dynamics of other SEAL teams may be different than those of SEAL Team One resulting in different approaches to physical training while on prolonged submarine missions. Differences in command leadership as well as mission duration, objectives and requirements could have significant impact on the degree of deconditioning following a prolonged submarine mission. Furthermore, differences in the submarine platform could significantly affect the working and living conditions of the SEAL team and lead to different outcomes on mission-related performance. For example, when compared to a converted ballistic missile submarine the fast attack submarine contains significantly less space for berthing and exercise equipment. Consequently, prolonged missions aboard a fast attack submarine could potentially have more of a negative impact on aerobic performance than that noted in the present study.

5. Conclusions

In conclusion, running performance and heart rate recovery profiles of US Navy SEALs were impaired following a 33-day deployment on a converted ballistic missile submarine. This decrement in aerobic performance is likely due to a combination of (1) deconditioning resulting from reduced levels of activity and alterations in training regimens during deployment, and (2) asynchrony of circadian rhythms and the effects of jet-lag resulting from their air flight travel immediately before conducting the post-deployment tests. Currently, it is unknown to what level the observed decrements in aerobic performance will affect the SEALs performance during an actual mission; it can be inferred from the results that mission-related performance will be impacted most during and immediately following extended periods of high intensity physical exertion. Mission elements that require the SEALs to recover quickly and react following periods of extended exertion may be compromised by the delay in recovery time associated with aerobic deconditioning. Since any factor that potentially affects mission success is a major concern for Special Operations Forces, countermeasures should be sought to maintain the aerobic fitness of SEALs during extended submarine deployments. Providing exercise equipment and structured training programmes that permit SEALs to attain the type and intensity of aerobic exercise equivalent to their land-based training regimens may prove to be the most fruitful avenue for avoiding aerobic deconditioning during prolonged submarine deployments.

Acknowledgements

This study was supported by the US Special Operations Command. The views expressed in this report are those of the authors and do not reflect the official policy or position of the Department of the Navy, the Department of Defense or the US Government. This work was undertaken by US Government employees as part of their official duties and therefore cannot be copyrighted. The authors wish to thank CDR Howard F. Reese, Commanding Officer of the USS Kamehameha (SSN 642), CDR Gerald Weer, Commanding Officer of SDV Team One and LCDR Sid Ellington, Executive Officer of SDV Team One for their co-operation in this study. The authors also wish to acknowledge the support of LCDR Perry Willette, CPT Mary A. Kautz, HMCS Thomas Bartholomew and HM2 Allen Miller during data collection, and the SEALs of SDV Team One for their participation. Design and statistical analysis of the exercise logs was kindly provided by Dr Christine Schlichting.

References

AMERICAN COLLEGE of SPORTS MEDICINE. 1998, American College of Sports Medicine Position Stand. The recommended quantity and quality of exercise for developing and maintaining cardiorespiratory and muscular fitness, and flexibility in healthy adults, *Medicine and Science in Sports and Exercise, 30, 975–991.*

ATKINSON, G. and NEVILL, A. M. 1998, Statistical methods for assessing measurement error (reliability) in variables relevant to sports medicine, *Sports Medicine*, 26, 217–238.

ATKINSON, G. and REILLY, T. 1996, Circadian variation in sports performance, *Sports Medicine*, 21, 292–312.

BARNES, L. and STRAUSS, R. H. 1986, The U.S. Navy SEAL team: Total commitment to total fitness, *Physician and Sports medicine, 14, 176–183.*

BENNETT, B. L., SCHLICHTING, C. L. and BONDI, K. R. 1985, Cardiorespiratory fitness and cognitive performance before and after confinement in a nuclear submarine, *Aviation Space and Environmental medicine, 56, 1085–1091.*

BRAUN, D. E., PRUSACZYK, W. K., GOFORTH, Jr., H. W. and PRATT, N. C. 1994, Personality profiles of U.S. Navy Sea-Air-Land (SEAL) personnel, Publication 94-8, Naval Health Research Center: San Diego, CA.

CARDUS, D., and SPENCER, W. A. 1976, Recovery time of heart frequency in healthy men: its relation to age and physical condition, *Archives of Physical Medicine and Rehabilitation, 21, 71–76.*

COOPER, K. H. 1968, A means of assessing maximal oxygen uptake, *Journal of the American Medical Association, 203, 201–204.*

COYLE, E. F., MARTIN, W. H. 3, SINACORE, D. R., JOYNER, M. J., HAGBERG, J. M. and HOLLOSZY, J. O. 1984, Time course of loss of adaptations after stopping prolonged intense endurance training, *Journal of Applied Physiology: Respiration, Environmental and Exercise Physiology, 57, 1857–1864.*

DARR, K. C., BASSETT, D. R., MORGAN, B. J. and THOMAS, D. P. 1988, Effects of age and training status on heart rate recovery after peak exercise, *American Journal of Physiology, 254, H340–343.*

DUETT, M. M. 1982, Aerobic exercise habits and motivation to exercise on board a deployed submarine. Unpublished dissertation, San Diego State University, San Diego, CA.

GRAVES, B. and WHITMAN, E. 1999, The Virginia class: America's next submarine, *Undersea Warfare, 1, 2–7.*

GREENLEAF, J. and KOZLOWSKI, S. 1982, Physiological consequences of reduced physical activity, in R. Terjung (ed.), *Exercise and Sport Science Reviews*, vol. 10 (Philadelphia, PA: Franklin Press Institute), 84–119.

HAGBERG, J. M., HICKSON, R. C., EHSANI, A. A. and HOLLOSZY, J. O. 1980, Faster adjustment to and recovery from submaximal exercise in the trained state, *Journal of Applied Physiology: Respiration, Environmental and Exercise Physiology, 48, 218–224.*

HICKSON, R. C., FOSTER, C., POLLOCK, M. L., GALASSI, T. M. and RICH, S. 1985, Reduced training intensities and loss of aerobic power, endurance and cardiac growth, *Journal of Applied Physiology, 58, 492–499.*
IMAI, K., SATO, H., HORI, M., KUSUOKA, H., OZAKI, H., YOKOYAMA, H., TAKEDA, H., INOUE, M. and KAMADA, T. 1994, Vagally mediated heart rate recovery after exercise is accelerated in athletes but blunted in patients with chronic heart failure, *Journal of the American College of Cardiology, 24, 1529–1535.*
KIRBY, T. and HARTUNG, G. H. 1980, Heart rate deceleration in conditioned and unconditioned men, *American Corrective Therapy Journal, 34, 161–163.*
KNIGHT, D. R., BONDI, K. R., DOUGHERTY, J. H., SHAMRELL, T. P., YOUNKIN, R. K., WRAY, D. D. and MOONEY, L. W. 1981, The effects of occupational exposure to nuclear submarines on human tolerance for exercise, *Federation Proceedings, 40, 497.*
LAMBERT, M. I., MBAMBO, Z. H., and ST CLAIR GIBSON, A. 1998, Heart rate during training and competition for long-distance running, *Journal of Sports Sciences, 16, S85–S90.*
MENEY, I., WATERHOUSE, J. ATKINSON, G. and REILLY, T. 1998, The effect of one night's sleep deprivation on temperature, mood, and physical performance in subjects with different amounts of habitual physical activity, *Chronobiology International, 15, 349-363.*
MILLAHN, H. P. and HELKE, H. 1968, Uber Beziehungen zwischen der Herzfrequenz wahrend Arbeitsleistung und in der Erholungsphase in Abhangikeit von der Leistung und der Erholungsdauer. Zugleich eine Betrachtung zum Verhalten der Herzfrequenz in der Erholungsphase. [The relationships between exercise and recovery heart rates; their dependence on work load and duration of recovery with special reference to the behaviour of the heart during recovery], *Internationale Zeitschrift Fur Angewandte Physiologie einschliesslich Arbeitsphysiologie, 26, 245–257.*
NANDI, P. S. and SPODICK, D. H. 1977, Recovery from exercise at varying work loads. Time course of responses of heart rate and systolic intervals, *British Heart Journal, 39, 958–966.*
PERINI, R., ORIZIO, C., COMANDE, A., CASTELLANO, M., BESCHI, M. and VEICSTEINAS, A. 1989, Plasma norepinephrine and heart rate dynamics during recovery from submaximal exercise in man, *European Journal of Applied Physiology, 58, 879–883.*
PIERCE, E. F., WELTMAN, A., SEIP, R. L. and SNEAD, D. 1990, Effects of training specificity on the lactate threshold and $\dot{V}O_2$ peak, *International Journal of Sports Medicine, 11, 267–272.*
PRUSACZYK, W. K., GOFORTH, Jr., H. W. and NELSON, M. L. 1991, Characteristics of physical training activities of West Coast U.S. Navy Sea-Air-Land personnel (SEALS), Publication 90-35, Naval Health Research Center: San Diego, CA.
PRUSACZYK, W. K., GOFORTH, Jr., H. W. and NELSON, M. L. 1994, Physical training activities of East Coast U.S. Navy SEALs, Publication 94-24, Naval Health Research Center: San Diego, CA.
PRUSACZYK, W. K., STUSTER, J. W., GOFORTH Jr., H.W., SOPCHICK SMITH, T. and MEYER, L. T. 1995, Physical demands of U.S. Navy Sea-Air-Land (SEAL) operations, Publication 95-24, Naval Health Research Center: San Diego, CA.
RAVEN, P. B., WELCH-O'CONNOR, R. M. and SHI, X. 1998, Cardiovascular function following reduced aerobic activity, *Medicine and Science in Sports and Exercise, 30, 1041–1052.*
SAVIN, W. M., DAVIDSON, D. M. and HASKELL, W. L. 1982, Autonomic contribution to heart rate recovery from exercise in humans, *Journal of Applied Physiology: Respiration, Environmental and Exercise Physiology, 53, 1572–1575.*
TANAKA, H. 1994, Effects of cross-training. Transfer of training effects on V·O2 max between cycling, running and swimming, *Sports Medicine, 18, 330–339.*
UNDERSEA WARFARE, 1999, http://www.chinfo.navy.mil/navpalib/cno/n87/usw/issue_3/ pull-out/speciall_warfare.htm
WRIGHT, J. E., VOGEL, J. A., SAMPSON, J. B. and KNAPIK, J. J. 1983, Effects of travel across time zones (jet-lag) on exercise capacity and performance, *Aviation Space and Environmental Medicine, 54, 132–137.*

6   Use of melatonin in recovery from jet-lag following an eastward flight across 10 time-zones

B. J. Edwards†, G. Atkinson†, J. Waterhouse†, T. Reilly†*, R. Godfrey‡ and R. Budgett‡

†Research Institute for Sport and Exercise Sciences, Liverpool John Moores University, Henry Cotton Campus, 15–21 Webster Street, Liverpool L3 2ET, UK

‡British Olympic Medical Centre, Northwick Park Hospital, Watford Road, Harrow, Middlesex HA1 3UJ, UK

Keywords: Body clock; Circadian rhythm; Grip strength; Body temperature; Eastward travel.

Subjective, physiological and physical performance variables are affected following travel across multiple time-zones (jet-lag). The objective of the study was to examine the effects of oral melatonin in alleviating jet-lag by investigating its effects on subjects who had flown from London to Eastern Australia, 10 time-zones to the east. Melatonin (5 mg day^{-1}) or placebo capsules were administered to 14 experimental (13 males and 1 female) and 17 control subjects (15 males and 2 females), respectively, in a double-blind study; the time of administration was in accord with the current consensus for maximizing its hypnotic effect. Grip strength and intra-aural temperature were measured on alternate days after arrival at the destination, at four different times of day (between the times 07:00–08:00 h, 12:00–13:00 h, 16:00–17:00 h and 19:00–20:00 h local time). In addition, for the first 6–7 days after arrival in Australia, subjective ratings of jet-lag on a 0–10 visual analogue scale and responses to a Jet-lag Questionnaire (incorporating items for tiredness, sleep, meal satisfaction and ability to concentrate) were recorded at the above times and also on retiring (at about midnight). Subjects continued normally with their work schedules between the data collection times. Subjects with complete data (13 melatonin and 13 placebo subjects), in comparison with published data, showed partial adjustment of the diurnal rhythm in intra-aural temperature after 6 days. A time-of-day effect was evident in both right and left grip strength during adjustment to Australian time; there was no difference between the group taking melatonin and that using the placebo. Right and left grip strength profiles on day 6 were adjusted either by advancing or delaying the profiles, independent of whether subjects were taking melatonin or placebo tablets. Subjects reported disturbances with most measures in the Jet-lag Questionnaire but, whereas poorer concentration and some negative effects upon sleep had disappeared after 3–5 days, ratings of jet-lag and tiredness had not returned to 'zero' (or normal values), respectively, by the sixth day of the study. Subjects taking melatonin showed no significant differences from the placebo group in perceived irritability, concentration, meal satisfaction, ease in getting to sleep and staying asleep, frequency of bowel motion and consistency of the faeces. These results suggest that, in subjects who, after arrival, followed a busy schedule which resulted in frequent and erratic exposure to daylight, melatonin had no benefit in alleviating jet-lag or the components of jet-lag, and it did not influence the process of phase adjustment.

*Author for correspondence. e-mail: t.p.reilly@livjm.ac.uk

1. Introduction

Most travellers suffer from 'jet-lag' after rapid transportation across multiple time-zones; this syndrome results from a temporary mismatch between the phase of the traveller's body clock and time in the new environment (Waterhouse et al. 1997). Until adjustment of the body clock has occurred, individuals feel tired during the new daytime, yet they are unable to sleep properly during the new night. For athletes this loss in sleep can affect mood and powers of concentration, increase irritability with coaches and fellow team members, and might result in poorer training performances and, even, sub-optimal competition results (Reilly et al. 1997b). However, detailed and rigorous studies of these changes as they affect athletes have not been performed (Youngstedt and O'Connor 1999).

For American and European athletes travelling to and competing in the Sydney Olympics in 2000, there is, therefore, a major problem while they are adjusting to the new time-zone. Current advice is that travellers should allow 1 day for each time-zone crossed in order to recover from the symptoms of jet-lag (Reilly et al. 1997b, Waterhouse et al. 1997). Athletes do not necessarily heed such advice owing to the interruptions to training that arriving earlier would cause, to cost and changes in diet, and to the discomfort and tensions that are reported by those who have lived in athletes' villages or unfamiliar accommodation.

Nevertheless, it is important to have methods of coping with jet-lag and its associated symptoms, especially strategies that minimize its detrimental effects and help to speed the process of adjustment. One method that received much attention in the 1990s was the ingestion of exogenous melatonin, a hormone normally secreted by the pineal gland during the late evening and throughout the night. There have been several reports that exogenous melatonin reduced subjective assessments of 'jet-lag' and its symptoms (Arendt et al. 1987, 1997, Petrie et al. 1989, Claustrat et al. 1992, Suhner et al. 1998), but also reports that exogenous melatonin was ineffective or only marginally effective (Nickelsen et al. 1991, Samel et al. 1991, Petrie et al. 1993, Spitzer et al. 1999). Part of the reason for these conflicting results might have been the differences in the lifestyles (associated with which would be differences in light exposure) of the subjects when in the new time-zone. One of the aims of this study, therefore, was to see if melatonin ingestion was effective in the particular case of subjects who, after travel from the UK to Australia, engaged in busy daily schedules that would involve erratic exposure to bright light outdoors.

There is another problem when considering 'jet-lag'. This arises because jet-lag is a global term covering several symptoms, and it has been shown elsewhere that these symptoms do not recover at the same rate (Waterhouse et al. 2000). This approach is now extended to investigate whether melatonin exerts an effect upon all of these symptoms.

Melatonin is often considered to be a natural hypnotic (Haimov and Arendt 1999), and its administration in the current study was timed so that its hypnotic effect in the new time-zone would be promoted (Arendt and Deacon 1997). Recent work (Lewy et al. 1992, 1998, Zaidan et al. 1994) indicates that melatonin might also shift the rhythm of a given biological variable in accord with its phase response curve. Details of this curve differ, but, considering the circadian rhythm of body temperature, melatonin given in the afternoon and evening (before the body temperature minimum) causes a phase advance, and in the hours after this minimum it causes a phase delay. Therefore, a second aim of the current study was to investigate if melatonin, taken in the evening by local time, influenced the direction of adjustment of the circadian system to the time-zone transition. In summary, the present study was concerned with assessing not only the efficacy of oral melatonin (5 mg day^{-1}) in alleviating 'jet-lag' and the components of jet-lag, but also its effects on adjustment of the body clock as inferred from the diurnal variations of physical performance and body temperature.

2. Methods

2.1. Subjects

The participants were 31 volunteers (28 males and 3 females), all travelling to the East Coast of Australia. They were either professional sports officials or sports scientists visiting Australia on business. Subjects gave written, informed consent to participate in the study, and the procedures were approved by the Human Ethics Committee of Liverpool John Moores University. The subjects had been given information regarding jet-lag in the form of a booklet produced by staff at the British Olympic Medical Centre (London).

Personal data comprising age, stature, body mass, chronotype score (Barton et al. 1990) and self-rated fitness (Barton et al. 1990) were collected before the flight. Initially there were 34 subjects who were divided into matched pairs on the basis of their physical characteristics (table 1).

Accompanying medical staff assigned numbers to these 17 matched pairs and then gave out numbered packages containing the sets of pills to be taken. The researchers were blind as to which of each pair of packages contained melatonin (Penn Pharmaceuticals Ltd, Gwent, UK) and which contained placebo; this information was not made available until after the end of the study.

Owing to illness, three of the subjects were obliged to drop out of the study; in the event, they were all from the melatonin group, but their illness was unrelated to the capsules taken. Also, owing to their schedules while in Australia, three of the placebo subjects could not provide a complete set of physiological data. Other observations of grip strength, body temperature and questionnaire data were also missed occasionally. These missing cases never exceeded 2 per variable (apart from one subject in each of the placebo and melatonin groups whose physiological data were discarded) and the missing data were estimated by the method of Winer et al. (1991). In summary, physiological data from 26 subjects were used (13 taking melatonin and 13 taking the placebo) and responses to the Liverpool Jet-lag Questionnaire (Winterburn et al. 1997, Waterhouse et al. 2000) from all 31 subjects.

Subjects had originally been matched for a number of factors; as three subjects from the melatonin group had dropped out, the two groups were re-tested for all factors. There was no significant difference between the melatonin and the placebo

Table 1. Physiological characteristics of subjects. Mean (SD) for age, stature, body mass, chronotype score and self-rated fitness (where a higher score indicates a higher rating of fitness) for those taking melatonin (13 males and 1 female) and those taking placebo (15 males and 2 females).

	Age (years)	Stature (m)	Body mass (kg)	Chronotype score	Self-rated fitness (1–5 scale)
Melatonin	40 (13)	1.78 (0.05)	80.5 (8.92)	35 (4.89)	3.6 (1.01)
Placebo	41 (12)	1.78 (0.05)	80.5 (15.29)	38 (5.23)	3.4 (1.18)

group with regard to age ($t_{29} = 0.217$, $p = 0.985$), self-rated fitness ($t_{29} = 0.455$, $p = 0.652$), transmeridian travel experience in both an eastward ($t_{29} = 1.024$, $p = 0.314$) or westward direction ($t_{29} = 0.585$, $p = 0.563$), or body mass ($t_{29} = 0.019$, $p = 0.958$).

2.2. *Measurements made*
The intensity of the daylight during the mornings of the 6 days was 10 000–16 000 lux (Actiwatch-AWL, CambridgeNeurotechnologies, Cambridge, UK).

Measurements were made from the day of arrival for up to 6 days afterwards. Subjects were familiarized with the equipment before leaving the UK. Grip strength for both right and left hands (Takei Kiki Kogyo, Tokyo) and intra-aural temperature (Genius 1000, Sherwood, Nottingham, UK) were measured at four different times (07:00–08:00 h, 12:00–13:00 h, 16:00–17:00 h and 19:00–20:00 h) on days 2, 4 and 6 after arrival in Australia. Giving subjects 1-h 'windows' in which to make their measurements was necessary to comply with their work schedules. The average of two measurements was taken for temperature and grip strength, where the right and left grip strength efforts were measured alternately to limit the effects of fatigue (Reilly *et al.* 1997a). If the difference between the two values for the variable was greater than 2% (grip strength) or 0.2°C (temperature) then a third measurement was taken.

Subjective ratings of jet-lag on a visual analogue scale (VAS) labelled 0 = no jet-lag to 10 = very bad jet-lag (Arendt and Deacon 1997, Reilly *et al.* 1997a) were obtained at the four times stated above and, in some of the subjects, also on retiring at about midnight. This rating of jet-lag was part of the Liverpool Jet-lag Questionnaire (Winterburn *et al.* 1997, Waterhouse *et al.* 2000). This questionnaire also incorporated self-assessments of tiredness, sleep latency, sleep quality, meal satisfaction, ability to concentrate and bowel activity. All these components used scales that ranged from −5 to +5, indicating marked changes compared with normal, and which had a central value of 0 indicating 'normal'. A positive value indicated worse feelings, 'normal' was defined as what the subjects would have felt habitually before the flight. The questionnaire was completed after the physical observations had been made and the subjects had eaten.

2.3. *Travel and drugs*
Travel details differed slightly between the sports administrators and sports scientists with regard to route and times of departure and arrival, but total travelling time was about 24 h and involved two flights separated by a stop-over of about 1 h in each case. Both groups were exposed to Australian daylight on arrival. On average, both melatonin and placebo groups managed a total sleep length of about 6 h duration during the flights, there being continual interruptions owing to noise and the cramped conditions; most of this sleep was taken in the first flight before the stopover. The timing and type of the meals were based on UK time until the stop-over, which hindered subjects from operating on local time in Eastern Australia during this part of the journey.

Following the recommendations of Arendt and Deacon (1997), a capsule containing placebo or 5 mg melatonin was taken by the subjects on the plane between 18:00–19:00 h according to the local time in Eastern Australia (EA). Subsequently, the same dose was given to subjects in the evening (22:00–23:00 h EA time) of the day of arrival, and for the next three evenings. The sports administrators arrived at the final destination early in the morning (05:30 h) and the sports scientists arrived in the evening (17:00–19:00 h).

2.4. Analysis of the results

The two groups were compared for the first 6 days after arrival. The data were analysed by means of the Statistical Package for Social Sciences (SPSS) for Windows, using a three-way ANOVA (with repeated measures) model: group [2 levels] × time-of-day [4 levels] × day-number [3 levels] for the physiological variables, and group [2 levels] × time-of-day [4 levels] × day-number [6 levels] for self-assessment of jet-lag and the other questionnaire components. The assumption of sphericity in the ANOVA model was examined using the Huyn-Feldt epsilon (ε), the degrees of freedom being corrected if $\varepsilon < 0.75$ (Vincent 1995). Not all subjects arrived in Australia at the same time, therefore in order for the time points used in the ANOVAs to be common to all subjects for jet-lag and tiredness only, the definition of 'day' was as follows. 'Day 1' was defined as from 16:00 h on the day of arrival to 12:00 h on the day after this (for the four subjects who arrived in the late evening at their destination, thereby missing the 16:00 h questionnaire session, the results were estimated by the method of Winer et al. 1991). 'Day 2' to 'day 6' were defined simply as starting and finishing at the same clock times as 'day 1'; as a result 'day 6' is from 16:00 h on the fifth day after arrival until 12:00 h on the sixth day after arrival (see top of figure 1).

The profiles of temperature and grip strength for each individual on day 6 were compared with those of adjusted circadian rhythms of body temperature (Reilly 1994) and grip strength (Reilly et al. 1997b). The delay or advance of these

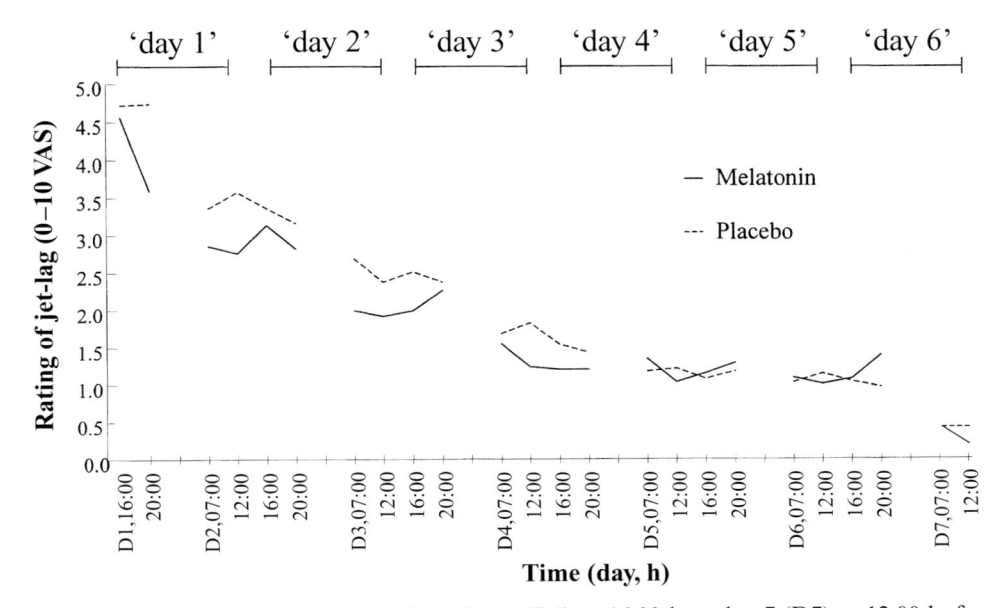

Figure 1. Pooled ratings of jet-lag from day 1 (D1) at 16:00 h to day 7 (D7) at 12:00 h, for melatonin and placebo groups. Bars at the top indicate 'days' used in the ANOVA (for more details see § 2.4). VAS = visual analogue scale.

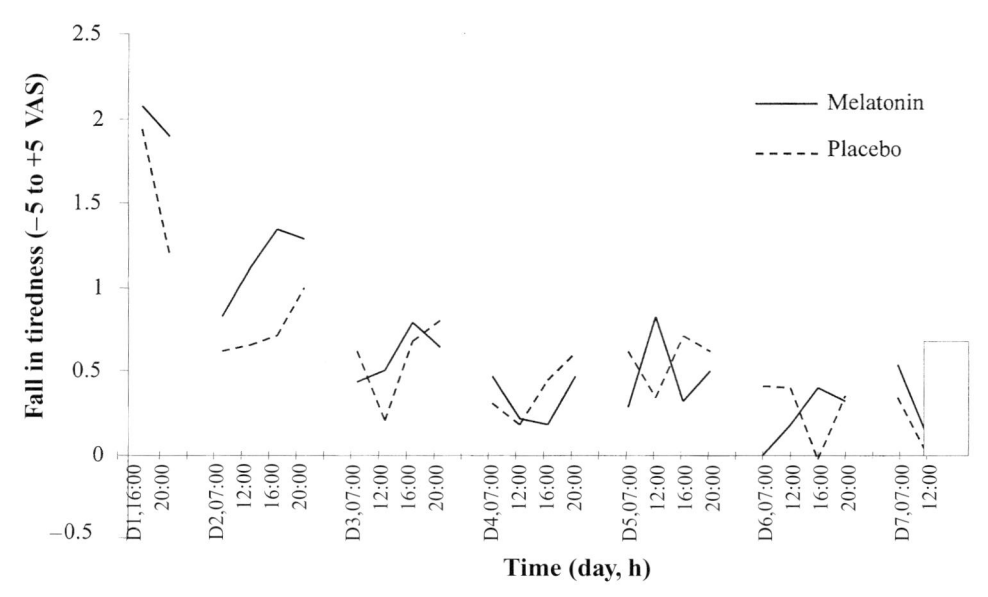

Figure 2. Fall in tiredness (the inverse of tiredness) from day 1 (D1) at 16:00 h to day 7 (D7) at 12:00 h, for melatonin and placebo groups. Bars at the top indicate 'days' used in the ANOVA (for more details see § 2.4). VAS = visual analogue scale.

normative profiles that best fitted the observed profile was used as a measure of the adjustment of the rhythms. Mann-Whitney tests compared the sizes of the shifts in the melatonin and placebo groups. The distribution of advances or delays in the groups was analysed by Chi-squared tests (χ^2).

3. Results

Both groups reported jet-lag during the first 6 days in Australia, with jet-lag being rated highest on day 1 and decreasing thereafter; the greatest amount of adjustment occurred in the first 3 days (figure 1). According to ANOVA, details of which are given in table 2, there was a significant day-number effect, and jet-lag was still significantly different from zero on day 6 ($t_{30} = 3.98$, $p = 0.0004$). There was also a significant time-of-day effect, jet-lag being higher in the afternoon and evening than in the morning and at noon. There was no significant group effect or significant interaction between group and the other ANOVA factors.

Other approaches have been used by other research groups to investigate whether the melatonin and placebo groups rated jet-lag differently, and these methods have been applied to the present data. For example, only the values obtained at 16:00–17:00 h were employed, using a two-way ANOVA with repeated measures model (Petrie et al. 1993). A significant effect for day-number was found, but no group effect.

In the studies of Arendt et al. (1987, 1997) and Nickelsen et al. (1991) the subjects were asked to rate jet-lag on a Likert scale 1 week after arrival at their destination. It is unclear to what extent the individuals' responses would have reflected their feeling at the time of recording or some sort of 'average'. Therefore, two further methods of analysis have been considered. First, the averages of individuals' daily ratings of jet-

Table 2. ANOVA results of *F*-ratio and degrees of freedom in parentheses (with corrected values) are shown for jet-lag, physiological variables and mood. Effects that are significant at $p<0.05$ are in bold. Effects that approach significance ($0.1>p>0.05$) are italicized.

	GP	DAY	DAY×GP	TOD	TOD×GP	DAY×TOD	GP×DAY×TOD
Jet-lag	0.20	**33.27**	0.63	**15.26**	0.36	**3.47**	0.67
	(1, 29)	**(2.12, 61.39)**	(2.12, 61.39)	**(3, 87)**	(3, 87)	**(8.66, 251.04)**	(8.66, 251.04)
Tiredness	0.29	**10.22**	1.07	**5.31**	0.32	**2.46**	0.57
	(1, 29)	**(3.16, 91.60)**	(3.16, 91.60)	**(3, 87)**	(3, 87)	**(15, 435)**	(15, 435)
Intra-aural temperature	0.74	2.19	0.38	**4.53**	0.18	**7.13**	0.80
	(1, 24)	(2, 48)	(2, 48)	**(1.58, 37.89)**	(1.58, 37.89)	**(6, 144)**	(6, 144)
Right grip strength	0.04	1.36	1.01	**4.66**	**2.21**	0.58	0.91
	(1, 25)	(2, 50)	(2, 5)	**(3, 75)**	**(3, 75)**	(6, 150)	(6, 150)
Left grip strength	0.000	2.21	0.52	**2.70**	1.10	1.49	0.67
	(1, 25)	(2, 50)	(2, 50)	**(3, 75)**	(3, 75)	(6, 150)	(6, 150)
Concentration	0.06	**14.80**	0.70	**30.78**	0.38	**7.98**	0.15
	(1, 29)	**(2.70, 78.40)**	(2.70, 78.40)	**(1, 29)**	(1, 29)	**(4, 116)**	(4, 116)
Motivation	0.16	**10.24**	0.84	**16.88**	0.430	**3.36**	0.57
	(1, 29)	**(2.79, 80.85)**	(2.79, 80.85)	**(1, 29)**	(1, 29)	**(4, 116)**	(4, 116)
Irritability	0.11	0.43	1.09	**10.26**	0.03	*2.36*	0.25
	(1, 29)	(2.82, 81.87)	(2.82, 81.87)	**(1, 29)**	(1, 29)	*(4, 116)*	(4, 116)
Headaches	1.10	0.33	0.30	1.16	0.11	0.73	0.34
	(1, 30)	(2.77, 83.14)	(2.77, 83.14)	(1, 30)	(1, 30)	(2.90, 87.08)	(2.90, 87.08)

GP = group; DAY = day number; TOD = time of day.

lag over the 6 days were calculated separately for the melatonin and placebo groups. The variances of the two groups were equal and a two-sample t-test established no significant difference between them ($t_{30} = 0.37$, $p = 0.741$). In the second of these methods, the subjects' average ratings of jet-lag on day 6 only were compared (Arendt et al. 1987); once again, the variances of the two groups were equal and a two-sample t-test failed to show a significant difference ($t_{29} = 0.22$, $p = 0.833$).

Subjects reported elevated levels of fatigue (tiredness) for the first 6 days in Australia; like jet-lag, this was highest on day 1 and fell thereafter (figures 1 and 2). The ANOVA indicated a significant day-number effect, but with ratings still above normal by day 6 ($t_{30} = 3.13$, $p = 0.004$). A main effect of time-of-day was found for the rating of fatigue, with subjects reporting higher values at 16:00 and 20:00 h than at 07:00 and 12:00 h, a result similar to that for jet-lag. There was no significant group effect or interaction between group and other ANOVA factors (table 2).

Some aspects of the sleep pattern were disturbed in Australia. Subjects reported it easier to get to sleep on their first night in Australia, having gone to bed slightly later than normal (partly because the new local time was 10 h in advance of UK time). There was no effect of day-number on being able to stay asleep, or on the time of waking. There was a trend for morning alertness (alertness 30 min after waking) to be worse than normal. By day 6, subjects still reported that their 'ease in getting to sleep' was significantly different from 'normal' ($t_{30} = 2.08$, $p = 0.042$), and they continued to go to bed at a slightly later time than normal ($t_{30} = 2.86$, $p = 0.006$). Subjects' ability to stay asleep and their alertness 30 min after waking up were not significantly different from normal by day 6. There were no significant differences between the melatonin or placebo groups in either sleep disturbances (table 3) or alertness 30 min after waking. There was a trend towards a significant group × day-number interaction for the time of waking, the melatonin group waking from their first night of sleep in Australia earlier than the placebo group.

With regard to consistency of stools, subjects reported that they were harder than normal but that there was no significant change in the frequency of defecation. The change in hardness lasted only for the first 2 days. There were no significant differences between the melatonin and placebo groups. There were no significant time-of-day, day-number or group effects for the aspects of hunger and appetite that were measured.

Table 3. ANOVA for sleep components, appetite and bowel action. Effects that are significant at $p < 0.05$ are in bold. Effects that approach significance (0.17 $p > 0.05$) are italicized. Degrees of freedom are shown with the F-values in parentheses.

	GP	DAY	DAY × GP
Ease in getting to sleep	0.03 (1, 29)	**4.38 (4, 116)**	1.10 (4, 116)
Retiring time	2.07 (1, 29)	*2.19 (4, 116)*	0.45 (4, 116)
Ability to stay asleep	2.04 (1, 29)	0.92 (4, 116)	0.44 (4, 116)
Waking time	0.00 (1, 29)	0.63 (4, 116)	*2.02 (4, 116)*
Morning alertness	0.08 (1, 29)	*2.14 (4, 116)*	0.99 (4, 116)
Hunger before meal	0.57 (1, 29)	0.83 (5, 145)	0.59 (5, 145)
Meal satisfaction	0.01 (1, 29)	1.66 (5, 145)	0.70 (5, 145)
Post-meal hunger	0.036 (1, 29)	1.285 (5, 145)	0.877 (5, 145)
Frequency of bowel	0.32 (1, 28)	1.48 (5, 140)	0.51 (5, 140)
Consistency of stools	0.01 (1, 29)	**2.92 (5, 145)**	0.53 (5, 145)

GP = group; DAY = day number.

For mood states, there were main effects of day-number for concentration and motivation (table 2), both being negatively affected; by contrast, irritability and reports of headaches did not show consistent effects of day-number (table 2). By day 6, motivation ($t_{30} = 1.93$, $p = 0.063$) and concentration ($t_{30} = 1.45$, $p = 0.157$) were no longer significantly different from normal. There were effects of time-of-day for motivation, concentration and irritability (table 2), but not for reports of headaches. There were effects for day-number × time-of-day for concentration, motivation and a trend, which did not reach significance, for irritability. Concentration was lower at 16:00 h on day 1 than at 12:00 h on day 2, motivation had improved by 12:00 h on day 2, but this was not reflected until 16:00 h on day 3; and there was a transient decrease in irritability at 12:00 h on day 2 which was worse than normal at 16:00 h on the same day. For none of these variables was there a significant group effect or interaction between group and the other ANOVA factors (table 2).

There were no main effects of group or day-number for body temperature or grip strengths (table 2), but there were significant time-of-day effects for all three variables. The process of adjustment of body temperature resulted in changed daily profiles, and this was supported statistically by significant interactions between day-number and time-of-day. Such changes in profile with day-number were not statistically significant for left or right grip strength. Again, there were no significant differences between melatonin and placebo groups (table 2). There was a marginally significant interaction between time-of-day and group for right grip strength with grip strength tending to be higher in the melatonin than in the placebo group at 20:00 h.

With regard to the direction of shift of the physiological variables, when all three days (day 2, day 4 and day 6) and all three variables (body temperature, right grip strength and left grip strength) were considered, 50–69% of the results showed delays, 20–38% showed advances, and 11–19% showed changes that could not be interpreted as shifts in a particular direction. On day 6, the distributions of shifts of temperature, right grip strength and left grip strength did not differ significantly between the melatonin and placebo groups ($\chi^2 = 0.810$, $p = 0.667$; $\chi^2 = 4.089$, $p = 0.129$; and $\chi^2 = 0.627$, $p = 0.730$, respectively).

Any hypnotic effect of melatonin could be monitored only in those subjects who recorded fatigue and jet-lag on retiring (1–2 h after taking the pill); 16 subjects (8 taking melatonin and 8 taking placebo capsules) made such recordings. In this small sample, the jet-lag and fatigue ratings on retiring for each of the 6 days were subjected to a 2-way ANOVA with group and day-number being the main factors. There were no significant differences between the two groups in fatigue on retiring ($F(1, 12) = 0.137$, $p = 0.894$) or jet-lag on retiring ($F(5, 70) = 1.01$, $p = 0.741$).

During the study all subjects were asked to list any minor medical problems. Six subjects reported an increased number of headaches, four subjects reported dizziness and six subjects described a disorientating 'rocking' feeling as though they were on a boat. Some of these adverse effects have previously been described as occurring after having taken melatonin (Reilly et al. 1998, Suhner et al. 1998). In the current study, there was no significant difference between the melatonin and placebo groups for headaches and dizziness ($p > 0.05$, χ^2 tests), but five out of the six subjects who reported a 'rocking' sensation were in the melatonin group ($\chi^2 = 4.377$, $p = 0.036$). That is five out of 14 subjects in the melatonin group reported this sensation compared to one out of the 17 taking placebo tablets.

After the study, subjects were asked to speculate as to which treatment they had been on. For those in the melatonin group, 9 had no idea, 3 guessed correctly and 5 incorrectly; for the placebo group, the figures were 8, 4 and 5, respectively.

4. Discussion

No differences between those taking melatonin and those taking placebo tablets were found in subjective ratings of jet-lag or its symptoms during the course of the first 6 days after arrival in Australia. This can be inferred from the lack of significant group effects and the lack of significant interactions between group and the other ANOVA factors. The study employed methods that had previously indicated significant differences between subjects taking melatonin or placebo (Arendt et al. 1987, 1997, Claustrat et al. 1992, Suhner et al. 1998), but the results agree instead with those of other authors who had found melatonin to be ineffective, or only marginally effective (Nickelsen et al. 1991, Samel et al. 1991, Spitzer et al. 1999).

Authors who have previously contributed to this controversy have not always addressed the reasons for this conflict. Haimov and Arendt (1999) recently described two possible reasons for studies that had shown no effect of melatonin in the alleviation of jet-lag. These reasons were complications in pre-flight melatonin administration (citing the study of Petrie et al. 1993) and lack of pre-flight synchronization of the subjects to the local environment (citing the study of Spitzer et al. 1999). These reasons would not seem to apply to the study of Nickelsen et al. (1991) or to the present study. In both cases the subjects had experienced no transmeridian travel or irregular hours of work and sleep for the last month and so would have been adjusted to UK time before the flight.

There are other ways of considering the possible efficacy of melatonin following time-zone transitions. One, based on the fact that 'jet-lag' is a global term encompassing several symptoms, is to investigate if melatonin reduces these symptoms when considered individually. The present authors did not find any significant effects of melatonin when the various symptoms were compared in the melatonin and placebo groups; a similar result was found by Spitzer et al. (1999), who used a syndrome-specific scale of jet-lag severity derived from their questionnaire.

Of particular interest was any hypnotic effect of melatonin. With regard to this aspect, the present authors found no evidence for a significant effect of melatonin on the ease of getting to sleep or the number of waking episodes. Such results agree with the findings from field studies (Nickelsen et al. 1991, Claustrat et al. 1992, Spitzer et al. 1999) and a laboratory-based study (Samel et al. 1991), but are in contrast with other studies (Arendt et al. 1987, Suhner et al. 1998). Again the issue is unresolved therefore.

It might be that melatonin works only in those individuals in whom fatigue is high and motivation is low; in the current study, all subjects were motivated to be active in the new environment, and many were determined to 'throw off' any negative effects due to sleep loss, for example. Moreover, subjects were allowed to drink alcohol in the evening, although most individuals chose not to do so. In those that reported alcohol intake, this was not excessive (1−2 bottles of beer per night), probably as subjects were following strict schedules that involved rising early. However, the hypnotic effects of ethanol taken before bed (both by melatonin and placebo groups) could have 'swamped' any benefits from melatonin ingestion.

There might also be a parallel here with some of the results of Reinberg's group (Reinberg *et al.* 1988) who described the amount of 'tolerance' of shift-work in individuals. They found that the shift-worker who showed a higher amplitude of the circadian rhythm of temperature appeared to suffer less from the negative effects of shift-work. By analogy, it is possible that individuals (like the present subjects) who are strongly motivated in the new time-zone might be less affected by (or less prepared to admit to suffering from) the negative effects that follow transmeridian flights; if so, measures designed to alleviate the effects of jet-lag would be of less value and appear to be less effective in these subjects.

Another approach is to consider melatonin as a chronobiotic (Dawson and Armstrong 1996), that is, a drug that shifts circadian rhythms. The advantage of such a substance is that it might promote the rate of resynchronization of the body clock to the new time-zone and, since it is this desynchrony between the body clock and the environment that is responsible for many of the symptoms of jet-lag, so ameliorate the condition.

The present subjects showed a progressive adjustment of the circadian rhythms of temperature and grip strength to Australian time, but there was no difference between the placebo and melatonin groups in the rates of adjustment. Moreover, both field studies and laboratory simulations (Mills *et al.* 1978, Colquhoun 1979, Klein and Wegmann 1980, Gander *et al.* 1989, Nickelsen *et al.* 1991, Spencer *et al.* 1994, Deacon and Arendt 1996) have indicated that, after a flight eastwards through about 10 time-zones, adjustment of the circadian system in different individuals can be by phase advance or phase delay. Adjustment of all three variables occurred in the present study by both phase advances and delays (although there were more phase delays), but the proportions of advances and delays were the same in melatonin and placebo groups.

These results do not support the view that melatonin had acted as a chronobiotic, but there are other factors that complicate the position. First, the size and direction of shifts in the circadian system depend upon the time of administration of melatonin, in accord with a phase response curve. Two such curves for melatonin have been reported, but they are not identical (Zaidan *et al.* 1994, Lewy *et al.* 1998), and this causes some difficulties of interpretation. For example, in the current study, melatonin was ingested at the time recommended by Arendt and Deacon (1997) to promote the hypnotic effect of melatonin; this was at about 22:00 – 23:00 h local time, and corresponds to 12:00 – 13:00 h on unadjusted 'body time'. This time appears to fall either in the phase-delay portion of the phase response curve (Zaidan *et al.* 1994), or in the phase-advance portion (Lewy *et al.* 1992, 1998). Second, giving melatonin at 18:00 – 19:00 h on the plane was equivalent to giving it at 08:00 – 09:00 h UK time. This would promote either a delay (Zaidan *et al.* 1994) or have no effect or a slight phase-advancing effect (Lewy *et al.* 1998). Third, there is the problem that timing of administering melatonin on the days after arrival does not change; this would be appropriate for a hypnotic effect, but is inappropriate for a chronobiotic effect upon a body clock that is adjusting to local time. Finally, the subjects' exposure to, and avoidance of, bright light must be controlled in such a way that any phase-shifting effects due to this *zeitgeber* (Waterhouse *et al.* 1997) act synergistically with those promoted by melatonin. No such control was exerted in the current study, both on the day of arrival and on subsequent days. Therefore, it is plausible that any effect of melatonin on the process of readjustment of the body clock would have been 'swamped' by the effects of the subjects' haphazard exposure to the light-dark cycle.

5. Conclusion

In a field study in which the habits and lifestyles of individuals were unrestrained (as a result of which light exposure was erratic), and when alcohol could be drunk in the evening, exogenous melatonin had no significant effect on the process of adjustment of circadian rhythms to the new time-zone (10 time-zones to the east) or on sleep hygiene. Also, there were no effects of the melatonin treatment on the subjective assessment of 'jet-lag' or on the symptoms that comprise jet-lag syndrome. On this evidence, therefore, the use of melatonin by travellers from the UK to the East coast of Australia cannot be recommended.

Acknowledgements

The authors would like to thank all of the subjects and the British Olympic Association for support with this project.

References

ARENDT, J. and DEACON, S. 1997, Treatment of circadian rhythm disorders—melatonin, *Chronobiology International*, **14**, 185–204.

ARENDT, J., ALDOUS, M., ENGLISH, J., MARKS, V., ARENDT, J. H., MARKS, M. and FOLKARD, S. 1987, Some effects of jet-lag and their alleviation by melatonin, *Ergonomics*, **30**, 1379–1393.

ARENDT, J., SKENE, D. J., MIDDLETON, B., LOCKLEY, S. W. and DEACON, S. 1997, Efficacy of melatonin treatment in jet lag, shift work, and blindness, *Journal of Biological Rhythms*, **12**, 604–617.

BARTON, J., FOLKARD, S., SMITH, L. R., SPELTON, E. R. and TOTTERDELL, P. A. 1990, Standard shiftwork index manual, SAPU Memo No. 1159, MRC/ESRC Social and Applied Psychology Unit, Department of Psychology, University of Sheffield, Sheffield S10 2TN.

CLAUSTRAT, B., BRUN, J., DAVID, M., SASSOLAS, G. and CHAZOT, G. 1992, Melatonin and jet-lag: confirmatory result using a simplified protocol, *Biological Psychiatry*, **32**, 705–711.

COLQUHOUN, W. P. 1979, Phase shift in temperature rhythm after transmeridian flight, as related to pre-flight phase angle, *International Journal of Occupational Medicine and Environmental Health*, **42**, 149–157.

DAWSON, D. and ARMSTRONG, S. M. 1996, Chronobiotics—drugs that shift rhythms, *Pharmacology and Therapeutics*, **69**, 15–36.

DEACON, S. and ARENDT, J. 1996, Adapting to phase-shifts. II. Effects of melatonin and conflicting light treatment, *Physiology and Behaviour*, **59**, 675–682.

GANDER, P. H., MYHRE, G., GRAEBER, R. C., ANDERSEN, H. T. and LAUBER, J. K. 1989, Adjustment of sleep and the circadian temperature rhythm after flights across nine time zones, *Aviation, Space and Environmental Medicine*, **64**, 733–743.

HAIMOV, I. and ARENDT, J. 1999, The prevention and treatment of jet-lag, *Sleep Medicine Reviews*, **3**, 229–240.

KLEIN, K. E. and WEGMANN, H. M. 1980, Significance of circadian rhythms in aerospace operations, AGARDOGRAPH No. 247, NATO-AGARD, Neuilly-sur-Seine, France.

LEWY, A. J., AHMED, S., JACKSON, J. M. L. and SACK. R. L. 1992, Melatonin shifts circadian rhythms according to a phase response curve, *Chronobiology International*, **9**, 380–392.

LEWY, A. J., BAUER, V. K., AHMED, S., THOMAS, K. H., CUTLER, N. L., SINGER, C. M., MOFFIT, M. T. and SACK, R. L. 1998, The human phase response curve (PRC) to melatonin is about 12 hours out of phase with the PRC to light, *Chronobiology International*, **15**, 71–83.

MILLS, J., MINORS, D. and WATERHOUSE, J. 1978, Adaptation to abrupt time shifts of the oscillator(s) controlling human circadian rhythms, *Journal of Physiology (Lond.)*, **285**, 455–470.

NICKELSEN, T., LANG, A. and BERGAU, L. 1991, The effect of 6, 9 and 12 hour time shifts on circadian rhythms: adaptation to sleep parameters and hormonal patterns following the intake of melatonin or placebo, in J. Arendt and P. Pévet (eds), *Advances in Pineal Research*, vol. 5 (London: John Libbey), 303–306.

PETRIE, K., CONAGLEN, J. V., THOMPSON, L. and CHAMBERLAIN, K. 1989, Effect of melatonin on jet-lag after long haul flights, *British Medical Journal*, **298**, 705–707.
PETRIE, K., DAWSON, A., THOMPSON, L. and BROOK, R. 1993, A double-blind trial of melatonin as a treatment for jet-lag in international cabin crew, *Biological Psychiatry*, **33**, 529–530.
REILLY, T. 1994, Circadian rhythms, in M. Harries, C. Williams, W. D. Stanish and L. J. Micheli (eds), *Oxford Textbook of Sports Medicine* (New York: Oxford University Press), 238–254.
REILLY, T., ATKINSON, G. and BUDGETT, R. 1997a, Effects of temazapam on physiological and performance variables following a westerly flight across five time-zones, *Journal of Sports Sciences*, **15**, 62.
REILLY, T., ATKINSON, G. and WATERHOUSE, J. 1997b, *Biological Rhythms and Exercise* (Oxford: Oxford University Press).
REILLY, T., MAUGHAN, R. and BUDGETT, R. 1998, Melatonin: a position statement of the British Olympic Association, *British Journal of Sports Medicine*, **32**, 99–100.
REINBERG, A., MOTOHASHI, Y., BOURDELEAU, P., ANDLAUER, P., LÉVI, F. and BICAKOVA-ROCHER, A, 1988, Alteration of period and amplitude of circadian rhythms in shift workers. With special reference to temperature, right and left grip strength, *European Journal of Applied Physiology*, **57**, 15–25.
SAMEL, A., WEGMANN, H. M., VEJOVODA, M., MAASS, H., GUNDEL, A. and SCHÜTZ, M. 1991, Influence of melatonin treatment on human circadian rhythmicity before and after simulated 9-hr shift, *Journal of Biological Rhythms*, **6**, 235–248.
SPENCER, M. B., NICHOLSON, A. N., PASCOE, P. A. and RODGERS, A. S. 1994, Adaptation of the temperature rhythm to a 10 hour eastwards time zone change, *Society for Research on Biological Rhythms*, **4**, 17.
SPITZER, R., TERMAN, M., WILLIAMS, J., TERMAN, J., MALT, U., SINGER, F. and LEWY, A. 1999, Jet lag: clinical features, validation of a new syndrome-specific scale, and lack of response to melatonin in a randomised, double-blind trial, *American Journal of Psychiatry*, **156**, 1392–1396.
SUHNER, A., SCHLAGENHAUF, P., JOHNSON, R., TSCHOPP, A. and STEFFEN, R. 1998, Comparative study to determine the optimal melatonin dosage form for the alleviation of jet lag, *Chronobiology International*, **15**, 655–666.
VINCENT, W. J. 1995, *Statistics in Kinesiology* (Champaign, IL: Human Kinetics).
WATERHOUSE, J., REILLY, T. and ATKINSON, G. 1997, Jet-lag, *The Lancet*, **350**, 1611–1616.
WATERHOUSE, J., EDWARDS, B. J., ATKINSON, G., REILLY, T., NEVILL, A., DAVIES, P. and GODFREY, R. 2000, Do subjective symptoms predict our perception of jet-lag?, *Ergonomics*, **43**, 1514–1527.
WINER, B. J., BROWN, D. R. and Michels, K. M. 1991, *Statistical Procedures in Experimental Design*, 3rd ed. (New York: McGraw-Hill Kougakusha).
WINTERBURN, S., ATKINSON, G., WATERHOUSE, J. and REILLY, T. 1997, Monitoring jet-lag in a national sports team, *Chronobiology International*, **14 (Suppl. 1)**, 183.
YOUNGSTEDT, S. D. and O'CONNOR, P. J. 1999, The influence of air travel on athletic performance, *Sports Medicine*, **28**, 197–207.
ZAIDAN, R., GEOFFRIAU, M., BRUN, J., TAILLARD, J., BUREAU, C., CHAZOT, G. and CLAUSTRAT, B. 1994, Melatonin is able to influence its secretion in humans; description of a phase-response curve, *Neuroendocrinology*, **60**, 105–112.

7 Do subjective symptoms predict our perception of jet-lag?

J. WATERHOUSE†*, B. EDWARDS†, A. NEVILL†, G. ATKINSON†, T. REILLY†, P. DAVIES‡ and R. GODFREY‡

†Research Institute for Sport and Exercise Sciences, Liverpool John Moores University, Henry Cotton Campus, 15–21 Webster Street, Liverpool L3 3ET, UK

‡British Olympic Medical Centre, Northwick Park Hospital, Northwick Park, Middlesex, UK

Keywords: Circadian rhythms; Fatigue; Mood; Sleep.

A total of 39 subjects were studied after a flight from the UK to either Sydney or Brisbane (10 time-zones to the east). Subjects varied widely in their age, their athletic ability, whether or not they were taking melatonin, and in their objectives when in Australia. For the first 6 days after arrival, subjects scored their jet-lag five times per day and other subjective variables up to five times per day, using visual analogue scales. For jet-lag, the scale was labelled 0 = no jet-lag to 10 = very bad jet-lag; the extremes of the other scales were labelled −5 and +5, indicating marked changes compared with normal, and the centrepoint was labelled 0 indicating 'normal'. Mean daily values for jet-lag and fatigue were initially high (+3.65 ± 0.35 and +1.55 ± 0.22 on day 1, respectively) and fell progressively on subsequent days, but were still raised significantly ($p<0.05$) on day 5 (fatigue) or day 6 (jet-lag). In addition, times of waking were earlier on all days. By contrast, falls in concentration and motivation, and rises in irritability and nocturnal wakings, had recovered by day 4 or earlier, and bowel activity was less frequent, with harder stools, on days 1 and 2 only. Also, on day 1, there was a decrease in the ease of getting to sleep (−1.33 ± 0.55), but this changed to an increase from day 2 onwards (for example, +0.75 ± 0.25 on day 6). Stepwise regression analysis was used to investigate predictors of jet-lag. The severity of jet-lag at all the times that were measured was strongly predicted by fatigue ratings made at the same time. Its severity at 08:00 h was predicted by an earlier time of waking, by feeling less alert 30 min after waking and, marginally, by the number of waking episodes. Jet-lag at 12:00 and 16:00 h was strongly predicted by a fall of concentration at these times; jet-lag at mealtimes (12:00, 16:00 and 20:00 h) was predicted by the amount of feeling bloated. Such results complicate an exact interpretation that can be placed on an assessment of a global term such as jet-lag, particularly if the assessment is made only once per day.

1. Introduction

After a rapid flight across several time-zones, a general malaise, commonly known as jet-lag, appears transiently. The exact symptoms, their severity and duration depend upon the individual, the direction of flight and the number of time-zones crossed (Graeber 1982, 1989, Graeber *et al.* 1986, Moline *et al.* 1992, Arendt *et al.* 1995,

---

*Author for correspondence. e-mail: waterhouseathome@hotmail.com

Waterhouse et al. 1997, Lowden and Akerstedt 1998). These symptoms are believed to arise because, for some days after arrival in the new time-zone, the internal 'body clock' has not adjusted its phase in accord with the new local time (Graeber 1982, 1989, Reilly et al. 1997). The symptoms include some or all of the following: fatigue during the new daytime and yet inability to sleep satisfactorily at night time; loss of appetite, indigestion and feeling bloated after food intake; decrease in the ability to concentrate and loss of motivation; increases in irritability and the frequency of headaches.

In spite of such a mixture of symptoms—or, perhaps because of it—it is common for the amount of jet-lag being experienced by an individual to be recorded on a visual analogue scale, the decision as to what exactly constitutes jet-lag at the time of assessment being left to that individual (Arendt et al. 1986, Arendt and Aldhous 1988). In nearly all studies other variables are measured, including those that are believed to act as markers of the circadian body clock—core temperature and some hormones—and those that reflect some of the characteristics of jet-lag—mood, sleep, mental and physical performance, for example. The common finding from such studies (Arendt et al. 1987, Monk et al. 1988, Petrie et al. 1989, Samel et al. 1991) is that different variables, including those that are estimated subjectively, adjust to a time-zone transition at different rates. Whatever the explanation of such a result, it implies that the symptoms that constitute an individual's perception of jet-lag might change during the course of a study.

For reasons of practicality, assessments of jet-lag and some of the variables that are believed to contribute to it can sometimes be made only once per day (Arendt et al. 1986, Petrie et al. 1993, Suhner et al. 1998). This might raise a problem since some of the symptoms of jet-lag are believed to show circadian rhythms and to be linked to the (unadjusted) body clock (Graeber 1982, 1989, Graeber et al. 1986, Moline et al. 1992, Reilly et al. 1997, Waterhouse et al. 1997). Thus, except for the inability to sleep, many of the symptoms would be predicted to be worst in the new time-zone at a time coincident with the middle of the night on home time.

For sleep, the position is more complex (Graeber et al. 1986, Moline et al. 1992). For example, falling asleep is easier when plasma adrenaline and core temperature are falling (evening in the home time-zone), staying asleep is easier during the night on home time (when plasma adrenaline and core temperature are low), and waking up is easier when plasma adrenaline and core temperature are rising most quickly (about 05:00–09:00 h on home time). As a result, after a flight to the west across 6 time-zones, fatigue in the evening is high and sleep initiation (at 23:00 h local time, 05:00 h body time) is comparatively easy (particularly since sleep has also been delayed) but waking up is often premature owing to the rising plasma adrenaline and core temperature; by contrast, after a flight to the east through 6 time-zones, sleep initiation (at 17:00 h body time) would be difficult (particularly since the waking period has also been curtailed) but premature waking is less likely, as 07:00 h local time coincides with the night on home time. That is, after a flight to the east, one of the symptoms of jet-lag would be an inappropriate *decrease* of fatigue at bedtime in the new time-zone. Also, after flights to the east crossing more than 9 time-zones, there is the added complication that adjustment of the body clock might be by phase delay rather than by phase advance (Gundel and Wegmann 1989).

Whatever the direction of flight and number of time-zones crossed, it is reasonable to suppose that, *in so far as they reflect the process of adjustment of the body clock to the new time-zone*, the time courses of manifestation of jet-lag and its components would be related to this process and to each other; however, *in so far as they reflect different relative contributions from the body clock and the direct effects of the new time-zone*, the manifestations of jet-lag and its components would show different rates of adjustment to the new time-zone.

The aim of the current study was to explore the relationship between assessments of jet-lag and other variables generally believed to be associated with it. Estimates of jet-lag and subjective measures of sleep, physiology, mood and mental performance were made several times each day for 6 days after a flight to the east crossing 10 time-zones. The analysis has focused on a comparison of the between-day and within-day changes in these variables in a group of subjects who had considerable differences in their activities after the time-zone transition. Apart from the recent study by Lowden and Akerstedt (1998), there does not seem to be any other study that has focused on predictors of jet-lag in this way, but the use of the Columbia jet-lag scale (Spitzer *et al.* 1999) would certainly enable replication of this type of approach.

2. Methods

2.1. Subjects and general protocol

A total of 39 subjects (10 females, 29 males) took part in the study, divided into three groups (15, 10 and 14 subjects). They varied widely in their ages (23–63 years), their athletic ability and in their aims when in Australia; some were elite athletes who underwent training sessions and others were administrators for the British Olympic Association or national sports governing bodies. All were taking part in a double-blind study to investigate the effect of melatonin on the process of adjustment to Australia, which is 10 time-zones to the east of their home time-zone, the UK (Edwards *et al.* 2000). Journeys differed slightly between the three groups with regard to route and times of departure and arrival, but took about 24 h in all cases. The day of arrival in Australia was called day 1, this being early in the morning (local time) for two groups disembarking in Brisbane and about mid-day for the third, who had flown to Sydney.

Subjects had been instructed on times of taking a pill containing placebo or melatonin, and advised to guard against cramp and dehydration during the flights, but otherwise they were free to choose their own schedule with regard to sleep times and activities, both during the flight and in the days immediately before departure. After arrival in Australia, all adopted local time for sleep, activities and meals. During the first 6 days after arrival, subjects were required to answer the Jet-lag Questionnaire as close to 08:00, 12:00, 16:00, 20:00 and 24:00 h as their commitments would allow.

2.2. Assessments of jet-lag and other variables

A Jet-lag Questionnaire for field use has been developed at Liverpool John Moores University. This is similar to, but was developed independently of, that produced at Columbia University, USA. In the authors' questionnaire, individuals assessed their subjective responses to a series of questions by means of visual analogue scales (VAS), which were reported in their own personal record. All VAS except that for jet-lag had the same format, namely a line divided into 10 parts with the centrepoint being labelled 0 and the extremes being labelled +5 and −5. In addition, these three points were labelled with brief descriptions of the sensations to be associated with them. For jet-lag, the extremes of the scale were labelled 0 and 10. The questions asked, and details of the scales used, are presented in the appendix.

Not all questions were given at each test time. The times of assessment were as follows.

Jet-lag: 08:00, 12:00, 16:00, 20:00, 24:00 h; *Sleep*: 08:00 h; *fatigue*: 08:00, 12:00, 16:00, 20:00, 24:00 h; *meals*: 12:00, 16:00, 20:00 h; *mental performance and mood*: 12:00, 16:00 h; *bowel activity*: 24:00 h.

Before the journey, subjects were shown the questionnaire in order to become familiar with its contents and the way to record their answers. Specific advice about the meaning of jet-lag or any of the other questions asked was not given. After having the nature of the protocol explained fully to them and having been given any clarification requested, the subjects signed informed consent forms. The protocol was approved by the Human Ethics Committee of Liverpool John Moores University.

2.3. *Statistical analysis*

Standard statistical methods have been used to describe and investigate the results. The correlations and stepwise regression analyses (see § 3) were performed on a Minitab package (Minitab Inc., State College, PA, USA).

3. Results

The demanding commitments upon the subjects during their stay in Australia meant that some times of testing were missed. On a very few occasions, up to 50% of the possible data were not obtained, but generally the response rate was greater than 90% of the maximum possible (39 sets of results). All results from all subjects—regardless of age, gender, activities in Australia, and whether taking melatonin or placebo—were pooled in the following analyses, since the authors did not investigate if differences between sub-groups, due to any one or more of these factors, affect the between-day and within-day relationships between the severity of jet-lag and changes in the other variables.

Figure 1 shows the mean daily values for the measured variables during the first 6 days in the new time-zone, and table 1 shows the mean number of days taken for each variable to return to its pre-flight value. (The graphical results have been expressed such that all curves tend to show the largest positive change on day 1 and then to decline over successive days; in some cases, this requires a presentation of the opposite of what was assessed. For examples, comparing the appendix and figure 1: ability to concentrate was assessed but 'Fall in concentration' was plotted; and looseness of bowel motion was assessed, but 'Stool hardness' was presented graphically).

Figure 1 shows that jet-lag was most marked on day 1 and decreased thereafter; it was still significantly greater than zero ($p<0.05$, Student's *t* test) on day 6. Increased fatigue and an earlier time of waking behaved similarly (figure 1 and table 1), except that fatigue was not raised significantly, and waking earlier only marginally so ($p>0.10$ and $0.10>p>0.05$, respectively, Student's *t* tests) on day 6.

By contrast, other variables appear to have behaved differently. Some (number of waking episodes, falls of concentration and motivation, increased irritability, decreased bowel frequency and increased hardness of stools) appeared to have become normal by day 4 or earlier; others either showed no reliable changes (lateness of getting to sleep, alertness 30 min after waking) or changes that did not appear to be related to the process of adjustment (assessments associated with meals). The difficulty of getting to sleep (the reverse of ease of getting to sleep, see the appendix) was different from the other variables in that it was significantly greater on day 1 but significantly *less* on subsequent days.

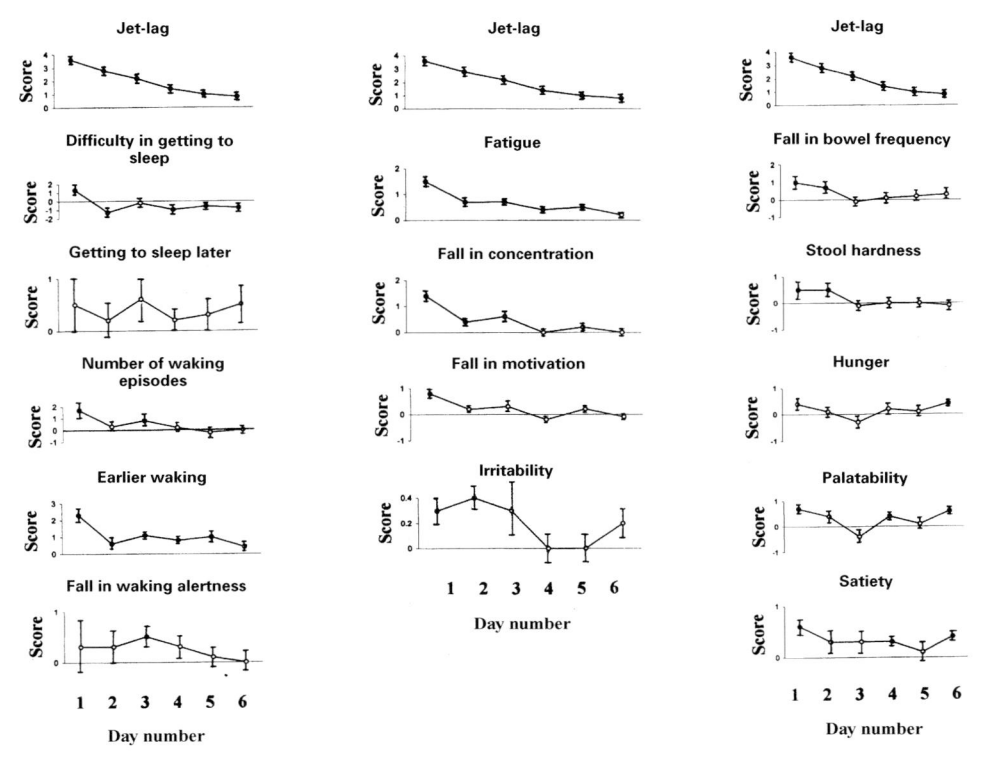

Figure 1. The mean (±SE) daily values of jet-lag and associated variables on the 6 days after the flight from the UK to Australia. Ordinates presented so that all variables tended to fall over successive days. For clarity, the jet-lag results have been presented three times. Day number applies to all plots.

While such results support the view that there are many negative effects after a time-zone transition, the observation that they do not all appear to adjust at the same rate as each other or as the global measure jet-lag indicates that the situation is complex, and that a close link between jet-lag and some of its symptoms is not likely to exist. This point is confirmed when the daily time courses of jet-lag and fatigue, the two variables that were measured most frequently (see § 2) and which showed similar time courses of the daily means (figure 1) are compared (figure 2). Thus, even though the correlation between jet-lag and fatigue was highly significant when the results from all times and days were considered ($r = +0.534$, $N=1170$), figure 2 shows that there is no reproducible within-day trend for jet-lag but that fatigue tends to increase throughout the daytime, at least on days 1–4 when it is most marked. Some support for this difference comes from a comparison of the results from the first 4 days. For jet-lag there was no significant within-day effect ($\chi^2 = 1.2$, $p > 0.30$; Friedman's ANOVA by ranks); for fatigue the effect was close to statistical significance ($\chi^2 = 9.2$, $0.10 > p > 0.05$; Friedman's ANOVA).

If jet-lag and the other variables do not appear to show the same time courses, it then becomes appropriate to investigate to what extent variables within the subgroups (sleep/meals/mood and mental performance/bowel activity) correlate with

Table 1. Day number after the flight by which a variable was no longer significantly different from zero.

Variable	Day number	Comment
Jet lag	>6	
Fatigue	6	
Sleep		
Ease of getting to sleep	>6	Day 1 differs from the others
Lateness of getting to sleep	?	Non-significant delay
Unbroken sleep	4	
Earlier waking	?6	$0.10 > p > 0.05$ on day 6
Alertness 30 min after waking	?	Non-significant fall
Meals		
Hunger	?	Tend to
Palatability	?	be raised
Satiety	?	throughout
Mood and mental performance		
Fall in connection	4	
Fall in motivation	2	
Increase in irritability	?3	$0.10 > p > 0.05$ on day 3
Increase in headaches	?	No significant change
Bowel activity		
Decreased frequency	3	
Increased hardness	3	

each other, and also to what extent they act as predictors of jet-lag. The correlation coefficients between variables within the sub-groups are shown in table 2. This table shows that, within the Sleep variables, there was a general positive association between the variables 'Ease of getting to sleep', 'Unbroken sleep', 'Lateness of waking' and 'Alertness 30 min after waking'. For the variables associated with Meals, the highest correlation was a positive one between 'Hunger' and 'Palatability'. For the Mood and mental performance variables, there was a strong association between 'Concentration' and 'Motivation' as well as negative correlations between 'Irritability' and both 'Concentration' and 'Motivation'. Finally, for the Bowel activity variables, there was a strong negative correlation between 'Frequency of defecation' and 'Hardness of the stools'.

To investigate the extent to which these variables acted as predictors of jet-lag, stepwise regression analyses were performed using jet-lag on all 6 days as the dependent variable and the other variables measured at the same time (see § 2) as predictors. No variable was forced into the analysis. From these analyses were extracted the β-coefficients that were significantly different from zero. The results are shown in table 3, from which the following conclusions are drawn.

    (1) Jet-lag at all times of testing was strongly predicted by the rating of fatigue at the same time.
    (2) Jet-lag at 08:00 h was predicted by earlier waking and by a decreased alertness 30 min after waking; the number of waking episodes had a

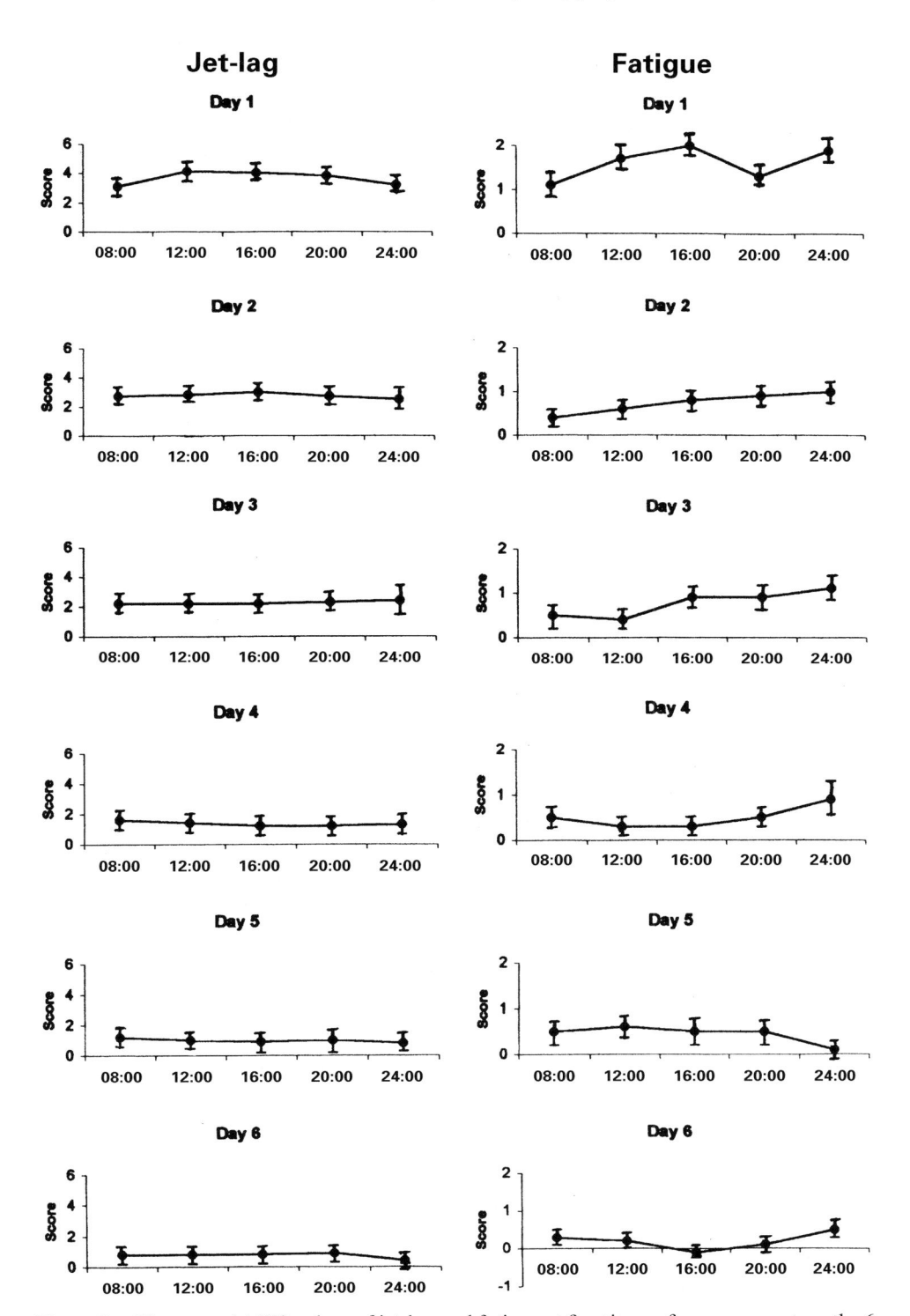

Figure 2. The mean ($\pm$SE) values of jet-lag and fatigue at five times of measurement on the 6 days after the flight from the UK to Australia.

Table 2. Correlations within sub-groups.

Sub-group (time(s); maximum number of values)		Correlation coefficient
Sleep (08:00 h; 234)		
Ease of getting to sleep	vs. Lateness of getting to sleep	−0.185
Ease of getting to sleep	vs. Unbroken sleep	+0.486
Ease of getting to sleep	vs. Lateness of waking	+0.279
Ease of getting to sleep	vs. Alertness 30 min after waking	+0.054
Lateness of getting to sleep	vs. Unbroken sleep	−0.075
Lateness of getting to sleep	vs. Lateness of waking	−0.041
Lateness of getting to sleep	vs. Alertness 30 min after waking	+0.044
Unbroken sleep	vs. Lateness of waking	+0.296
Unbroken sleep	vs. Alertness 30 min after waking	+0.257
Lateness of waking	vs. Alertness 30 min after waking	+0.132
Meals (12:00, 16:00, 20:00 h; 702)		
Hunger vs. Palatability		+0.385
Hunger vs. Satiety		−0.191
Palatability vs. Satiety		+0.139
Mood and mental performance (12:00, 16:00 h; 468)		
Concentration vs. Motivation		+0.714
Concentration vs. Irritability		−0.206
Motivation vs. Irritability		−0.290
Bowel activity (20:00 h; 234)		
Frequency of defecation vs. Hardness of the stools		−0.505

Table 3. Significant predictors of jet-lag; summary of results from regression analyses.

Variable (times; max. number of values)	β-coefficient	SD	P-value	%Variance*
Fatigue (08:00, 12:00, 16:00, 20:00, 24:00 h; 1170)	+0.83	±0.04	<0.001	29
Sleep (08:00 h; 234)				
Ease of getting to sleep	n.s.			
Lateness of getting to sleep	n.s.			
Number of waking episodes	+0.12	±0.07	0.08	
Lateness of waking	−0.45	±0.08	<0.001	22
Alertness 30 min after waking	−0.47	±0.09	<0.001	
Meals (12:00, 16:00, 20:00 h; 702)				
Hunger	n.s.			
Palatability	n.s.			
Satiety	+0.22	±0.09	0.009	2
Mood and mental performance (12:00, 16:00 h; 468)				
Concentration	−0.96	±0.08	<0.001	28
Motivation	n.s.			
Irritability	n.s.			
Bowel activity (20:00 h; 234)				
Frequency of defecation	n.s.			
Looseness of the stools	n.s.			

*Percentage of total variance accounted for by regression equation.

marginally predictive value. The predictive value of these sleep variables upon jet-lag measured at this time was 22% of the total variance.
(3) Jet-lag in the daytime was predicted by feeling bloated after meals and also by the fall in the ability to concentrate.
(4) Jet-lag at the end of the day was not predicted significantly by either the frequency of defecation or the consistency of the stools.

Attention is drawn also to the other variables shown in table 3 that did *not* act as predictors of jet-lag; these were the ease of getting to sleep and getting to sleep later, changes in hunger and the palatability of food, and the fall in motivation and rise of irritability.

4. Discussion

The decline in the mean daily ratings of jet-lag indicates that the malaise would have been over after about 7 days (figure 1). Given that the rates of phase advance and delay of the body clock are estimated to be about 1 and 2 hours per day, respectively (Graeber 1982, 1989, Reilly *et al.* 1997) the observed time course would be about that predicted, whether phase adjustment was by a 10-h phase advance or a 14-h phase delay (Gundel and Wegmann 1989).

4.1. *A comparison of the time courses of adjustment*

When considering other subjective assessments that might act as predictors of jet-lag, one approach was to compare the time course of jet-lag with those of other variables (figure 1). This approach focused attention towards increased fatigue and an earlier time of waking.

(1) While the mean daily ratings for fatigue and jet-lag were similar, the within-day time courses of the two variables differed (figure 2) indicating that some other factor(s) peculiar to one of these variables must exist also (see below). Another variable where change persisted throughout the 6 days of measurement was an early time of waking from sleep. Early morning (06:00 h) local time coincided initially with 20:00 h home (UK) time, and so it is likely that raised temperature and adrenaline levels at this time would have promoted waking. For the next day or so, 06:00 h local time would have corresponded to falling—but still quite high—evening temperatures and adrenaline levels on home time if adjustment had been by phase advance, or high afternoon values had it been by phase delay; in both cases, there exists a possible explanation for the continuation of premature waking. The present data do not enable the amount of adjustment on any day, and whether it was by phase advance or delay, to be established, but previous work (Mills *et al.* 1978, Gundel and Wegmann 1989) has shown that both possibilities exist following a time-zone transition of 10 h in an eastward direction.
(2) The mean daily adjustments of the components of mood and mental performance, and of bowel activity, both showed more rapid adjustment than did jet-lag. Such differences in the time-courses of adjustment of different variables to a time-zone transition have been shown before (see reviews by Graeber 1982, 1989, Reilly *et al.* 1997, Waterhouse *et al.* 1997), but they render it impossible for all, if not any, of them to act as a marker of the process of adjustment of the body clock. In attempting to account for the

more rapid adjustment of some variables, it seems possible that the personal objectives of the subjects when in Australia enabled them to maintain a positive attitude to their tasks in spite of the jet-lag that they all acknowledged they were suffering from. The rapid adjustment of bowel activity could have followed from their increased appetites throughout the stay (figure 1).

(3) The persistent increase (after day 1) in the ease of getting to sleep deserves comment. Since the time of retiring would have coincided with 14:00 h on home time, and then, as adjustment proceeded, to the afternoon or late morning (depending upon whether adjustment was by phase advance or phase delay, respectively), a difficulty with sleep initiation would have been predicted if core temperature and plasma adrenaline had been the determinants (Graeber et al. 1986, Graeber 1989, Moline et al. 1992, Reilly et al. 1997). Clearly, other factors must have been involved; these could have included the earlier waking time (see above) and the concerted effort, both physical and mental, that was being expended by the subjects during their stay in Australia.

In summary, the different symptoms that have been associated with jet-lag do not all adjust at the same rate as each other or as jet-lag itself. These findings are not new, but rather confirm the results of many others (see Arendt et al. 1987, Monk et al. 1988, Samel et al. 1991, for example, and reviews in Graeber et al. 1986, Graeber 1989, Reilly et al. 1997). A consideration of these mean daily rates of adjustment suggests that increased fatigue and an early time of waking from sleep are likely to be the best predictors of jet-lag. Such predictors can be assessed more directly by the correlation and regression analyses, particularly if the time of day when the measurements were made is taken into account.

4.2. *The correlation and regression analyses*
The following conclusions can be drawn from the correlation and regression analyses.

(1) Fatigue correlated strongly with jet-lag and, as table 3 indicates, acted as a highly significant predictor of jet-lag.
(2) Some, but not all, components of sleep were correlated with each other (table 2), and there was evidence that the amount of jet-lag at 08:00 h was predicted by some of them (table 3). However, 'Lateness of getting to sleep', in particular, did not seem to correlate highly with the other assessments of sleep and did not act as a significant predictor of jet-lag. Taken together, these assessments of sleep seem to indicate that, when assessing their jet-lag in the morning, individuals attach more importance to staying asleep, waking up on time and feeling alert after having woken up.
(3) The increase in 'Hunger' (figure 1) possibly reflects the busy work schedule that the subjects had. Even though there was evidence that 'Hunger' and 'Palatability of food' correlated with each other, neither acted as a predictor of jet-lag during the daytime. By contrast, an increase in 'Satiety' above normal—that is, feeling bloated—did act as a predictor. There was also a clear change in bowel activity, with stools initially being harder and produced less frequently; but recovery from these changes was rapid and neither acted

as a predictor of jet-lag. That is, there was evidence that some, but not all, aspects of gastrointestinal tract function were slow to adjust fully to the time-zone transition, as can also be inferred from the time-courses of adjustment (figure 1).

(4) Even though the different assessments of 'Mood and mental performance' formed a cluster that correlated with each other (table 2), only the fall in 'Concentration' was a highly significant predictor of daytime jet-lag (table 3), 'Motivation' and 'Irritability' being non-significant predictors.

5. Conclusions

The main conclusions from the present data are that, after a flight to the east across 10 time-zones the amount of perceived jet-lag at any time is predicted by the amount of fatigue at that time (increased fatigue predicting more jet-lag); the amount of jet-lag in the morning is predicted by the time of waking from sleep (earlier times predicting more jet-lag) and by a decreased alertness 30 min after waking; and the amount of jet-lag in the daytime is predicted by the fall in the perceived ability to concentrate. Other subjective variables are adversely affected by time-zone transitions, but their time course of adjustment differs from that of jet-lag, and they are less reliable as predictors of jet-lag. Such findings stress the importance of sleep and of not feeling fatigued. They compare closely with some of the findings from the recent study of aircrew by Lowden and Akerstedt (1998), who found that jet-lag was related mainly to sleepiness ratings and the number of awakenings.

These results imply that methods to reduce jet-lag might concentrate on reducing fatigue and extending sleep in the morning. One method to achieve this could be by the use of a hypnotic, but the question remains if this alone would be sufficient or whether promotion of the adjustment of the body clock to the new time-zone would be required. (Immediately after the time-zone transition, the unadjusted body clock would preclude maximum mental and physical performance during the daytime.) Short-acting hypnotics of the benzodiazepine group have been used with success (Nicholson 1984), as has ingestion of capsules containing melatonin (Arendt *et al.* 1987, 1995, Arendt and Aldhous 1988, Claustrat *et al.* 1992, Petrie *et al.* 1993, Comperatore *et al.* 1996, Sack *et al.* 1996, Suhner *et al.* 1998). However, since both substances can phase shift the body clock (Van Reeth and Turek 1989, Lewy *et al.* 1998), it is not clear to what extent any beneficial effects have been due to the hypnotic effect, the clock-shifting effect, or both. Particularly if the important effect is that of promoting adjustment of the body clock, the time of administration becomes crucial, and the evening in the new time-zone (when administration would be required for the hypnotic effect) might not be the correct time (Waterhouse *et al.* 1997). An assessment of the amount of adjustment of the body clock requires the use of well-established markers of it; such an analysis is currently being performed by the present authors.

Acknowledgements

The authors would like to thank all of the subjects and the British Olympic Association for support with this project.

References

ARENDT, J. and ALDHOUS, M. 1988, Further evaluation of the treatment of jet-lag by melatonin: a double-blind crossover study, *Annual Review of Chronopharmacology*, **5**, 53–56.

ARENDT, J., ALDHOUS, M. and MARKS, V. 1986, Alleviation of jet lag by melatonin: preliminary results of controlled double blind trial, *British Medical Journal*, **292**, 1170.

ARENDT, J., DEACON, S., ENGLISH, J., HAMPTON, S. and MORGAN, L. 1995, Melatonin and adjustment to phase shift, *Journal of Sleep Research*, **4** (Suppl. 2), 74–79.

ARENDT, J., ALDHOUS, M., ENGLISH, J., MARKS, V., ARENDT, J. H., MARKS, M. and FOLKARD, S. 1987, Some effects of jet-lag and their alleviation by melatonin, *Ergonomics*, **30**, 1379–1393.

CLAUSTRAT, B., BRUN, J., DAVID, M., SASSOLAS, G. and CHAZOT, G. 1992, Melatonin and jet lag: confirmatory results using a simplified protocol, *Biological Psychiatry*, **32**, 705–711.

COMPERATORE, C., LIEBERMAN, H., KIRGY, A., ADAMS, B. and CROWLEY, J. 1996, Melatonin efficacy in aviation missions requiring rapid deployment and night missions, *Aviation, Space and Environmental Medicine*, **67**, 520–524.

EDWARDS, B. J., ATKINSON, G., WATERHOUSE, J., REILLY, T., GODFREY, R. and BUDGETT, R. Use of melatonin in recovery from jet-lag following an eastward flight across 10 time-zones, *Ergonomics*, **43**, 1501–1513.

GRAEBER, R. 1982, Alterations in performance following rapid transmeridian flight, in F. Brown and R. Graeber (eds), *Rhythmic Aspects of Behavior* (Hillsdale, NJ: Lawrence Erlbaum), 173–212.

GRAEBER, R. 1989, Jet lag and sleep disruption, in M. Kryger, T. Roth and W. Dement (eds), *Principles and Practice of Sleep Medicine* (Phildelphia, PA: W. Saunders), 324–331.

GRAEBER, R., DEMENT, W., NICHOLSON, A., SASAKI, M. and WEGMANN, H. 1986, International cooperative study of aircrew layover sleep, operational summary, *Aviation, Space and Environmental Medicine*, **57** (Suppl.), B10–B13.

GUNDEL, A. and WEGMANN, H. 1989, Transition between advance and delay responses to eastbound transmeridian flights, *Chronobiology International*, **6**, 147–156.

LEWY, A., BAUER, V., AHMED, S., JACKSON, J. and SACK, R. 1998, The human phase response curve (PRC) to melatonin is about 12 hours out of phase with the PRC to light, *Chronobiology International*, **15**, 71–83.

LOWDEN, A. and AKERSTEDT, T. 1998, Retaining home-base sleep hours to prevent jet lag in connection with a westward flight across nine time zones, *Chronobiology International*, **15**, 365–376.

MILLS, J., MINORS, D. and WATERHOUSE, J. 1978, Adaptation to abrupt time shifts of the oscillator(s) controlling human circadian rhythms, *Journal of Physiology (London)*, **285**, 455–470.

MOLINE, M., POLLACK, C., MONK, T., LESTER, L., WAGNER, D., ZENDALL, S., GRAEBER, R., SALTER, C. and HIRSCH, E. 1992, Age-related differences in recovery from simulated jet lag, *Sleep*, **15**, 28–40.

MONK, T., MOLINE, M. and GRAEBER, R. 1988, Inducing jet lag in the laboratory: patterns of adjustment to an acute shift in routine, *Aviation, Space and Environmental Medicine*, **59**, 703–710.

NICHOLSON, A. 1984, Long periods of work and disturbed sleep, *Ergonomics*, **27**, 629–630.

PETRIE, K., CONGLEU, J., THOMPSON, L. and CHAMBERLAIN, K. 1989, Effect of melatonin on jet lag after long haul flights, British Medical Journal, **298**, 705–707.

PETRIE, K., DAWSON, A., THOMPSON, L. and BROOK, R. 1993, A double-blind trial of melatonin as a treatment for jet lag in international cabin crew, *Biological Psychiatry*, **33**, 526–530.

REILLY, T., ATKINSON, G. and WATERHOUSE, J. 1997, *Biological Rhythms and Exercise* (Oxford: Oxford University Press).

SACK, R., LEWY, A., HUGHES, R., MCARTHUR, A. and BLOOD, M. 1996, Melatonin as a chronobiotic drug, *Drug News and Perspectives*, **6**, 325–332.

SAMEL, A., WEGMANN, H., VEJVODA, M., MAAS, H., GUNDEL, A. and SCHUTZ, M. 1991, Influence of melatonin treatment on human circadian rhythmicity before and after a simulated 9-hr time shift, *Journal of Biological Rhythms*, **6**, 235–248.

SPITZER, R., TERMAN, M., WILLIAMS, J., TERMAN, J., MALT, U. and LEWY, A. 1999, Jet lag: clinical features, validation of a new syndrome-specific scale, and lack of response to melatonin in a randomized, double-blind trial., *American Journal of Psychiatry*, **156**, 1392–1396.

Suhner, A., Schlagenhauf, P., Johnson, R., Tschopp, A. and Steffen, R., 1998, Comparative study to determine the optimal melatonin dosage form for the alleviation of jet lag, *Chronobiology International*, **15**, 655–666.

Van Reeth, O. and Turek, F. 1989, Administering triazolam on a circadian basis entrains the activity rhythm of hamsters, *American Journal of Physiology*, **256**, R639–R645.

Waterhouse, J., Reilly, T. and Atkinson, G. 1997, Jet-lag, *Lancet*, **350**, 1611–1616.

Appendix. The self-assessment visual analogue scales.

1. Jet-lag:
How much jet-lag do you have?
0 ———————————————————————————— 10
(insignificant jet-lag)   (very bad jet-lag)

2. Last night's sleep. When compared with normal:
a. How easily did you get to sleep?
−5 ————————— 0 ——————————— +5
(less)   (normal)   (more)

b. What time did you get to sleep?
−5 ————————— 0 ——————————— +5
(earlier)   (normal)   (later)

c. How well did you sleep?
−5 ————————— 0 ——————————— +5
(more waking episodes)   (normal)   (fewer waking episodes)

d. What was your waking time?
−5 ————————— 0 ——————————— +5
(earlier)   (normal)   (later)

e. How alert did you feel 30 min after rising?
−5 ————————— 0 ——————————— +5
(less)   (normal)   (more)

3. Fatigue:
In general, compared to normal, how tired do you feel at the moment?
−5 ————————— 0 ——————————— +5
(more)   (normal)   (less)

4. Meals. Compared with normal: *
a. How hungry did you feel before your meal?
−5 ————————— 0 ——————————— +5
(less)   (normal)   (more)

b. How palatable (appetising) was the meal?
−5 ————————— 0 ——————————— +5
(less)   (normal)   (more)

c. After your meal, how do you now feel?
−5 ————————— 0 ——————————— +5
(still hungry)   (satisfied)   (bloated)

5. Mental performance and mood. Compared with normal:
a. How well have you been able to concentrate?
−5 ————————— 0 ——————————— +5
(worse)   (normal)   (better)

b. How motivated do you feel?
−5 — ———————————— 0— ————————————— +5
(less) (normal) (more)

c. How irritable do you feel?
−5 — ———————————— 0— ————————————— +5
(less) (normal) (more)

6. **Bowel activity today. Compared with normal:** **
a. How frequent have your bowel motions been?
−5 — ———————————— 0— ————————————— +5
(less) (normal) (more)

b. How has the consistency been?
−5 — ———————————— 0— ————————————— +5
(harder) (normal) (looser)

*If no meal was taken at a particular time, then the results were treated as missing data for the purpose of analysis.
**If there was no bowel motion on a particular day then its consistency was treated as missing data.

Part III
Ergonomics in Competitive Sports

8 Effect of carbohydrate supplementation on simulated exercise of rugby league referees

D. P. M. MacLaren* and G. L. Close

Research Institute for Sport and Exercise Sciences, Liverpool John Moores University, Henry Cotton Campus, Webster Street, Liverpool L3 2ET, UK

Keywords: Energy; Maltodextrin; Perceived exertion; Referee; Sprint; Endurance run.

This investigation evaluated the effectiveness of supplementing eight elite rugby league referees with a 6% maltodextrin (Md) solution whilst undertaking a simulated rugby league game. The simulation was based on motion analysis of six rugby league matches. Subjects undertook two trials of repetitive 20-m shuttle activity on an indoor track. During one trial 200 ml of Md was ingested at eight time points and in the other trial a similarly tasting placebo (Pl) was administered. A single-blind, counterbalanced design was employed. The simulation involved subjects performing four, 10-min blocks of shuttle activity before a 10-min break was instigated. Three further 10-min blocks of shuttle activity were also performed before a performance test to volitional exhaustion involving 20-m shuttles at paces varying between 55 and 95% of a pre-determined $O_{2\ max}$ was undertaken. Timed 15-m sprints took place during each of the 10-min blocks. The rating of perceived exertion (RPE), and blood glucose and lactate concentrations were also determined throughout. The mean number of shuttles to exhaustion was significantly greater with Md ingestion than with Pl ingestion (57 ± 19 vs. 43 ± 15; $p<0.05$), while the mean 15-m sprint times were significantly shorter for the Md than the Pl condition (2.40 ± 0.09 s vs. 2.51 ± 0.14 s; $p<0.01$). The mean RPE was ~5.2% lower during Md than Pl ingestion, the values being significantly different (Md: 12.07 ± 0.32; Pl: 12.73 ± 0.28; $p<0.01$). Maltodextrin ingestion significantly elevated blood glucose levels compared with placebo ($F(1,7) = 18.07$; $p<0.01$), although no significant differences were apparent for blood lactate levels ($F(1,7) = 4.39$; $p>0.05$). These results highlight the beneficial effects of maltodextrin ingestion on work-rates of rugby league referees in a simulation of a game's activity. The improvement may be related to higher circulating concentrations of blood glucose.

1. Introduction

The game of rugby league consists of two, 40-min halves separated by a 10-min break. The majority of the game (approximately 90%) entails low-level activities such as walking, jogging and standing (Brewer and Davis 1995), although there are a high number of physical confrontations such as tackling and running hard with the ball in hand. The total distance covered by players during a game is 5000–7000 m, and the activity may be described as high intensity intermittent exercise. There have been no studies of the distance covered and type of activities undertaken by rugby

*Author for correspondence. e-mail: d.p.maclaren@livjm.ac.uk

league referees, although it may be surmised that whereas players have some degree of control over their level of participation, the referee must keep pace with the game. In a study on association football referees, the total distance covered was found to be 8000–10 000 m over 90 min, during which 47% of the time was spent jogging, 23% walking, 18% running backwards, and 12% sprinting (Catterall et al. 1993). Similar results were found for association football referees by Johnston and McNaughton (1994).

The effect of exercise intensity on muscle glycogen depletion is well documented. Not only prolonged exercise but also high intensity exercise results in significant depletion of muscle glycogen stores in various muscle fibres (Gollnick et al. 1974, Vollestad and Blom 1985). Since muscle glycogen depletion has been associated with fatigue during exercise (Bergstrom et al. 1967) and sport (Saltin 1973), ingesting carbohydrate before and/or during exercise may be of significant benefit (Coyle 1992, MacLaren et al. 1994). A recent study demonstrated the positive effect of ingesting a carbohydrate-electrolyte solution on endurance running capacity during intermittent high-intensity shuttle running (Nicholas et al. 1995). In a follow-up study Nicholas et al. (1999) concluded that muscle glycogen utilization was reduced by 22% after ingesting a carbohydrate-electrolyte drink when compared with placebo during such activity.

In spite of the wealth of evidence relating muscle glycogen depletion to fatigue during prolonged exercise, depletion of liver glycogen during intense, prolonged exercise has been shown to result in hypoglycaemia (Coggan and Coyle 1987). This is probably due to the fact that once liver glycogen is depleted, the process of gluconeogenesis is unlikely to provide glucose at a sufficient rate for use by tissues such as the central nervous system (CNS) and muscle. The resultant hypoglycaemia brings about 'central' fatigue, since the brain requires glucose as a metabolic fuel.

Although there has been a recent review of the physiological demands of playing rugby league (Brewer and Davis 1995), and two studies have reported the physiological demands on association football referees (Catterall et al. 1993, Johnston and McNaughton 1994), there have been no physiological or movement analyses of elite rugby league referees. The aim of this investigation was initially to establish the motion activities of elite rugby league referees and then determine the effectiveness of carbohydrate supplementation on a rugby league simulated activity.

2. Pilot study

2.1. Motion analysis

Videotapes of six premiership rugby league matches were obtained from BSB 'Sky Sports'. Aerial views of the matches were selected for analyses since the position and movement patterns of the referees could be seen throughout. A manual notation system was employed to record the time (in seconds) that the referee spent either standing, walking, jogging, striding, jogging backwards or sideways, and sprinting.

2.2. Results

Table 1 presents the findings from this pilot investigation. The percentages of time spent engaged in the activities were 18% standing, 22% walking, 32% jogging, 10% striding, 17% jogging backwards and sideways, and 1% sprinting. The mean data (i.e. percentage of time spent performing the activity) was then used in the main study in which a laboratory simulation was carried out.

Table 1. Total time (min) in which referees were engaged in various activities during the six competitive televised games.

	Total (min) spent in the activity						
	Stand	Walk	Jog	Stride	Reverse	Sprint	Total
Game 1	16.82	16.4	27.35	8.83	14.31	0.42	84.13
Game 2	12.99	17.85	27.59	9.73	12.18	0.81	81.15
Game 3	14.49	20.53	23.74	6.01	14.48	1.16	80.47
Game 4	11.36	14.99	29.97	6.48	17.01	1.24	81.01
Game 5	16.62	20.02	25.98	11.08	9.37	2.13	85.21
Game 6	17.26	20.14	21.36	6.58	13.97	1.19	80.51
Mean	14.92	18.32	28.00	8.12	13.55	1.16	82.08
±SD	±2.39	±2.28	±3.06	±2.07	±2.57	±0.57	±2.05

3. Methods

Eight elite rugby league referees with a mean ± SD age of 28.4 ± 1.5 years, height of 1.77 ± 0.03 m, body mass of 83.5 ± 4.4 kg, and estimated $O_{2\,max}$ of 52.2 ± 3.1 ml kg^{-1} min^{-1} gave informed consent both verbally and in writing to take part in the study. All the referees were members of the Wigan and District Referee Society and had a minimum of 3 years of refereeing experience at senior level. At the time of the study, the referees were engaged in training for 6–8 h per week. Approval for the study was obtained from the Ethics Committee of Liverpool John Moores University.

Subjects visited the laboratory on three occasions, the first of which was to establish their $O_{2\,max}$ while the second and third visits were to perform simulated rugby league activities after ingesting either an experimental drink (6% maltodextrin) or a placebo. An adapted version of the Loughborough Intermittent Shuttle Test (LIST) was used (Nicholas et al. 1997). A 1-week period separated the trials, and each condition was counterbalanced to account for any bias. A single-blind cross-over design was employed.

Maximal oxygen uptake was estimated using a progressive 20-m multi-stage shuttle run test (Ramsbottom et al. 1988) on an indoor running track. Following the test, individual audio tapes were produced so that 'bleeps' were recorded to elicit running speeds of 40, 55 and 95% of the final running speed in the multi-stage shuttle test. These individualized tapes were then used in the two subsequent visits.

Subjects were tested at the same time of day to take circadian variations into account, and had not consumed food or drinks (other than water) for 4 h prior to testing. No exercise was allowed in the 24 h prior to the test, and all subjects monitored and kept their dietary intake constant during this period.

On entering the laboratory subjects rested before providing a capillary blood sample for the determination of glucose concentration. A 10-min standardized warm-up, devised by the physical trainer of the Rugby Football League, ensued before the subjects ingested 200 ml of either an orange-flavoured test drink (6% maltodextrin) or orange-flavoured placebo (water). Both drinks were indistinguishable in taste to the subjects, and neither contained electrolytes. The simulated performance then took place on an indoor track, and consisted of repeated 8-min bouts of activity followed by a 2-min rest period (i.e. standing) as determined from the motion analysis. Table 2 illustrates a typical activity pattern undertaken by subjects. Four bouts of activity with rest represented the first half of a match, which was followed by a 10-min half-time break and then a further three simulated bouts of activity. During each of the seven activity bouts a sprint was timed using electronic timing gates (Retro-reflective Photo Relay System, Model WL27, Sensick, Germany) set 15 m apart. After the final bout, subjects performed a run to fatigue at intensities varying between 55 and 95% of their estimated $O_{2\,max}$, i.e. two 20-m shuttles at 55% and then two 20-m shuttles at 95%. Fatigue was deemed to have occurred when subjects were unable to maintain the selected pace for two successive shuttles. Figure 1 is a schematic diagram of the overall procedures adopted.

Additional 200 ml drinks were consumed during the 2-min rest periods, at half-time, and immediately prior to the run to exhaustion. Capillary blood samples were taken at half-time, just before the run to exhaustion, and at the end of the trial. The rating of perceived exertion was assessed during the final minute of each 8-min block of exercise, and following every 10 shuttles during the run to exhaustion (Borg 1973).

Table 2. Activity pattern undertaken by one of the referees during an 8-min block of simulation. This pattern was followed by 2 min of standing and then repeated.

Activity	Number of shuttles	Time per shuttle (s)	Total number of shuttles in one 8-min bout
Walk	1	12.4	10
Jog	2	9.5	10
Reverse	1	9.0	10
Stride	1	5.5	10
Stand	1	5.0	10
Sprint	n/a	n/a	1

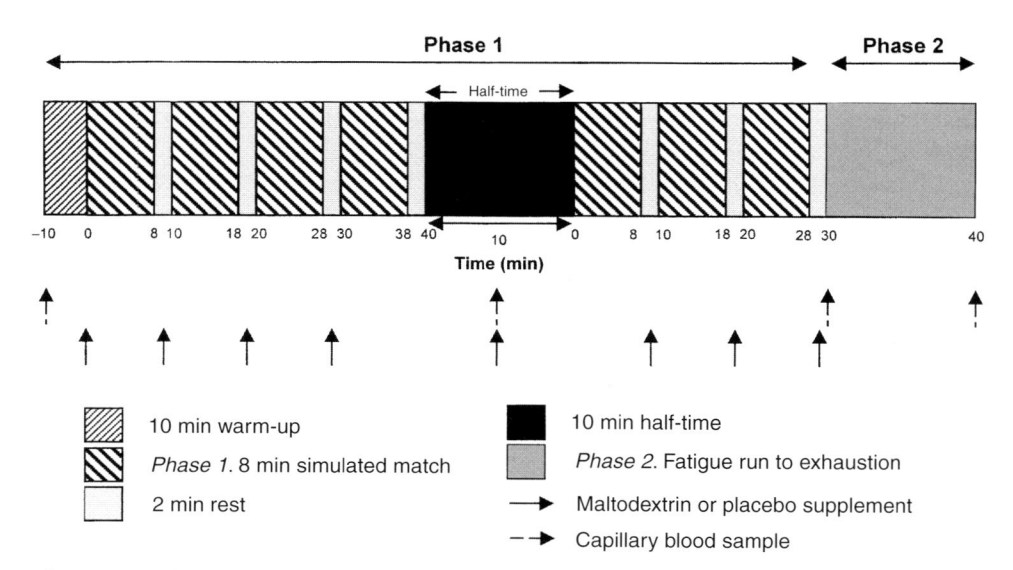

Figure 1. Schematic diagram of the protocol adopted (adapted from Nicholas et al. 1997).

Lysed capillary whole blood was later analysed in duplicate for glucose and lactate using an Analox Micro-stat GM7 analyser (Alpha Laboratories, London).

Data are presented as means ($\pm$SD); *t*-tests were employed on the number of shuttles to fatigue (after normality had been assessed using the Anderson-Darling normality plot) and to the mean sprint times between treatments. Multi-factorial ANOVAs with repeated measures were performed on the sprint times, RPE, and the blood glucose and lactate concentrations. Any significant findings from the ANOVAs were subjected to further analyses using a *post-hoc* Tukey test. Significance was accepted at $p<0.05$.

4. Results

4.1. *Number of 20-m shuttles to fatigue*

The mean number of 20-m shuttles was $\sim 25\%$ higher for the Md than for the Pl condition (57 ± 19 shuttles for Md vs. 43 ± 15 shuttles for Pl). A one-sample *t*-test revealed this to be significant ($p<0.05$).

4.2. *Sprint times*

The mean sprint times for the seven 15-m sprints were significantly faster in the Md condition compared with the Pl condition (2.40 ± 0.09 s for Md vs. 2.51 ± 0.14 s for Pl). A two-way ANOVA revealed both a significant time effect ($p<0.05$) and a treatment effect ($p<0.05$). The first sprint was the slowest in both treatments before becoming faster. However, whereas sprint speed was maintained in the Md condition, the times became longer in the Pl condition after the half-time break (figure 2). The significant differences between treatments was evident from the first sprint and maintained throughout.

4.3. *Rating of perceived exertion (RPE)*

The mean RPE was $\sim 5\%$ lower for the Md condition than for the Pl condition ($p<0.05$) and the values increased significantly during the seven activity bouts ($p<0.05$). In the Pl treatment, the RPE increased steadily throughout the bouts, whereas for Md ingestion the values increased to the fourth bout (i.e. after the half-time interval) and then were maintained (figure 3).

4.4. *Blood glucose concentrations*

There were no significant differences between treatments for the blood glucose concentrations at rest ($p>0.05$). There were significant increases in blood glucose at the three sample points following maltodextrin ingestion compared with placebo ($p<0.05$). Figure 4 illustrates these findings.

4.5. *Blood lactate concentrations*

No significant differences were found between Pl and Md ingestions for blood lactate concentrations ($p>0.05$) although there was a significant difference with respect to time ($p<0.05$). The differences with time occurred in the blood samples following half-time (figure 5).

5. Discussion

The activity patterns determined for rugby league referees in the pilot study are not too dissimilar to those found for referees officiating in the Premier Division of the English

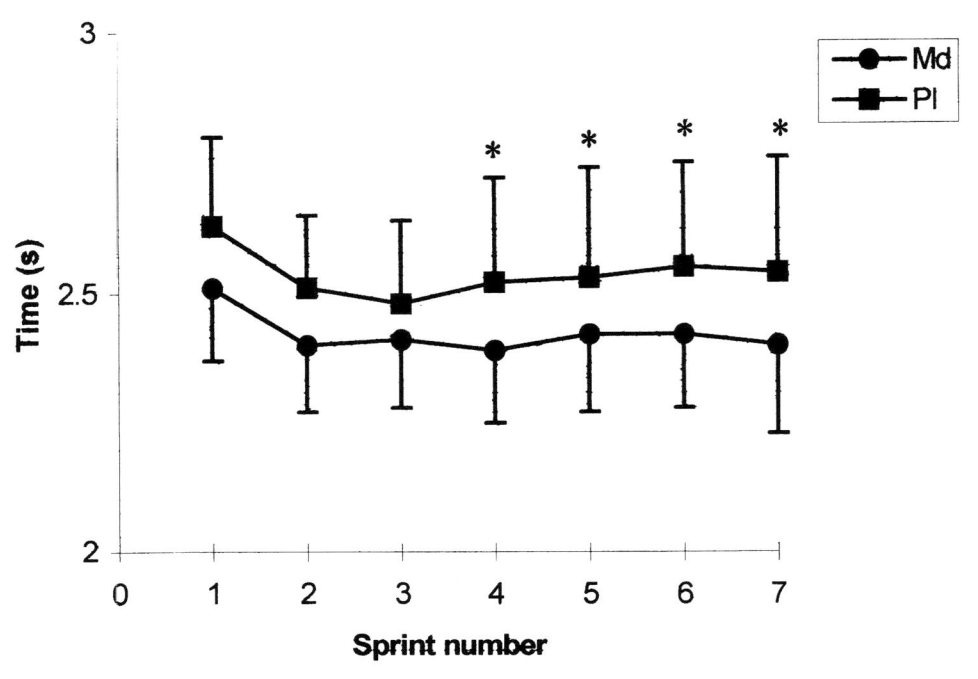

Figure 2. Mean (±SD) 15-m sprint times during the activity bouts under each of the two treatments, ingestion of 6% maltodextrin (Md) or placebo (Pl) (* indicates significant difference between treatments, $p < 0.05$).

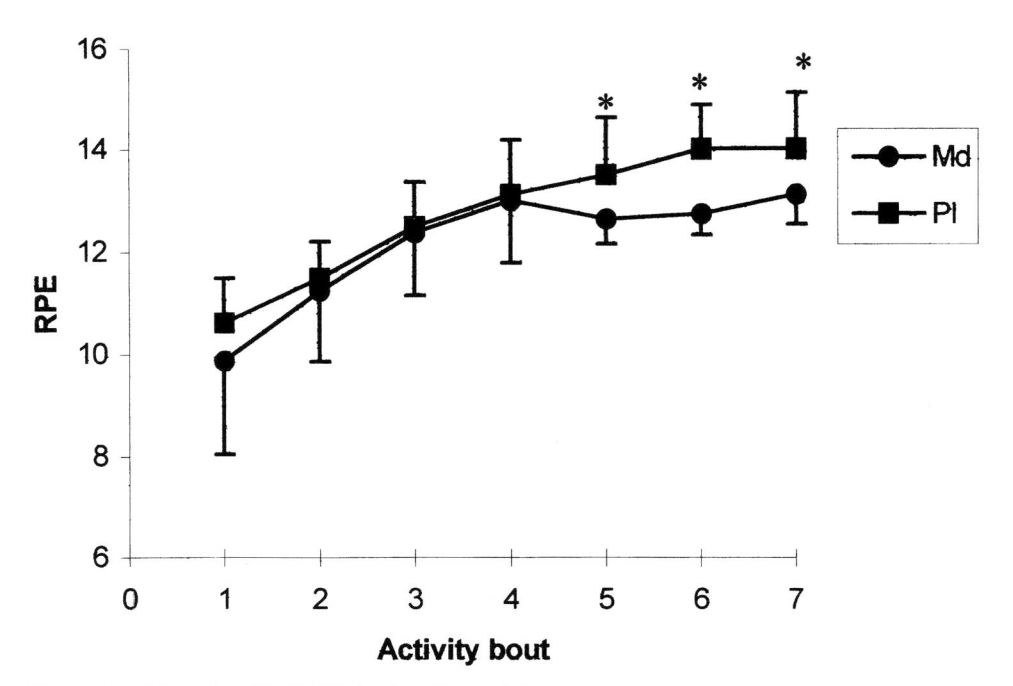

Figure 3. Mean (±SD) RPE during the activity bouts under each of the two treatments, ingestion of 6% maltodextrin (Md) or placebo (Pl) (*indicates significant difference between treatments, $p < 0.05$).

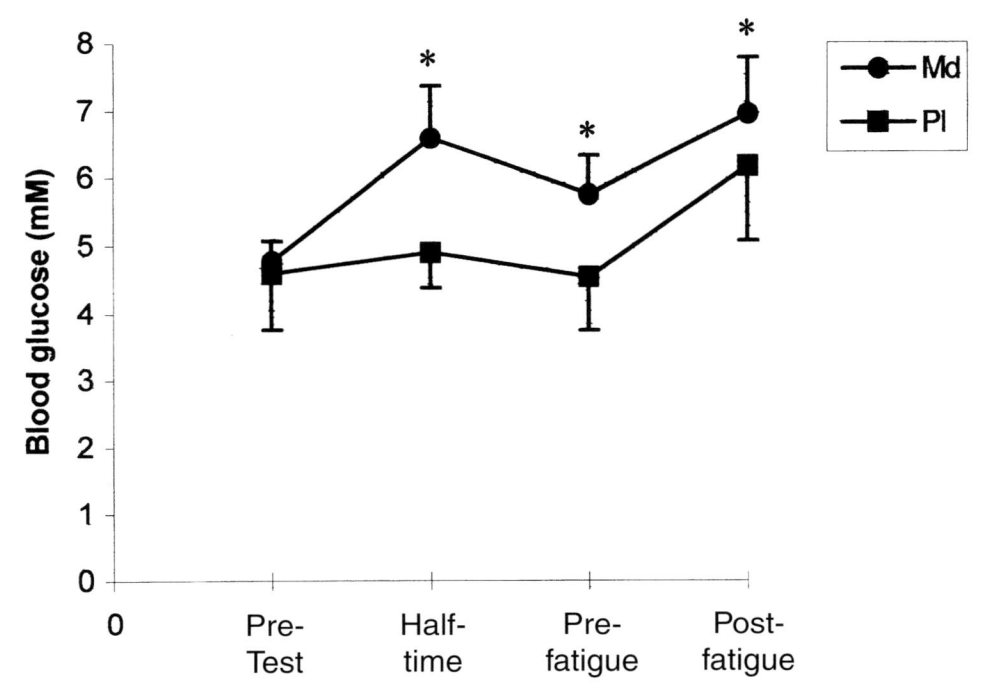

Figure 4. Mean (±SD) blood glucose concentrations during the activity bouts under each of the two treatments, ingestion of 6% maltodextrin (Md) or placebo (Pl) (* indicates significant difference between treatments, $p<0.05$).

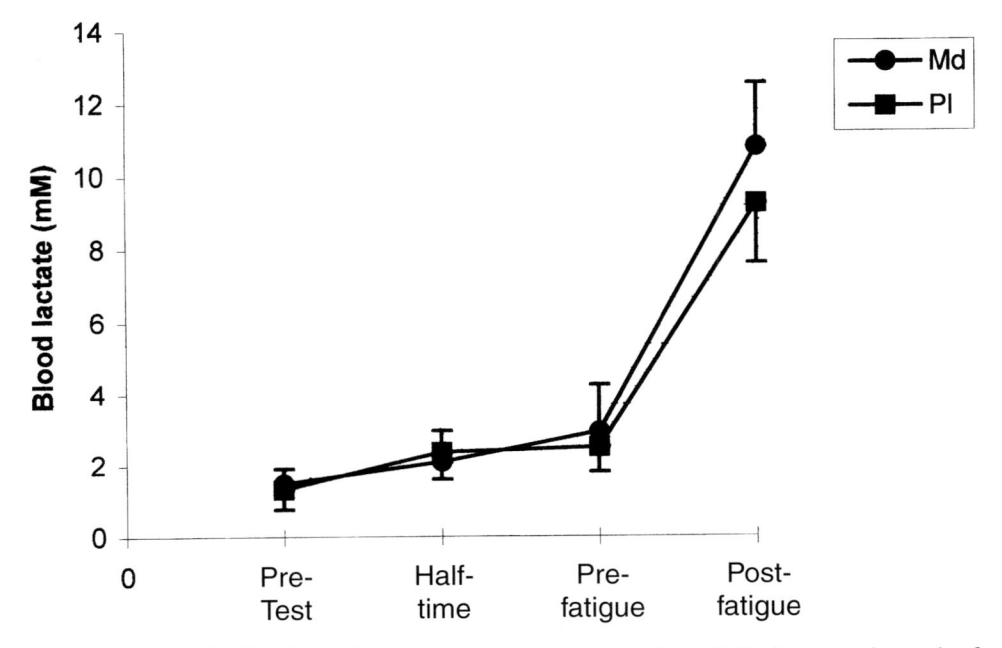

Figure 5. Mean (±SD) blood lactate concentration during the activity bouts under each of the two treatments, ingestion of 6% maltodextrin (Md) or placebo (Pl).

Football League (Catterall *et al.* 1993). The major difference between the present study and that of Catterall *et al.* (1993) was that rugby league referees were found to spend 1% of the total time sprinting as opposed to 12% by the association football referees. In the present study there was an extra criterion of 'striding', which accounted for 10% of the game demand and presented an activity proportional to 95% $O_{2\,max}$.

The intermittent high intensity bouts of activity during the simulation clearly resulted in fatigue since sprint times were elevated in the placebo treatment although the ability to sprint was maintained during Md ingestion. A surprising finding was that a significant difference in sprint time was apparent between treatments at the first sprint. It is difficult to interpret this observation since the maltodextrin would have been ingested a matter of 8 min before the sprint. This appears to be too short a time period for any significant effect to have accrued from the carbohydrate, although the effect of mild hyperglycaemia (which is evident 8 min after ingestion of maltodextrin) on 'central' factors remains a possibility.

Of particular interest was the observation that after half-time (i.e. sprints 5–7) the sprint times actually decreased for the Md condition whereas they increased for the Pl condition. Depletion of muscle glycogen stores has been shown to attenuate high intensity activities and an increase in muscle glycogen stores has been shown to promote high intensity exercise (Maughan and Poole 1981, Pizza *et al.* 1995). Since muscle glycogen concentration was not assessed in this study, one can only speculate that if ingestion of carbohydrate leads to muscle glycogen sparing (Tsintzas *et al.* 1995) then the greater differences in ability to sprint between the Pl and Md treatments in the second half could, in part, be attributed to that. Recently, Nicholas *et al.* (1999) reported a 22% reduction in muscle glycogen utilization when a 6.9% carbohydrate-electrolyte drink was compared with a control drink for trained games players performing the Loughborough Intermittent Shuttle Test. The approximate total amount of carbohydrate ingested during the study was 78 g, which compares with the 96 g ingested by subjects in the present study.

Since glycogenolysis provides energy during intense sprint running within 3 s (Cheetham *et al.* 1986), the possibility exists of a significant degree of muscle glycogen depletion in the fast glycolytic fibres during the repeated sprints. Fast glycolytic fibres generate the necessary power for maintenance of peak sprint speeds, and so any significant depletion of glycogen may require other muscle fibres to be recruited, leading to a decrement in speed. It should also be remembered that recruitment of fast glycolytic fibres takes place during steady-paced running. This too will contribute to the loss in muscle glycogen from these fibres.

The distance covered by the seven, 10-min bouts of activity was 7000 m (1000 m per bout), a distance expected to be covered by referees in a competitive game. Maltodextrin ingestion improved time to fatigue by 25% over the placebo treatment, which represents an approximate 300 m increase in distance covered. Furthermore, all but one subject increased his endurance after Md ingestion. The improvement in endurance is in agreement with previous findings (MacLaren *et al.* 1994, Nicholas *et al.* 1995, Davis *et al.* 1997, El-Sayed *et al.* 1997), although these studies investigated the effects of carbohydrate ingestion on activities other than rugby league. Nicholas *et al.* (1995) employed the LIST and found an improvement of 33% in time to fatigue when comparing a 6.9% carbohydrate-electrolyte solution with a control.

Although the present investigation was not designed to elucidate the exact mechanism as to how carbohydrate supplementation delayed fatigue, it is apparent that the role of elevating and maintaining mild hyperglycaemia could be a factor

(Coyle et al. 1986, Coggan and Coyle 1988). Recent studies have indicated that carbohydrate supplementation improves endurance performance during intermittent exercise (Davis et al. 1997, Dennis et al. 1997), and indeed Jacobs et al. (1982) reported that fatigue was delayed during a soccer match when carbohydrate was ingested.

The high level of exercise intensity over a protracted period of time has consequences not only for physical fitness of referees but also for mental judgement. Decrements in cognitive function and some psychomotor tasks have been found once the exercise intensity exceeds 55% $O_{2\,max}$ (Reilly and Smith 1986). The need to prevent hypoglycaemia is recognized if decision-making is required. Reilly and Lewis (1985) reported that carbohydrate ingestion during exercise improved mental function both in terms of the number of tasks carried out and fewer resultant errors. The RPE data in the present study clearly shows the benefit of carbohydrate provision in terms of perception of effort. The authors did not include decision-making tasks in this study, and so are unable to conclude whether the psychological benefits of 'feeling less tired' can extend to more correct decisions being made by the referees. Clearly there is scope for a further investigation.

6. Conclusions

This investigation was successful in exhibiting the ergogenic benefits of maltodextrin ingestion on simulated rugby league activity and number of shuttles to fatigue after such an activity. Supplementation with maltodextrin improved endurance and maintained sprint performance, possibly due to higher levels of blood glucose. The benefits of carbohydrate provision may also be realized in an attenuated RPE score. Recommendations can be made to rugby league referees to consume carbohydrate drinks before and during competitive games.

Acknowledgements

The help from Professor Alan Nevill with the statistical analyses of the data is very much appreciated.

References

BERGSTROM, J., HERMANSEN, L., HULTMAN, E. and SALTIN, B. 1967, Diet, muscle glycogen and physical performance, *Acta Physiologica Scandinavica*, **71**, 140–150.
BORG, G. A. V. 1973, Perceived exertion: a note on 'history' and methods, *Medicine and Science in Sports*, **5**, 90–93.
BREWER, J. and DAVIS, J. 1995, Applied physiology of rugby league, *Sports Medicine*, **20**, 129–135.
CATTERALL, C., REILLY, T., ATKINSON, G. and COLDWELLS, A. 1993, Analysis of the work rates and heart rates of association football referees, *British Journal of Sports Medicine*, **27**, 193–196.
CHEETHAM, M. E., BOOBIS, L. H., BROOKS, S. and WILLIAMS, C. 1986, Human muscle metabolism during sprint running, *Journal of Applied Physiology*, **61**, 54–60.
COGGAN, A. R. and COYLE, E. F. 1987, Reversal of fatigue during prolonged exercise by carbohydrate infusion or ingestion, *Journal of Applied Physiology*, **63**, 2388–2395.
COGGAN, A. R. and COYLE, E. F. 1988, Effect of carbohydrate feeding during high intensity exercise, *Journal of Applied Physiology*, **65**, 1703–1709.
COYLE, E. F. 1992, Carbohydrate feeding during exercise, *International Journal of Sports Medicine*, **13**, s126–s128.
COYLE, E. F., COGGAN, A. R., HEMMERT, M. K. and IVY, J. L. 1986, Muscle glycogen utilisation during prolonged strenuous exercise when fed carbohydrate, *Journal of Applied Physiology*, **61**, 165–172.

Davis, J. M., Jackson, D. A., Broadwell, M. S., Queary, J. L. and Lambert, C. L. 1997, Carbohydrate drinks delay fatigue during intermittent, high-intensity cycling in active men and women, *International Journal of Sports Nutrition*, **7**, 261–273.

Dennis, S. C., Noakes, T. D. and Hawley, J. A. 1997, Nutritional strategies to minimize fatigue during prolonged exercise: fluid, electrolyte and energy replacement, *Journal of Sports Sciences*, **15**, 305–313.

El-Sayed, M. S., Balmer, J. and Rattu, A. J. M. 1997, Carbohydrate ingestion improves endurance performance during 1 hour simulated cycling time trial, *Journal of Sports Sciences*, **15**, 223–230.

Gollnick, P. D., Piehl, K. and Saltin, B. 1974, Selective glycogen depletion patterns in human muscle fibers after exercise of varying intensity and at varying pedalling rates, *Journal of Physiology*, **241**, 45–57.

Jacobs, I., Westlin, N., Karlsson, J. and Rasmusson, M. 1982, Muscle glycogen and diet in elite soccer players, *European Journal of Applied Physiology*, **48**, 297–302.

Johnston, L. and McNaughton, L. 1994, The physiological requirements of soccer refereeing, *The Australian Journal of Science and Medicine in Sport*, Sept/Dec, 67–72.

MacLaren, D. P., Reilly, T., Campbell, I. T. and Frayn, K. N. 1994, Hormonal and metabolite responses to glucose and maltodextrin ingestion with or without the addition of guar gum, *International Journal of Sports Medicine*, **15**, 466–471.

Maughan, R. J. and Poole, D. C. 1981, The effects of a glycogen-loading regimen on the capacity to perform anaerobic exercise, *European Journal of Applied Physiology*, **46**, 211–219.

Nicholas, C. W., Green, P. A., Hawkins, R. D. and Williams, C. 1997, Carbohydrate intake and recovery of intermittent running capacity, *International Journal of Sports Nutrition*, **7**, 251–260.

Nicholas, C. W., Tsintzas, K., Boobis, L. and Williams, C. 1999, Carbohydrate-electrolyte ingestion during intermittent high-intensity running, *Medicine and Science in Sports and Exercise*, **31**, 1280–1286.

Nicholas, C. W., Williams, C., Lakomy, H. K. A., Phillips, G. and Nowitz, A. 1995, Influence of ingesting a carbohydrate-electrolyte solution on endurance capacity during intermittent, high-intensity shuttle running, *Journal of Sports Sciences*, **13**, 283–290.

Pizza, F. X., Flynn, M. G., Duscha, B. D., Holden, J. and Kubitz, E. R. 1995, A carbohydrate loading regimen improves high intensity, short duration exercise performance, *International Journal of Sports Nutrition*, **5**, 110–116.

Ramsbottom, R., Brewer, J. and Williams, C. 1988, A progressive shuttle run test to estimate maximal oxygen uptake, *British Journal of Sports Medicine*, **22**, 141–144.

Reilly, T. and Lewis, W. 1985, Effects of carbohydrate loading in mental functions during sustained physical work, in I. D. Brown, R. Goldsmith, K. Coombes and M. Sinclair (eds), *Contemporary Ergonomics 85* (London: Taylor & Francis), 700–702.

Reilly, T. and Smith, D. 1986, Effect of work intensity on performance in a psychomotor task during exercise, *Ergonomics*, **29**, 601–606.

Saltin, B. 1973, Metabolic fundamentals in exercise, *Medicine and Science in Sports*, **5**, 137–146.

Tsintzas, O.-K., Williams, C., Boobis, L. and Greenhaff, P. 1995, Carbohydrate ingestion and glycogen utilization in different muscle fibre types in man, *Journal of Physiology*, **489**, 243–250.

Vollestad, N. K. and Blom, P. C. S. 1985, Effect of varying exercise intensity on glycogen depletion in human muscle fibres, *Acta Physiologica Scandinavica*, **125**, 395–405.

9 Determining the protective function of sports footwear

MARK J. LAKE*

Research Institute for Sport and Exercise Sciences, Liverpool John Moores University, 15–21 Webster Street, Liverpool L3 2ET, UK

Keywords: Biomechanics; Injury; Impact; Stability; Traction.

To reduce the risk of injury associated with foot–ground interaction during sporting activities, there is a need for adequate assessment of the protective function of sports footwear. The present objectives are to review the typical biomechanical approaches used to identify protection offered by sports footwear during dynamic activities and to outline some of the recent methodological approaches aimed at improving this characterization. Attention is focused on biomechanical techniques that have been shown to best differentiate safety features of footwear. It was determined that subject tests would be used in combination with standard mechanical techniques to evaluate footwear protection. Impact attenuation characteristics of footwear during sporting activities were most distinguished by analysis of tibial shock signals in the frequency and joint time-frequency domains. It has been argued that lateral stability and traction properties of footwear are better assessed using game-like manoeuvres of subjects on the actual sporting surface. Furthermore, the ability of such tests to discriminate between shoes has been improved through methods aimed at reducing or accounting for variability in individual execution of dynamic manoeuvres. Advances in tools allowing measurement of dynamic foot function inside the shoe also aid our assessment of shoe protective performance. In combination, these newer approaches should provide more information for the design of safer sports footwear.

1. Introduction

Footwear can be designed to enhance the performance of all sports involving locomotion and dynamic foot–ground interaction. Shoe characteristics such as traction, stability, flexibility and weight can all be modified towards that goal. Much attention has also been focused on the protective role of footwear in helping to prevent either acute injury or chronic damage to the body during sporting activities. Although sometimes the professional elite performer may place slightly more emphasis on performance rather than safety features of footwear, for the majority of sports participants the protective function of sports shoes becomes increasingly important. A mixture of protective features for sports footwear, geared towards the specific demands of each sport, is often required. Adequate and durable cushioning systems in the shoe are necessary for safe participation in sports such as distance running where the lower limbs need to be protected from repetitive impact loading. For many team sports the body must also be protected against less frequent but

*e-mail: M.J.Lake@livjm.ac.uk

more severe impact forces of jump landings, and high shear forces generated during sudden stops and turns. In this latter case, compliant and shock absorbing shoes are essential but it is also necessary to have lateral stability and optimal traction. This combination of cushioning protection and support of the foot during forward and lateral dynamic manoeuvres presents a difficult challenge for sports shoe design engineers (Frederick 1995). The fit of the shoe and the amount of pressure the foot is exposed to at its plantar and dorsal surfaces are additional factors that may be associated with the long-term overloading and damage of the foot (Hennig 1998). Clearly there is a need for sufficient characterization and subsequent improvement of the safety aspects of sports shoes. There are some standard mechanical techniques available to examine sports shoe properties but there is a growing trend to combine such information with biomechanical measurements on subjects wearing the footwear during typical sporting manoeuvres. The present aims are to review the typical biomechanical approaches used to identify protection offered by sports footwear and to outline some of the methodological advances that may improve the characterization of footwear protective performance.

2. Impact attenuation characteristics

2.1. *Background*

During locomotor activities ground contact generates impact forces. These external reaction forces result in compressive loading of the lower extremity and cause a shock wave to travel through the body. Ground reaction forces (GRF) during running and jumping actions have high frequency impact peak magnitudes reaching from 2 to 10 times body weight in less than 35 ms (Cavanagh and Lafortune 1980, McClay *et al.* 1994b). Since they are transmitted to the musculoskeletal structures through the shoe sole, the sole has the opportunity to lower the rate of external loads acting on the body. Properly cushioned footwear may lead to a reduction of injuries associated with the repetitive impacts experienced during running and jump landings. It is reasonable to assume that some injuries which occur during running are related to phenomena that occur during the landing phase of a running stride (Bobbert *et al.* 1992). Speculated potential consequences of repetitive lower extremity loading of high magnitude and rate include degenerative joint disease (Simon *et al.* 1972, Radin *et al.* 1972, 1982, Collins and Whittle 1989) and low-back pain (Voloshin and Wosk 1982). Although the rate of lower extremity loading has been linked to joint deterioration and osteoarthritic changes in animals (Yang *et al.* 1989), the biological effects of different loading rates in humans are unknown. However, fast external force increases to the body during ground contact are very likely to be unfavourable. Different midsole material and design characteristics of footwear dictate the load and rate of loading to which specific foot structures will be exposed at contact with the ground. This impact attenuation function is crucial to protect the body from damage.

2.2. *Typical measurement techniques*

Measurement and analysis of high amplitude and short duration impacts typically occurring during locomotion are not straightforward. Mechanical tests and subject tests have been used to evaluate the cushioning properties of sports footwear. In mechanical testing, the shock attenuating capacity of a running shoe is quantified by dropping an instrumented weighted missile on the heel and fore part of the shoe (ASTM 1999). A lack of correlation between results from mechanical tests on running shoes and peak impact force results from force platform measurements during actual running has been reported (Clarke *et al.* 1983, Snel *et al.* 1985, Hennig and Milani 1995). These authors have indicated that such purely mechanical measurements are certainly not suited to make predictions with regard to the influence of shoe construction on the loading of the body during running. Typically, the simulation of the foot inside the shoe and the kinematic adaptations of the runner cannot be undertaken sufficiently well and, generally, most researchers have a preference for controlled subject tests to supplement data obtained from mechanical impact testers.

For footwear with midsole material having different properties, peak impact forces of similar magnitude for groups of individuals during running have been found using floor-mounted force platforms (Clarke *et al.* 1983, Snel *et al.* 1985, Nigg *et al.* 1987, Lafortune *et al.* 1993). This has been explained by a concurrent difference in internal forces (Nigg *et al.* 1987) and adaptations in the neuromuscular control system to changing shoe sole hardness (Snel *et al.* 1985). The apparent lack of sensitivity of the force platform to discern shoe cushioning properties may not be surprising given that the measured vertical ground reaction force reflects the acceleration of the total body's centre of mass. Such measures of the average acceleration of the body may mask some of the large differences in acceleration of the impacting leg segments owing to changes in the severity of foot–ground impact (Shorten and Winslow 1992). For walking, tibial axial acceleration has been found to be more sensitive at distinguishing between footwear cushioning than GRF measurement (Lafortune and Hennig 1992). Consequently, the impact attenuation properties of different footwear during locomotion have more recently been evaluated in different laboratories using tibial accelerometry (Hamill 1999).

2.3. *Tibial shock measures*

Tibial shock during foot–ground impact is typically measured using a miniature accelerometer attached to the distal shank with balsa wood and tight strapping (Valiant 1990). A low mass of the transducer assembly and good mechanical coupling with the underlying bone are crucial if the signals are to resemble the acceleration of the bone (Lafortune *et al.* 1996a). Although the magnitude and time course of tibial shock data during walking and running have been used to differentiate footwear cushioning properties (Lafortune and Hennig 1992, Lees and Jones 1994), it is very useful to examine the frequency content of these signals. Shorten and Winslow (1992) demonstrated that a frequency analysis of tibial shock signals during running could distinguish between the signal component associated with impact and that due to resonance of the accelerometer mounting on the skin. Significant differences in impact attenuation properties of both footwear and playing surfaces have been found using a spectral analysis (based on the Fourier transform (FT)) performed on tibial shock signals during locomotor activities (Johnson 1986, Shorten *et al.* 1986, Shorten and Himmelsbach 1999). Substantial reductions have been found in the higher frequency components of tibial shock signals (>20 Hz) during running by the use of specific midsole materials in footwear (Shorten *et al.* 1986). Higher mean power frequencies have been associated with more severe impacts (Johnson 1986), but the relationship of the higher frequencies to detrimental tissue effects remains to be determined. It has been shown in animal studies that bone tissue development is very sensitive to the frequency content of transmitted impact shock (Rubin *et al.* 1990). Valiant (1990) indicated that by characterizing tibial shock in the frequency domain, it may be possible to identify potentially harmful frequency components being transmitted through the body and generate a greater understanding of how to filter or dampen them with footwear material and design characteristics. Recent studies are now beginning to focus on shock transmission through the human body resulting from locomotor-type impacts, on how effective the body is at attenuating shock travelling to the head at specific frequencies and on the influence of different cushioning materials present in footwear (Lafortune et al. 1996b).

2.4. *Analysis techniques to improve the characterization of impact attenuation*
Classic algorithms such as the square of the Fourier transform (FT) indicate only the average of a signals power spectrum and are suited to the analysis of signals whose characteristics do not change over time. For transient signals such as tibial acceleration, the FT approach may cause some distortion of the frequency information that must be corrected with appropriate algorithms. To characterize the frequency content of a transient signal over time, researchers have begun to represent such signals in the joint time-frequency plane (Qian and Chen 1996). Rakotomamonjy et al. (1997a) demonstrated the benefit of a joint time-frequency domain (JTFD) analysis of tibial shock signals to illustrate the large improvements in impact attenuation during shod running compared to barefoot running. More recently, JTFD-derived impact severity variables, determined on tibial shock data from controlled locomotor-like impacts, were found to better differentiate sports shoe impact attenuation ability than severity variables calculated from either time domain or frequency domain (FT) signal analysis alone (Lake and Patritti 1999). Impact attenuation characteristics in the latter study were clearly distinguished between two types of footwear cushioning by subtracting the tibial shock signal energy spread over the time-frequency plane (figure 1).

Some JTFD analysis techniques, such as the windowed Fourier transform, fix the time and frequency resolutions once the time window is selected. However, biomechanical measurements associated with locomotor activities and foot-ground impact have a short time period at high frequency and relatively long time periods at low frequency. This suggests that the window should have high time resolution at high frequency. A promising JTFD approach that has good time resolution at high frequencies and good frequency resolution at low frequencies is the discrete wavelet transform (Aldroubi and Unser 1996). Wavelet analysis can be considered to be a filtering method that extracts specific frequency bands from the signal. Impact signals can be decomposed into time series of different frequency bands and this represents a powerful technique with which to characterize the data (Burn et al. 1997). Plots similar to figure 1 can provide a highly visual representation of time and frequency localization of power during impact. Another advantage of the JTFD approach is that noise-corrupted signals are easier to recognize and correct than with separate analyses of time-domain and frequency domain information. Noise reduction is a powerful JTFD tool that has been previously used to minimize the influence of the skin mounting of accelerometers on estimates of tibial acceleration signals during lower limb impacts (Rakotomamonjy et al. 1997b).

These newer signal analysis approaches allow a more thorough characterization of the impact attenuation properties of footwear in both the time and frequency domains. Once this aspect of footwear protection has been properly identified then possible mechanisms for impact-related injury can be examined (Burn et al. 1997).

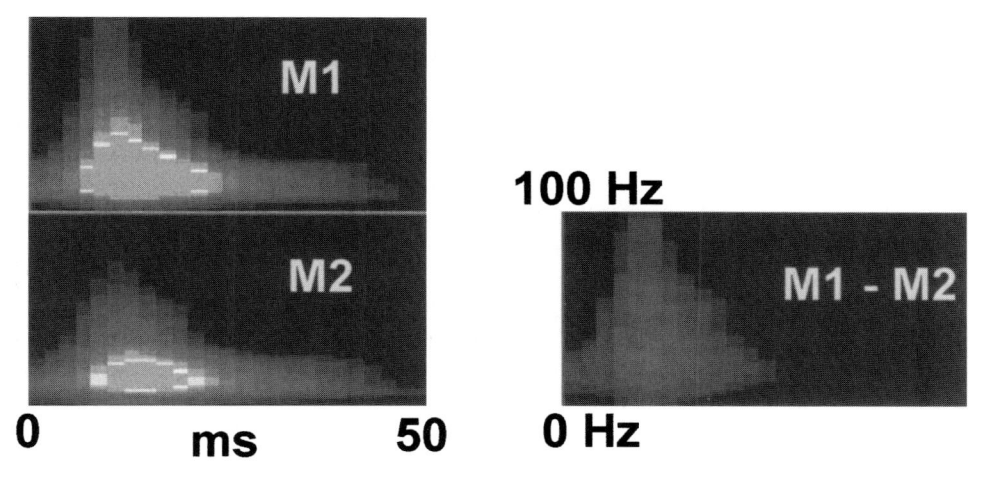

Figure 1. Tibial shock data in the joint time-frequency domain for controlled lower limb impacts in the same footwear with two different cushioning material inserts (M1 and M2). The y-axis represents frequency content (0 to 100 Hz) and the x-axis represents the time after initial foot impact (0 to 50 ms). The shading represents the level of signal energy (black being the lowest level). Impact attenuation differences between footwear are clearly illustrated by subtracting data in this time-frequency plane (M1–M2).

The aetiology of such musculoskeletal injury remains to be determined but adequate characterization of ground contact impact transients combined with more information concerning energy absorbing properties of the structures of the lower limb can help to further our understanding in this area.

3. Traction and lateral stability

3.1. Background

In a review of injury statistics for a number of sports activities where sideward cutting movements were common, Stacoff (1996) indicated that the ankle was the most frequently injured joint and accounted for between 15 and 30% of all reported injuries. Most of these were ankle sprains occurring during cutting, stopping, landing and rotating manoeuvres. Higher cut sports shoes can reduce the risk of these types of sprain injury (Stacoff 1996), but at the expense of full ankle mobility and perhaps performance. As mentioned previously, elite athletes may choose to wear low-cut shoes and thereby compromise protection for enhanced performance. There is evidence to suggest that injuries of the ankle and knee are related to high levels of translational and rotational traction demanded by such dynamic movements (Torg et al. 1974, Bowers and Martin 1975, Lambson et al. 1996). Friction characteristics of the soccer boot–surface interface are assumed to be important in both performance and injury risk. Grip must be sufficient to prevent slip and facilitate turning manoeuvres but excessive fixation has been implicated in non-contact injuries that occur during turning and cutting when excessive torque can develop on lower extremity joint structures (Lambson et al. 1996).

Cutting manoeuvres in ball sports are frequent and considerable shear components of the ground reaction force are observed. Peak shear forces greater than 1.5 times body mass in anterior-posterior and medio-lateral directions have been reported (McClay et al. 1994b). These shear forces produce dual requirements, placing a demand upon the lateral stability as well as the traction characteristics of sports shoes. Lafortune (1997) stated that footwear must provide a stable base for the initiation and execution of cutting manoeuvres and protect the athlete from injuries resulting from high forces developed at the shoe–ground interface. However, the acceptable limits for the generation of shear forces and moments of rotation about a vertical axis during dynamic turning manoeuvres remain to be established. Nigg (1990) observed that moments of rotation seem to be kept below a limit of 25 N m in subject rotational manoeuvres on different surfaces. In order to design footwear with appropriate lateral stability and traction properties to minimize injury risk for a specific shoe–sports surface interaction, shoe manufacturers need much more information regarding 'safe' boundaries for the magnitudes of moments and shear forces that can be generated by the body.

3.2. Typical measurement techniques

Mechanical test feet have been used to assess sports footwear–surface traction characteristics in a controlled manner (Schlaepfer et al. 1983, Valiant 1994). Nigg (1990) criticized such friction test devices because they did not provide meaningful comparisons between athletic footwear as the applied loads were not representative of the actual situation during sporting activities. He recommended that mechanical tests and subject tests be used in combination for the assessment of traction characteristics. Therefore, once again the use of subject tests has been endorsed for assessing an important safety feature of sporting footwear. Stucke et al. (1984) developed a series of controlled subject tests to examine the translational and rotational traction characteristics of footwear on a given surface. Subjects in their study performed stopping movements and standardized rotations of 180° on surfaces mounted onto a force platform. They also determined shoe traction properties using more realistic dynamic turning manoeuvres although shoe lateral stability was not assessed.

The lateral stability of different footwear designs during sideward cutting moves has been determined using standard two-dimensional film analysis of the rearfoot movement (Luthi et al. 1986, Stacoff 1996). Stacoff (1996) also measured the movement of the heel inside the shoe during turning using small windows cut in the heel counter. He demonstrated the importance of shoe sole design and upper design (high cut) in reducing the risk of ankle injury. Some investigators have measured rearfoot motion and shear forces at the same time during cutting manoeuvres by having subjects land with their pivot foot on the force platform (Luthi et al. 1986, McClay et al. 1994a, b). These studies have provided some insight into the interaction between lateral stability and shear force development. Depending on the type of playing surface and the kind of movement, traction characteristics may vary considerably for the same shoe. Therefore biomechanical measurements should be performed in game-like situations on the actual surface of play (Hennig 1998). However, very few studies have determined the protective function of footwear using this approach (Coyles et al. 1998).

3.3. Approaches to improve subject tests of traction and lateral stability

The main drawbacks with the subject testing of footwear traction and lateral stability properties are: (1) the lack of tight control of experimental conditions for comparative shoe evaluations, and (2) the adaptation of the athlete to the shoe–

ground frictional characteristics and shoe support by alterations in their movement patterns and shear forces and rotational moments generated (Nigg 1990). Although some researchers might correctly argue that such adaptive effects are important to examine, for the comparative evaluation of footwear they limit the ability of subject tests to identify small differences in the protective function of the shoe during high-risk cutting and turning movements.

Adaptation of the athlete can be minimized by having the traction and support performance of the test footwear unknown to the athlete. Coyles *et al.* (1998) developed a field test methodology that allowed the evaluation of footwear traction during rigorous game-like manoeuvres that placed maximal demand on the footwear–surface interaction. In their study, knowledge of the footwear condition was kept from the subject for each manoeuvre by concealing shoe markings and changing the footwear each time. Although some differences were detected in actual and perceived traction performance of their tested footwear, in general differences were quite small. Coyles *et al.* (1998) attributed this observation to intra-subject variability in the initial landing kinematics of the foot although average speed approaching each turn was controlled. Larger differences between shoe traction properties have recently been observed by these authors when the control of initial turning conditions for each footwear type was improved by measuring both the continuous running velocity profile (using a laser system, Laveg-Sport, Jenoptik, Germany) and the adjustments in step length into the turn (Coyles 1999). Such precautions are necessary to allow differences in traction and support to be primarily the result of the change in footwear and not due to differences in execution of the manoeuvre. Lafortune (1997) used an alternative approach to minimize the influence of differences in the subject's execution of movements on the comparative evaluation of footwear. He simultaneously recorded high-speed video of the rearfoot and in-shoe pressure to quantify the lateral support and stability offered by court shoes during side-stepping moves. Lateral displacement of the heel counter (stability) was combined with measurement of in-shoe lateral heel pressure (support) to provide a heel control index. This index was not appreciably modified by individual differences in execution of the manoeuvre and also allowed a high discrimination of footwear rearfoot control. These types of methodological advances can improve the determination of the protective properties of footwear using subject tests.

The lateral heel counter pressure in the study by Lafortune (1997) was measured using three small discrete sensors. However, flexible arrays of force sensors can be attached on the medial and lateral side of the shoe (or foot) to estimate the shear forces during sudden stops and cutting movements in the field (Hennig 1998). Stacoff (1996) illustrated the importance of measuring foot movements inside footwear during lateral manoeuvres to determine fully their protective function. Techniques are now becoming available for quantifying the horizontal displacements of the foot relative to the shoe (Grau *et al.* 1997) and the actual shear forces at the foot–shoe interface (Christ *et al.* 1998). These developments in measurement tools that can be used inside the shoe will allow biomechanists to gain a better understanding of foot–shoe interaction during sporting activities.

4. Other protective aspects of footwear

Hennig and Milani (1995) measured in-shoe plantar pressures, GRF and tibial shock during running in 19 different commercially available running shoes. They found low correlations for the comparison of plantar pressures with GRF or shock-related variables and stated that the pressure measurements provided a different insight into the interaction between the human body, footwear and the ground as compared to recordings of GRF or tibial acceleration. The distribution of loads under the foot during sporting activities provides an additional indication of shoe protective performance. Areas of high pressure at the foot shoe interface can be detected and pressure sores of the foot caused by uncomfortable shoes can be avoided. In their review of in-shoe plantar pressure measurements, Cavanagh et al. (1992) pointed out that the same technology could be used for other areas of the foot. Since then investigators have begun to measure pressures on the dorsal as well as the plantar aspect of the foot while wearing different types of footwear (Jordan and Bartlett 1995, Kalpen and Seitz 1997). As well as providing a more complete assessment of foot loading due to shoe mechanics (Kalpen and Seitz 1997), the dorsal pressure measurements have also been related to shoe comfort during walking (Jordan and Bartlett 1995).

Another important aspect of footwear protection is the relationship between footwear construction features and foot structure. For those sports needing cleated footwear for traction, there are usually areas of localized high plantar pressure where studs are located (Shorten 1998, Coyles and Lake 1999). This is the probable cause of the high incidence of foot pain in areas of the foot where studs are typically located (Lafortune 1998). One example of this is the first metatarsal head, which is usually directly overlying a cleat position (figure 2). Coyles and Lake (1999) measured in-shoe

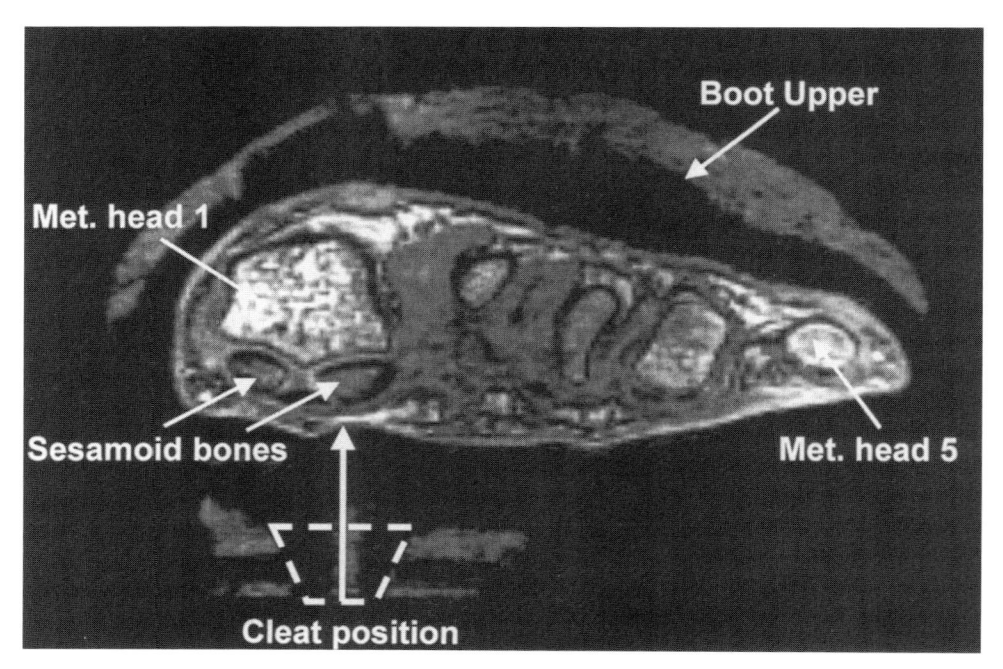

Figure 2. Magnetic resonance image of a foot inside a soccer boot while loaded. The two-dimensional slice is through the centre of the first metatarsal head (Met. head 1) with an oily paste used to mark the boot upper and stud screw threads. Met. head 1 (and the delicate sesamoid bones beneath it) is directly above a stud location where elevated plantar pressures are common. Consequently, pain as well as injury are very common at this location for many top-level participants in sports that utilize cleated footwear.

plantar pressure during running in cleated footwear and matched the pressure measures to loaded foot structure locations and cleat positions. Shoe design features such as outsole plate stiffness and cleat number were found to be important in reducing the in-shoe pressure hot spots and thereby decreasing the risk of foot injury.

One of the most recent tools that has been utilized to predict sports shoe response to external forces is mathematical modelling. Finite element models of shoe soles have recently been developed to estimate internal stresses and strains that cannot be measured practically (Shorten 1998). Such models have also facilitated the evaluation of footwear in relation to performance factors such as energy return and oxygen cost during running (Shorten 1993, Thomson *et al.* 1999). Shoe protective aspects can be optimized mathematically by using the finite element models to modify systematically the sole design and material properties and by examining the resulting plantar loading. In this manner, safety features of footwear can be predicted before prototypes are developed and then later confirmed using subject tests.

5. Overview

The protective requirements of sports footwear are complex and present a difficult challenge for shoe manufacturers. This review has highlighted the need for subject tests to supplement the mechanical assessment of footwear characteristics. The methods available for determining the protective role of sports footwear include force, acceleration and pressure measurements. Some safety aspects can be also be quantified effectively by using high-speed motion analysis systems. Most useful information is gained when these measurement approaches are employed simultaneously, e.g. force platform measures or in-shoe pressure combined with data for foot and leg motion. For the determination of impact attenuation performance of footwear, variables derived from tibial acceleration measures have been found to be preferable to those quantified from force platform ground reaction force information. This is particularly the case for tibial acceleration analysed in the frequency domain and more recently in the joint time-frequency domain. An exciting research task that lies ahead is to determine specific frequency and energy level combinations of the impact shock signal that might be damaging to structures of the human body. Traction and lateral stability performance of sports footwear were shown to be ideally determined during game-like manoeuvres on the actual playing surface. The main disadvantage with subject tests of this nature is that it is difficult to control completely the execution of manoeuvres in different footwear conditions. Recent studies have offered some methodological solutions either to reduce variability in movement execution or to allow for it. Advances in measurement tools that describe the motion of the foot inside the shoe will also permit a more thorough investigation of the shoe's protection function during dynamic stopping and turning moves. Finally, although some of the recent developments and modifications in techniques to determine protective aspects of sports footwear have been addressed, the full benefit of using the above analysis and measurement techniques to aid in the development of safer sports footwear remains to be demonstrated.

References

ALDROUBI, A. and UNSER, M. 1996, *Wavelets in Medicine and Biology* (Boca Raton: CRC Press).

ASTM 1999, F-1975 Standard test method for cushioning properties of athletic shoes using an impact test, American Society of Testing and Materials, West Conshohocken, PA.
BOBBERT, M. F., YEADON, M. R. and NIGG, B. M. 1992, Mechanical analysis of the landing phase in heel-toe running, *Journal of Biomechanics*, **25**, 223–234.
BOWERS, K. D. and MARTIN, B. 1975, Cleat-surface friction on new and old Astroturf, *Medicine and Science in Sports*, **7**, 132–135.
BURN, J. F., WILSON, A. and NASON, G. P. 1997, Impact during equine locomotion: techniques for measurement and analysis, *Equine Veterinary Journal*, **23**, 9–12.
CAVANAGH, P. R. and LAFORTUNE, M. A. 1980, Ground reaction forces in distance running, *Journal of Biomechanics*, **13**, 397–406.
CAVANAGH, P. R., HEWITT, F. G. and PERRY, J. E. 1992, In-shoe plantar pressure measurement: a review, *The Foot*, **2**, 185–194.
CHRIST, P., GENDER, M. and SEITZ, P. 1998, A 3-D pressure distribution measuring platform with 8 * 8 sensors for simultaneous measurement of vertical and horizontal forces, *Proceedings of the VI EMED Scientific Meeting*, Brisbane, Australia, 8–13 August, 17.
CLARKE, T. E., FREDERICK, E. C. and COOPER, L. B. 1983, Biomechanical measurement of running shoe cushioning properties, in B. Nigg and B. Kerr (eds), *Biomechanical Aspects of Sports Shoes and Playing Surface* (Calgary: University of Calgary Press), 25–33.
COLLINS, J. J. and WHITTLE, M. W. 1989, Impulsive forces during walking and their clinical implications, *Clinical Biomechanics*, **4**, 179–187.
COYLES, V. R. 1999, Unpublished document, Liverpool John Moores University.
COYLES, V. R. and LAKE, M. J. 1999, Forefoot plantar pressure distribution inside the soccer boot during running, *Proceedings of the Fourth Symposium on Footwear Biomechanics*, Canmore, Canada, 5–7 August, 30–31.
COYLES, V. R., LAKE, M. J. and PATRITTI, B. 1998, Comparative evaluation of soccer boot traction during cutting manoeuvres—methodological considerations for field testing, in S. Haake (ed.), *The Engineering of Sport* (London: Blackwell Science), 183–190.
FREDERICK, E. C. 1995, Biomechanical requirements of basketball shoes, *Proceedings of the Second Symposium on Footwear Biomechanics*, Cologne, Germany, 28–30 June, 18–19.
GRAU, S., SCHIEBL, F., KREBS, H. and SIMON, S. 1997, The linear position sensing system: a measurement device to measure horizontal movements in the shoe, *Proceedings of the Third Symposium on Footwear Biomechanics*, Tokyo, Japan, 21–23 August, 34.
HAMILL, J. 1999, Evaluation of shock attenuation, *Proceedings of the Fourth Symposium of Footwear Biomechanics*, Canmore, Canada, 5–7 August, 7–8.
HENNIG, E. M. 1998, Measuring methods for the evaluation of soccer shoe properties, *Proceedings of Soccer Player Oriented Science and Technology Congress*, Centre Technique Cuir Chaussure Maroquinerie, Lyons, France, 14–15 May.
HENNIG, E. M. and MILANI, T. L. 1995, In-shoe pressure distribution for running in various types of footwear, *Journal of Applied Biomechanics*, **11**, 14–15 May, 299–310.
JOHNSON, G. R. 1986, The use of spectral analysis to assess the performance of shock absorbing footwear, *Engineering in Medicine*, **15**, 117–122.
JORDAN, C. and BARTLETT, R. M. 1995, A method of measuring pressure distribution on the dorsum of the foot, *Proceedings of the Second Footwear Biomechanics Meeting*, Cologne, Germany, 28–30 June, 40–41.
KALPEN, A. and SEITZ, P. 1997, Influence of the shoe mechanics on the load of the foot during the push-off phase, *Proceedings of the Third Symposium on Footwear Biomechanics*, Tokyo, Japan, 21–23 August, 53.
LAFORTUNE, M. A. 1997, A new approach to assess in vivo rearfoot control of court footwear during side-stepping moves, *Journal of Applied Biomechanics*, **13**, 197–204.
LAFORTUNE, M. A. 1998, Performance, protection and education of soccer players, *Proceedings of Soccer Player Oriented Science and Technology Congress*, Centre Technique Cuir Chaussure Maroquinerie, Lyons, France, 14–15 May.
LAFORTUNE, M. A. and HENNIG, E. M. 1992, Cushioning properties of footwear during walking: accelerometer and force platform measurements, *Clinical Biomechanics*, **7**, 181–184.
LAFORTUNE, M. A., HENNIG, E. and VALIANT, G. A. 1996a, Tibial shock measured with bone and skin mounted transducers, *Journal of Biomechanics*, **28**, 989–993.

Lafortune, M., Lake, M. J. and Hennig, E. M. 1996b, Differential shock transmission response of the human body to impact severity and lower limb posture, *Journal of Biomechanics*, **29**, 1531–1537.

Lafortune, M. A., Lake, M. J. and Valiant, G. A. 1993, In-shoe heel pressure distribution and impact ground reaction force during running, *Proceedings of 17th Annual Meeting of the American Society of Biomechanics*, Iowa City, Iowa, 21–23 October, 57–58.

Lake, M. J. and Patritti, B. 1999, Signal analysis approaches to evaluate the performance of shock absorbing inserts for footwear, *Proceedings of the Fourth Symposium of Footwear Biomechanics*, Canmore, Canada, 5–7 August, 70–71.

Lambson, R. B., Barnhill, B. S. and Higgins, R. W. 1996, Football cleat design and its effect on anterior cruciate ligament injuries, *American Journal of Sports Medicine*, **24**, 155–159.

Lees, A. and Jones, H. 1994, The effect of shoe type and surface type on peak shank deceleration, *Journal of Sports Sciences*, **12**, 173.

Luthi, S., Frederick, E. C., Hawes, M. R. and Nigg, B. M. 1986, Influence of shoe construction on lower extremity kinematics and load during lateral movements in tennis, *International Journal of Sports Biomechanics*, **2**, 166–174.

McClay, I .S., Robinson, J. R., Andriacchi, T. P., Frederick, E. C., Gross, T., Martin, P., Valiant, G., Williams, K. R. and Cavanagh, P. R. 1994a, A kinematic profile of skills in professional basketball players, *Journal of Applied Biomechanics*, **10**, 205–221.

McClay, I. S., Robinson, J. R., Andriacchi, T. P., Frederick, E. C., Gross, T., Martin, P., Valiant, G., Williams, K. R. and Cavanagh, P. R. 1994b, A profile of ground reaction forces in professional basketball players, *Journal of Applied Biomechanics*, **10**, 222–236.

Nigg, B. M. 1990, The validity and relevance of tests used for the assessment of sports surfaces, *Medicine and Science in Sports and Exercise*, **22**, 131–139.

Nigg, B. M., Bahlsen, H. A., Leuthi, S. M. and Stokes, S. 1987, The influence of running velocity and midsole hardness on external impact forces in heel-toe running, *Journal of Biomechanics*, **20**, 951–959.

Qian, S. and Chen, D. 1996, *Joint Time-frequency Analysis* (Englewood Cliffs, NJ: Prentice-Hall).

Radin, E. L., Paul, I. L. and Rose, R. M. 1972, Role of mechanical factors in pathogenesis of primary osteoarthritis, *Lancet*, **4**, 519–522.

Radin, E. L., Paul, I. L. and Rose, R. M. 1982, The effect of prolonged walking on concrete on the knee of sheep, *Journal of Biomechanics*, **15**, 487–492.

Rakotomamonjy, A., Barbaud, M., Tronel, M. and Marche, P. 1997a, Time-frequency analysis of impact shock during running, *Proceedings of the Third Symposium on Footwear Biomechanics*, Tokyo, Japan, 21–23 August, 14–15.

Rakotomamonjy, A., Barbaud, M., Tronel, M. and Marche, P. 1997b, Attenuation of the effect of skin mounting technique in tibial acceleration measurement based on the wavelet shrinkage, *Proceedings of the Third Symposium on Footwear Biomechanics*, Tokyo, Japan, 21–23 August, 35.

Rubin, C., McLeod, K. and Bain, S. 1990, Osteoregulatory nature of mechanical stimuli, Journal of Biomechanics, **23**(Suppl.), 43–54.

Schlaepfer, F., Unold, E. and Nigg, B. M. 1983, The frictional characteristics of tennis shoes, in B. M. Nigg and B. A. Kerr (eds), *Biomechanical Aspects of Sport Shoes and Playing Surfaces* (Calgary: University Printing), 153–160.

Shorten, M. 1993, The energetics of running and running shoes, *Journal of Biomechanics*, **26**, 41–51.

Shorten, M. 1998, Finite element modeling of soccer shoe soles, *Proceedings of the Soccer Player Oriented Science and Technology Congress*, Lyons, France, 14–15 May.

Shorten, M., and Himmelsbach, J. 1999, Impact shock during controlled landings on natural and artificial turf, *Proceedings of the International Society of Biomechanics Meeting*, Calgary, Canada, 8–13 August, 783.

Shorten, M. R. and Winslow, D. S. 1992, Spectral analysis of impact shock during running, *International Journal of Sport Biomechanics* **8**, 288–304.

Shorten, M., Valiant, G. and Cooper, L. 1986, Frequency analysis of the effects of shoe cushioning on dynamic shock in running, *Medicine and Science in Sports and Exercise*, **18**(Supplement), S80–S81.

SIMON, S. R., RADIN, E. L., PAUL, I. L. and ROSE, R. M. 1972, The response of joints to impact loading. II. *In vivo* behaviour of subchondral bone, *Journal of Biomechanics*, **5**, 267–272.

SNEL, J. G., DELLMANN, N. J., HEERKENS, Y. F. and VAN INGEN SCHENAU, G. J. 1985, Shock-absorbing characteristics of running shoes during actual running, in D. A. Winter, R. W. Norman and R. P. Wells (eds), *Biomechanics IX-A* (Champaign, IL: Human Kinetic Publishers), 133–138.

STACOFF, A. 1996, Lateral stability in sideward cutting movements, *Medicine and Science in Sports and Exercise*, **28**, 350–358.

STUCKE, H., BAUDZUS, W. and BAUMANN, W. 1984, On friction characteristics of playing surfaces, in E. C. Frederick (ed.), *Sport Shoes and Playing Surfaces* (Champaign, IL: Human Kinetic Publishers), 87–97.

THOMSON, R. D., BIRKBECK, A. E., TAN, W. L., McCAFFERTY, L. F., GRANT, S. and WILSON, J. 1999, The modelling and performance of training shoe cushioning systems, *Sports Engineering*, **2**, 109–120.

TORG, J. S., QUEDENFELD, T. C. and LANDAU, S. 1974, The shoe-surface interaction and its relationship to football knee injuries, *Journal of Sports Medicine*, **2**, 261–269.

VALIANT, G. A. 1990, Transmission and attenuation of heelstrike accelerations, in P. R. Cavanagh (ed.), *Biomechanics of Distance Running* (Champaign, IL: Human Kinetic Publishers), 225–247.

VALIANT, G. A. 1994, Evaluating outsole traction of footwear, in W. Herzog, B. M. Nigg and T. van den Bogert (eds), *Proceedings of the Eighth Biennial Conference of the Canadian Society for Biomechanics*, Calgary, Canada, 19–20 August, 326–327.

VOLOSHIN, A. and WOSK, J. 1982, An in vivo study of low back pain and shock absorption in the human locomotor system, *Journal of Biomechanics*, **15**, 21–27.

YANG, K. H., BOYD, R. D., KISH, V. L., BURR, D. B., CATERSON, D. and RADIN, E. L. 1989, Differential effect of load magnitude and rate on the initiation and progression of osteoarthrosis, *Proceedings of Orthopaedic Research Society Meeting*, Las Vegas, Nevada.

10 How the free limbs are used by elite high jumpers in generating vertical velocity

A. Lees[†]*, J. Rojas[‡], M. Cepero[‡], V. Soto[‡] and M. Gutierrez[‡]

[†]Research Institute for Sport and Exercise Sciences, Liverpool John Moores University, 15–21, Webster Street, Liverpool L3 2ET, UK

[‡]Department of Physical Education, University of Granada, Granda, Spain

Keywords: High jump; Biomechanical analysis; Relative momentum; Elite athletes

The aim of this study was to quantify how elite high jumpers used their free limbs in a competitive high jump and to estimate the contribution that these made to vertical take-off velocity. This was achieved by analysing the competitive performances of six elite male high jumpers using 3D motion analysis and assessing limb function using the relative momentum method. The mean peak relative momentum of the arm nearest to the bar at take-off was 9.4 kg m s^{-1}, while that of the arm furthest away from the bar was 11.3 kg m s^{-1} and these did not differ significantly. The free (lead) leg reached a mean peak relative momentum of 20.9 kg m s^{-1}. At touch-down the free leg had a large positive relative momentum that was offset by the negative relative momentum of the arms, although their combined value still remained positive. The mean combined free limbs' relative momentum at touch-down was 13.8 kg m s^{-1} and reached a peak of 37.6 kg m s^{-1}. The difference between these two values amounted to 7.1% of whole-body momentum, which was judged to be the amount by which the free limbs contributed to performance. The arms had a greater influence on performance than had the lead leg. This was because the lead leg increased its relative momentum little during the contact period while the arms had an initial negative value that increased markedly after touch-down. The compressive force exerted by the motion of the free limbs, estimated by the change in the combined free limbs' relative momentum, reached a mean peak of 366 N and was greatest at 37% of the contact period. It was concluded that to maximize the contribution the free limbs can make to performance, given the restraints imposed on technique by other performance requirements, the arms should have a vigorous downward motion at touch-down to make the most use of the high (but little changing) relative momentum of the lead leg.

1. Introduction

In athletic jumping events the arms and free leg are used purposefully during the take-off phase in an attempt to improve performance and to control the motion of the body. In high jumping there is a need to produce both the appropriate angular momentum of the body in order to clear the bar successfully as well as to generate a

*Author for correspondence.

high vertical velocity at take-off. It is known that the vertical velocity at take-off is a major determinant of performance outcome in high jumping and it is thought that the limbs will make a greater contribution to this vertical velocity the more vigorously they are used (Dapena 1993). This effect is explained in different ways but a common explanation is that as the arms are thrown upwards into the air they create an impulse that is transmitted through the body to the ground, generating a reaction impulse that propels the centre of mass (CM) upwards (Dapena 1993, Yu and Andrews 1998).

The role of the free limbs during competitive high jump performances are best determined by kinematic methods. There have been two major approaches used to quantify the effect that the limbs may have on performance. The first is the determination of 'arm activeness', defined by Dapena (1987) as the difference between minimum and maximum vertical velocity of the centre of mass of the arm relative to the centre of mass of the trunk during take-off. He reported arm activeness values for elite male high jumpers ranging from 0 to 11 m s^{-1}, with the greater value associated with a greater effect of the arms. Although the concept can be applied equally well to the leg, the concept of 'leg activeness' seems not to have been used despite the fact that the leg is used with similar vigour during the take-off phase. Arm activeness gives a single maximal measure of arm use but does not allow its specific contribution to the generation of vertical velocity to be established. Further, it cannot identify how the limb action is timed in relation to the jump, or identify the interaction between one arm and another, or between the arms and free leg. The second approach is the relative momentum method proposed by Ae and Shibukawa (1980). The vertical relative momentum of a limb is defined as the product of the mass of the limb and its vertical velocity relative to the proximal joint. This is similar to the arm activeness described above, but because it quantifies the extra momentum generated by the limb during its upward motion it can, through Newton's Second Law, be related to the impulse generated at the proximal joint and, using rigid body modelling, the reactive impulse from the support surface. The relative momentum of a limb is defined over the whole of the movement and as such can be summed with other limb relative momenta in order to identify their specific contribution to the whole-body vertical momentum at take-off.

The relative momentum method would seem to have some merit as a means of identifying the contribution that the limbs make to performance, but there have been no studies in which this method has been applied to athletic high jumping. There have been two studies (Ae *et al.* 1983, Vint and Hinrichs 1996) in which this approach has been used in an activity related to the high jump, that of a running one-legged take-off vertical jump for height, but subjects were not asked to perform a bar clearance and were not high-level athletes. Therefore it was the aim of this study to identify the individual characteristics of each free limb motion, their co-ordination (how they move in relation to each other), their relative importance and how they might combine to affect vertical velocity of the CM at take-off in high jumping. This aim was achieved through an analysis of elite male high jumpers.

2. Methods

2.1. *Data collection*

Six elite male high jumpers were filmed during their competitive performances in the high jump finals during the 1991 Gran Premio Ciudad de Granada held in Granada (Spain). The anthropometric measurements of body height, mass and the performance of each high jumper are presented in table 1. All of the high jumpers used a left foot take-off.

The jumps were filmed simultaneously with two cine cameras. The first, a Bealieu R-16 (Maison Brandt Freres, Charenton-le-Pont, France), used a sample rate of 60 Hz, and was placed perpendicular to the bar at a distance of 31 m. The second camera, a Photo-Sonics 16-1PL (Photo-Sonics, Burbank, California), used a sample rate of 50 Hz, and was placed at a distance of 36 m parallel to the bar. The field of view enabled the foot contact during touch-down of the last stride and the total clearance of the bar to be seen. This ensured that sufficient film frames were available for analysis before and after the take-off phase. The best jump of each finalist recorded on film was used for analysis.

2.2. Data reduction

A 14-segment biomechanical model defined by 21 body landmarks was used. The co-ordinates of each landmark were obtained by digitizing the projected images from each camera. The digitized data were smoothed, interpolated and synchronized to an equivalent sampling rate of 100 Hz using a fifth-order spline (Wood and Jennings 1979), modified for the purpose. The 3D co-ordinates were calculated using a direct linear transformation (DLT) procedure (Abdel-Aziz and Karara 1971). A separate computer programme was used to compute the kinematic data in this report using body segment parameters from Dempster (1955). Data were smoothed using a Butterworth fourth-order zero-lag filter with padded end-points (Smith 1989) and a cut-off frequency of 9 Hz based on a residual analysis and qualitative evaluation of the data. Derivatives were calculated by direct differentiation (Winter 1990).

Three free limbs were studied. The leg raised at take-off was termed the lead leg (LL). The arm on the same side of the body as the lead leg was termed the ipsilateral arm (IA) while that on the opposite side was termed the contralateral arm (CA). All of the segments of a limb are combined as one unit, although in principle these can be treated individually. The characteristics of individual free limb action were computed using the relative momentum method of Ae and Shibukawa (1980). This considers the vertical momentum of a free limb to be composed of a transfer and a relative momentum component. The total vertical momentum of a limb can therefore be expressed as

$$m_L V_L = m_L V_{PJ} + m_L V_{L/PJ} \qquad (1)$$

where m_L = mass of limb, V_L = vertical velocity of CM of limb, V_{PJ} = vertical velocity of the proximal joint, and $V_{L/PJ}$ = relative velocity of V_L to V_{PJ}. The term $[m_L V_{PJ}]$ represents the transfer momentum of the limb due to movement of the proximal joint, and the term $[m_L V_{L/PJ}]$ represents the individual free limb relative momentum.

Table 1. Anthropometric data for each high jumper.

Athlete	Height (m)	Mass (kg)	Personal best (m)
Austin	1.83	76	2.35
Sotomayor	1.93	83	2.44
Malchencho	1.91	82	2.38
Drake	1.91	77	2.32
Dakov	1.92	72	2.36
Ortiz	1.93	74	2.34

The relative momentum component is used to assess the contribution that an individual limb makes to whole-body momentum.

The relative momentum (RM) of each free limb was termed CARM, IARM and LLRM for the contralateral arm, ipsilateral arm and lead leg relative momentum, respectively, and defined by

$$\text{CARM} = m_{\text{CA}} V_{\text{CA/CS}} \tag{2a}$$

$$\text{IARM} = m_{\text{IA}} V_{\text{IA/IS}} \tag{2b}$$

$$\text{LLRM} = m_{\text{LL}} V_{\text{LL/H}} \tag{2c}$$

where the subscripts refer to the limb (CA = contralateral arm, IA = ipsilateral arm, LL = lead leg) and proximal joint (CS = contralateral shoulder, IS = ipsilateral shoulder, H = Hip joint of the lead leg).

As the positive relative momentum of one limb can cancel the negative relative momentum of another, the combined effect of all three free limbs is given by the sum of each free limb relative momentum and termed the Combined Free Limb Relative Momentum (CFLRM) defined as

$$\text{CFLRM} = \text{CARM} + \text{IARM} + \text{LLRM} \tag{3}$$

The force applied by the free limbs was obtained by direct differentiation of CFLRM. The whole-body vertical momentum (WBVM) was computed from the vertical velocity of the whole-body CM and body mass.

2.3. Data analysis

The definition of touch-down (TD) was taken as the first clear frame in which the foot was in contact with the ground. Similarly the definition of take-off (TO) was taken as the first frame in which the foot was clear of the ground. Data are presented as means and standard deviations for whole-body momentum, individual and combined free limb relative momentum, temporal occurrence of the peak relative momentum for each free limb with respect to the period of ground contact, touch-down and take-off vertical velocities and vertical force due to free limb motions.

The whole-body vertical momentum at take-off (WBVM_{TO}) was evaluated. The contribution of CFLRM to whole-body momentum was obtained in the following way. The change in the relative momentum of a limb is equated to the impulse acting on the proximal joint and it is assumed that this generates an equal and opposite reactive impulse at the ground and in turn is responsible for the change in vertical momentum of the whole-body CM. This provides a means whereby the contribution of free limbs to whole-body motion can be assessed. With regard to this contribution, the interpretation described by Lees and Barton (1996) was used. Briefly, the extra impulse applied by the free limbs is measured by the change in the CFLRM. This change is assessed as the positive increase (PI) in CFLRM (CFLRM_{PI}) from its lowest positive value after touch-down ($\text{CFLRM}_{\text{LPV}}$) to its peak value ($\text{CFLRM}_{\text{PEAK}}$). A negative value of CFLRM is ignored as this cannot contribute to positive motion and in these cases $\text{CFLRM}_{\text{LPV}}$ is equated to zero. In addition, any reduction in CFLRM after $\text{CFLRM}_{\text{PEAK}}$ is ignored as it is assumed that this represents momentum sharing of the free limbs with the rest of the body. Thus, the positive increase in CFLRM and hence its contribution to whole-body momentum at take-off is defined as

$$\text{CFLRM}_{\text{PI}} = \text{CFLRM}_{\text{PEAK}} - \text{CFLRM}_{\text{LPV}} \qquad (4)$$

The maximum value that the combined free limbs could generate would be when all free limbs reached their relative momentum peak at the same time, which is termed the limbs potential and is defined as

$$\text{Limbs potential} = \text{CARM}_{\text{PEAK}} + \text{IARM}_{\text{PEAK}} + \text{LLRM}_{\text{PEAK}} \qquad (5)$$

Where appropriate, data are presented as percentage ratios in order to relate to specific aspects of performance. Specifically these are: (1) Limb contribution%, which is defined as the percentage ratio of CFLRM_{PI} to WBVM_{TO}, and is taken as an indicator of the contribution made by the combined free limbs to vertical velocity; and (2) Limb effectiveness%, which is defined as the percentage ratio of the CFLRM_{PI} to limbs potential, and is taken as an indicator of how well the limb actions are used and co-ordinated with respect to maximizing the influence of the free limbs.

Statistical procedures used were the Shapiro-Wilk's test for establishing the non-normality of data sets, Pearson's product-moment correlation coefficient for establishing relationships and Student's t-test for establishing differences. A level of significance of $P<0.05$ was used to establish statistical significance.

3. Results

A graph of the individual free limbs relative momentum together with the combined free limbs relative momentum is given in figure 1 for athlete Austin. This shows that the relative momentum of the arms are negative at touch-down as the arms are being driven downwards while that of the leg is positive as the leg is flexing at the knee giving its CM a net upward velocity; however the sum of all three limbs is slightly positive at touch-down. The relative momentum of the leg continues to increase as the contact phase progresses and it is flexed and elevated; that of the arms becomes positive as they reach the bottom of their swing and are elevated. The relative momentum profiles of each arm are similar and they reach their peak at approximately the same point in time but that of the lead leg reaches its peak earlier than the arms. The combined free limb relative momentum (CFLRM) reaches a peak just after the lead leg but just before the arms. The relative momentum of all free limbs reduces at take-off, illustrating the transfer of each segment momentum to the rest of the body.

Table 2 shows the whole-body vertical momentum at take-off, the peak relative momentum for each limb during the touch-down to take-off phase, $\text{CFLRM}_{\text{PEAK}}$, CFLRM_{TD} and CFLRM_{PI}. This table also shows limbs potential, limbs contribution% and limbs effectiveness%, which are computed as described in § 2.3. The limbs contribution% suggests that the free limbs contributed 7.1% to the whole-body vertical momentum at take-off. The limbs effectiveness% expresses the ability of athletes to utilize the relative momentum they generate through their technique and the process of co-ordination. This ranged from 36–93% showing wide variation in the way in which the limbs are used. A graph of the individual free limbs relative momentum together with the combined free limbs relative momentum is given in figure 2 for athlete Ortiz who demonstrated the lowest value of limbs effectiveness%. This athlete performed poorly with regard to limb effectiveness% due to a high value of lead leg relative momentum at touch-down.

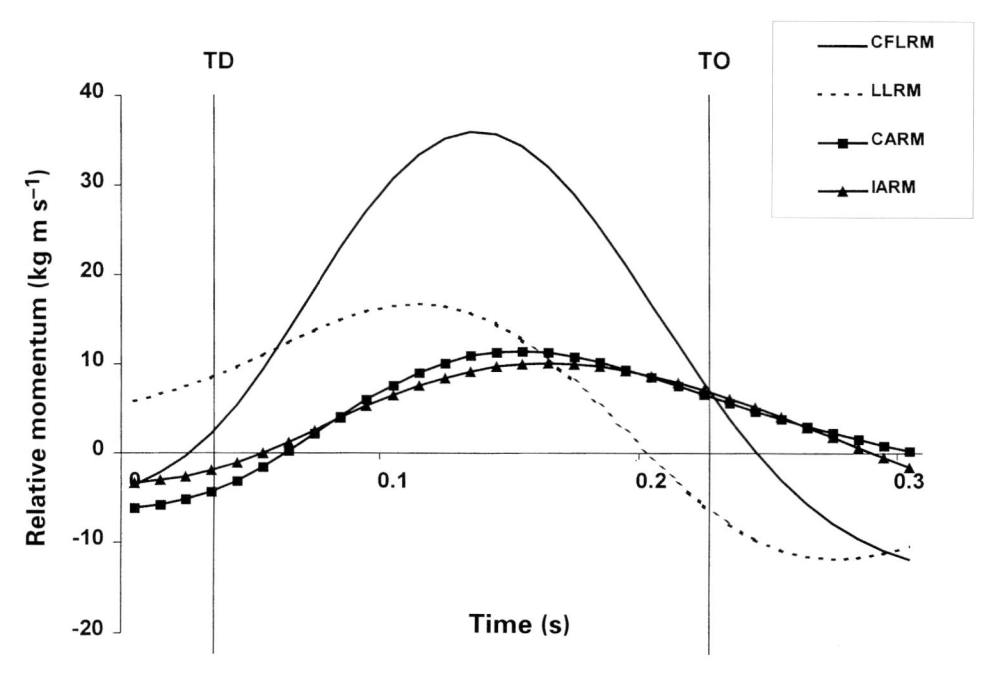

Figure 1. Vertical relative momentum for individual free limbs (CARM = contralateral arm relative momentum, IARM = ipsilateral arm relative momentum, LLRM = lead leg relative momentum) and combined free limbs relative momentum (CFLRM) for athlete Austin.

Table 3 shows the official height jumped, the duration of contact and the percentage of the contact period at which the CARM, IARM, LLRM and CFLRM peaks were reached. The LLRM always reached its peak before CARM and IARM, and CFLRM reached its peak at around 63% of the contact phase. The table also shows the vertical velocity of the CM at touch-down and take-off. The values at touch-down are positive, which is an unusual finding but those at take-off are similar to values reported in the literature (e.g. Dapena (1988) reported a mean of 4.37 m.s^{-1}).

It was of interest to analyse the detailed motion of the CM and the touching-down leg during the contact period to investigate how the action of the free limbs relate to touch-down leg kinematics. Typical graphs of the radial distance of the CM from the ankle joint illustrated in figure 3 show that the minimum radial distance consistently occurs at 0.1 s before MKF, which itself occurs at about 50% of the contact period. The earlier trough in the radial distance is probably due to the raising of the limbs, while MKF represents the end of the musculoskeletal compression phase. The vertical force acting as a result of free limb motion is given typically in figure 4 and illustrates the peak force and it temporal occurrence in relation to touch-down and take-off. This is positive at touch-down, which would be unhelpful to the athlete in coping with the impact at touch-down, although it is lower than it otherwise would be due to the negative relative momentum of the arms at that point. It reaches a peak at the same time as minimum radial distance, suggesting that the additional force generated by the action of the free limb could be a contributing

Table 2. Whole body vertical momentum, peak individual free limb relative momentum (CARM, IARM, LLRM), combined free limbs relative momentum (CFLRM) values together with derived variables.

Athlete	WBVM$_{TO}$ (kg m s^{-1})	CARM$_{PEAK}$ (kg m s^{-1})	IARM$_{PEAK}$ (kg m s^{-1})	LLRM$_{PEAK}$ (kg m s^{-1})	CFLRM$_{PEAK}$ (kg m s^{-1})	CFLRM$_{TD}$ (kg m s^{-1})	CFLRM$_{PI}$ (kg m s^{-1})	Limbs potential (kg m s^{-1})	Limbs contribution %	Limbs effectiveness %
Austin	335.4	11.4	10.1	16.6	35.9	0.3	35.6	38.1	10.6	93.4
Sotomayor	376.6	13.5	11.0	21.9	44.3	16.1	28.2	46.4	7.5	60.8
Malchencho	355.2	10.4	9.9	24.7	41.6	15.2	26.4	45.0	7.4	58.7
Drake	315.2	11.1	11.9	23.0	41.7	21.0	20.7	46.0	6.6	45.0
Dakov	288.3	10.7	7.8	19.5	34.9	16.0	18.9	38.0	6.6	49.7
Ortiz	306.4	10.5	5.5	19.7	27.0	14.1	12.9	35.7	4.2	36.1
Mean	329.5	11.3	9.4	20.9	37.6	13.8	23.8	41.5	7.1	57.3
SD	32.7	1.2	2.3	2.9	6.3	7.0	8.0	4.8	2.1	19.9

WBVM$_{TO}$, whole body vertical momentum at take-off; CARM$_{PEAK}$, contralateral arm relative momentum at peak; IARM$_{PEAK}$, ipsilateral arm relative momentum at peak; LLRM$_{PEAK}$, leading leg relative momentum at peak; CFLRM$_{PEAK}$, combined free limb relative momentum at peak; CFLRM$_{TD}$, combined free limb relative momentum at touch-down; CFLRM$_{PI}$, combined free limb relative momentum positive increase.

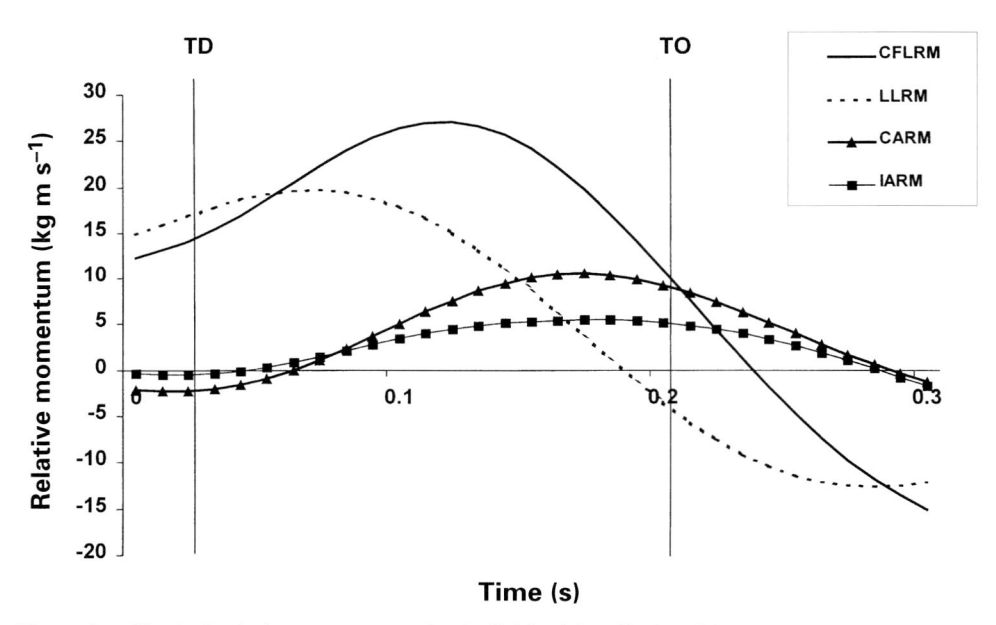

Figure 2. Vertical relative momentum for individual free limbs (CARM = contralateral arm relative momentum, IARM = ipsilateral arm relative momentum, LLRM = lead leg relative momentum) and combined free limbs relative momentum (CFLRM) for athlete Ortiz.

factor to the extent of body compression. The peak force for each athlete is given in table 4, together with the percentage of the contact period at which it is reached.

The data were tested for non-normality using the Shapiro-Wilk's statistic W, which was non-significant ($p < 0.01$). The contralateral arm was no more dominant than the ipsilateral arm ($t = 2.23$, $p = 0.10$). A significant relationship was found ($r = 0.86$, $p < 0.05$) between the vertical velocity of the CM at take-off and the official height jumped, confirming the importance of this variable to performance. The relationship between $CFLRM_{PI}$ and vertical velocity at take-off did not reach the required level of significance ($r = 0.77$, $p < 0.10$) but its value suggests that some importance should still be attached to the correct use of the free limbs.

4. Discussion

The interpretation of relative momentum used in the literature has proved to be equivocal. With specific regard to activities such as high jumping there have been two reports of interest. Ae et al. (1983) and Vint and Hinrichs (1996) have both investigated activities in which subjects were asked to take a running jump off one leg in order to attain maximum height. As this activity is similar to high jumping, and both authors used the relative momentum approach, their results provide a relevant comparison. Ae et al. (1983) found that the relative momentum of the arms was negative at touch-down and reached a peak shortly after the mid-point of the contact phase, similar to that found in this study. The relative momentum of the free leg was close to zero at touch-down and reached a peak slightly earlier than the arms but still after the mid-point of the contact phase. Although no absolute values for relative momentum of the arms and legs were given, from their graphical data the peak

Table 3. Official height jumped, contact time and relative occurrence in relation to contact period of the peak individual free limb relative momentum, and vertical velocities (V_v) at touch-down and take-off.

Athlete	Official height jumped (m)	Contact time (s)	CARM$_{PEAK}$ (%)	IARM$_{PEAK}$ (%)	LLRM$_{PEAK}$ (%)	CFLRM$_{PEAK}$ (%)	V_v (TD) (m s^{-1})	V_v (TO) (m s^{-1})
Austin	2.33	0.19	73.7	79.0	52.6	63.1	0.64	4.41
Sotomayor	2.30	0.17	88.2	84.2	52.9	76.5	0.61	4.53
Malchencho	2.30	0.16	68.8	75.0	37.5	62.5	0.89	4.32
Drake	2.25	0.17	72.2	83.3	44.4	61.1	0.49	4.09
Dakov	2.20	0.16	70.6	64.7	35.3	58.8	0.49	4.00
Ortiz	2.20	0.17	88.2	88.2	29.4	58.8	0.62	4.13
Mean	2.26	0.17	77.0	79.1	42.0	63.5	0.62	4.25
SD	0.05	0.01	8.1	7.6	8.8	6.1	0.13	0.19

CARM$_{PEAK}$, contralateral arm relative momentum at peak; IARM$_{PEAK}$, ipsilateral arm relative momentum at peak; LLRM$_{PEAK}$, leading leg relative momentum at peak; CFLRM$_{PEAK}$, combined free limb relative momentum at peak; V_v (TD), vertical velocity at touch-down; V_v (TO), vertical velocity at take-off.

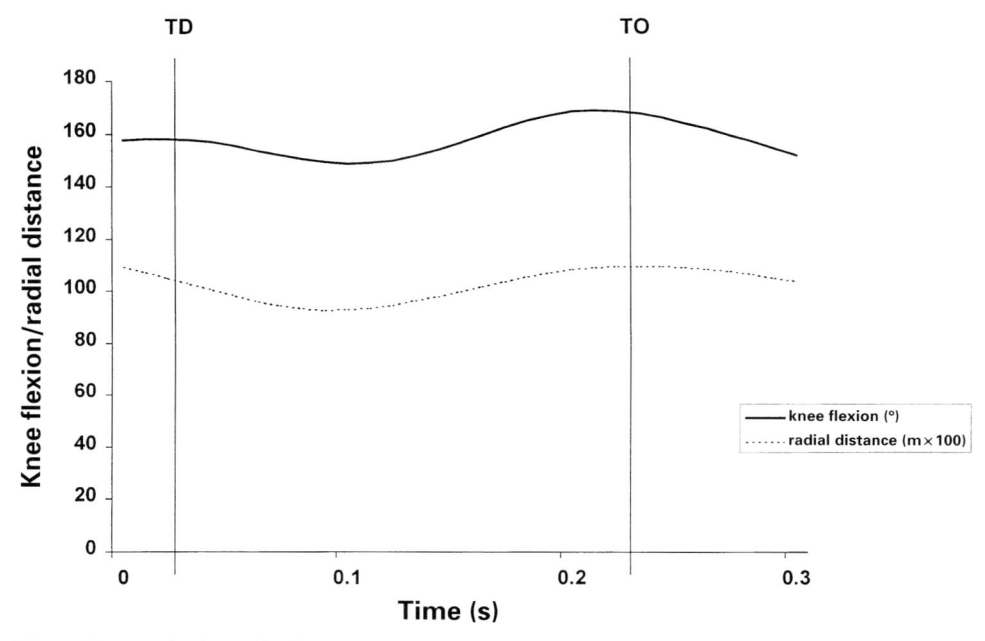

Figure 3. Typical graph of the radial distance of the centre of mass from the ankle joint and the angle of flexion at the knee.

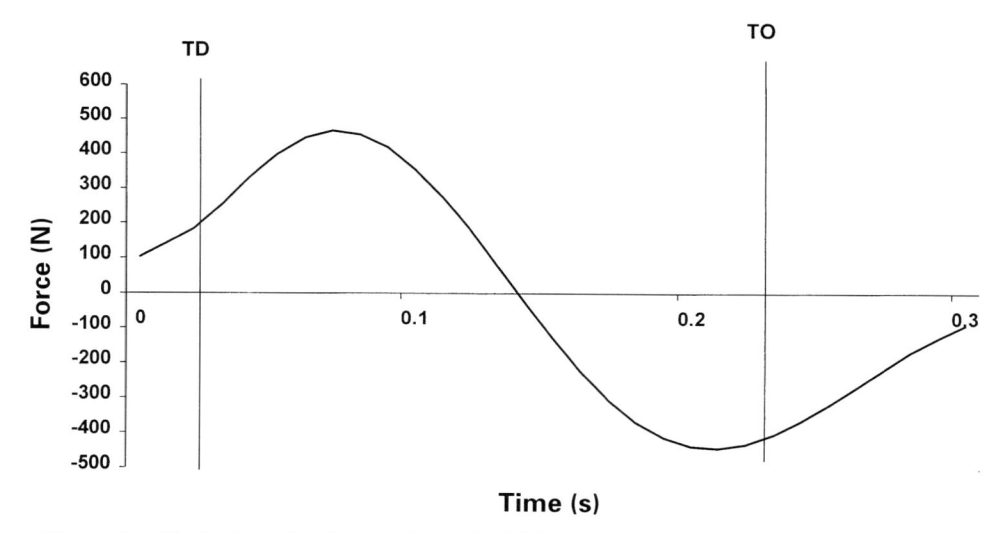

Figure 4. Typical graph of the estimated additional force generated as a result of limb motion.

relative momentum of the arms appeared to be about 6% of the total body momentum, while for the free leg this was about 10%. Their estimates of the contribution of the arms and free leg to the change in whole-body momentum were 15% and 10%, respectively. No attempt was made to investigate the combined effect

Table 4. The estimated additional force generated as a result of limb motion together with the percentage of contact tine at which it occurs.

Athlete	Peak force (N)	Proportion of contact time (%)
Austin	462	50
Sotomayor	441	31
Malchencho	412	44
Drake	373	23
Dakov	327	27
Ortiz	180	47
Mean	366	37
SD	94	10

of the limbs and unfortunately the authors did not give sufficiently clear information as to how these percentages were computed and so their usefulness is limited. Vint and Hinrichs (1996) reported a relative momentum profile of the arms and legs similar to that of Ae et al. (1983). At touch-down the relative momentum of both the arms and the free leg were close to zero and became negative before reaching a peak at about 60% of the contact phase. The peak relative momentum appeared to be about 20 kg m s^{-1} for the arms and 25 kg m s^{-1} for the leg, which corresponded to 8% and 10%, respectively, of the total body momentum. Using the method described by Hinrichs et al. (1987) to assess the contribution of the limbs to the total body vertical velocity (which took the difference between the take-off and touch-down values), they estimated that the arms contributed 6.6% and the free leg −3.4% to total body vertical momentum. Again no attempt was made to quantify the total free limbs contribution. Surprisingly, they estimated that the leg made a negative contribution to the total vertical velocity when one might expect that it would make a positive contribution. This led these authors to comment that 'care be taken when interpreting results from the Ae and Shibukawa (1980) relative momentum methodology' (Vint and Hinrichs, 1996: 354). The interpretation suggested by Lees and Barton (1996) overcomes these difficulties and offers clear guidance on how this might be done.

There was no significant difference between the peak relative momentum produced by each arm in high jumping. The arm furthest from the bar (the contralateral arm) generated a mean peak relative momentum of 11.3 kg m s^{-1}, while the arm nearest the bar generated a mean peak relative momentum of 9.4 kg m s^{-1}. These results are in contrast to those of Dapena (1988), who reported differences in arm activeness between the arms for similar level high jumpers. He reported arm activeness values of 5.42 m s^{-1} for the arm furthest from the bar and 2.97 m s^{-1} for the arm nearest the bar. If these were peak vertical velocities of the CM of the arm, then the equivalent relative momentum value for a 75 kg athlete would be 24.4 and 13.4 kg m s^{-1}, respectively. These values are considerably higher than those found in the present study and suggest that the data used to compile arm activeness was not directly comparable to that used for the relative momentum, even though there appears to be great similarity between their respective definitions. This is probably caused by the inclusion of negative velocities in the computation of arm activeness. If this were the case then the results from 'arm activeness' computations cannot be directly compared to those from the relative momentum method. Dapena

(1987) has reported that elite jumpers do use different arm techniques, but in this study all subjects retracted both arms simultaneously to assist the jump, leading to similar contributions from each arm.

The timing of arm and leg action was an important factor determining their combined effect. The arms appeared to influence strongly the total free limb contribution to the vertical velocity at take-off largely because the free leg has quite high positive relative momentum at touch-down and increased only slightly as the jump progressed. This was explained by the need to flex the leg early to enable it to be brought through when the CM was in its lowest position at touch-down. The positive relative momentum of the free leg at touch-down was initially off-set by the negative relative momentum of the arms, thus the arms have an important role at touch-down in creating sufficient negative relative momentum to maximize the effect of the combined free limbs and to reduce the impact force at touch-down. An important aspect of high jump technique would appear to be a vigorous downward motion of the arms at touch-down and this has not been highlighted in the literature before. Further, this effect is enhanced if both arms are retracted and can generate downward velocity at touch-down. All athletes in this study used a double-arm retraction technique but in some athletes this was more asymmetrical, which would not appear to be the most beneficial technique to use from the point of view of generating vertical momentum. Once the relative momentum of the arms became positive, there was only a small increase in the relative momentum of the free leg, indicating that it made little further additional contribution to the generation of vertical velocity. The interaction between the arms and free leg is critical to maximize the total free limb contribution with the leg dominating the magnitude of the CFLRM but the arms dominating its change over the contact period. These data suggest that the free leg has a role other than in generating vertical velocity. The most likely explanation is that its early generation of relative momentum is necessary to enable it to contribute significantly to the creation of angular momentum of the body thus enabling the athlete to clear the bar successfully.

Studies investigating jumping behaviour typically have been concerned with isolating the specific contribution of each limb to the performance. While the peak relative momentum of each limb can be readily quantified, these values have little merit on their own as the positive value of one limb can be offset by the negative value of another. The relative momentum method uses the CFLRM to estimate the influence of the limbs on jump performance. Based on the data reported in this study, and using the interpretation of Lees and Barton (1996) for estimating the contribution of the limbs to the generated vertical velocity, it was found that the mean sum of individual limb peak relative momentum (the limbs potential) was 41.5 kg m s^{-1} but due to the interaction between the limbs, only 23.8 kg m s^{-1} of this was estimated to contribute to the whole-body's vertical momentum. This represents a considerable ineffectiveness of the use of the free limbs and must in some way be related to other technical requirements of the event as noted above. Using this latter figure, it was estimated that the free limbs contributed a mean of 7.1% to the whole-body vertical momentum at take-off. This value is lower than might be expected from observations of the way in which athletes use their limbs, but the explanation for this low value is the poor co-ordination of use of limbs imposed by the technical requirements of high jumping, the main ones being a very low CM at touch-down and the need to generate angular momentum for bar clearance. This was also lower than that calculated from the investigations of Ae *et al.* (1983) and Vint and Hinrichs (1996) (16% and 18%, respectively, interpreted from their peak relative momentum data), although in both of these studies it is reasonable to assume that the free limbs may provide a greater percentage contribution due to the lower total body vertical velocity at take-off, and they were less restricted in the technique used. One athlete (Ortiz) had a particularly low total value for relative momentum. This stemmed from a lack of speed of one arm, but mainly by poor timing of the free leg. There was a positive but non-significant correlation between the vertical velocity of the CM at take-off and $CFLRM_{PI}$. This suggests that while the limbs are a contributing element to total vertical velocity there will be other factors that are important.

The errors involved in estimating the percentage contribution come from the errors in the kinematic data, the choice of frame identifying touch-down and measurement artefacts. The errors in velocity data for studies investigating jumping in competition have been estimated by Lees et al. (1994) to be about 5%. It is thought that the errors in the kinematic data in this study would be no less than this although no alternative velocity measures could be used in the competitive environment to check this. The error associated with choice of frame of touch-down would lead to a small change in the combined free limbs' total relative momentum value, but as both the combined free limb and total body relative momentum increase over the first few data samples the effect on the estimated percentage contribution of the free limbs to total vertical velocity gain will be small. There will be no error introduced by the choice of frame for touch-down, as the relative momentum reaches a peak well before the point of take-off. The vertical velocity of the CM in this study had a small positive value at touch-down, in contrast to a range of published data (for example Dapena 1988) with the CM having a slightly negative velocity at touch-down. For the early frames of the analysis during the last stride, the digitized body landmarks were outside the calibrated volume. There is a likelihood that these data were less reliably reproduced, and the subsequent smoothing used on the data was unable to reproduce the sharp change in conditions associated with touch-down. For this reason it was assumed that the data representing the touch-down conditions of the whole-body CM were in error. The values of the relative momentum of the limbs before touch-down could also be in error for the same reason, but these data seem more reasonable as they agree with that produced by other authors (Ae et al. 1983, Vint and Hinrichs 1996). Any small discrepancy in these data would have a small effect on the estimate of the percentage contribution, but would not affect the conclusions drawn regarding the interaction between the limbs and their timing within the general movement.

The relative momentum method allows an estimate to be made for the compressive force produced due to the upward acceleration of the free limbs. The illustrative data in figure 4 indicate that the additional compressive force is well established at touch-down thus contributing to the impact force at touch-down, rising to a peak within 37% of the contact period, to reach a value of around 350 N. The negative velocity of the arms at touch-down serves to keep this force low thus reducing the severity of impact. The additional compressive force diminishes gradually until about 63% of the contact period when it becomes negative. At this point the force begins to lift the trunk. This would serve to reduce the compressive force on the leg muscles allowing them to contract rapidly producing an increase in the knee extension velocity.

5. Summary and conclusion

With regard to high jumping technique, the general interaction between the arms and the leg used by elite high jumpers would appear to be non-optimal and this is probably determined by certain technical requirements of performance such as the need to create vertical velocity and angular momentum during take-off. The main features of high jumping technique are the lowering of the CM and the planting of the touch-down leg markedly in front of the body, which helps the body to 'pivot' over the support leg in order to generate vertical velocity. In order to bring the lead leg through during this low position of the CM the free leg must begin flexing at an early stage. The flexing of the lead leg gives it positive relative momentum and so it is inevitable that it will have substantial positive relative momentum at touch-down. This motion of the lead leg continues so that it reaches its peak relative momentum early (42%) in the contact period. It is advantageous for the arms to have a negative relative momentum at touch-down to off-set that of the lead leg, and this is produced by the forward and downward motion of the arms from their retracted position. The change in relative momentum from touch-down onwards is influenced by the vigour and timing of each limb. All limbs must be moved upwards vigorously and this becomes an essential requirement of jumping technique. The free limb action would need to be completed before take-off and so it is unlikely, given the need for the relative momentum of the arms at touch-down to be negative, that the timing of arm action can be advanced from its peak relative momentum at 78% of the contact period to match more closely that of the free leg. Equally the timing of the free leg may be difficult to delay as its motion is determined by the need to bring the leg through while the CM is close to the ground. It may also have an importance to the balance and rotation of the overall action, which would make any alteration in its relative timing inadvisable. It therefore means that the key factor affecting the contribution of the limbs to the generation of vertical velocity is the initial downward motion of the arms and the vigour with which each limb is used as it is driven upwards. Those athletes who show a low $CFLRM_{PI}$ value, as illustrated in the data of table 2, use an early leg action and a low movement speed of the arms after touch-down. It should be noted that these subjects have achieved a high vertical velocity of their free leg, but have not increased it by much during the contact period. It is therefore not the speed of the limb *per se* that defines vigour but the change in its vertical velocity while the jumper is in contact with the ground. It is concluded that to maximize the contribution that the free limbs make to performance, given the restraints imposed on technique by other performance requirements, the arms should have a vigorous downward motion at touch-down to make the most use of the high (but little changing) relative momentum of the lead leg.

References

ABDEL-AZIZ, Y. I. L. and KARARA, H. M. 1971, Direct linear transformation from comparator coordinates into the object space coordinates in close range photogrammetry, *ASP Symposium on Close Range Photogrammetry*, Falls Church, VA, American Society of Photogrammetry.

AE, M. and SHIBUKAWA, K. 1980, A biomechanical analysis of the segmental contribution to the take off of the one-legged running jump for height, in H. Matsui and K. Kobayashi (eds), *Biomechanics VIII-B* (Champaign, IL: Human Kinetic Pubishers).

AE, M., SHIBUKAWA, K., TADA, S. and HASHIHARA, Y. 1983, A biomechanical method for the analysis of the contribution of the body segments in human movement, *Japanese Journal of Physical Education*, **25**, 233–243.

Dapena, J. 1987, Basic and applied research in the biomechanics of high jumping, in B. Van Gheluwe and J. Atha (eds), *Current Research in Sports Biomechanics* (Basle: Karger).

Dapena, J. 1988, Biomechanical analysis of the Fosbury Flop, *Track Technique*, **104,** 3307–3317.

Dapena, J. 1993, Biomechanical studies in the high jump and implications for coaching, *Modern Athlete and Coach*, **31,** 7–12.

Dempster, W. T. 1955, Space requirements of the Seated Operator. Report WADCTR 55-159, Wright Patterson Air Force Base, Dayton, OH.

Hinrichs, R. N., Cavanagh, P. R. and Williams, K. R. 1987, Upper extremity function in running 1. Centre of mass and propulsion considerations, *International Journal of Sport Biomechanics*, **3,** 222–241.

Lees, A. and Barton, G. 1996, The interpretation of relative momentum data to assess the contribution of the free limbs to the generation of vertical velocity in sports activities, *Journal of Sports Sciences*, **14,** 503–511.

Lees, A., Graham-Smith, P. and Fowler, N. 1994, An analysis of the touch-down and take-off characteristics of the men's long jump, *Journal of Applied Biomechanics*, **10,** 61–68.

Smith, G. 1989, Padding point extrapolation techniques for the Butterworth digital filter, *Journal of Biomechanics*, **22,** 967–971.

Vint, P. F. and Hinrichs, R. N. 1996, Differences between one foot and two foot vertical jump performances, *Journal of Applied Biomechanics*, **12,** 338–358.

Winter, D. A. 1990, *Biomechanics and Motor Control of Human Movement* (New York: John Wiley).

Wood, G. A. and Jennings, L. S. 1979, On the use of spline functions for data smoothing, *Journal of Biomechanics*, **12,** 477–479.

Yu, B. and Andrews, J. G. 1998, The relationship between free limb motions and performance in the triple jump, *Journal of Applied Biomechanics*, **14,** 223–227.

11 Kinematic adjustments in the basketball jump shot against an opponent

F. J. Rojas[†*], M. Cepero[‡], A. Oña[†] and M. Gutierrez[†]

[†]Faculty of Physical Activity and Sports Sciences, University of Granada, Crtra. de Alfacar s/n 18071 Granada, Spain

[‡]Faculty of Human Sciences, University of Jaen, Spain

Keywords: Biomechanical analysis; Shooting; Opposition; Training.

The aim of this study was to analyse the adjustments in technique made by a basketball player when shooting against an opponent. The subjects used consisted of 10 professional basketball players of the Spanish First Division League. Three-dimensional motion analysis based on video recordings (50 Hz) was used to obtain the kinematic characteristics of basketball jump shots with and without an opponent. It was found that when performing against an opponent the release angle of the ball increased, the flight time was reduced and postural adjustments as determined by the angles at the knee and shoulder increased, all significantly. There were several other non-significant differences that helped to interpret the changes in technique imposed by the presence of an opponent. It was suggested that when shooting with an opponent, players attempted to release the ball more quickly and from a greater height. This strategy will lessen the chance of the opponent intercepting the ball. It was concluded that the differences noted in the technical execution of the skill had implications for practice. It was suggested that training would benefit from practice with an opponent for at least some of the time to condition players to the demands which they were more likely to meet in the game situation.

1. Introduction

Shooting is the principal method used to score points in basketball and for this reason it is the most frequently used technical action (Hay 1994). The jump shot is distinguished as the most important of all the shooting actions (Hess 1980), and in the Spanish Basketball League it is the one most often used successfully, since 41% of all points are scored by using this technique (Asociación Clubs Baloncesto (ACB) 1997). Efficacy in shooting is identified with the ability to perform well in this sport and consequently it is extensively practised.

In basketball studies biomechanical research has focused on various aspects including basic shooting techniques (Brancazio 1981, Hay 1994), differences in play between the sexes (Elliott and White 1989) and the characteristics of players at different skill levels (Hudson 1985). Some of these studies have analysed the jump shot and the variables studied are mainly those that determine the flight characteristics of the ball. The principal factors determining the flight characteristics of the ball (and therefore outcome) are release speed, release angle and release height (Hay 1994).

---

*Author for correspondence. e-mail: fjrojas@platon.ugr.es

Some of these studies have also included analysis of the jump shot under different conditions, as the variability in the performance of the shot is determined by a number of factors (Sáenz and Ibáñez 1995) such as arm action (standard, hook and lay-up), previous technical action (dribble, reception fake), previous movement of the legs (stationary or running), final movement of the legs (with or without jump), body orientation, height and distance of the shot, and opposition. For example, Elliott and White (1989), Walters et al. (1990), Miller and Bartlett (1993) and Satern (1993) studied the effects of increased shooting distance in the jump shot, whilst Gabbard and Shea (1980) and Chase et al. (1994) analysed the effects of equipment modifications on children and jump shot performance.

Of these influencing factors, no research group has attempted to establish the effects of opposition on the movement characteristics of the jump shot. As the technical performance of the shot may be expected to change with the presence of opposition, then practising the jump shot skill without realistic opposition may be less beneficial to skill development and maintenance. Therefore, the aim of this study was to determine the influence of the presence of an opponent on jump shot technique. This aim was met by investigating the biomechanical characteristics of jump shot technique with and without an opponent.

2. Methods

2.1. Subjects

The subjects used were 10 male, active professional basketball players from the First Division of the Spanish Basketball League (ACB) who volunteered to take part. All were right-handed and specialists in mid- and long-distance shooting. The mean age was 23.36 (± 2.87) years with a mean height of 1.95 (± 0.09) m and a mean mass of 90.43 (± 12.40) kg.

2.2. Data collection

The execution of the jump shot is subject to all types of stimuli, external contingencies and attentional mechanisms. For this reason, and in order to control these variables, it was necessary to analyse the action using a protocol similar to that encountered in competition, where the variables manipulated are controlled and those that influence it are kept constant. The manipulated variable was the presence or absence of opposition, while the controlled variables were the previous technical action (running and stop), body orientation and distance of the jump shot.

Two video cameras were used at 50 Hz to record the performance of the shots. The first was placed at a distance of 10 m from where the shot was to be made with an orientation of 45° to the direction of the shot, and the second was situated 11 m from the shot with an orientation of 45° to the direction of the shot and 90° to the orientation of the first camera. The cameras were started approximately 3 s prior to the beginning of each shot and were not switched off until the ball passed through the hoop to ensure the recording of a sufficient portion of the performance to permit analysis of release variables. After positioning the cameras, and before filming the shots, a reference object was filmed. The reference object was so oriented that the x-axis was in line with the direction of the shot, the z-axis was perpendicular and horizontal to the direction of the shot and the y-axis was perpendicular to the plane of the floor.

Once the warm-up was conducted, the subject completed the experimental protocol (figure 1). The starting position was in the central zone of the court

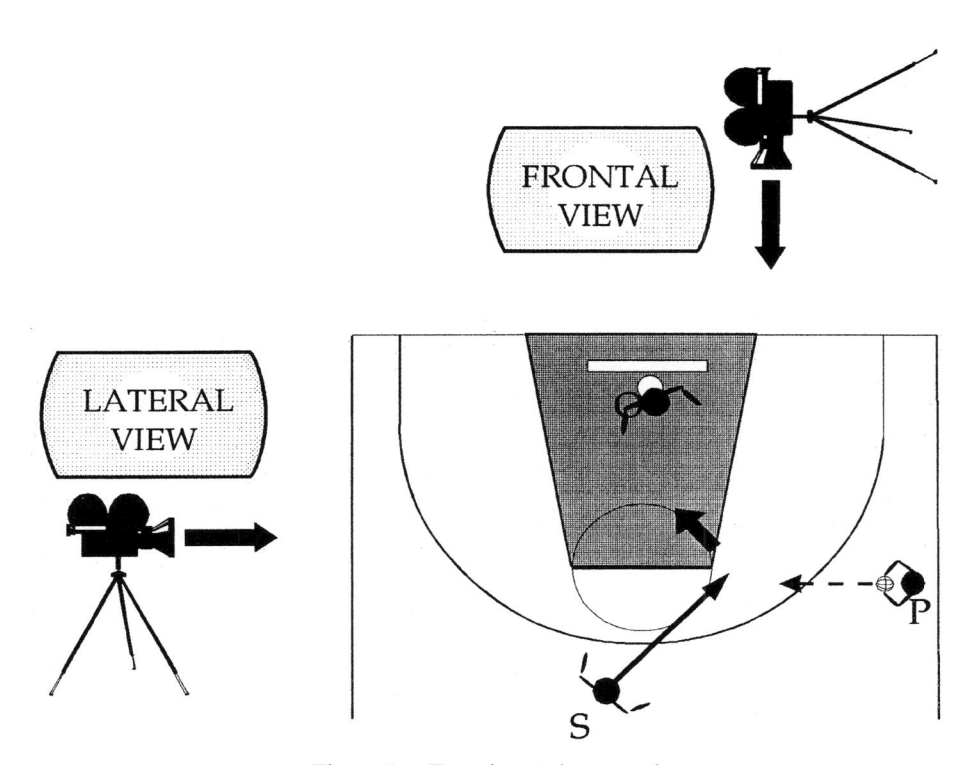

Figure 1. Experimental protocol.

(Position S of figure 1), and from that static position the player ran along a line as shown in figure 1. During his run, the player received a ball from player P at a point 2 m before reaching the shooting position. At the instant of receiving the ball, the player stopped and he finally made the shot. The opponent, situated in the horizontal projection of the hoop, O, remained in that position until the moment in which the ball left the passer's hands. The opponent, at that moment, at random, either remained in that position or moved to intercept the ball, sometimes succeeding in doing so. This protocol was continued until each player had performed 15 successful shots.

Eight shots by each player (four with and four without opposition) were selected for analysis, the criterion being those where the ball passed through the hoop without touching either it or the backboard.

2.3. *Data analysis*

The human model used for the analysis consisted of 14 segments plus an implement, in this case the ball, which was considered as a sphere. Twenty-three points, using the inertial parameters given by De Leva (1996), defined these segments. The co-ordinates of each point were obtained by digitizing the images of both cameras. The digitized data were smoothed, interpolated and synchronized to an equivalent sampling rate of 100 Hz, using a fifth order spline (Wood and Jennings 1979). The three-dimensional co-ordinates were obtained through the direct linear transformation (DLT) procedure (Abdel-Aziz and Karara 1971) and subsequently each of the biomechanical variables that defined the characteristics of the action were calculated.

2.4. *Selection of the variables*

The dependent variables were selected because they had been studied extensively in the literature and/or because basketball coaches focus on them in training sessions to improve the player's technique.

The variables were grouped according to Hudson (1985) who postulated two classes: (1) product variables, which determine the final result of the action and which here correspond to the angle, speed and height of the release of the ball and are based on the mechanical inter-relationships of projectile motion and are illustrated in figure 2, and (2) process variables, which are the most significant causes in determining the efficacy of the action during its execution. The process variables were classified into three groups and are described in tables 1, 2 and 3 and illustrated in figure 2. The temporal variables (table 1) were obtained from the key moments defined by the spatial positions adopted by the player and the ball during the course of the performance of the jump shot. The positional variables (table 2) were obtained from the spatial positions adopted, choosing the most significant positions in relation to the performance of the jump shot. The velocity variables (table 3) refer to the velocities developed during different phases of the shot.

2.5. *Statistical analysis*

As the aim of this study was to examine the effect of one independent variable (opposition) on the dependent variables mentioned above, a one-way analysis of variance was used. A value of $p < 0.05$ was used to indicate significance.

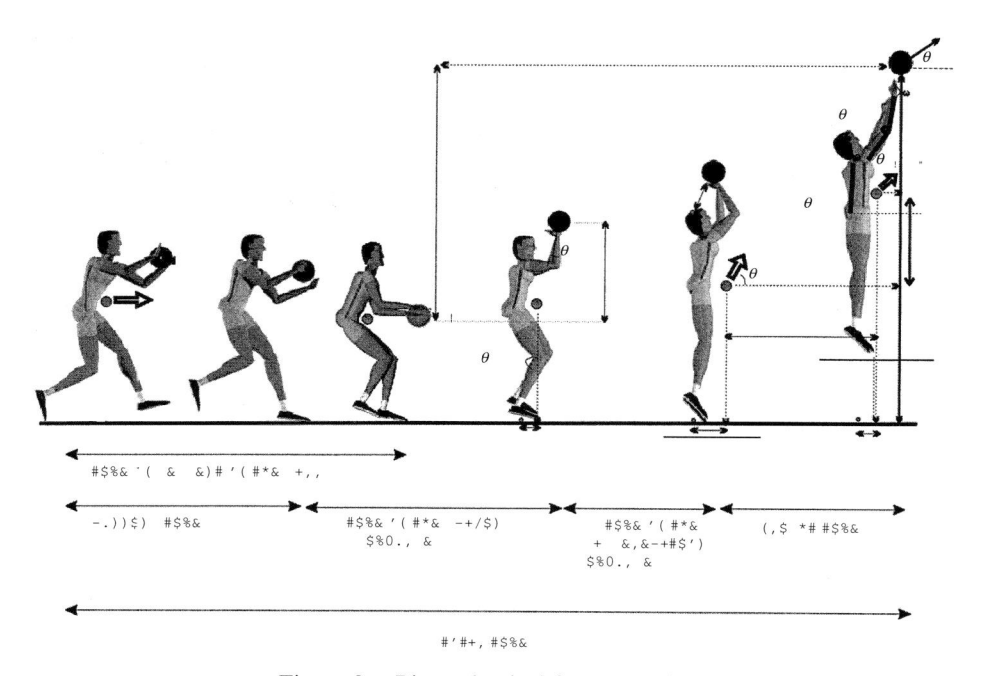

Figure 2. Biomechanical factors analysed.

Table 1. Temporal variables in the jump shot.

T_1	Moment of receiving the ball.
T_2	Moment when one or both feet made contact with the ground in order to jump.
T_3	Instant when the ball was at its lowest point.
T_4	Instant in which the centre of gravity (CG) of the player and the ball reached its lowest vertical point.
T_5	Instant in which the player took both feet off the ground.
T_6	Moment when the ball left the player's hands.

Table 2. Positional variables in the process of shooting.

DCG-cb. (T_4)	Distance between horizontal projection of CG and the centre of the support base and the lowest position of CG.
DCG-cb. (T_5)	Distance between horizontal projection of CG and the centre of the support base and take-off.
DCG-cb. (T_6)	Distance between horizontal projection of CG and the centre of the support base and ball release.
DCG (T_5-T_6)	Distance between the horizontal projection of CG at take-off and ball release.
Ball-vertex (T_5)	Ball-vertex distance at take-off.
$Sy_{(ball)}$	Vertical displacement of the ball, from its lowest position to height of release.
θ_{knee} (T_4)	Knee angle at the beginning of the acceleration impulse phase, the three-dimensional included angle formed by a line joining the hip, knee and ankle joint centres.
θ_{elbow} (T_4)	Elbow angle at the lowest position of the CG, the three dimensional included angle formed by a line joining the wrist, elbow and shoulder joint centres.
θ_{CG} (T_5)	Take-off angle of CG, the two-dimensional included angle formed by the projection, onto the x,z (sagittal) plane, of the line joining the positions of the CG at take-off and one frame after take-off and the forward horizontal.
θ_{trunk} (T_6)	Trunk angle at ball release, the two-dimensional included angle formed by the projection, onto the x,y (sagittal) plane, of the line joining the mid-points of those lines joining the right and left shoulder, and right and left hip joints, and the forward horizontal.
$\theta_{shoulder}$ (T_6)	Shoulder angle at ball release, the three-dimensional included angle formed by a line joining the centres of the right shoulder and elbow joints and the line joining the mid-points of those lines joining the right and left shoulder, and right and left hip joints.
θ_{elbow} (T_6)	Elbow angle at ball release.

Table 3. Velocity variables in the jump shot.

$V_{x,z}$ (T_1)	Horizontal velocity of the CG at the moment reception of the ball.
$V_{x,z}$ (T_5)	Horizontal velocity of the CG at the moment take-off.
$V_{x,z}$ (T_6)	Horizontal velocity of the CG at the moment ball release.
V_{ball} (T_3-T_4)	Mean vertical velocity of the ball from its lowest vertical point (T_3) until the beginning of the acceleration phase (T_4).
ω_{wrist} (T_6)	Angular velocity of the wrist at ball release.

3. Results and discussion

3.1. Product variables

The release angle of the ball (θ_{ball}) increased significantly in the presence of an opponent (table 4) and this helped the player to avoid the possible interception of the ball by the opponent's hand. The mean release angle of the ball in this study was 45° in contrast to an angle of 48° found in the studies carried out by Mortimer (1951), Brancazio (1981), Hudson (1985), Satern (1988), Walters et al. (1990) and Miller and Bartlett (1996). This may be due to the greater height of the release of the ball caused by the greater height of the subjects in the present study, 1.95 m, against the mean of 1.83 m reported in the studies above. The distance of the ball from the basket in this study was similar to the distance in the studies referred to above, although the player's position on court was different.

The velocity of ball release (Vs_{ball}) was not significantly different between the opponent and non-opponent conditions. The mean value (6.33 m s^{-1}) was similar to data reported by Walters et al. (1990), where the jump shots taken from the free throw line were between 6.6 and 6.9 m s^{-1}, and by Miller and Bartlett (1993, 1996) who reported values of around 6.2 m s^{-1}, but it was lower than that reported in other studies where shots were taken from the same distance, such as Mortimer (1951) and Hudson (1985), who reported values of 9.95 and 7.13 m.s^{-1}, respectively.

The height the ball reached at release (Hs_{ball}) was higher when shooting against an opponent than without an opponent even though the height reached by the centre of gravity (CG) from the take-off position of the player to the release of the ball (SCG_{ball}) was lower, but neither of these were significantly different. The greater ball release height could be related to the greater release angle as several authors (Yates and Holt 1982, Toyoshima et al. 1985, Satern 1988, Miller and Bartlett 1996) have reported a close relation between these two variables, which reflects a more vertical orientation of the arm at release. The lower height reached by the CG may be related to the requirements of releasing the ball quickly with the presence of opposition.

3.2. Process variables

The temporal variables are shown in table 5. The significant reduction in flight time (T_5-T_6) when there was an opponent suggests that the player is jumping more

Table 4. Results of product variables.

	Without opponent ($N=40$) Mean	SD	With opponent ($N=40$) Mean	SD	F	p
θ_{ball} (°)	44.7	2.3	47.0	1.7	4.592	0.036*
Vs_{ball} (m s^{-1})	6.30	0.57	6.36	0.50	0.194	0.666
Hs_{ball} (m)	2.85	0.16	2.88	0.16	0.559	0.4651
SCG_{ball} (m)	0.32	0.09	0.31	0.09	0.264	0.614

*$p<0.05$.

rapidly and with less effort. Consequently, he will achieve a lower peak elevation of the CG. Although the latter variable was not significantly different between the two conditions (as noted above), the temporal data confirm the importance of this observation.

Given that the height of release of the ball was greater with opposition than when unopposed, the explanation may be that the player adopts changes in the position of his joints during shooting so that the height of the release of the ball is greater even though the jump elevation of the CG is lower in the presence of opposition. These postural adjustments are detailed as position variables (table 6) in which the angle at the knee at time T_4 and the angle at the shoulder at time T_6 were significantly different between the two conditions. These findings indicate that where players face an opponent they begin the propulsive phase of the jump shot with a larger angle of the knee ($\theta_{knee}(T_4)$) giving them a more elevated position of the CG. In addition, at release the player increases shoulder flexion, ($\theta_{shoulder}(T_6)$) and the angles of inclination of the trunk ($\theta_{trunk}(T_6)$) and the elbow angle ($\theta_{elbow}(T_6)$), although these two latter variables were not significantly different. It is probable that the sum of all the joint angles taken together would reveal significant differences between the two situations, since the tendency for each was for an increase when facing opposition.

Table 5. Results of process variables: temporal.

	Without opponent		With opponent		F	p
	Mean	SD	Mean	SD		
Running time, (T_2-T_2) (s)	0.14	0.10	0.17	0.12	1.183	0.280
Time of descent of the ball, (T_1-T_3) (s)	0.14	0.07	0.12	0.04	0.163	0.692
Time of braking impulse, (T_2-T_4) (s)	0.30	0.09	0.30	0.08	0.048	0.828
Time of acceleration impulse, (T_4-T_5) (s)	0.16	0.03	0.16	0.03	0.349	0.563
Flight time, (T_5-T_6) (s)	0.26	0.04	0.24	0.03	3.984	0.04*
Total time, (T_1-T_6) (s)	0.86	0.10	0.87	0.12	0.233	0.636

*$p < 0.05$.

Table 6. Results of process variables: positional.

	Without opponent		With opponent		F	p
	Mean	SD	Mean	SD		
$\theta_{CG}(T_5)$ (°)	77.86	6.54	77.17	5.80	0.227	0.640
DCG-cb.(T_4) (m)	0.16	0.07	0.15	0.07	0.416	0.528
DCG-cb.(T_5) (m)	0.14	0.05	0.13	0.05	0.997	0.332
DCG-cb.(T_6) (m)	0.07	0.04	0.08	0.04	1.438	0.235
DCG (T_5-T_6) (m)	0.12	0.04	0.12	0.04	0.041	0.841
$\theta_{knee}(T_4)$ (°)	107.01	8.36	110.1	7.14	6.351	0.014*
$\theta_{elbow}(T_4)$ (°)	73.00	15.43	71.81	12.45	0.131	0.722
Ball-vertex (T_5) (m)	0.26	0.05	0.26	0.05	0.136	0.718
Sy_{ball} (m)	1.89	0.20	1.92	0.20	0.664	0.427
$\theta_{trunk}(T_6)$ (°)	82.68	2.87	85.26	2.64	0.782	0.389
$\theta_{shoulder}(T_6)$ (°)	136.95	3.92	138.79	2.64	2.494	0.032*
$\theta_{elbow}(T_6)$ (°)	123.81	9.89	126.42	10.54	0.064	0.803

*$p < 0.05$.

These results agree with the findings of authors such as Ryan and Holt (1989), White and Elliott (1989), Satern (1993) and Miller and Bartlett (1996) who reported that when the shooting distance was increased the joint angles of the propelling arm also increased at release. It is therefore suggested that the positions adopted by the upper arm segments at the end of the shot must be considered as relevant factors that are affected by the presence of opposition. These differences are illustrated in figure 3.

Without opponent		With opponent
−	θ_{ball}	+
−	θ_{elbow}	+
−	$\theta_{shoulder}$	+
+	$SCG_{(ball)}$	−
=	$Hs_{(ball)}$	=

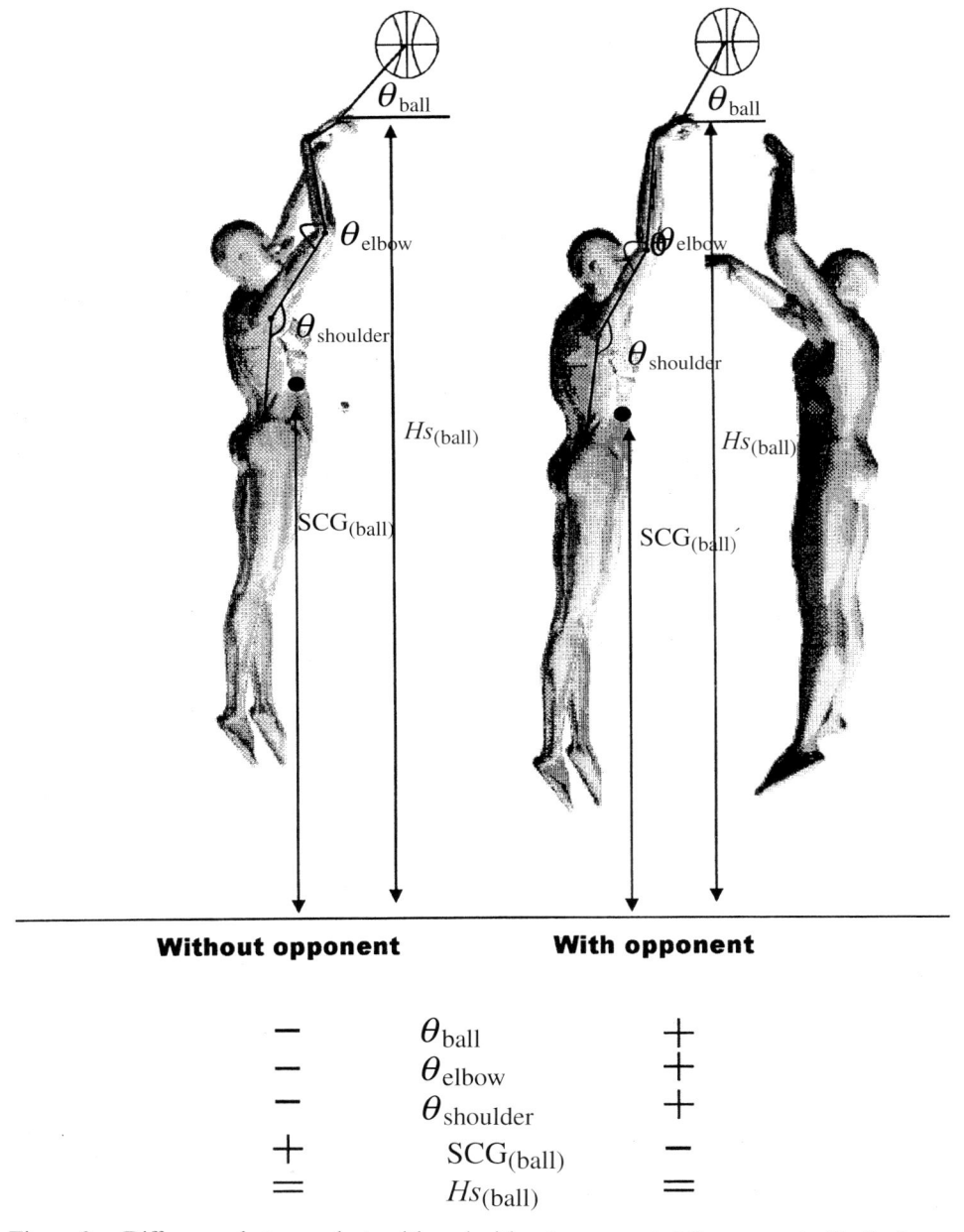

Figure 3. Differences between shots with and without opponent at the moment of ball release.

With regard to the velocity variables (table 7) there appears to be no difference in the horizontal velocity characteristics of the CG during the movement although there was a tendency for the CG to have a slightly greater horizontal velocity at the beginning of the movement (probably reflecting the urgency imposed by the presence of an opponent) and a lower horizontal velocity at release (reflecting the more upward projection angle as noted above in order to gain extra height). The vertical velocity of the ball from its lowest point until its release was significantly greater with an opponent, which confirms the earlier suggestion that when faced with opposition players move more rapidly in the initial phase of the shot. The wrist angular velocities were not significantly different between the two situations, but the lower value against opposition might reflect the emphasis placed on joint extension to gain height and in so doing also use this as a means to propel the ball.

4. Conclusions

In conclusion, it can be stated that players attempt to release the ball more quickly and from a greater height when confronted with an opponent. This strategy lessens the chance of the opponent intercepting the ball. Players realize this strategy by approaching more rapidly and positioning the body in a more upright position at the initiation of the upward movement of the ball. This manoeuvre gives players greater initial height but also a more stable base for generating a greater initial velocity of the ball. The greater initial knee position restricts the ability of the player to jump and therefore he performs a quicker but less powerful jump, while the more rapid upward movement of the ball helps to increase the joint angles at shoulder and elbow at release and this, combined with a more upright trunk, helps the ball to attain a greater height and a more vertical angle of projection. This interpretation is supported by significant differences and trends in the biomechanical data collected.

The differences in technical execution of the skill have implications for practice. Although the differences between the two situations are small, it is likely that they lead to significantly different demands on the neuromuscular co-ordination requirements for situations with and without opposition. This implies that training would benefit from practice with an opponent for at least some of the time to condition the players to the demands that they are more likely to meet in the game situation. It is unknown at this time what proportion of practice would best be done in this way, or the effects that variability of practice has on performance of the basketball jump shot.

Table 7. Results of process variables: velocity.

	Without opponent		With opponent		F	p
	Mean	SD	Mean	SD		
$V_{x,z}(T_1)$ (m s^{-1})	2.11	0.71	2.25	0.73	0.721	0.408
$V_{x,z}(T_5)$ (m s^{-1})	0.55	0.29	0.58	0.25	0.084	0.776
$V_{x,z}(T_6)$ (m s^{-1})	0.64	0.25	0.61	0.29	0.211	0.653
$V_{ball}(T_3-T_4)$ (m s^{-1})	4.07	0.90	4.45	0.91	6.483	0.021*
$\omega_{wrist}(T_6)$ (rad s^{-1})	26.22	8.19	24.03	5.10	1.864	0.176

*$p < 0.05$.

Acknowledgements

This study was funded by the Centro de Alto Rendimiento y de Investigación en Ciencias del Deporte, Proyecto no. 19/UNI21/97. The authors wish to express their appreciation to Adrian Lees for his help in preparing this paper.

References

ABDEL-AZIZ, Y. I. and KARARA, H. M. 1971, Direct linear transformation from comparator coordinates into object space coordinates in close-range photogrammetry, in *Proceedings of the ASP/UI Symposium on Close-Range Photogrammetry*, Falls Church, VA (Urbana, IL: American Society of Photogrammetry), 1–18.

ASOCIACIÓN CLUBS BALONCESTO (ACB) 1997, *Servicio de Estadísticas. Estadísticas de la temporada 1996–97. [Statistics service. Season Statistics 1996–97.]* (Barcelona: ACB).

BRANCAZIO, P. J. 1981, The physics of basketball, *American Journal of Physics*, **49**, 356–365.

CHASE, M., EWING, M., LIRGG, C. and GEORGE, T. 1994, The effects of equipment modification on children's self-efficacy and basketball shooting performance, *Research Quarterly for Exercise and Sport*, **65**, 159–168.

COHEN, L. and HOLLIDAY, M. 1982, *Statistics for Social Sciences* (London: Harper & Row).

DE LEVA, P. 1996, Adjustments to Zatsiorsky-Seluyanolv's segment inertia parameters, *Journal of Biomechanics*, **29**, 1223–1230.

ELLIOTT, B. and WHITE, E. 1989, A kinematic and kinetic analysis of the female two point and three point jump shots in basketball, *Australian Journal of Science and Medicine in Sport*, **21**(2), 7–11.

GABBARD, C. P. and SHEA, C. H. 1980, Effects of varied goal height practice on basketball foul shooting performance, *Coach and Athlete*, **42**, 10–11.

HAY, J. G. 1994, *The Biomechanics of Sports Techniques* (Englewood Cliffs, NJ: Prentice-Hall).

HESS, C. 1980, Analysis of the jump shot, *Athletic Journal*, **61**(3), 30–33, 37–38, 58.

HUDSON, J. L. 1985, Prediction of basketball skill using biomechanical variables, *Research Quarterly for Exercise and Sport*, **56**, 115–121.

MILLER, S. and BARTLETT, R. M. 1993, The effects of increased shooting distance in the basketball jump shot, *Journal of Sports Sciences*, **11**, 285–293.

MILLER, S. and BARTLETT, R. M. 1996, The relationship between basketball shooting kinematics, distance and playing position, *Journal of Sports Sciences*, **14**, 243–253.

MORTIMER, E. M. 1951, Basketball shooting, *Research Quarterly*, **22**, 234–243.

RYAN, P. and HOLT, L. E. 1989, Kinematic variables as predictors of performance, in W. E. Morrison (ed.), *Proceedings of the Seventh International Symposium of the Society of Biomechanics in Sports* (Melbourne, Victoria: Footscray Institute of Technology), 79–88.

SÁENZ, P. and IBÁÑEZ, S. 1995, El Tiro: Clasificación, evaluación y su entrenamiento en cada categoría [The shot: Classification, evaluation and training at each age], *Clinic*, **3**, 29–34.

SATERN, M. N. 1988, Basketball: shooting the jump shot, *Strategies*, **1**(4), 9–11.

SATERN, M. N. 1993, Kinematics parameters of basketball jump shots projected from varying distances, in Hamill, J. (ed) *et al.*, *Biomechanics in Sport XI: Proceedings of the XI International Symposium of Biomechanics in Sports*, Amherst, Mass., 313–317.

TOYOSHIMA, S., HOSHIKAWA, T. and IKEGAMI, Y. 1985, Effects of initial ball velocity and angle of projection on accuracy in basketball shooting, in H. Matsui and K. Kobayashi (eds), *Biomechanics*, vol. VII-B (Champaign, IL: Human Kinetics Books), 525–530.

WALTERS, M., HUDSON, J. M. and BIRD, M. 1990, Kinematics adjustments in basketball shooting at three distances, in Nosek, M. (ed) *et al., Proceedings of the VIIIth International Symposium of the Society of Biomechanics in Sports*, Prague, 219–224.

WHITE, L. and ELLIOTT, B. C. 1989, A comparison of the female jump shot technique for the two point and three point goals in basketball, *Sports Coach*, **12**(4), 33–35.

WOOD, G. A. and JENNINGS, L. S. 1979, On the use of spline functions for data smoothing, *Journal of Biomechanics*, **12**, 477–479.

YATES, G. and HOLT, L. E. 1982, The development of multiple linear regression equations to predict accuracy in basketball jump shooting, in J. Terauds (ed.) *Biomechanics in Sports* (Del Mar, CA: Academic Publications), 103-109.

Part IV
Ergonomics in Special Populations

12 Decreased submaximal oxygen uptake during short duration oral contraceptive use: a randomized cross-over trial in premenopausal women

M. GIACOMONI†‡* and G. FALGAIRETTE‡

†Research Institute for Sport and Exercise Sciences, Liverpool John Moores University, Henry Cotton Campus, 15–21 Webster Street, Liverpool L3 2ET, UK

‡Unité d'Ergonomie Sportive et Performance, UFR STAPS, Université de Toulon et du Var, Avenue de l'Université, BP 132, 83957 La Garde Cedex, France

Keywords: Contraceptive steroids; Oxygen uptake; Aerobic exercise; Running economy; Premenopausal women.

Long-term oral contraceptive (OC) use is known to be associated with changes in haemostasis, cardiovascular dynamics, and carbohydrate and lipid metabolism. Less well documented are the short-term variations in cardiorespiratory responses to exercise during the menstrual cycle of OC users. In this study the short-term effects of the usage of OC on cardiorespiratory and ventilatory responses to submaximal exercise were examined.

Ten women (age = 23 ± 3 years) on monophasic OC were tested at three different times during their cycle: during menstruation, off OC use (off OC: days 2–4), early on OC use (EOC: days 7–9) and late on OC use (LOC: days 19–21). Times of testing were assigned randomly. On each occasion participants performed a continuous 12-min run exercise on a treadmill at three submaximal intensities (averaging 7, 8 and 9 km h^{-1}), each for 4 min. Heart rate, ventilation ($\dot{V}$E), oxygen uptake ($\dot{V}O_2$), carbon dioxide output ($\dot{V}CO_2$), respiratory exchange ratio and running economy were assessed in the last minute of each stage of exercise.

No significant variations were observed between the different times for heart rate, $\dot{V}$E, and $\dot{V}CO_2$ irrespective of the stage of exercise ($p > 0.05$). Using two-way analysis of variance (ANOVA) with repeated measures on both factors (three stages and three times), $\dot{V}O_2$ (ml kg^{-1} min^{-1}) was lower by 3% to 5.8% when participants were on early and late OC use compared to off OC regardless of the stage of the exercise ($F(2,18) = 6.3$; $p = 0.008$). Running economy (ml O$_2$ kg^{-1} km^{-1}) was significantly improved (lower values) when women were on late OC use compared to off OC regardless of the stage of exercise. No significant interaction effect between stage of exercise and time of pill usage was demonstrated in any of the parameters studied. Results suggest that oral contraceptive users may expect lower $\dot{V}O_2$ and better running economy during the pill ingestion phase and consequently have implications for exercise performances.

1. Introduction

Many exercising women are prescribed oral contraceptive pills for contraception, regulation of the menstrual cycle, control of perimenstrual symptoms and skeletal

*Author for correspondence. e-mail: m.giacomoni@livjm.ac.uk

protection in case of amenorrhoea. The majority of the studies to date have dealt with the long-term effect of different oestrogen/progestogen preparations on haemostasis, cardiovascular dynamics or carbohydrate and lipid metabolism (Littler et al. 1974, Linder et al. 1989, Heintz et al. 1996, Kluft and Lansink 1997). The influence of oral contraceptive (OC) use on exercise responses of women has been studied mostly by comparing subjects on OC to non-OC users or engaging participants as their own control before and after contraceptive pill medication (Lehtovirta et al. 1977, Bemben et al. 1992, Rahkila and Laatikainen 1992).

In normally menstruating women, fluctuations in endogenous ovarian steroid concentrations along the menstrual cycle have been shown to modify respiratory drive, substrate utilization and thermoregulatory responses. High levels of progesterone during the luteal phase are associated with an increase in resting minute ventilation ($\dot{V}E$), ventilatory responses to hypoxia (Schoene et al. 1981), and higher core body temperature (Pivarnic et al. 1992). Increased concentrations in oestradiol-17β during the luteal phase of the menstrual cycle have been shown to spare muscle glycogen content (Hackney 1990) and have a stimulating effect on lipolysis (Bunt 1990).

Very few studies have focused on exercise responses of OC users in different times relative to cyclic OC use. During submaximal exercise, the period of OC use has been reported to induce an enhancement of human growth hormone concentrations (Bonen et al. 1991), and an upward shift in the threshold for heat loss responses, resulting in elevated core body temperature (Rogers and Baker 1997). Moreover, Bonen et al. (1991) demonstrated that a high steroid status, either endogenous or exogenous, increased insulin concentrations during exercise by about 12%. These results could have implications for metabolic responses to submaximal exercise and consequently for oxygen uptake ($\dot{V}O_2$) and energy cost of exercise in OC users. It is not well-known how cardiorespiratory and ventilatory responses to exercise are affected by short-term OC usage (21 days on OC vs 7 days off OC). Reilly and Whitley (1994) observed, in four women on monophasic pills, a rise in the ventilatory equivalent for oxygen, suggesting a lesser ventilatory efficiency or a decreased $\dot{V}O_2$ for a given $\dot{V}E$ when participants were taking OC.

The purpose of the present study was to analyse the ventilatory and cardiorespiratory responses to submaximal exercise in women taking monophasic OC at different times during cyclic OC use.

2. Methods

Ten female physical education students (age: 23±3 years, height: 1.67±0.05 m, body mass: 56.6±6.8 kg, sum of skinfolds: 38.6±10.9 mm) participated in the study after giving written informed consent. The experiments received the ethical approval of the Nice Committee of Human Experimentation. All participants had been using monophasic OC (constant doses of oestrogen and progestogen for 21 days) for at least 18 months before entering the study. The OC medication consisted of low-dose combined contraceptive pills, containing ethinylestradiol (0.02–0.03 mg day^{-1}, 21 days month^{-1}) associated with 0.150 mg day^{-1} desogestrel (4 cases), or 0.075 mg day^{-1} gestodene (6 cases). The cycle of OC users consisted of 21 days of OC ingestion, days 5–25 from the first day of menstrual bleeding, followed by 7 days of non-OC consumption. Participants were tested during three different times of OC usage: during menstruation, off OC use (off OC: days 2–4), early on OC use (EOC: days 7–9) and late on OC use (LOC: days 19–21). Times for testing were assigned randomly and all participants first performed a pre-study trial session to counter balance and minimize learning effects. All participants were tested three times on each occasion: in the morning (08:00 h), in the early afternoon (13:00 h) and in the late afternoon (17:00 h) and the mean value of the three measurements was retained for analyses. This procedure was used to offset inter-individual differences in chronopharmacological effects of drugs, and the differences between participants in the time of pill intake. Moreover, this procedure was used in order to eliminate a likely effect of circadian rhythm modifications of biological parameters (e.g. cortisol secretion, heart rate, blood pressure) which could be induced by OC use (Reinberg *et al.* 1996), and consequently might interfere with the effect of OC usage period. Participants were asked to maintain a stable level of physical activity throughout the experimental period, and were requested to avoid vigorous activity during the 24 h preceding the testing sessions. Height and body mass were measured with a medical scale (accuracy 0.1 cm) and a medical calibrated balance (accuracy 0.1 kg), respectively. Sum of skinfolds was calculated from measures taken at four different sites (biceps, triceps, subscapular, suprailiac) using a Harpenden caliper (British Indicators Ltd, Luton, UK), according to Durnin and Rahaman (1967).

All participants performed a submaximal aerobic exercise in the laboratory ($21 \pm 2°C$, 760 ± 6 mmHg) on a treadmill (Powerjog, Birmingham, UK). The exercise consisted of a continuous 12-min run at 3 submaximal intensities (6.6 ± 0.9, 7.7 ± 0.7 and 8.7 ± 0.7 km h^{-1}) each lasting for 4 min. The running speeds were estimated from the heart rate response during the pre-study trial session. The heart rate targets were 140, 150 and 160 beats min^{-1}, and were chosen to elicit approximately 50, 60 and 65% of the $\dot{V}O_2$ max in the last minute of each stage of exercise (Åstrand and Ryhming 1954).

Ventilation ($\dot{V}E$), expired fraction of oxygen (FE O_2) and carbon dioxide (FE CO_2) were measured with a computerized open-circuit respiratory test system (Sensor Medics 2900, Anaheim, USA). Expired fractions of O_2 and CO_2 were sampled using a zirconia analyser and an infra-red analyser, respectively. The analysers were calibrated automatically before each test using standard gas mixtures ($16 \pm 0.17\%$ for O_2 and $4 \pm 0.03\%$ for CO_2). The heart rate (HR) was continuously recorded by short-range radio telemetry (Sport Tester PE 4000, Polar Electro, Kempele, Finland). The following variables were calculated: oxygen uptake ($\dot{V}O_2$), CO_2 output ($\dot{V}CO_2$), respiratory exchange ratio (RER), $\dot{V}E/\dot{V}O_2$ and $\dot{V}E/\dot{V}CO_2$. Mean values were obtained every 20 s pre-exercise (1 min), during exercise (12 min), and during the first 2 min of the recovery period. Exercise responses were assessed in the last minute of each stage of exercise (steady state). The rate of oxygen consumption during a submaximal steady-state run has been defined as the running economy (Morgan *et al.* 1989). The higher the $\dot{V}O_2$ at a given running speed the less economical is the run. The running economy (RE) was analysed for each submaximal steady-state running velocity. Respiratory exchange ratio and $\dot{V}O_2$ were used to estimate the gross energy expenditure in kJ kg^{-1} km^{-1} from the table that provides thermal equivalents of oxygen utilization based on RER (McArdle *et al.* 1996). It was assumed that RER reflected the non-protein respiratory quotient.

The data were analysed using a two-way analysis of variance (ANOVA) with repeated measures on both factors (stage and time of pill usage). Tukey's Honestly Significant Differences *post hoc* analysis was used to locate the differences. Normality of distributions was checked and the F values have been corrected for violation of the assumption of sphericity using Huynh-Feldt adjustment when epsilon (ε) values were $\varepsilon < 0.75$, according to Vincent's (1995) guidelines. Statistical significance was chosen as $p < 0.05$.

3. Results

3.1. *Main effect for stage of exercise*

Mean values and standard deviations of the studied cardiorespiratory and ventilatory responses are given for each stage of exercise and each testing time in table 1.

As expected, the stage of exercise had a significant main effect on HR, $\dot{V}$E, $\dot{V}O_2$, $\dot{V}CO_2$, RER, RE (ml O_2 kg^{-1} km^{-1}) and gross energy expenditure (kJ kg^{-1} km^{-1}), irrespective of the OC usage time (table 2). Heart rate, $\dot{V}$E, $\dot{V}O_2$, and $\dot{V}CO_2$ were significantly increased between each stage of exercise (table 2). Respiratory exchange ratio was lower in the first stage compared to the second and third stages of exercise ($p < 0.05$ and 0.01, respectively). Running economy and gross energy expenditure were significantly higher at the first intensity of exercise compared to the intermediate and last intensities of exercise (table 2). No significant main effect for stage of exercise was observed on $\dot{V}E/\dot{V}O_2$ and $\dot{V}E/\dot{V}CO_2$ (table 2).

3.2. *Main effect for time of pill usage*

The body mass did not vary significantly between off OC, EOC and LOC (56.8 ± 6.8 kg, 56.8 ± 6.6 kg, 56.9 ± 6.8 kg, respectively; $F(2,18) = 1.3$; $p = 0.3$). No significant variations were observed between the three occasions for HR, $\dot{V}$E, and $\dot{V}CO_2$ regardless of the stage of exercise (table 2). In contrast, $\dot{V}O_2$ (l min^{-1}) was significantly lower when participants were using OC compared to off OC use irrespective of exercise intensity ($p < 0.05$; table 2). Oxygen uptake corrected for body mass also displayed a systematic decrease during both OC use times compared to off OC use regardless of the stage of exercise ($p < 0.01$; figure 1). A significant increase in RER was noted early on OC use compared to off OC use ($p < 0.05$; table 2). A main effect of the time of pill usage was also observed for $\dot{V}E/\dot{V}O_2$ irrespective of the stage of exercise ($p < 0.05$; table 2). The ratio $\dot{V}E/\dot{V}CO_2$ did not vary significantly between the three occasions (table 2).

The decrease in oxygen demand between the off OC use time and both times on OC induced a decrease in the energy cost of running (ml O_2 kg^{-1} km^{-1}) irrespective of stage of exercise when participants were on OC use ($p < 0.05$; figure 2). The differences were significant between off OC use and late on OC use ($p < 0.05$; table 2).

When thermal equivalents of oxygen for the non-protein respiratory quotient were taken into account, a main effect of time of pill usage was observed on energy expenditure (kJ kg^{-1} km^{-1}). Energy expenditure was lower when participants were late on OC use compared to off OC use ($p < 0.05$; table 2).

3.3. *Interaction effect between stage of exercise and time of pill usage*

No significant interaction effects (stage of exercise × time of pill usage) were demonstrated in any of the parameters studied (table 2).

4. Discussion

The major finding emerging from the present study was the decrease in oxygen uptake during the period of oral contraceptive consumption for a given exercise intensity.

Table 1. Means and standard deviations for cardiorespiratory and ventilatory responses to submaximal exercise of monophasic oral contraceptive users ($n = 10$) in the off OC use time (off OC), early on OC use time (EOC) and late on OC use time (LOC).

Variables	Stage 1 off OC	Stage 1 EOC	Stage 1 LOC	Stage 2 off OC	Stage 2 EOC	Stage 2 LOC	Stage 3 off OC	Stage 3 EOC	Stage 3 LOC
HR (beats min^{-1})	137±13	141±4	137±5	150±12	152±8	150±9	158±12	160±10	158±12
$\dot{V}$E (l min^{-1})	38.0±5.3	38.3±5.7	37.3±6.0	44.5±6.0	44.3±7.1	44.5±6.6	50.9±8.8	52.0±8.5	51.0±8.3
$\dot{V}O_2$ (l min^{-1})	1.37±0.23	1.31±0.25	1.30±0.25	1.57±0.19	1.51±0.25	1.50±0.24	1.73±0.20	1.69±0.23	1.68±0.23
$\dot{V}O_2$ (ml kg^{-1} min^{-1})	24.1±2.1	23.0±2.9	22.7±2.3	27.6±1.2	26.4±2.3	26.2±2.1	30.6±1.5	29.7±1.9	29.4±1.8
$\dot{V}CO_2$ (l min^{-1})	1.21±0.17	1.21±0.18	1.17±0.18	1.43±0.20	1.42±0.22	1.39±0.20	1.61±0.26	1.63±0.25	1.59±0.24
RER	0.89±0.07	0.92±0.07	0.90±0.06	0.92±0.08	0.95±0.08	0.94±0.08	0.93±0.10	0.96±0.11	0.95±0.10
$\dot{V}E/\dot{V}O_2$	28.1±3.1	29.7±4.2	29.1±4.3	28.6±3.1	29.8±4.5	30.1±4.6	29.5±4.4	31.1±5.2	30.7±5.3
$\dot{V}E/\dot{V}CO_2$	31.6±2.7	32.0±3.3	32.1±3.7	31.2±2.5	31.3±3.6	32.4±4.4	31.6±2.3	32.0±3.3	32.3±3.7

Stage 1, 6.6±0.9 km h^{-1}; Stage 2, 7.7±0.7 km h^{-1}; Stage 3, 8.7±0.7 km h^{-1}. HR, heart rate; $\dot{V}$E, minute ventilation; $\dot{V}O_2$, oxygen uptake; $\dot{V}CO_2$, carbon dioxide output; RER, respiratory exchange ratio.

Table 2. Summary of the two-way analysis of variance with repeated measures on both factors (stage of exercise and time of pill usage) for each variable.

Variables	Effect of stage F	Effect of stage p	Effect of stage Tukey's post hoc comparison	Effect of phase F	Effect of phase p	Effect of phase Tukey's post hoc comparison	Interaction effect (stage × phase) F	Interaction effect (stage × phase) p
HR (beats min^{-1})	40.9	***	S1<S2 & S3 (**) S2<S3 (*)	1.1	NS		0.6	NS
$\dot{V}E$ (l min^{-1})	27.3	***	S1<S2 & S3 (**) S2<S3 (**)	0.3	NS		1.7	NS
$\dot{V}O_2$ (l min^{-1})	90.5	***	S1<S2 & S3 (**) S2<S3 (**)	5.6	*	off OC>EOC & LOC (*)	0.3	NS
$\dot{V}CO_2$ (l min^{-1})	40.7	***	S1<S2 & S3 (**) S2<S3 (**)	3.2	NS		0.5	NS
RER	6.8	**	S1<S2 (*) S1<S3 (**)	4.7	*	off OC<EOC (*)	0.9	NS
$\dot{V}E/\dot{V}O_2$	2.4	NS		3.8	*	off OC<EOC (*)	1.5	NS
$\dot{V}E/\dot{V}CO_2$	0.3	NS		1.4	NS		1.5	NS
RE (ml O$_2$ kg^{-1} km^{-1})	9.2	**	S1>S2 (*) S1>S3 (**)	5.0	*	off OC>LOC (*)	1.5	NS
Energy expenditure (kJ kg^{-1} km^{-1})	8.7	**	S1>S2 (*) S1>S3 (**)	4.7	*	off OC>LOC (*)	1.7	NS

S1, Stage 1 (6.6±0.9 km h^{-1}); S2, Stage 2 (7.7±0.7 km h^{-1}); S3, Stage 3 (8.7±0.7 km h^{-1}). HR, heart rate; $\dot{V}E$, minute ventilation; $\dot{V}O_2$, oxygen uptake; $\dot{V}CO_2$, carbon dioxide output; RER, respiratory exchange ratio; RE, running economy; off OC, off oral contraceptive use; EOC, early on oral contraceptive use; LOC, late on oral contraceptive use. P, probability of null hypothesis, *$p<0.05$; **$p<0.01$; ***$p<0.001$. NS; $p>0.05$.

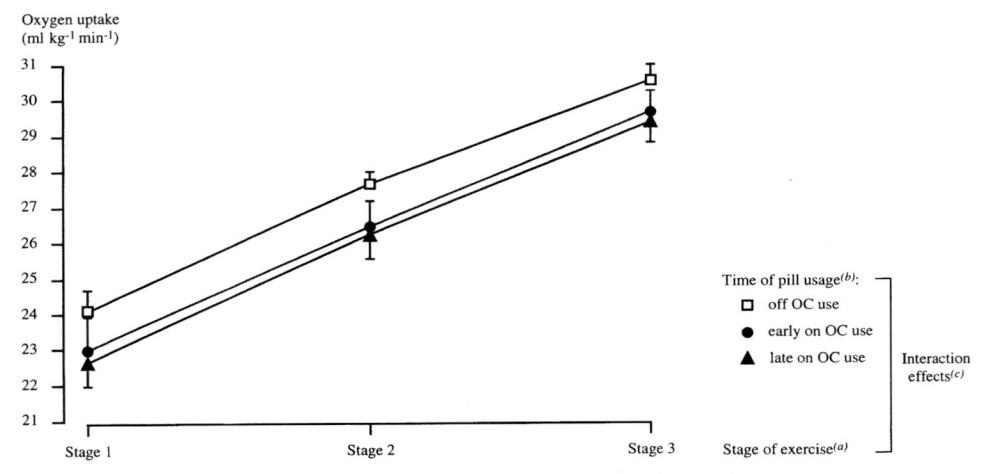

Figure 1. Variations in submaximal oxygen uptake (ml kg^{-1} min^{-1}) between the different times of testing in monophasic oral contraceptive users ($n = 10$). Values are means ± SEM. (a), main effect for stage of exercise: $F(2,18) = 91.1$, $p < 0.001$; (b), main effect for time of pill usage: $F(2,18) = 6.3$, $p < 0.01$; (c), interaction effect between stage of exercise and time of pill usage: $F(4,36) = 0.3$, $p > 0.05$.

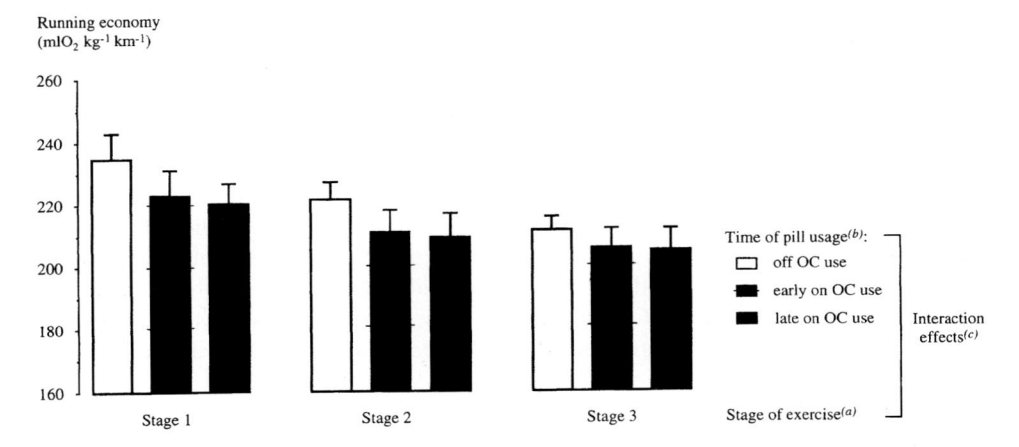

Figure 2. Variations in running economy at each stage of exercise between the different times of testing in monophasic oral contraceptive users ($n = 9$). Values are mean ± SEM. (a), main effect for stage of exercise: $F(2,16) = 9.2$, $p < 0.01$; (b), main effect for time of pill usage: $F(2,16) = 5.0$, $p < 0.05$; (c), interaction effect between stage of exercise and time of pill usage: $F(4,32) = 1.7$, $p > 0.05$.

The influence of OC use on aerobic capacity has been mostly studied by comparing subjects on OC to non-OC users or using participants as their own control before and after contraceptive pill medication. Comparisons between studies are difficult owing to the wide range of oral contraceptive preparations (monophasic, biphasic, triphasic; type and dosage of oestrogen and progestogen) and the intensity-duration configuration of the exercise performed. However, maximal aerobic power

($\dot{V}O_2$ max) has been shown to decrease by 5% to 8% during long-term oral contraceptive medication compared to pre-treatment and/or placebo values, possibly through cellular changes including decrement in muscle mitochondrial citrate (Lebrun 1993).

In normally menstruating women, maximal aerobic capacities have been shown to be slightly lower during the luteal phase compared to the follicular phase (Lebrun *et al.* 1995). The majority of the studies to date failed to demonstrate any changes in submaximal $\dot{V}O_2$ along the menstrual cycle (Jurkowski *et al.* 1981, Eston and Burke 1984, Dombovy *et al.* 1987, De Souza *et al.* 1990). In some studies, high ovarian steroid status observed during the luteal phase has been shown to be associated with higher resting and submaximal $\dot{V}O_2$ values than those observed at low endogenous hormone concentrations during the early or mid-follicular phase (Hessemer and Brück 1985, Williams and Krahenbuhl 1997). The concentrations of synthetic steroids were higher during the 21 days on OC use than during the off OC week. Therefore, the present study suggested specific cardiorespiratory responses to changes in synthetic steroid levels.

The present results partially contrast with the observations of Reilly and Whitley (1994) who failed to show significant differences in $\dot{V}O_2$ between the period off OC use and the period on OC use in four women using monophasic oral contraceptives when exercise was conducted at 70% of $\dot{V}O_2$ max until exhaustion. These conflicting observations could be partly related to the small sample size as well as to the exercise intensity. The decrease in oxygen consumption between off OC use and the two periods on OC in the present study ranged from 3% to 5.8%, the lowest variations being observed for the highest intensity. However, Reilly and Whitley (1994) demonstrated a significant increase in the ventilatory equivalent for oxygen during the period on OC use, suggesting a slight increase in the minute ventilation for a given oxygen uptake or, as observed in the present study, a reduced $\dot{V}O_2$ for a given value of $\dot{V}E$.

Many factors are likely to affect oxygen consumption during submaximal exercise. A modification of exercise ventilation constitutes one of these factors. In normally menstruating women, Williams and Krahenbuhl (1997) observed higher submaximal $\dot{V}O_2$ and $\dot{V}E$ values during the mid-luteal phase (high progesterone status) compared to the early follicular phase (low progesterone status). The increase in ventilation for a given submaximal exercise intensity during the mid-luteal phase has been attributed to the influence of progesterone. The stimulatory effect of progesterone and/or progestogens on ventilatory function has been demonstrated in women during pregnancy (Lyons and Antonio 1959), during the luteal phase of a natural menstrual cycle (Schoene *et al.* 1981, Dombovy *et al.* 1987), and in men or women supplemented with progesterone or synthetic progestogen (medroxyprogesterone acetate). These ventilatory adaptations were expressed by an increase in ventilatory responses to hypercapnia and/or hypoxia and would be due to an increased sensitivity of chemoreceptors to carbon dioxide (Lyons and Antonio 1959, Zwillich *et al.* 1978). In normally menstruating women, the increase in submaximal $\dot{V}O_2$ with high progesterone levels could be partly a consequence of a greater exercise hyperpnea and of the increase in the energy cost of ventilation. In the present study, $\dot{V}E$ did not vary significantly between off OC and both EOC and LOC use suggesting that the third-generation progestogens contained in the oral contraceptive pills (desogestrel and gestodene) might not have the same effect on ventilatory function as natural progesterone at least at the concentrations used. As

$\dot{V}$E did not vary significantly between the different cycle phases, the decrease in $\dot{V}O_2$ observed in this study was not linked to a reduction in the energy cost of ventilation.

The reduction in $\dot{V}O_2$ observed when women were on OC could have been induced by a decrease in HR, in stroke volume and/or in arterio-venous difference in oxygen. However, HR was not altered between the different phases in any stage of exercise. Moreover, a reduction in the stroke volume between off OC use time and times on OC use seemed unlikely. Actually, as far as the long-term effect of OC use could be transposable to its short-term effect, the stroke volume has been demonstrated to increase after 2 months of oestrogen/progestogen administration compared with pre-treatment values (Lehtovirta et al. 1977). Then, reduction in exercise $\dot{V}O_2$ in EOC and LOC might be related to a decrease in the arterio-venous difference in oxygen and in muscle oxygen demand. This reduction in muscle oxygen consumption for a given running speed might be linked to substrate utilization (i.e. metabolic factors) and/or to a greater running efficiency (i.e. biomechanical factors).

The decrease in $\dot{V}O_2$ during the OC use time for a given submaximal intensity could be partly related to a shift in substrate utilization. A significant increase in RER was observed between times off OC use and times on OC use irrespective of the stage of submaximal exercise. These results might suggest an increase in carbohydrate metabolism as an energy source when participants were on OC or a shift from carbohydrate to lipids when they were off OC. More oxygen is needed to provide the same quantity of energy from lipids than from carbohydrate, but this suggestion is not bolstered by the literature. Bonen et al. (1991) demonstrated that oral contraceptive users had higher blood concentrations of free fatty acids (FFA) than non-users, during 30 min of light exercise, but they did not report any significant OC usage-phase effect on blood glucose or FFA concentrations in pill users. The glycogen-sparing property of oestradiol-17β has been demonstrated in animals (Ahmed-Sorour and Bailey 1981) as well as in humans during the luteal phase of the menstrual cycle (Hackney 1990). Moreover, Bemben et al. (1992) reported lower blood glucose and lower carbohydrate oxidation during prolonged exercise (90 min) in women on OC pills compared to non-OC users, suggesting a carbohydrate-sparing effect of OC. These authors also demonstrated an enhancement of growth hormone responses to exercise, a hormone implicated in the breakdown and release of triglycerides from adipose tissue. Thus, the decrease in $\dot{V}O_2$ during OC use phase did not seem to be associated with a shift in substrate utilization. Moreover, when the thermal equivalents of oxygen for the non-protein respiratory quotient were taken into account, the energy expenditure (kJ kg^{-1} km^{-1}) still decreased when participants were on OC use compared to off OC use time. The improvement of running economy during the period on OC might not have been induced by metabolic effects of OC.

Biomechanical factors are known to affect running economy (Williams and Cavanagh 1987). More precisely in running the stretch-shortening cycle activation is very important as an amount of mechanical energy is stored in the series elastic components of active muscles and tendons during the eccentric phase and released during the concentric phase (Asmussen and Bonde-Petersen 1974, Komi 1984). The amount of elastic energy stored in the series elastic components depends on the degree of muscle and tendon stiffness (Komi 1984). Tendon stiffness could have been affected by structural and functional modifications induced by exogenous hormone usage. Both tendons and ligaments are comprised of collagen structural proteins. Liu et al. (1997) observed an inhibition of collagen synthesis of the anterior cruciate ligament with increasing local oestradiol concentrations inducing consequently greater ligament stiffness. Assuming a similar effect of ethinylestradiol on tendon collagen proteins, higher ethinylestradiol concentrations during the OC usage time could have increased the tendon stiffness and consequently the amount of energy stored during the stretching phase of running. Moreover, treatment with low-dose combined oral contraceptives has been shown (Kuhl et al. 1985, Walden et al. 1986) to be associated with an oestrogen-mediated increase in thyroid (thyroxine, triiodothyronine) hormone concentrations known to affect the tendon reflex. The Achilles tendon reflex time has been frequently used as a test of thyroid function and a shortening of the contraction time has been observed in hyperthyroid patients (Kissel et al. 1964). Therefore, in the present study, the greater running economy during the phases of OC usage seemed more likely to be related to biomechanical than to metabolic factors. Further studies are needed to assess these suggestions more accurately.

In conclusion, the present study demonstrated that in users of low-dose combined contraceptive pills the oxygen uptake was decreased during the period of exogenous hormone supplementation compared to the period off oral contraceptive use for a given submaximal exercise intensity. A greater running economy could be expected in these women during the oral contraceptive usage phase possibly through biomechanical factors.

References

AHMED-SCROUR, H. and BAILEY, C. J. 1981, Role of ovarian hormones in the long-term control of glucose homeostasis, glycogen formation and gluconeogenesis, *Annals of Nutrition and Metabolism*, **25**, 208–212.

ASMUSSEN, E. and BONDE-PETERSEN, F. 1974, Apparent efficiency and storage of elastic energy in human muscles during exercise, *Acta Physiologica Scandinavica*, **92**, 537–545.

ÅSTRAND, P. O. and RYHMING, I. 1954, A nomogram for calculation of aerobic capacity (physical fitness) from pulse rate during submaximal work, *Journal of Applied Physiology*, **7**, 218–221.

BEMBEN, D. A., BOILEAU, R. A., BAHR, J. M., NELSON, R. A. and MISNER, J. E. 1992, Effects of oral contraceptives on hormonal and metabolic responses during exercise, *Medicine and Science in Sports and Exercise*, **24**, 434–441.

BONEN, A., HAYNES, F. J. and GRAHAM, T. E. 1991, Substrate and hormonal responses to exercise in women using oral contraceptives, *Journal of Applied Physiology*, **70**, 1917–1927.

BUNT, J. C. 1990, Metabolic actions of estradiol: significance for acute and chronic exercise responses, *Medicine and Science in Sports and Exercise*, **22**, 286–290.

DE SOUZA, M. J., MAGUIRE, M. S., RUBIN, K. R. and MARESH, C. M. 1990, Effects of menstrual phase and amenorrhea on exercise performance in runners, *Medicine and Science in Sports and Exercise*, **22**, 575–580.

DOMBOVY, M. L., BONEKAT, H. W., WILLIAMS, T. J and STAATS, B. A. 1987, Exercise performance and ventilatory response in the menstrual cycle, *Medicine and Science in Sports and Exercise*, **19**, 111–117.

DURNING, J. V. G. A. and RAHAMAN, M. M. 1967, The assessment of the amount of fat in the human body from measurements of skinfold thickness, *British Journal of Nutrition*, **21**, 681–689.

ESTON, R. G. and BURKE, E. J. 1984, Effects of the menstrual cycle on selected responses to short constant-load exercise, *Journal of Sports Sciences*, **2**, 145–153.

HACKNEY, A. C. 1990, Effects of the menstrual cycle on resting muscle glycogen content, *Hormone and Metabolism Research*, **22**, 647.

HEINTZ, B., SCHMAUDER, C., WITTE, K., BREUER, I., BALTZER, K., SIEBERTH, H. G. and LEMMER, B. 1996, Blood pressure rhythm and endocrine functions in normotensive women on oral contraceptives, *Journal of Hypertension*, **14**, 333–339.

HESSEMER, V and BRÜCK, K. 1985, Influence of menstrual cycle on thermoregulatory, metabolic, and heart rate responses to exercise at night, *Journal of Applied Physiology*, **59**, 1911–1917.

JURKOWSKI, J. E. H., JONES, N. L., TOEWS, C. J. and SUTTON, J. R. 1981, Effects of menstrual cycle on blood lactate, O_2 delivery, and performance during exercise, *Journal of Applied Physiology*, **51**, 1493–1499.

KISSEL, P., HARTEMANN, P., DUC, M. and DUC, M. L. 1964, Le réflexogramme achilléen dans les dysthyroïdies et dans les troubles du métabolisme électrolytique, *La Presse Médicale*, **72**, 2201–2204.

KLUFT, C. and LANSINK, M. 1997, Effect of oral contraceptives on haemostasis variables, *Thrombosis and Haemostasis*, **78**, 315–326.

KOMI, P. V. 1984, Physiological and biomechanical correlates of muscle function: effects of muscle structure and stretch-shortening cycle on force and speed, in R. L. Terjung (ed.), *Exercise and Sport Sciences Reviews* (Lexington: The Collamore Press), 81–121.

KUHL, H., GAHN, G., ROMBERG, G., ALTHOFF, P. H. and TAUBERT, H. D. 1985, A randomized cross-over comparison of two low-dose oral contraceptives upon hormonal and metabolic serum parameters: II. Effects upon thyroid function, gastrin, STH, and glucose tolerance, *Contraception*, **32**, 97–107.

LEBRUN, C. M. 1993, Effect of the different phases of the menstrual cycle and oral contraceptives on athletic performance, *Sports Medicine*, **16**, 400–430.

LEBRUN, C. M., MCKENZIE, D. C., PRIOR, J. C. and TAUNTON, J. E. 1995, Effects of menstrual cycle phase on athletic performance, *Medicine and Science in Sports and Exercise*, **27**, 437–444.

LEHTOVIRTA, P., KUIKKA, J. and PYÖRÄLÄ, T. 1977, Hemodynamic effects of oral contraceptives during exercise, *International Journal of Gynaecology and Obstetrics*, **15**, 35–37.

LINDER, C. W., DURANT, R. H., JAY, S. and BRYANT-PITTS, N. 1989, The influence of oral contraceptives and habitual physical activity on serum lipids in black adolescents and young women, *Journal of Adolescent Health Care*, **10**, 275–282.

LITTLER, W. A., BOJORGES-BUENO, R. and BANKS, J. 1974, Cardiovascular dynamics in women during the menstrual cycle and oral contraceptive therapy, *Thorax*, **29**, 567–570.

LIU, S. H., AL-SHAIKH, R. A. PANOSSIAN, V., FINERMAN, G. A. M. and LANE, J. M. 1997, Estrogen affects the cellular metabolism of the anterior cruciate ligament. A potential explanation for female athletic injury, *The American Journal of Sports Medicine*, **25**, 704–709.

LYONS, H. A. and ANTONIO, R. 1959, The sensitivity of the respiratory center in pregnancy and after the administration of progesterone, *Transactions of the Association of American Physicians*, **72**, 173–180.

MCARDLE, W. D., KATCH, F. I. and KATCH, V. L. 1996, *Exercise Physiology. Energy, Nutrition and Human Performance*, 4th ed. (Baltimore: Williams & Wilkins).

MORGAN, D. W., MARTIN, P. E. and KRAHENBUHL, G. S. 1989, Factors affecting running economy, *Sports Medicine*, **7**, 310–330.

PIVARNIK, J. M., MARICHAL, C. J., SPILLMAN, T. and MORROW, J. R. 1992, Menstrual cycle phase affects temperature regulation during endurance exercise, *Journal of Applied Physiology*, **72**, 543–548.

RAHKILA, P. and LAATIKAINEN, T. 1992, Effect of oral contraceptives on plasma β-endorphin and corticotropin at rest and during exercise, *Gynecological Endocrinology*, **6**, 163–166.

REILLY, T. and WHITLEY, H. 1994, Effects of menstrual cycle phase and oral contraceptive use on endurance exercise, *Journal of Sports Sciences*, **2**, 150.

REINBERG, A. E., TOUITOU, Y., SOUDANT, E., BERNARD, D., BAZIN, R. and MECHKOURI, M. 1996, Oral contraceptives alter circadian rhythm parameters of cortisol, melatonin, blood pressure, heart rate, skin blood flow, transepidermal water loss, and skin amino acids of healthy young women, *Chronobiology International*, **13**, 199–211.

ROGERS, S. M. and BAKER, M. A. 1997, Thermoregulation during exercise in women who are taking oral contraceptives, *European Journal of Applied Physiology & Occupational Physiology*, **75**, 34–38.

SCHOENE, R. B., ROBERTSON, H. T., PIERSON, D. J. and PETERSON, A. P. 1981, Respiratory drives and exercise in menstrual cycles of athletic and nonathletic women, *Journal of Applied Physiology*, **50**, 1300–1305.

VINCENT, W. J. 1995, *Statistics in Kinesiology* (Champaign, IL: Human Kinetics).
WALDEN, C. E., KNOPP, R. H., JOHNSON, J. L., HEISS, G., WAHL, P. W. and HOOVER, J. J. 1986, Effect of estrogen/progestin potency on clinical chemistry measures. The lipid research clinics program prevalence study, *American Journal of Epidemiology*, **123**, 517–531.
WILLIAMS, K. R. and CAVANAGH, P. R. 1987, Relationship between distance running mechanics, running economy, and performance, *Journal of Applied Physiology*, **63**, 1–9.
WILLIAMS, T. J. and KRAHENBUHL, G. S. 1997, Menstrual cycle phase and running economy, *Medicine and Science in Sports and Exercise*, **29**, 1609–1618.
ZWILLICH, C. W., NATALINO, M. R., SUTTON, F. D. and WEIL, J. V. 1978, Effects of progesterone on chemosensitivity in normal men, *Journal of Laboratory and Clinical Medicine*, **92**, 262–269.

13 Promoting children's physical activity in primary school: an intervention study using playground markings

GARETH STRATTON

Centre for Physical Education, Sport and Dance, I. M. Marsh Campus, Liverpool John Moores University, Barkhill Road, Liverpool L17 6BD, UK

Keywords: Children; Exercise; Heart rate; Playground markings; School playtime.

The physical activity levels of 47, 5- to 7-year-old children were assessed before and after a school playground was painted with fluorescent markings. Children's physical activity was measured using heart rate telemetry during three playtimes before and after the markings were laid down. Children in the experimental and control groups spent 27 and 29 min, respectively, in moderate to vigorous physical activity (MVPA) before the intervention, increasing to 45 and 36 min, respectively, during the intervention period. MVPA, vigorous physical activity (VPA) and mean heart rate remained relatively stable in the control group compared to respective increases of 10 and 5% of playtime and 6 beats min^{-1} in the experimental group during the intervention period. The ANCOVA analysis revealed significant interactions and main effects for the intervention for MVPA, VPA and mean heart rate. Conversely there were no main effect differences between groups. These results suggest that while playground markings had a significant and positive influence on children's physical activity, factors other than playground markings may also influence children's physically active play.

1. Introduction

Children's physical inactivity has been deemed to be a modifiable risk factor for lifestyle-related diseases such as osteoporosis (Bailey 1995) and coronary heart disease (Berenson 1986). Low levels of physical activity during childhood may compromise the current and future health and well-being of the population (Center for Disease Control and Prevention 1997, Biddle *et al.* 1998). Moderate to vigorous physical activity (MVPA) for 60 min each day has been recommended as the optimal level, compared to 30 min as the minimal level for 5- to 18-year-olds (Center for Disease Control and Prevention 1997, Biddle *et al.* 1998). Some British 6- to 7-year-old schoolchildren failed to meet the optimal criterion (Welsman and Armstrong 1997). Schools have long been recognized as key settings in promoting physical activity recommendations (Iverson *et al.* 1985, Center for Disease Control and Prevention 1997, Biddle *et al.* 1998). Furthermore, within the school day, physical education lessons and playtime represent the two main contexts in which children have the opportunity to be physically active. The advantage of school playtime over physical education is that 'all' children have the opportunity to take 'daily' physical activity and recent evidence suggests that young children are more likely to participate in MVPA during unstructured play environments than in structured settings (Pate *et al.* 1996).

Few investigators have succeeded in implementing sustainable interventions aimed at increasing children's participation in physically active behaviour

(McKenzie et al. 1997). Welk (1999) suggested that factors which 'enable' children to be physically active are essential in a successful health promotion model. With this in mind, Health Promotion Wales (1997) piloted a project where playgrounds were painted with multi-coloured markings aimed at stimulating physically active play. Video evidence revealed that more children were playing on the area of the markings 15 months after compared to before they were first painted. Unfortunately, this study did not include a control group, only the number of children occupying the area of one new marking was assessed, individual behaviour was not recorded, and MVPA was not quantified at any time during the study. Rather than investigate the effects of interventions, other researchers measured the effects of prompts such as encouragement (McKenzie et al. 1997) and play equipment/environment (Whitehurst et al. 1996) on children's physical activity. McKenzie et al. (1997) investigated physical activity levels and prompts (encouragement) in a bi-ethnic sample of 287 children over 2 years. Children in elementary school spent more time walking during playtime than their pre-school counterparts (25% versus 20%, respectively), whereas both groups were 'very active' for approximately 20% of playtime. The investigation by Whitehurst et al. (1996) into physical activity in commercial play centres revealed that the mean heart rate of 5- to 10-year-old children was 158 beats min^{-1}, 20 to 30 beats min^{-1} higher than children's heart rates in school playgrounds (Bradfield et al. 1971, Stratton 1999). These differences may be attributed to the fact that this study took place in a multi-level, 525 m^2 area that was well equipped with stimulating and colourful apparatus.

Up to now scant attention has been given to the contribution that primary school playtime makes to physical activity goals for children (McKenzie et al. 1997). Moreover, there is no controlled study in the extant literature of the effects of an intervention that promotes children's physical activity during school playtime. Enabling factors are central to a health promotion model for children (Welk 1999). In Britain, children of primary school age will experience up to 600 playtimes per year (3 times a day, 5 days per week, 39 weeks per year). Playgrounds that stimulate physical activity also have the potential to promote children's health because of the amount of time that children spend in the playground. The effects on the MVPA of primary school children of painting a school playground with bright and colourful markings were investigated in this study.

2. Methods

2.1. Subjects and settings

Two schools situated in an urban industrialized area in north-west England took part in the study. Sixty children and their parents gave written informed consent to participate in the investigation. The experimental group (school 1) consisted of 18 boys and 18 girls aged 5 to 7 years, randomly selected from a stratified sample of reception, first and second year infant school children. These included 6 boys and 6 girls from each school year. The control group (school 2) consisted of 12 boys and 12 girls (4 from each school year) of similar age, stature and body mass to children in the experimental group. Both schools were located in the same geographical area and had similar playground space (30 × 30 m for the experimental group; 40 × 20 m for the control group). Neither school had playground markings at the beginning of the study. During morning and afternoon playtimes approximately 200 children occupied the playground. This varied greatly during the lunch break when numbers in the playground were between 50 and 200.

The experimental period included a phase where children in school 1 ('the experimental school') designed a series of markings that were painted in bright fluorescent colours (Magical Markings, Castleford, England) on the tar macadam playground surface. The experimental school had a surface with 10 markings, each linked to a school curricular theme. A castle, dragon, pirate ship, clock face, flower maze, fun trail and dens, hopscotch, letter squares, snakes and ladders, and a circular maze were evenly spaced throughout the playground area. With the exception of a single football, other play equipment was not allowed in the playground area. School 2 ('the control school') had no playground markings, but allowed limited equipment into the playground.

2.2. *Heart rate monitoring*

There are over 30 available methods for assessing physical activity and the lack of a reference technique presents a persistent problem for physical activity research (Freedson and Melanson 1996). Heart rate radio telemeters have been found to be valid (Treiber *et al.* 1989) and reliable when used as a measure of physical activity in young children throughout the day (Durant *et al.* 1992) and during playtime (Whitehurst *et al.* 1996). Heart rate also has a significant relationship with energy expenditure and has been widely used in studies of physical activity in children (Eston *et al.* 1997). In this study, short-range radio telemeters (Electro-Polar, Kempele, Finland) were attached to the children by fitting a lightweight chest strap (transmitter) and wristwatch (receiver). Heart rates during sleep were recorded minute by minute over one night. The three lowest consecutive heart rates were used as baseline. Using 200 beats min^{-1} as maximum heart rate (Rowland 1996) individual heart rate reserves (HRR) at 50, 60 and 75% thresholds were calculated for each individual. These heart rate reserve thresholds represent moderate (MPA), moderate to vigorous (MVPA) and vigorous physical activity (VPA), respectively (Stratton 1996). Heart rates were recorded once every 5 s during a morning, lunch and afternoon playtime before and after marking the playground. Telemeters started recording just prior to the children entering the playground and were stopped when children returned from the playground. The recording time on the telemeter was used as a measure of playtime duration. The percentages of playtime and actual time in minutes spent in heart rate zones and the overall mean heart rate were calculated for each individual. The sum of HRR50, HRR60 and HRR75 represented the total time in MVPA. The measurement of physical activity lasted for 4 weeks (20 weekdays) before and 4 weeks (20 weekdays) immediately after the playground was painted. Data were collected only from safe and dry playgrounds.

Three children from each school (1 from each year group) were monitored on the same day. This procedure was repeated more or less every weekday (depending upon the weather) for 20 days. Of the initial 36 children in the experimental school, 14 boys (4 from reception, 5 from year 1 and 5 from year 2) and 13 girls (4 from reception, 4 from year 1 and 5 from year 2) had complete data-sets for further analysis. Children in the control school followed the same procedure. Of the initial 24 children in the control school, 20 complete sets of data were collected. These data included the heart rates of 9 boys (3 from reception, 2 from year 1 and 4 from year 2) and 11 girls (3 from reception, 4 from year 1 and 4 from year 2).

2.3. Data analysis

The data for girls and boys were combined to increase statistical power. Statistics were subsequently computed for group (experimental and control school) and intervention (before and after). Moderate to vigorous physical activity, mean heart rate and HRR75 were included for analysis because of their relationship with health, data reported in other studies, and exercise intensity that has been reported to increase aerobic fitness (Payne and Morrow 1993), respectively. A series of 2×2 ANOVA (group × intervention) analyses were computed for each dependent variable as well as the duration of playtime. Subsequently, data were analysed using ANCOVA to adjust for changes in the duration of playtime (SPSS-PC). Alpha was set at $p \leqslant 0.05$.

3. Results

Body mass and stature (table 1) were within the normal range for children of this age. These values changed little over the whole of the programme.

3.1. Calculating heart rate thresholds

Heart rates during sleep were slightly lower for girls than for boys. The mean ($\pm$SD) heart rates for boys in the experimental and control schools were 72 (± 6) and 71 (± 6) beats min^{-1}, respectively. These were not different from the values for girls in the experimental and control schools, which were 69 (± 5) and 68 (± 6) beats min^{-1}, respectively. Heart rate thresholds for moderate (HRR50), moderate to vigorous (HRR60) and vigorous (HRR75) physical activity were 134 and 136, 147 and 149, and 167 and 168 beats min^{-1} for boys and girls, respectively. These results equated favourably with heart rates for easy to moderate, moderate to fast, and fast or hard translocation in children of a similar age (Puhl et al. 1990).

3.2. Heart rate

3.2.1. MVPA (>HRR50):
Before the intervention, children in the experimental school spent approximately 5% less time in MVPA than children in the control school. Differences in MVPA between schools increased after the intervention period (table 2). The ANOVA analysis revealed a significant interaction for MVPA ($F(1,278) = 9.71$; $p \leqslant 0.01$) as well as a significant main effect for intervention ($F(1,278) = 5.73$; $p \leqslant 0.05$) but not for groups ($F(1,278) = 0.18$; $p > 0.05$).

After adjusting for duration of playtime ANCOVA analysis revealed a significant interaction for MVPA after the intervention ($F(1,272) = 9.39$; $p \leqslant 0.01$) and the main effect of the intervention ($F(1,272) = 5.06$; $p \leqslant 0.05$). There was no significant main effect difference between groups ($F(1,272) = 0.33$; $p > 0.05$).

Table 1. Mean (SD) body mass and stature of experimental and control groups.

	Experimental group ($n = 27$)		Control group ($n = 20$)	
	Before intervention	After intervention	Before intervention	After intervention
Body mass (kg)	22.9 (4.2)	23.3 (4.3)	22.4 (3.8)	22.9 (4.1)
Stature (m)	1.18 (0.08)	1.18 (0.08)	1.17 (0.07)	1.18 (0.08)

Table 2. Mean (SD) heart rate, physical activity and duration of play in experimental and control groups.

	Experimental group ($n=27$)		Control group ($n=20$)	
	Before intervention	After intervention	Before intervention	After intervention
Heart rate				
Mean (beats min^{-1})	126.1 (9.5)	132.8 (11.6)	128.9 (9.5)	129.4 (11.4)
MVPA (HRR50) (% playtime)	35.1 (17.7)	46.2 (18.1)	40.5 (14.8)	39.1 (15.8)
VPA (HRR75) (% playtime)	5.3 (6.8)	10.0 (10.8)	7.0 (7.8)	6.8 (8.3)
Play time				
Duration (min)	25.3 (12.08)	32.0 (9.82)	23.4 (11.19)	29.8 (8.95)

3.2.2. *Vigorous physical activity (>HRR75)*: Vigorous physical activity almost doubled in the experimental group compared to a very small decrease in the control group. Analysis of variance revealed significant interaction for HRR75 ($F_{(1,278)} = 6.0$; $p \leqslant 0.01$). Main effects for HRR75 revealed significant differences before and after the intervention ($F_{(1,278)} = 4.92$; $p \leqslant 0.05$), but not between groups ($F_{(1,278)} = 0.50$; $p > 0.05$). The ANCOVA analysis revealed a significant interaction for HRR75 after the intervention ($F_{(1,272)} = 6.28$; $p \leqslant 0.01$) but not for the main effects of intervention ($F_{(1,272)} = 2.85$; $p < 0.05$) or group ($F_{(1,272)} = 0.48$; $p < 0.05$).

3.2.3. *Mean heart rate*: The mean heart rate of the experimental group increased by 7 beats min^{-1} during the intervention compared to little change in the control group. There was a significant interaction ($F(1,278) = 6.00$; $p \leqslant 0.01$) and a significant main effect for the intervention ($F(1,278 = 8.09$; $p \leqslant 0.01$). There was no significant main effect for groups ($F(1,278) = 0.05$; $p > 0.05$). After adjusting for duration of playtime by means of ANCOVA, a significant interaction for mean heart rate ($F(1,272) = 5.69$; $p \leqslant 0.05$) and a main effect for intervention were evident ($F(1,272) = 6.11$; $p \leqslant 0.01$). There was no significant main effect between groups ($F(1,272) = 0.05$; $p > 0.05$).

3.2.4. *Duration of playtime*: The duration of play during the intervention phase exceeded that of the pre-intervention phase in the control and experimental school by almost 7 min and 6 min, respectively. After analysis, a significant effect for duration of playtime was evident (group × intervention; $F(1,272) = 9.51$; $p \leqslant 0.01$), with significant main effect differences in play duration before and after the intervention ($F(1,272) = 25.67$; $p \leqslant 0.01$). There were no significant main effect differences in play duration between schools ($F(1,272) = 2.4$; $p > 0.05$).

Prior to the intervention, children in the experimental and control groups, respectively, spent a mean of 8.9 and 9.5 min in MVPA during each play period. This equated to approximately 27–29 min of playtime spent in MVPA per school day. During playtime after intervention, children in the experimental and control groups increased the amount of time spent in MVPA by a mean of 6 min and 2 min, respectively. During the intervention phase children in the experimental group spent nearly 45 min of playtime per day in MVPA. This equated to an increase of 18 min per day compared to pre-intervention values.

4. Discussion

The aim of this study was to measure the effect of painting markings in the playground on the physical activity of 5- to 7-year-old primary school children. Mean heart rates for the boys in this study were 6 and 10 beats min^{-1} higher in the experimental and control groups before the intervention than those reported by Bradfield et al. (1971). These differences increased to 13 beats min^{-1} during the intervention period. Furthermore, even though there was a significant increase in mean heart rate (to 133 beats min^{-1}) during the intervention period, values were substantially lower than the mean heart rate of 158 beats min^{-1} reported for children in commercial play centres (Whitehurst et al. 1996). The investigation by Whitehurst et al. (1996) highlights the potential for further increases in children's mean heart rates during playtimes given the appropriate environment and enabling factors.

With some exceptions (Mota and Stratton 1999, Stratton 1999), there are few studies that have reported MVPA using heart rate zones during play. Stratton (1999) revealed that girls and boys (aged 10–11 years) spent between 15 and 30%, respectively, of playtime in MVPA. Children in this study were more active and engaged in excess of 35 and 40% of MVPA during playtime before the intervention compared to 45 and 40% after the intervention. The significant interaction and main effect for intervention on MVPA resulted from marking the playground and this effect remained when data were adjusted for the duration of playtime. The absence of a main effect difference between groups suggested that factors other than playground markings were responsible for higher levels of physical activity in the control school before the intervention. These factors may be attributed to the availability of skipping ropes and balls allowed in the playground of the control school. Some investigators have found that children engaged in physical activity for approximately 60% of playtime (Hovell et al. 1978, Kraft 1989) or 40 to 50% in pre-school and elementary school (McKenzie et al. 1997). Kraft's observation of behaviour revealed children engaged in minimal, moderate and vigorous activity for 18.5, 21.5 and 17.4% of playtime, respectively. In comparison with Kraft's investigation (1989), children in the control group engaged in similar amounts of MVPA during the intervention whereas children in the experimental group engaged in 4% less MVPA before the intervention and 7% more MVPA during the intervention period.

Vigorous physical activity at 75%HRR is thought to increase cardiorespiratory fitness in pre-pubertal children (Payne and Morrow 1993). The significant interaction observed supported the positive influence of the playground markings on the percentage of time spent in VPA in the experimental school. However, ANCOVA results revealed a significant interaction but no significant main effects for group or intervention. The significant interaction suggests that the playground markings stimulated greater levels of VPA in children in the experimental group when compared to a small decrease in the control group. During the intervention vigorous physical activity increased two-fold in the experimental group and was similar to data revealed by Parcel et al. (1987). However, during the intervention phase children in the experimental school engaged in 3 min of VPA, only half that reported by Kraft (1989).

Increases in MVPA caused by the intervention were also magnified by an apparent Hawthorn effect on the duration of playtime. There were no requests made to the schools to increase the duration of playtime but both schools significantly increased the time children spent in the playground. There were no significant differences between schools in the number of minutes given to play. The increase in duration of playtime may be due to awareness within the schools of involvement in a project whose aim was to measure physical activity. This in turn may have prompted a slow but significant increase in the duration of playtime. The increase in play duration enhanced the increase in children's MVPA in both experimental and control schools. The ANCOVA results indicated that the increase in length of playtime did not significantly change the effect of the intervention programme on children's levels of MVPA. Whereas the percentage of playtime that children spent in MVPA is the key variable to consider when assessing the effects of the playground markings independent of time, a measure of time is essential when quantifying the contribution that playtime makes to physical activity guidelines (Biddle *et al.* 1998). Before the intervention the mean time spent in MVPA ranged between 27 and 29 min, slightly short of the 30 min recommended as the minimum amount of daily physical activity for children of this age. After the intervention, this duration increased to approximately 45 min in the experimental group and 35 min in the control group. The change in the time spent in MVPA during the intervention was almost 70% in the experimental group compared to 22% in the control group. The time spent in MVPA may be inflated because of a possible Hawthorn effect. Nevertheless, evidence suggests that children's engagement in MVPA during playtime can make a substantial contribution to the 60 min of daily MVPA recommended for children. Moreover, engagement in MVPA is probably even more significant for all children during the winter months and for some children who are transported to and from school (Sallo *et al.* 1997).

5. Conclusion

The aim in this study was to use a sustainable physical activity intervention (playground markings) to increase MVPA in 5- to 7-year-old schoolchildren. To this end, the aim was realized by increasing children's mean heart rate and levels of MVPA and VPA. The significant interaction for all three variables also suggests that in schools where physical activity levels may be low, playground markings can have a positive influence. Alternatively there were no main effect differences between groups suggesting that factors other than playground markings stimulated physical activity in the control school. Furthermore, during the first weeks of the intervention, there may have been a novelty effect that disappears with time. With these issues in mind it is recommended that future study designs attempt to match schools for physical activity levels and include medium- and long-term follow-up periods.

There may be further potential for increasing physical activity during school playtime as mean heart rates reported in this study fell short of those achieved by children in commercial play centres. This study was limited by its sample size of only 47 children from two schools and while initial results were promising, further investigations are necessary into the effects of painting school playgrounds with brightly coloured markings on children's physical activity.

Acknowledgements

The author would like to thank the two schools and all the children that took part in the study. The author would also like to thank Janine Leonard and Sue Byers for their assistance with data collection, and Jean Phillis at Magical Markings for marking the playground.

References

BAILEY, D. A. 1995, The role of mechanical loading in the regulation of skeletal development during growth, in C. J. R. Blimkie and O. Bar-Or (eds), *New Horizons in Pediatric Exercise Science* (Champaign, IL: Human Kinetics), 97–108.

BERENSON, G. S. 1986, Evolution of cardiovascular risk factors in early life. Perspectives on causation, in G. S. Berenson (ed.), *Causation of Cardiovascular Risk Factors in Children* (New York: Raven Press), 1–26.

BIDDLE, S. J. H, SALLIS, J. and CAVILL, N. 1998, *Young and Active; Physical Activity Guidelines for Young People in the UK* (London: Health Education Authority).

BRADFIELD, R. B., CHAN, H., BRADFIELD, N. E. and PAYNE, P. R. 1971, Energy expenditure and heart rates of Cambridge boys at school, *American Journal of Clinical Nutrition*, **244**, 1461–1466.

CENTER FOR DISEASE CONTROL AND PREVENTION 1997, Guidelines for school and community programs to promote lifelong physical activity among young people, *Morbidity and Mortality Weekly Report*, **46** (No. RR-6).

DURANT, R. H., BARANOWSKI, T., DAVIS, H., THOMPSON, W. O, PUHL, J., GREAVES, K. A and RHODES, T. 1992, Reliability and variability of heart rate monitoring in 3 year old, 4 year old, or 5 year old children, *Medicine and Science in Sports and Exercise*, **24**, 265–271.

ESTON, R. G., ROWLANDS, A. V. and INGLEDEW, D. K. 1997, Validation of the Tritrac-R3D activity monitor during typical children's activities, in N. Armstrong, B. Kirby and J. Welsman (eds), *Children and Exercise XIX: Promoting Health and Well Being* (London: E. & F. N. Spon), 132–138.

FREEDSON, P. S. and MELANSON, E. L. 1996, Measuring physical activity, in D. Docherty (ed.), *Measurement in Pediatric Exercise Science* (Champaign, IL: Human Kinetics), 261–281.

HEALTH PROMOTION WALES 1997, *The Health Promoting Playground Project: Evaluation Findings*, Research Summary, Issue 1, Cardiff.

HOVELL, M. F., BURSICK, J. H., SHARKEY, R. and MCCLURE, J. 1978, Evaluation of elementary students' voluntary physical activity during recess, *Research Quarterly for Exercise and Sport*, **49**, 460–474.

IVERSON, D. C, FIEILDING, J. E., CROW, R. S. and CHRISTENSO, G. M. 1985, The promotion of physical activity in the United States population: the status of programs in medical, worksite, community, and school settings, *Public Health Reports*, **100**(2), 212–224.

KRAFT, R. E. 1989, Children at play; behaviour of children at recess, Journal of Physical Education, *Recreation and Dance*, **60**, 21–24.

MCKENZIE, T. L., SALLIS, J. F., ELDER, J. P., BERRY, C. C., HOY, P. L., NADER, P. R., ZIVE, M. M. and BROYLES, S. L. 1997, Physical activity levels and prompts in young children at recess: a two-year study of a bi-ethnic sample, *Research Quarterly for Exercise and Sport*, **68**, 195–202.

MOTA, J and STRATTON, G. 1999, Individual differences in physical activity during primary school recess: a preliminary investigation on Portuguese and English children, *Pediatric Exercise Science*, **11**, 297–298.

PARCEL, G. S., SIMONS-MORTON, B. G., O'HARA, N. M., BARANOWSKI, T., KOLBE, L. J., and BEE, D. E. 1987, School promotion of healthful diet and exercise behaviour: an integration of organisational change and social learning theory interventions, *Journal of School Health*, **57**, 150–156.

PATE, R. R., BARANOWSKI, T., DOWDA, M. and TROST, S. G. 1996, Tracking of physical activity in young children, *Medicine and Science in Sports and Exercise*, **28**, 92–96.

PAYNE, V. G. and MORROW, J. R. 1993, The effect of physical training on prepubescent VO2max; a meta-analysis, *Research Quarterly for Exercise and Sport*, **64**, 305–313.

PUHL, J., GREAVES, K., HOYT, M. and BARANOWSKI, T. 1990, Children's activity rating scale: description and calibration, *Research Quarterly for Exercise and Sport*, **60**, 42–47.

ROWLAND, X. 1996, Personal communication.

SALLO, M., HARRO, J. and VIRU, A. 1997, Moderate to vigorous physical activities in preadolescent school children, in N. Armstrong, B. Kirby and J. Welsman (eds), *Children and Exercise XIX: Promoting Health and Well Being* (London: E. & F. N. Spon), 151–156.

STRATTON, G. 1996, Children's heart rates during physical education lessons: a review, *Pediatric Exercise Science*, **6**, 215–233.

STRATTON, G. 1999, A preliminary study of children's physical activity in one urban primary school playground: differences by sex and season, *Journal of Sport Pedagogy*, **5**, 71–81.

TREIBER, F. A., MUSANTE, L., HARTDAGAN, S., DAVIS, H., LEVY, M. and STRONG, W. B. 1989, Validation of a heart rate monitor with children in laboratory and field settings, *Medicine and Science in Sports and Exercise*, **21**, 338–342.

WELK, G. J. 1999, The youth physical activity promotion model: a conceptual bridge between theory and practice, *Quest*, **51**, 5–23.

WELSMAN, J. R. and ARMSTRONG, N. 1997, Physical activity patterns of 5 to 11 year old children, in N. Armstrong, B. Kirby and J. Welsman (eds), *Children and Exercise XIX: Promoting Health and Well Being* (London: E. & F. N. Spon), 139–144.

WHITEHURST, M., GROO, D. R. and BROWN, L. E. 1996, Prepubescent heart rate response to indoor play, *Pediatric Exercise Science*, **8**, 245–250.

14 Exercise testing in children: an alternative approach

D. B. CLAXTON†*, J. H. CHAPMAN†, N. V. CHALLIS‡ and M. L. FYSH†

†Sports Science Research Institute and ‡School of Science and Mathematics, Sheffield Hallam University, Collegiate Crescent Campus, Sheffield S10 2BP, UK

Keywords: Exercise testing; Children; Oxygen uptake kinetics; PRBS; Frequency domain analysis.

In recent years there has been a call for new methods of evaluating the cardiorespiratory responses of children to exercise that complement their everyday exercise patterns. One potential method would be to use a sub-maximal, intermittent, pseudo-random binary sequence (PRBS) exercise test protocol to measure oxygen uptake kinetics ($\dot{V}O_{2\ kinetics}$). Ten children of mean (SD) age 10.8 (± 1.5) years completed a 20–50 W cycle ergometer protocol of 17-min duration. An estimate of alveolar oxygen uptake ($\dot{V}O_2$) was calculated on a breath-by-breath basis. The $\dot{V}O_2$ kinetic parameters were expressed in the frequency domain as amplitude ratio and phase delay using standard Fourier techniques. Analysis was restricted to the frequency range 2.2 to 8.9 mHz. The mean (SD) amplitude ratio responses decreased from 10.33 (± 0.73) to 7.42 (± 0.99) ml min^{-1} W^{-1} and the mean phase delay increased from $-26.78°$ ($\pm 6.37°$) to $-81.93°$ ($\pm 10.45°$) over the frequency range 2.2–8.9 mHz. Significant correlations ($p<0.05$) were found between chronological age and amplitude ratio ($r=0.68$ and 0.62), and chronological age and phase delay ($r=-0.62$ and -0.69) at the frequencies of 2.2 and 4.4 mHz, respectively. No significant correlations were found between $\dot{V}O_{2\ kinetics}$ and stature or $\dot{V}O_{2\ kinetics}$ and body mass. The observations demonstrated the use of the PRBS technique to measure $\dot{V}O_{2\ kinetics}$ in the frequency domain in children. This approach may be a useful addition to the tests that are used to quantify the oxygen uptake responses to exercise in children.

1. Introduction

Since the work of Robinson (1938), significant progress has been made towards the understanding of how the processes of growth and maturation affect both the chronic and acute responses to exercise. Many studies have focused on the issues of establishing suitable criteria for determining maximum oxygen uptake ($\dot{V}O_{2\ max}$) (Rivera-Brown *et al*. 1992, Duncan *et al*. 1996), on establishing non-invasive indices for the determination of the anaerobic threshold (Washington 1993, Chicharro *et al*. 1995, Pfitzinger and Freedson 1997) and finally on scaling these parameters to make appropriate comparisons between individuals of different body size (Rowland 1998). There has yet to be a consensus on any of these issues. Traditional methods of investigation involve protocols that may not provide the most appropriate exercise stimuli for assessing the child's physiological responses to exercise (Armstrong and Welsman 1994, Cooper 1995). In recent years there has been a call for paediatric researchers not to be fixated with the approach to exercise testing routinely employed

---

*Author for correspondence.

for adult populations (Rowland 1996), but to search out ways to look more realistically at how children exercise in real life (Cooper 1995).

A useful way of quantifying cardiorespiratory responses to exercise is to measure the rate of change of oxygen uptake to exercise transitions ($\dot{V}O_2$ kinetics). As intermittent activity characterizes the exercise patterns of children (Cooper 1994), the measurement of $\dot{V}O_2$ kinetics using dynamic exercise perturbations such as the pseudo-random binary sequence (PRBS) test (Bennett et al. 1981, Eßfeld et al. 1987) may provide a more appropriate measure of the child's acute response to exercise. The PRBS exercise test provides a description of $\dot{V}O_2$ kinetics for a range of sinusoidal input signals in a single sub-maximal assessment (Hoffmann et al. 1994b). The rationale behind the PRBS exercise test is the principle of superposition. That is, in a linear system, the response to the PRBS signal may be considered as a sum of its individual harmonic (sinusoidal) components.

Using Fourier analysis, the $\dot{V}O_2$ response is transformed from the time domain into the frequency domain, essentially breaking down the input signal (work-rate) and the output signal ($\dot{V}O_2$) into their constituent harmonic components. The relationship between the work-rate input and the $\dot{V}O_2$ response, expressed in terms of amplitude ratio and phase delay, provides a description of frequency domain $\dot{V}O_2$ kinetics. These terms are calculated using the variables described in figure 1 and equations 1 and 2.

$$\text{Amplitude ratio (ml min}^{-1}\text{W}^{-1}) = \frac{\dot{V}O_2 \text{ amplitude}}{\text{Work-rate amplitude}} \qquad (1)$$

$$\text{Phase delay (}°\text{)} = \text{Work-rate phase angle} - \dot{V}O_2 \text{ phase angle} \qquad (2)$$

If the $\dot{V}O_2$ responses to the work-rate transitions are below the ventilatory threshold (T_{vent}), the same absolute exercise intensities can be used for all

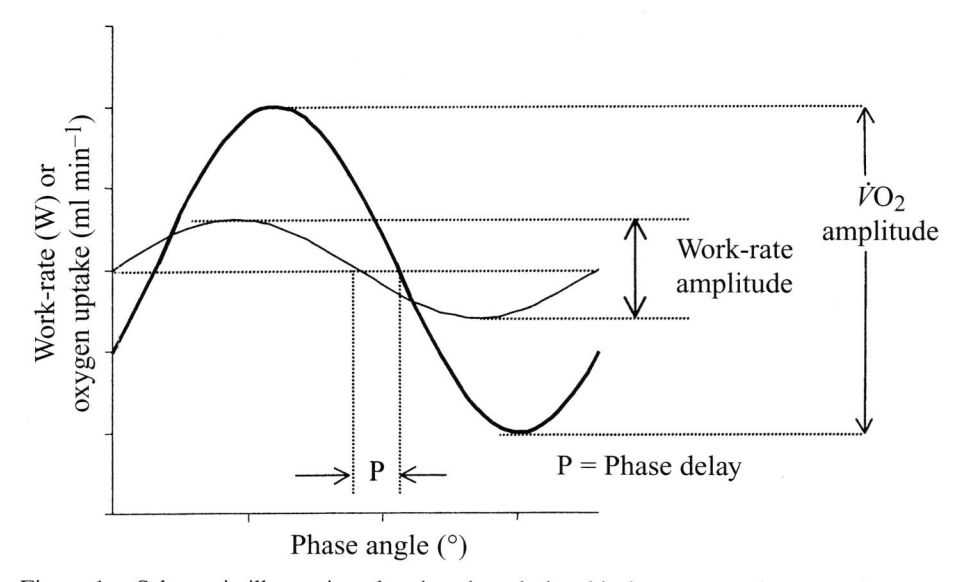

Figure 1. Schematic illustration showing the relationship between work-rate and oxygen uptake after Fourier analysis.

subjects. This is because in the aerobic range the rate of change in $\dot{V}O_2$ is independent of the magnitude of the work-rate change (Whipp and Ward 1990, Hoffmann et al. 1994b). The PRBS protocol is a simple, sub-maximal exercise test that has been shown to be well tolerated by children (Kusenbach et al. 1994) and to provide a reliable measurement of $\dot{V}O_{2\ kinetics}$ (Hughson et al. 1990, Hoffmann et al. 1994a).

The aim of this study was to measure $\dot{V}O_{2\ kinetics}$ in children using a PRBS exercise test in which the upper work-rate was set using a prediction based on stature. In a separate exercise test, T_{vent} was determined in order to verify that the predicted work-rate was below the T_{vent}. Maximal aerobic power ($\dot{V}O_{2\ max}$) was also measured in order to characterize the group. Subsequently, preliminary analyses were carried out to identify any relationships between chronological age or body size and $\dot{V}O_{2\ kinetics}$.

2. Methods

2.1. Subjects

Ten children (7 males and 3 females) were drawn from a group of 20 children who had been selected as control subjects for a previously published evaluation (Marven et al. 1998). Sexual maturity status was established by a qualified auxologist using the criteria of Tanner (1962). The test population consisted of 7 subjects at pubertal stage 1, two subjects at stage 2 and one subject at stage 3. The Tanner stage was generated using the mean score for pubic hair rating and genitalia (boys) or breast rating (girls). The subject profiles are shown in table 1.

The study was approved by the South Sheffield Ethics Committee. Informed consent was obtained from the children and their parent/guardian. At all stages of exercise testing a paediatrician and a parent/guardian or their representative was present.

2.2. Exercise protocols

Following a familiarization period, the subjects completed two exercise tests, a PRBS test followed by a $\dot{V}O_{2\ max}$ test on an electrically braked cycle ergometer (550 ERG, Bosch, Berlin, Germany). Both tests were carried out at a pedalling frequency of 1 Hz. The cycle ergometer was calibrated using a dynamic calibration rig (Maxwell et al. 1998). The familiarization period consisted of cycling at 1 Hz at both 20 W and 50 W (including work transitions) and also allowed the child to become accustomed to the breathing apparatus.

2.2.1. *Pseudo-random binary sequence exercise test*: The PRBS protocol consisted of two identical 450-s sequences of 15, 30-s units (Eßfeld et al. 1987). The test was preceded by a 2-min warm up consisting of the last 2-min of the PRBS sequence. The PRBS test signal was generated using a four-bit digital shift register with modulo-2

Table 1. Chronological age, stature and mass of the children ($n = 10$).

	Range	Mean	Standard deviation
Age (years)	8.4–13.0	10.8	1.5
Stature (m)	1.30–1.57	1.43	0.08
Mass (kg)	28.2–48.9	35.9	7.2

adder feedback (Kerlin 1974). Throughout the test the work-rate alternated between 20 W and 50 W in the manner described in table 2.

The work-rates were selected in order to maximize the $\dot{V}O_2$ signal without compromising the dynamic linearity of the system (Hoffmann et al. 1994b). The selection of the highest intensity of 50 W was calculated as 50% of the predicted maximal work-rate based on the stature of the smallest subject. The maximal work-rate prediction was calculated as an increase of 2.5 W for every centimetre increase in stature from a baseline of 25 W at a stature of 1.1 m (Godfrey 1974). Previous work undertaken by the present authors (unpublished) indicated that this approach was likely to result in an exercise intensity below the T_{vent} during the PRBS test. This was confirmed *post hoc*.

2.2.2. *The maximal exercise test*: A 15 W min^{-1} continuously increasing, incremental ramp protocol was used for the determination of T_{vent} and $\dot{V}O_{2\ max}$. Stature was used as the criterion for determining the protocol (Godfrey 1974) in the expectation that the maximal work intensity would be attained in approximately 10 min of cycling (Wasserman et al. 1987).

2.3. *Gas analysis*
The inspired and expired gas compositions were measured breath-by-breath using a respiratory mass spectrometer (Marquette MGA 1100, Marquette Electronics Inc., Milwaukee, Wisconsin). The calibration of the mass spectrometer was checked immediately before and after each test with standard calibration gases. Inspired and expired air flows were measured using a low dead space (90 ml) bi-directional turbine (Alpha Technologies VMM-2A, Interface Associates, Laguna Niguel, CA). A 3-l syringe was used to calibrate the volume of turbine using flow rates in the range 1.5–2.0 l s^{-1}. Gas exchange was calculated at the alveolar level with corrections for changes in lung volume and lung gas composition using the algorithm of Beaver et al. (1981). This method of calculating gas exchange corrects for breath-to-breath changes in lung gas stores and requires an estimate of functional residual capacity (FRC). The initial value for FRC was predicted from height, age group and gender (Taylor et al. 1989).

2.4. *Data analyses*
Prior to any analyses, anomalous (non-physiologic) $\dot{V}O_2$ data (± 4 SD greater than the mean breath-to-breath variation) were identified and removed from the breath-by-breath data record.

2.4.1. *Maximum aerobic power*: The breath-by-breath data for $\dot{V}O_2$ and carbon dioxide production ($\dot{V}CO_2$) were averaged on a 30-s basis. Respiratory exchange ratio (RER) was calculated as $\dot{V}CO_2/\dot{V}O_2$. The highest $\dot{V}O_2$ obtained at the termination of the progressive test to exhaustion was accepted as a maximal index if

Table 2. The protocol for a single 450-s pseudo-random binary sequence.

	PRBS protocol							
Duration at each intensity (s)	60	30	90	120	30	30	30	60
Work-rate (W)	20	50	20	50	20	50	20	50

the subject showed signs of intense effort and additionally exhibited one or more of the following: a plateau in $\dot{V}O_2$ with respect to increasing work-rate (Rowland and Cunningham 1992); a cardiac frequency of at least 190 beats min^{-1}; or a RER of at least unity (Rowland et al. 1997).

2.4.2. *Ventilatory threshold*: The breath-by-breath data were averaged every three breaths and T_{vent} was identified from individual graphs of $\dot{V}E/\dot{V}O_2$, $\dot{V}E/\dot{V}CO_2$, and end-tidal O_2 and CO_2 tensions (PETO$_2$ and PETCO$_2$, respectively) plotted against $\dot{V}O_2$. Ventilatory threshold was identified by visually locating the nadir of $\dot{V}E/\dot{V}O_2$ and PETO$_2$ without either a concomitant increase in $\dot{V}E/\dot{V}CO_2$ or a concomitant decrease in PETCO$_2$ (Whipp et al. 1981).

2.4.3. *Pseudo-random binary sequence exercise test*: The breath-to-breath variability of the $\dot{V}O_2$ was minimized by substituting the effective lung volume (ELV) calculated *post hoc* for the estimated FRC in the algorithm of Beaver et al. (1981). Using the estimated FRC as an initial value, an iterative process (Excel Solver, Micosoft Excel 97 SR-1, Microsoft Corporation, Redmond, WA) was used to calculate the ELV which minimizes the sum of the squared differences between each pair of successive $\dot{V}O_2$.

The $\dot{V}O_2$ values were linearly interpolated between breaths to yield values every 1 s. The auto-correlation function (ACF) of the PRBS work-rate input and the cross-correlation function (CCF) of the $\dot{V}O_2$ output with respect to the work-rate input were then calculated (figure 2).

Fourier analysis was performed on the ACF and CCF to provide phase angle and amplitude terms for both work-rate and $\dot{V}O_2$. The relationship between work-rate and $\dot{V}O_2$ was expressed as the amplitude ratio and the phase delay. Only those parameters in the frequency range 2.2 – 8.9 mHz were considered to be suitable for analysis (Hoffmann et al. 1994a).

2.5. Statistical analysis

The relationship between $\dot{V}O_{2\ kinetics}$ and chronological age, and $\dot{V}O_{2\ kinetics}$ and anthropometric variables was assessed using a Pearson's correlation coefficient test. Statistical significance was accepted at $p < 0.05$.

3. Results

3.1. Oxygen uptake kinetics

The frequency domain $\dot{V}O_{2\ kinetic}$ parameters of amplitude ratio and phase delay at the first four harmonics are shown in table 3.

3.2. The maximal exercise test

The mean results from the continuous cycle-ergometer ramp protocol to maximal exertion are displayed in table 4.

The work-rate transitions during the PRBS exercise elicited $\dot{V}O_2$ values that were below the ventilatory thresholds of all the children tested.

3.3. Correlation of oxygen uptake kinetics with chronological age and body mass

Significant correlations ($p < 0.05$) were found between chronological age and amplitude ratio ($r = 0.68$ and 0.62) and chronological age and phase delay ($r = -0.62$ and -0.69) at the frequencies of 2.2 and 4.4 mHz, respectively. No

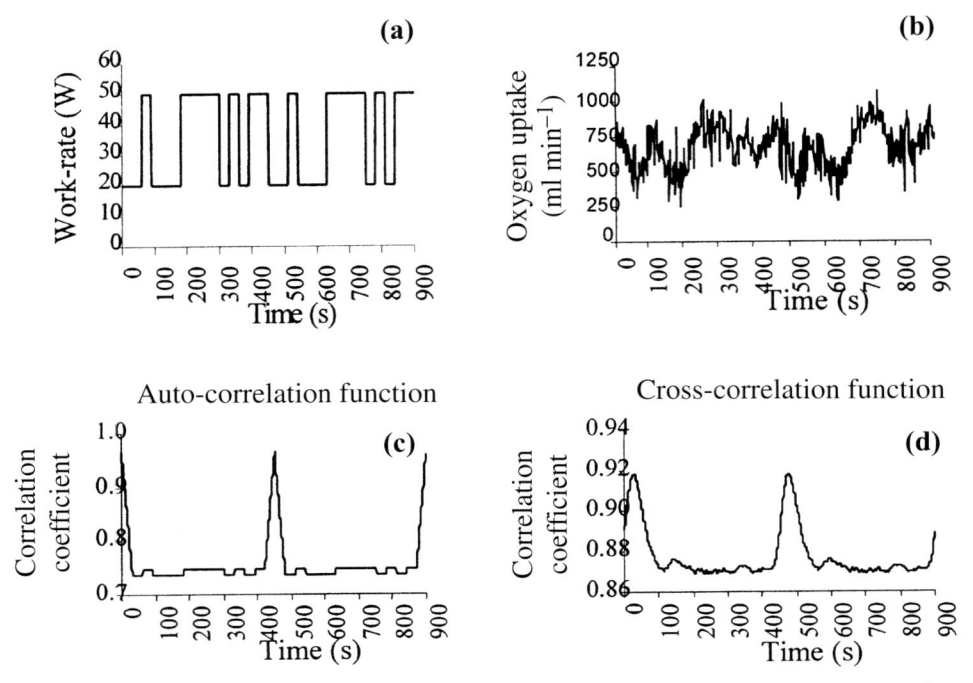

Figure 2. (a) Two consecutive pseudo-random binary sequences (PRBS), (b) the $\dot{V}O_2$ response in a 9-year-old child, (c) the auto-correlation function of the PRBS, (d) the cross-correlation function of the $\dot{V}O_2$ response with respect to the PRBS work-rate.

Table 3. Oxygen uptakes in children ($n=10$) determined by a 15, 30-s unit PRBS exercise test. Values expressed as mean (standard deviation).

	Frequency (mHz)			
	2.2	4.4	6.7	8.9
Amplitude ratio (ml min^{-1} W^{-1})	10.33 (0.73)	9.42 (0.99)	7.46 (1.14)	7.42 (0.99)
Phase delay (°)	−26.78 (6.37)	−46.22 (2.49)	−62.67 (10.60)	−81.93 (10.45)

Table 4. Maximum oxygen uptake ($\dot{V}O_{2\,max}$), ventilatory threshold (T_{vent}), maximum work-rate and test duration in children ($n=10$). Values expressed as mean (standard deviation).

$\dot{V}O_{2\,max}$ (ml min^{-1})	T_{vent} (ml min^{-1})	Maximum Work-rate (W)	Duration of incremental test (min)
1606 (391)	1115 (285)	139 (32)	9.36 (2.20)

significant correlations were found between $\dot{V}O_{2\ kinetics}$ and stature or $\dot{V}O_{2\ kinetics}$ and body mass.

4. Discussion

The speed at which oxygen uptake reaches a steady-state value following the onset of exercise reflects the ability of the circulatory system to deliver and the working muscles to utilize oxygen (Whipp and Wasserman 1972, Tschakovsky and Hughson 1999). The initial exercise stimulus results in an immediate increase in cardiac output, pulmonary blood flow and a consequent abrupt increase in $\dot{V}O_2$ that is described as Phase I of the response. The subsequent Phase II of the response reflects the changing mixed venous blood composition that is a result of increased extraction of O_2 by the working muscles and any residual component of the cardiac output increase that extends beyond Phase I (Whipp 1987). Phase II is characterized by the exponential increase in $\dot{V}O_2$ that terminates in steady state (Phase III). The traditional approaches of fitting mathematical models to describe these dynamic $\dot{V}O_2$ phases have not been validated in children. The use of a PRBS protocol and frequency domain analysis, which does not require explicit modelling, provides an alternative approach to the measurement of Phase II $\dot{V}O_{2\ kinetics}$ (Hughson *et al.* 1990).

In order to keep the work-rate perturbations within the aerobic, dynamically linear range, children can only undertake relatively small work-rate transitions. The problem associated with these small exercise transitions is that the underlying $\dot{V}O_2$ response may be obscured by breath-to-breath noise. If total lung gas exchange is measured (Beaver *et al.* 1973), the calculation of oxygen uptake is sensitive to the naturally occurring irregularities in breathing pattern (Lamarra *et al.* 1987, Potter *et al.* 1999) and changes in pulmonary gas stores (Swanson 1980). This noise has recently been characterized in children as being non-Gaussian and uncorrelated and concern has been raised with regard to the confidence with which the $\dot{V}O_{2\ kinetic}$ parameters can be interpreted after fitting explicit models to characterize the Phase II $\dot{V}O_2$ response in children (Potter *et al.* 1999). Although the effect of noise has not yet been characterized in the frequency domain analysis of $\dot{V}O_{2\ kinetics}$ in either children or adults, the measurement and analytical techniques that the present authors have employed have been shown to reduce the influence of noise (Beaver *et al.* 1981, Swanson and Sherrill 1983, Eßfeld *et al.* 1987). In the study of exercise transients, where breath-to-breath changes contain significant information, it is important to calculate gas exchange using an algorithm that has a low sensitivity to irregular changes in breathing pattern. One such algorithm proposed by Auchincloss *et al.* (1966) and adapted by Beaver *et al.* (1981) has been shown to provide a more precise description of breath-by-breath oxygen uptake in comparison to methods that calculate total lung gas exchange. The algorithm of Beaver *et al.* (1981) provides an estimate of breath-to-breath alveolar gas exchange by accounting for changes in both FRC and alveolar gas concentrations. This approach is considered to provide a reasonable estimate of alveolar gas exchange but only if the FRC reflects the pulmonary gas stores taking part in gas exchange (Swanson and Sherrill 1983). The *post hoc* iterative process to calculate the ELV further minimized the $\dot{V}O_2$ breath-to-breath variability in this study. As a further noise reduction technique the CCF was calculated. This leads to an attenuation of all $\dot{V}O_2$ components not correlated with the PRBS signal, thereby providing an improvement in signal to noise ratio (Eßfeld *et al.* 1987).

The children all tolerated the test well and were easily able to maintain a constant pedal frequency, for the duration of the test, without lapses in concentration. The highest intensity used in the PRBS exercise protocol elicited $\dot{V}O_2$ responses that were below T_{vent} for all of the subjects. In recently published research using the PRBS technique in children, individually predicted maximal working capacities were based on leg muscle volume (Kusenbach et al. 1999). The upper work-rate for the PRBS protocol was set to 40% of the individually predicted maximal working capacity rather than 50% as in the present study. Although this would give a greater degree of confidence that the upper work-rate would be below T_{vent}, particularly if the cohort contained subjects who were likely to have a reduced working capacity, it would reduce the signal to noise ratio.

In this small cohort, the younger children demonstrated faster $\dot{V}O_{2\ kinetics}$ than the older children. Unlike the traditional measurement of $\dot{V}O_{2\ max}$, the $\dot{V}O_{2\ kinetic}$ responses were not significantly influenced by stature and body mass. The slowing of $\dot{V}O_{2\ kinetics}$ with chronological age may indicate maturational changes, however, given the small size of the sample, these findings would need to be confirmed in a larger cohort of subjects where the effect of biological age could be controlled for. The question of whether the responses at different frequencies reflect different physiological processes has not been resolved. Therefore, at present, there is no physiological explanation for the significant correlation between chronological age and $\dot{V}O_{2\ kinetics}$ found at the frequencies of 2.2 and 4.4 mHz.

In summary, this study has demonstrated the use of the PRBS technique to measure $\dot{V}O_{2\ kinetics}$ in the frequency domain in children. This approach may be a useful addition to the tests that are used to quantify the $\dot{V}O_2$ responses to exercise in children.

References

ARMSTRONG, N. and WELSMAN, J. R. 1994, Assessment and interpretation of aerobic fitness in children and adolescents, *Exercise and Sport Sciences Reviews*, **22**, 435–476.

AUCHINCLOSS, H. L., GILBERT, R. and BAULE, G. H. 1966, Effect of ventilation on oxygen transfer during early exercise, *Journal of Applied Physiology*, **21**, 810–818.

BEAVER, W. L., LAMARRA, N. and WASSERMAN, K. 1981, Breath-by-breath measurement of true alveolar gas exchange, *Journal of Applied Physiology*, **51**, 1662–1675.

BEAVER, W. L., WASSERMAN, K. and WHIPP, B. J. 1973, On-line computer analysis and breath-by-breath graphical display of exercise function tests, *Journal of Applied Physiology*, **34**, 128–132.

BENNETT, F. M., REISCHL, P., GRODINS, F. S., YAMASHIRO, S. M. and FORDYCE, W. E. 1981, Dynamics of ventilatory response to exercise in humans, *Journal of Applied Physiology*, **51**, 194–203.

CHICHARRO, J. L., CALVO, F., ALVAREZ, J., VAQUERO, A. F., BANDRES, F. and LEGIDO, J.C. 1995, Anaerobic threshold in children: determination from saliva analysis in field tests, *European Journal of Applied Physiology and Occupational Physiology*, **70**, 541–544.

COOPER, D. M. 1994, Evidence for and mechanisms of exercise modulation of growth—an overview, *Medicine and Science in Sports and Exercise*, **26**, 733–740.

COOPER, D. M. 1995, Rethinking exercise testing in children: a challenge, *American Journal of Respiratory and Critical Care Medicine*, **152**, 1154–1157.

DUNCAN, G. E., MAHON, A. D., HOWE, C. A. and DEL CORRAL, P. 1996, Plateau in oxygen uptake at maximal exercise in male children, *Pediatric Exercise Science*, **8**, 77–86.

EßFELD, D., HOFFMANN, U. and STEGEMANN, J. 1987, $\dot{V}O_2$ kinetics in subjects differing in aerobic capacity: investigation by spectral analysis, *European Journal of Applied Physiology*, **56**, 508–515.

GODFREY, S. 1974, *Exercise Testing in Children: Applications in Health and Disease*, (London: W. B. Saunders).

Hoffmann, U., Eßfeld, D., Wuderlich, H. and Stegemann, J. 1994a, $\dot{V}O_2$ kinetics determined by PRBS-technique and sinusoidal testing, *Zeitschrift fur Kardiologie*, **3**, 57–60.
Hoffmann, U., Eßfeld, D., Leyk, D., Wuderlich, H. and Stegemann, J. 1994b, Prediction of individual oxygen uptake on-step transients from frequency responses, *European Journal of Applied Physiology and Occupational Physiology*, **69**, 93–97.
Hughson, R. L., Winter, D. A., Patla, A. E., Swanson, G. D. and Cuervo, L. A. 1990, Investigation of $\dot{V}O_2$ kinetics in humans with pseudorandom binary sequence work rate change, *Journal of Applied Physiology*, **68**, 796–801.
Kerlin, T. W. 1974, Properties of important test signals, *Frequency Response Testing in Nuclear Testing* (New York: Academic Press), 52–82.
Kusenbach, G., Wieching, R., Barker, M., Hoffmann, U. and Eßfeld, D. 1999, Effects of hyperoxia on oxygen uptake kinetics in cystic fibrosis patients as determined by pseudo-random binary sequence exercise, *European Journal of Applied Physiology*, **79**, 192–196.
Kusenbach, G., Wieching, R., Barker, M., Hoffmann, U., Eßfeld, D. and Heimann, G. 1994, Exercise testing with pseudo-random binary sequences of work load in children and adolescents, *European Respiratory Journal*, **7**, 233S.
Lamarra, N., Whipp, B. J., Ward, S. A. and Wasserman, K. 1987, Effect of interbreath fluctuations on characterising exercise gas exchange kinetics, *Journal of Applied Physiology*, **62**, 2003–2012.
Marven, S. S., Smith, C. M., Claxton, D. B., Chapman, J. H., Davies, H. A., Primhak, R. A. and Powell, C. V. E. 1998, Pulmonary function, exercise performance, and growth in survivors of congenital diaphragmatic hernia, *Archives of Diseases in Childhood*, **78**, 137–142.
Maxwell, B. F., Withers, R. T., Ilsley, A. H., Wakim, M. J., Woods, G. F. and Day, L. 1998, Dynamic calibration of mechanically, air- and electromagnetically braked cycle ergometers, *European Journal of Applied Physiology and Occupational Physiology*, **78**, 346–352.
Pfitzinger, P. and Freedson, P. 1997, Blood lactate responses to exercise in children: Part 2. Lactate threshold, *Pediatric Exercise Science*, **9**, 299–307.
Potter, C. R., Childs, D. J., Houghton, W. and Armstrong, N. 1999, Breath-to-breath 'noise' in the ventilatory and gas exchange responses of children to exercise, *European Journal of Applied Physiology and Occupational Physiology*, **80**, 118–124.
Rivera-Brown, A. M., Rivera, M. A. and Frontera, W. R. 1992, Applicability of criteria for $\dot{V}O_2$ max in active adolescents, *Pediatric Exercise Science*, **4**, 331–339.
Robinson, S. 1938, Experimental studies of physical fitness in relation to age, *Internationale Zeitschrift fur Angewandte Physiologie Einschliesslich Arbeitphysiologie*, **10**, 251–323.
Rowland, T. W. 1996, On buying a used car, or, reflections on Dr. Cooper's opus, *Pediatric Exercise Science*, **8**, 189–192.
Rowland, T. W. 1998, The case of the elusive denominator, *Pediatric Exercise Science*, **10**, 1–5.
Rowland, T. W. and Cunningham, L. N. 1992, Oxygen uptake plateau during maximal treadmill exercise in children, *Chest*, **101**, 485–489.
Rowland, T. W., Vanderburgh, P. M. and Cunningham, L. 1997, Body size and the growth of maximal aerobic power in children: a longitudinal analysis, *Pediatric Exercise Science*, **9**, 262–274.
Swanson, G. D. 1980, Breath-to-breath considerations for gas exchange kinetics, in P. Cerretelli and B. J. Whipp (eds), *Exercise Bioenergetics and Gas Exchange* (Amsterdam: North-Holland Medical Press Elsevier), 211–222.
Swanson, G. D. and Sherrill, D. L. 1983, A model evaluation of estimates of breath-to-breath alveolar gas exchange, *Journal of Applied Physiology*, **55**, 1936–1941.
Tanner, J. M. 1962, *Growth at Adolescence*, 2nd ed. (Oxford: Blackwell Scientific).
Taylor, A. E., Rehder, K., Hyatt, R. E. and Parker, J. C. 1989, *Clinical Respiratory Physiology* (Philadelphia, PA: Saunders).
Tschakovsky, M. E. and Hughson, R. L. 1999, Interaction of factors determining oxygen uptake at the onset of exercise, *Journal of Applied Physiology*, **86**, 1101–1113.
Washington, R. L. 1993, Anaerobic threshold, in T. W. Rowland (ed.), *Pediatric Laboratory Exercise Testing* (Champaign, IL: Human Kinetics Books), 115–129.
Wasserman, K., Hansen, J. E., Sue, D. J. and Whipp, B. J. 1987, *Principles of Exercise Testing and Interpretation*, 1st ed. (Philadelphia, PA: Lea and Febiger).

WHIPP, B. J. 1987, Dynamics of pulmonary gas exchange, *Circulation*, **76** (Suppl. VI), 18–28.
WHIPP, B. J. and WARD, S. A. 1990, Physiological determinants of pulmonary gas exchange kinetics during exercise, *Medicine and Science in Sports and Exercise*, **22**, 62–71.
WHIPP, B. J. and WASSERMAN, K. 1972, Oxygen uptake kinetics for various intensities of constant load work, *Journal of Applied Physiology*, **33**, 351–356.
WHIPP, B. J., DAVIS, J. A., TORRES, F. and WASSERMAN, K. 1981, A test to determine parameters of aerobic function during exercise, *Journal of Applied Physiology*, **50**, 217–221.

15 Walking in visually handicapped children and its energy cost

VÁCLAV BUNC*, JARMILA SEGETOVA and LUCIE SAFARIKOVA

Faculty of Physical Education and Sports, Charles University, J.Martiho 31, CZ 162 52 Prague 6, Czech Republic

Keywords: Visually handicapped; Children; Walking; Energy cost; Conditioning.

Walking is a basic activity in visually handicapped subjects, and often it is used as a general means of improving physical fitness. The level of adaptation to walking may be assessed by means of energy cost, c. The variable c was studied during walking on a treadmill in two groups of visually handicapped children (international classification of vision of 5/200 or less). The two groups were comprised of 15 boys (mean age = 11.8 ± 2.1 years) and 13 girls (mean age = 11.6 ± 3.1 years). The mean energy cost in boys was found to be 3.79 ± 0.31 J kg^{-1} m^{-1} and in girls it was 3.77 ± 0.36 J kg^{-1} m^{-1}. Both these values were not significantly higher than the energy cost in untrained non-handicapped children of the same age. There was a U-shaped dependence of c on increased speed of walking. The minimum was about 3.6 km h^{-1} in both groups of handicapped children, which was similar to that for non-handicapped subjects. It is concluded that in visually handicapped children the energy cost of walking, and thus adaptation to walking, is the same as in the healthy children. The visually handicapped individuals show a 'normal' response to exercise, to which they are adapted, with increases in both cardiovascular and muscular fitness.

1. Introduction

Physical disability often significantly limits the capacity for work and may restrict a person from certain activities. Nevertheless, most people with disabilities benefit from regular and appropriate physical activity, with results including improved stamina and function for activities involved in daily living, enhanced self-esteem and confidence through the perception of improved appearance, control and function, and improved tolerance to stress. A major factor also is an improvement in the subject's independence (Hopkins *et al.* 1987, Lockette and Keyes 1994).

Some individuals with visual impairments or blindness have significantly lower physical fitness levels compared to their peers with unaffected vision. The low physical work capacity in young visually handicapped subjects is due to the lack of physical activity, which contributes to age-related obesity, muscular weakness and low tolerance of exercise (Hopkins *et al.* 1987).

Walking is a basic locomotory activity both in non-handicapped subjects and in visually handicapped subjects. Walking is also often used as a means of improving physical fitness, and thus may contribute to the independence of handicapped subjects (Bernbaum *et al.* 1989). For evaluating conditioning effects of exercise and for monitoring exercise, it is relevant to assess the energy cost of exercise (Åstrand and Rodahl 1986).

*Author for correspondence. e-mail: bunc@ftvs.cuni.cz

The energetics of several forms of locomotion on land and in water can be appropriately described by specifying the amount of energy required above the resting level to transport the subject's body (plus whatever goes with it, e.g. bicycle) over one unit of distance (di Prampero et al. 1986, Bunc and Heller 1989). This quantity is defined as energy cost of transport, or locomotion, and is often expressed per unit body weight, or body surface area.

The energy requirement per unit of time (mainly 1 min), with metabolic power requirement, E, for moving at speed v is given by

$$E = cv + E_0 \tag{1}$$

where c, defined as the energy cost of moving, is expressed in J kg^{-1} m^{-1} and the speed of motion is expressed in m s^{-1}. E_0 is the resting energy requirement to cover basal metabolic functions.

By rearranging equation (1) and applying it to maximal conditions

$$v_{max} = (E_{max} - E_0)c^{-1} \tag{2}$$

it can be shown that the maximal speed of motion depends on both the maximal metabolic power of the subject and on the energy cost of locomotion at that speed. Since E is about the same for different groups of subjects, the maximal speed attained in different forms of locomotion (e.g. running vs. cycling or swimming) is set essentially by the value of c, being higher in those forms in which c is low and vice versa (di Prampero et al. 1986, Bunc and Heller 1989).

Equation (2) applies regardless of the sources, aerobic and/or anaerobic, supplying the energy for work performance, the only constraint being that the sum total of metabolic power from different sources is expressed in appropriate units. This equation shows that the theoretical maximal speed can be calculated from the appropriate values of E_{max} and c. This last measure, however, is a function of the speed (at submaximal intensities c is practically independent of speed of movement), whereas E_{max} depends on the duration of the effort and on the speed of motion. For an exercise with a higher speed of movement like running or cycling c is proportional to the air density, to the frontal area of the subject and thus to the body dimensions of the subject, as well as to the speed of running (di Prampero et al. 1986, Brisswalter et al. 1996).

Direct measurement of E during physical activity is difficult. For practical reasons E is often expressed as oxygen uptake ($\dot{V}O_2$) for a particular activity. In these cases it has been convenient to express c in ml kg^{-1} m^{-1} and/or J kg^{-1} m^{-1} and speed of motion in m min^{-1}, to obtain E in more customary units, ml kg^{-1} min^{-1}.

Thus, the relationship for the energy cost of movement may be rearranged as follows:

$$\dot{V}O_2 = cv + VO_{20} \tag{3}$$

The energy cost of motion may be characterized by the coefficient of energy cost of movement c, which indicates how much energy is needed to carry a body mass of 1 kg over a distance of 1 m (di Prampero et al. 1986, Bunc and Heller 1989). The physiological and biomechanical factors affecting c have recently been reviewed by Morgan et al. (1989) and by Bailey and Pate (1991). From the biomechanical point of view the energy demands during human transport are increased with the subject's body mass. Kinetic, potential and rotational parts of total energy demands during human movement are proportional to the subject's body mass.

Since walking is a basic activity in visually handicapped children, and often it is used for enhancing their fitness, the purpose of this study was to determine the energy cost of walking at different speeds in visually handicapped boys and girls.

2. Methods

The energy cost of walking was studied on a treadmill in two groups of visually handicapped children. The groups were classified according to international classification 3/200 to 5/200 for motion perception—the ability to see at 3 to 5 ft (1–1.5 m) what a person with normal vision sees at 200 ft (60 m), and less than 3/200 for light perception—the ability to distinguish a strong light at a distance of 3 ft from the eye, but inability to detect movement of a hand at the same distance (Lieberman and Cowart 1996). The groups consisted of 15 boys (mean age = 11.8 ± 2.1 years, stature = 1.461 ± 0.047 m, body mass = 35.2 ± 6.3 kg, percentage of body fat = $17.3 \pm 3.2\%$, $\dot{V}O_2$max = 42.3 ± 5.8 ml kg^{-1} min^{-1}), and 13 girls (age = 11.6 ± 3.1 years, stature = 1.475 ± 0.047 m, body mass = 36.2 ± 6.3 kg, percentage of body fat = $20.4 \pm 3.7\%$, $\dot{V}O_2$max = 35.4 ± 4.9 ml kg^{-1} min^{-1}). Two groups of untrained non-handicapped children of the same age were assessed as controls (for boys: $N = 17$; age = 11.7 ± 2.4 years; stature = 1.469 ± 0.041 m; body mass = 35.0 ± 5.7 kg; percentage of body fat = $16.4 \pm 3.2\%$; $\dot{V}O_2$max = 44.5 ± 5.3 ml kg^{-1}; and for girls: $N = 15$, age = 11.8 ± 3.5 years; stature = 1.472 ± 0.047 m; body mass = 35.8 ± 6.3 kg; percentage of body fat = $20.0 \pm 2.7\%$; $\dot{V}O_2$max = 36.7 ± 5.2 ml kg^{-1} min^{-1}).

Approval of the experimentation was granted by the Ethics Committee of Charles University. Before participation in the experiment the subjects' parents signed an informed consent form and the subjects were medically screened.

Maximal oxygen uptake ($\dot{V}O_2$max) was determined by using stepwise increases in running speed to subjective exhaustion on a treadmill at a slope of 5%. After a treadmill familiarization (warm up) at 5 and 7 km h^{-1} and 0% grade, for a duration of 4 min, the speed of running was increased by 1 km h^{-1} until subjects reached volitional exhaustion. The initial speed on the treadmill was 6 km h^{-1}. The duration of exercise to exhaustion ranged from 4 to 6 min.

Respiratory parameters were measured using an open system (Ergooxyscreen, Jaeger, Würzburg, Germany or TEEM 100, Aerosport, Ann Arbor, USA—the differences in maximal oxygen uptake between these two systems were smaller than 3%). The subjects breathed into a two-way valve with a small dead space. The gas analysers were checked before and after each test by using a calibration gas mixture of a known concentration, a pneumotachometer was tested by means of a mechanical pump. A computer printed the values every 20 or 30 s. The maximal values were calculated from the two and/or three consecutive highest values.

Energy costs of walking, c, were calculated for different walking speeds within submaximal range of intensities (the speed of walking ranged from 2 to 6 km h^{-1}). Body fat was determined by means of bioimpedance measurement with the lying position being on the right side of the body, by tetrapolar electrode configuration (Bodystat 500, Douglas, Isle of Man) using the prediction equation for children of this age.

Means and standard deviations were calculated according to standard methods. The Student's t-test and ANOVA were used for assessment of differences between groups. Statistical significance was accepted at the $p < 0.05$ level.

3. Results

The basic anthropometric data in the visually handicapped children were practically the same as in the subjects without handicap. Body mass was not significantly higher in handicapped children in either group. Maximal oxygen uptake was not significantly lower in the handicapped children than in the children without handicap.

The mean heart rate (HR) value at the highest speed of walking was 175 ± 4 beats min^{-1} in boys, and 177 ± 6 beats min^{-1} in girls, and both these values were lower than HR at the 'anaerobic threshold' (AT) levels (in boys $HR_{AT} = 184 \pm 5$ beats min^{-1} and in girls $HR_{AT} = 186 \pm 4$ beats min^{-1}), which were determined from the dependence of ventilation on oxygen uptake during a maximal treadmill test (Bunc and Heller 1989).

The mean energy cost of walking in the range of 3–6 km h^{-1} was 3.79 ± 0.31 J kg^{-1} m^{-1} in visually handicapped boys and 3.77 ± 0.36 J kg^{-1} m^{-1} in visually handicapped girls. The differences in c between visually handicapped boys and girls were not significant. Both these values were not significantly higher than c in untrained healthy children of the same age (for boys $c = 3.65 \pm 0.33$ J kg^{-1} m^{-1} and for girls $c = 3.62 \pm 0.37$ J kg^{-1} m^{-1}).

The dependence of the energy cost of walking on increasing walking speed is presented in figure 1 for boys with and without visual handicaps and in figure 2 for girls with and without visual handicaps. The same dependence in visually handicapped boys and girls only is presented in figure 3. The differences between data of handicapped children and children without a handicap at different speeds were not significant. Similar to other researchers, the authors have found a U-shaped dependence of coefficient c on increased speed of walking.

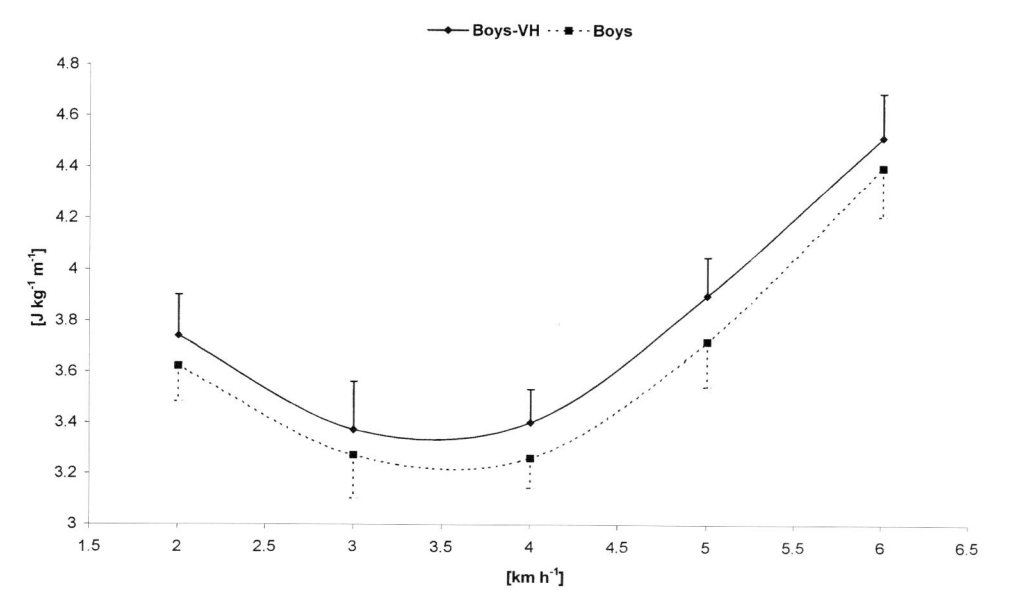

Figure 1. Dependence of energy cost of walking, c, on speed of walking, v, in visually handicapped boys (Boys-VH) and boys without handicap (Boys).

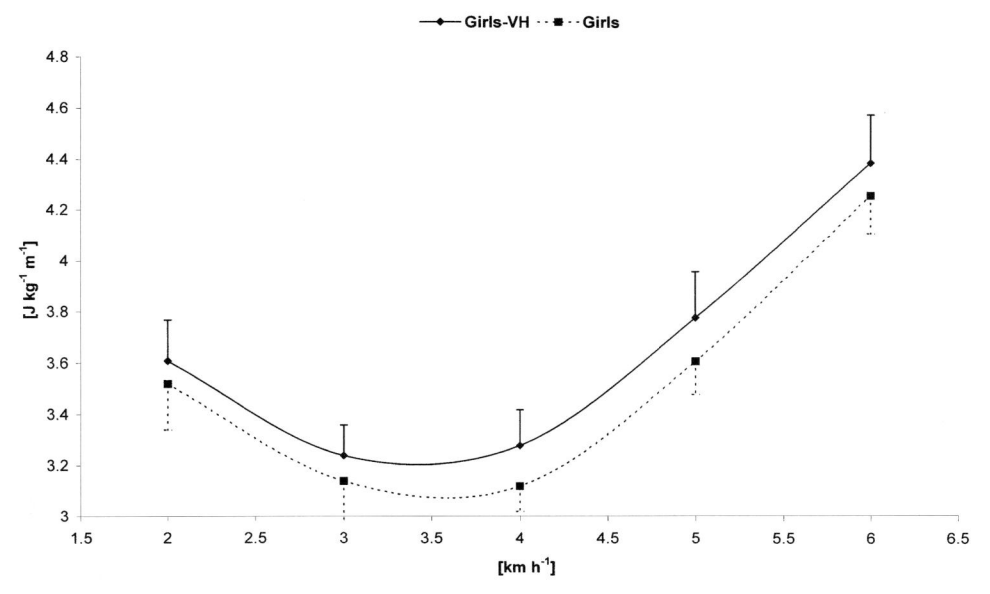

Figure 2. Dependence of energy cost of walking, c, on speed of walking, v, in visually handicapped girls (Girls-VH) and girls without handicap (Girls).

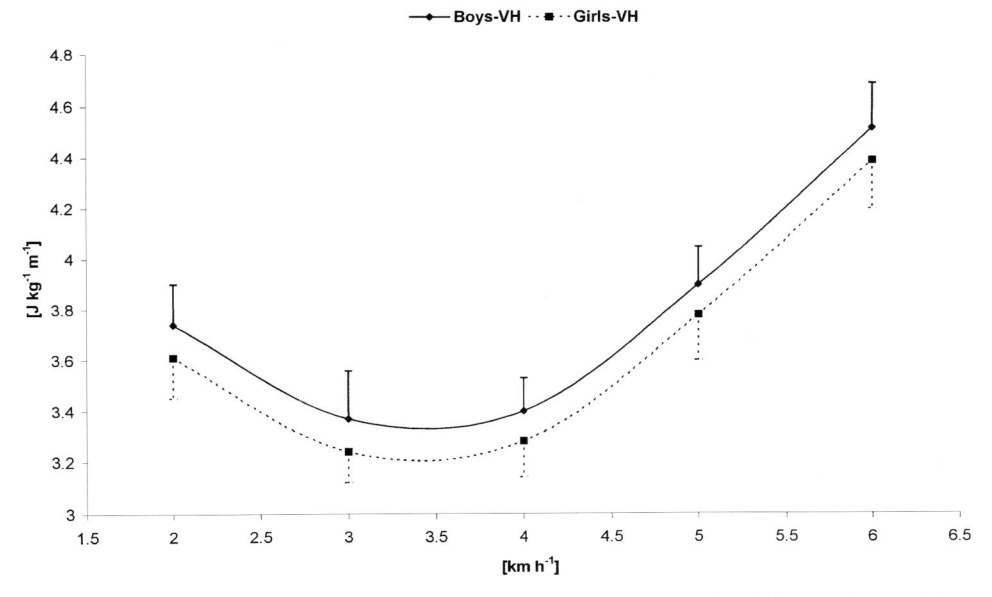

Figure 3. Dependence of energy cost of walking, c, on speed of walking, v, in visually handicapped boys (Boys-VH) and girls (Girls-VH).

The minimum energy cost of walking was found at a speed of about 3.6 km h^{-1} in both groups of visually handicapped children (for boys $v = 3.52 \pm 0.11$ km h^{-1}; for girls $v = 3.50 \pm 0.09$ km h^{-1}). This is similar to what was found in the non-handicapped children (for boys $v = 3.62 \pm 0.10$ km h^{-1}; for girls $v = 3.58 \pm 0.08$ km

h^{-1}). This minimum value of c was not significantly higher in the handicapped boys and girls (for boys $c = 3.36 \pm 0.05$ J kg^{-1} m^{-1}, and for girls $c = 3.30 \pm 0.07$ J kg^{-1} m^{-1}) than in the children without handicap (for boys $c = 3.27 \pm 0.08$ J kg^{-1} m^{-1} and for girls $c = 3.20 \pm 0.06$ J kg^{-1} m^{-1}).

4. Discussion

One methodological factor must be considered in the interpretation of the results. The validity of the measure of energy cost of running depends on the assumption that energy expenditure can be determined from oxygen uptake. During the submaximal test within range of intensities lower than the 'anaerobic threshold', the energy expenditure may be identified using oxygen consumption (Åstrand and Rodahl 1986, Bunc and Heller 1989).

Mean values of maximal oxygen uptake are in good agreement with the results of many previous studies in children (Åstrand and Rodahl 1986, Rowland 1989, Wilmore and Costill 1994). The level of adaptation to exercise may be assessed using energy cost. Generally, the higher the adaptation to exercise, the lower is the energy cost of exercise (di Prampero et al. 1986, Bunc and Heller 1989). The higher energy cost of walking in handicapped children is probably caused by: (1) a lower adaptation to walking as a result of lower volume of exercise in handicapped children compared to non-handicapped children; and (2) continuous muscle tension in these subjects.

The U-shaped dependence of energy cost of walking on speed of movement supports a suggestion of Pearce et al. (1983) that 'efficient' walking and/or a 'comfortable' speed of walking may be definable in terms of the least metabolic cost of walking per unit distance travelled. This may be connected with problems of maintaining balance while walking. Maintaining balance seems to be a major consideration in low-velocity bipedal locomotion (Workman and Armstrong 1986, Bunc and Dlouhá 1997). It is surmised that it is the need to maintain balance while walking that imposes curvilinearity on the relationship between ground speed and the step frequency that is freely chosen (Zarrugh and Radcliffe 1978).

The metabolic cost of walking is the product of frequency multiplied by the metabolic cost of a step (Workman and Armstrong 1986). Thus the minimum observed in the U-shaped curves of c is not artefactual but reflects the real situation (physical and metabolic) for walking.

The non-significantly lower values of c in the visually handicapped girls compared to the visually handicapped boys of the same age may be explained by a higher training state in males than in females. Second, Bosco et al. (1980) showed that under equivalent training conditions and thus practically in the same training state there are, in women, better predispositions for higher economy of exercise in the submaximal ranges, i.e. lower values of c, which result from a higher proportion of slow twitch fibres in their muscles than in males.

The present results are similar to data from children without any handicap. It can therefore be concluded that in visually handicapped children the energy cost of basic locomotion, and thus adaptation to walking, is the same as in healthy children. The visually handicapped individuals showed a 'normal' response to exercise.

Acknowledgements

This study was supported by a grant from the Czech Ministry of Education No. VS 97131.

References

ASTRAND, P. O. and RODAHL, K. 1986, *Textbook of Work Physiology*, 3rd ed. (New York: McGraw-Hill).

BAILEY, S. and PATE, R. 1991, Feasibility of improving running economy, *Sports Medicine*, **12**, 228–236.

BERNBAUM, M., ALBERT, S. G. and COHEN, J. D. 1989, Exercise training in individuals with diabetic retinopathy and blindness, *Archives of Physical Medicine and Rehabilitation*, **70**, 605–611.

BOSCO, C., KOMI, P. V. and SINKKONEN, K. 1980, Mechanical power, net efficiency, and muscle structure in male and female middle distance runners, *Scandinavian Journal of Sports Science*, **2**, 47–51.

BRISSWALTER, J., LEGROS, P. and DURAND, M. 1996, Running economy, preferred step length correlated to body dimensions in elite middle distance runners, *Journal of Sports Medicine and Physical Fitness*, **36**, 7–15.

BUNC, V. and DLOUHÁ, R. 1997, Energy cost of treadmill walking, *Journal of Sports Medicine and Physical Fitness*, **37**, 103–109.

BUNC, V. and HELLER, J. 1989, Energy cost of running in similarly trained men and women, *European Journal of Applied Physiology*, **59**, 178–183.

DI PRAMPERO, P. E., ATCHAU, G., BRUECKNER, J.C. and MOIA, C. 1986, The energetics of endurance running, *European Journal of Applied Physiology*, **55**, 256–266.

HOPKINS, W. G., GAETA, H., THOMAS, A. C. and MCHILL, P. 1987, Physical fitness of blind and sighted children, *Journal of Applied Physiology*, **56**, 69–73.

LIEBERMAN, L. J. and COWART, J. F. 1996, *Games for People with Sensory Impairments* (Champaign, IL: Human Kinetics).

LOCKETTE, K. F. and KEYES, A. M. 1994, *Conditioning with Physical Disabilities* (Champaign, IL: Human Kinetics).

MORGAN, D., MARTIN, P. and KRAHENBUHL, G. 1989, Factors affecting running economy, *Sports Medicine*, **7**, 310–330.

PEARCE, M. E., CUNNINGHAM, D. A., DONNER, A. P., RECHNITZER, P. A., FULLERTON, G. H. and HOWARD, J. H. 1983, Energy cost of treadmill and floor walking at self-selected paces, *European Journal of Applied Physiology*, **52**, 115–119.

ROWLAND, T. E. 1989, Oxygen uptake and endurance fitness in children: a developmental perspective, *Pediatric Exercise Science*, **1**, 313–328.

WILMORE, J. H. and COSTILL, D. L. 1994, *Physiology of Sport and Exercise*, 1st ed. (Champaign, IL: Human Kinetics).

WORKMAN, J. M. and ARMSTRONG, B. W. 1986, Metabolic cost of walking: equation and model, *Journal of Applied Physiology*, **61**, 1369–1374.

ZARRUGH, M. Y. and RADCLIFFE, C. W. 1978, Predicting metabolic cost of level walking, *European Journal of Applied Physiology*, **38**, 215–223.

16 Touch-down and take-off characteristics of the long jump performance of world level above- and below-knee amputee athletes

Lee Nolan* and Adrian Lees

Research Institute for Sport and Exercise Sciences, Liverpool John Moores University, Henry Cotton Campus, Webster St, Liverpool L3 2ET, UK

Keywords: Above-knee amputees; Below-knee amputees; Long jump; Biomechanical analysis.

The aims of this study were to establish the take-off characteristics of long jump performance of disabled amputee athletes, and to establish to what extent amputee athletes conform to a model of performance defined for elite able-bodied athletes. The jumps of 8 male below-knee (trans-tibial) and 8 male above-knee (trans-femoral) amputee athletes who competed in the finals of the long jump at the 1998 World Disabled Championships were recorded in the sagittal plane on video (50 Hz). Approach speed was measured using a laser Doppler system. The best jump for each athlete was digitized, and kinematic data from the key instants of touch-down (TD), maximum knee flexion (MKF) and take-off (TO) were obtained. Amputees demonstrated a lower approach speed and jumped less far than able-bodied athletes although below-knee amputees performed better than above-knee amputees. For each amputee group there was a significant ($p < 0.05$) linear relationship between approach speed and distance jumped. With the exception of their slower horizontal speed and greater negative vertical speed at touch-down, below-knee amputees demonstrated characteristics of technique that were similar to elite able-bodied long jumpers. Above-knee amputees at touch-down had a more upright trunk, smaller hip and knee angles and consequently a smaller leg angle. This was attributed to the difficulty of taking off on the last stride on the prosthetic limb. Consequently, above-knee amputees were less able to gain vertical velocity during the compression (TD–MKF) phase, but were able to compensate for this by using a greater hip range of motion during the extension (MKF–TO) phase. It was concluded that below-knee amputees displayed the same basic jumping technique as elite able-bodied long jumpers, but above-knee amputees did not. These findings have implications for the training and technical preparation of amputee long jumpers.

1. Introduction

Disabled athletes now have the opportunity to perform on the world stage in competitions such as the Paralympics and World Disabled Athletics Championships. In these championships athletes who lack of one or more segments of their lower limb compete in events such as the long jump. Their performances have rarely been the subject of scientific analysis and it is not known whether they exhibit similar or differing techniques to able-bodied athletes, or in detail how their disability affects their performance.

*Author for correspondence. e-mail: L.Nolan@Livjm.ac.uk

Performance in the long jump is determined by a small number of key variables that can be identified through a mechanical analysis of performance. The long jump is essentially a projectile event in which the distance jumped is determined by the flight distance and this in turn is dependent on the height, horizontal and vertical speed of the centre of mass (CM) at take-off (Hay 1986). It is known that athletes try to maximize their horizontal speed by making a long approach run, and the linear relationship between approach speed and distance jumped is one of the most robust relationships emanating from scientific analyses of long jumping (Hay 1993, Lees et al. 1993, 1994). On the other hand, the vertical speed and height of the CM at take-off is almost completely determined by the actions athletes make while in contact with the take-off board. These actions are defined by several key instants and key phases. Key instants include touch-down (touch-down, defined as the instant of first contact by the take-off leg with the ground), maximum knee flexion (MKF, defined as the instant when the knee of the take-off leg reaches its smallest angle of flexion) and take-off (take-off, defined as the instant of last contact of the take-off leg with the ground). Key phases include compression (the period from touch-down to MKF), extension (the period from MKF to take-off) and contact (the period from touch-down to take-off). Previous research on able-bodied elite long jumpers has suggested that in the compression phase the athlete pivots over the support leg to gain vertical velocity and that over 65% of the take-off vertical velocity at take-off is reached by MKF (Lees et al. 1993, 1994). The pivot comes to an end at around MKF when the CM is over the support foot. In order to perform these actions during take-off, able-bodied athletes prepare themselves in advance of touch-down. As long jumpers approach the final few strides they are observed to lower their CM. This is particularly noticeable as athletes take off from their last stride (TOLS) before touch-down, and has several advantages. The low CM position at TOLS ensures its downward vertical speed at touch-down is as small as possible so that effort does not have to be expended in reversing it for take-off. It also enables the touch-down leg to be planted well in front of the body so that the CM is able to pivot over the foot using the body's horizontal speed to create vertical speed. The pivot action is enhanced by a large horizontal speed and large leg angle at touch-down. It is hindered by a flexion at the major joints of the support leg, which occurs during the compression phase. Consequently, as performance increases, a greater load is exerted on the touch-down leg and so high eccentric strength and a tolerance to high impact forces are additional requirements for preventing excessive joint flexion. The extension phase is concerned with increasing the height and speed of the CM through rapid support leg extension and free limb elevation. Thus, as well as approach speed and projectile take-off variables, the additional variables that define a model of long jump performance are height and speed of the CM, and lower limb joint angles at the key instants, and changes in these variables over the key phases.

There have been few investigations of long jump technique in lower limb amputees. Williams et al. (1997) and Simpson et al. (1998) have previously looked at visual targeting and toe-to-board placement prior to take-off in above- and below-knee amputees, but no researchers have looked at the touch-down and take-off technique of lower limb amputees. Problems that limit amputee performance may be related to their asymmetrical gait. Amputees spend longer in stance and have a shorter swing phase on their intact limb than their prosthetic limb possibly due to discomfort, reduced prosthetic mass, and inertial characteristics of the prosthesis (Breakey 1976, Murray et al. 1983). They also spend longer stepping onto their prosthetic limb than onto their intact limb (Nolan *et al.* 1998) and exhibit a shorter stride on their prosthetic limb compared to their intact limb. Enoka *et al.* (1982) reported that below-knee amputees tended to modify step frequency rather than step length in order to increase running speed. Therefore, the adjustments seen over the last few strides in able-bodied long jumpers may not be demonstrated by amputees. Simpson *et al.* (1998) reported that elite above-knee amputees increased intact limb step lengths and decreased intact limb step times throughout the approach, while little change was found for the prosthetic limb. Thus, the velocity increase throughout the approach appeared to be regulated by the intact limb. Owing to the modifications seen in amputee gait, there may be modifications to their long jump technique in comparison to able-bodied athletes.

Consequently the aims of this study were to (1) establish the take-off characteristics of long jump performance of disabled amputee athletes following the model of performance defined by elite able-bodied athletes, and (2) establish to what extent amputee athletes conform to such a model.

2. Methods

2.1. Data collection

Eight male below-knee (trans-tibial) and eight male above-knee (trans-femoral) amputees were filmed during the finals of the long jump competition at the 1998 World Disabled Athletics Championships. All athletes except one below-knee amputee jumped off their intact limb. Two Panasonic VHS video cameras (50 Hz) were used to film each jump in the frontal and sagittal planes. They were placed so that the athlete was visible at touch-down last stride (touch-down LS) through to 1 m after the take-off board. During the competition, the cameras were left running and the judges agreed to place themselves so that they were not in view of the cameras. All the finalists' jumps were recorded and the best (greatest official distance) for each competitor was selected for detailed kinematic analysis using the sagittal plane recording only. Approach velocity for each jump was measured by means of a laser Doppler system (Jenoptik™, Jena, Germany), which gave a continuous estimate of speed at 0.02 s intervals. No data were available for athletes' height and mass.

2.2. Data reduction and analysis

The video was analysed using a 9-segment biomechanical model defined by 18 points. The segmental data used were those proposed by Dempster (1955) for adult males, and modified for each subject to account for the prosthetic limb. This modification was based on stump length and prosthetic segment mass and CM location. Stump length was taken from measurements on video and expressed as a percentage of the corresponding intact segment, also taken from video measurements. For each of the amputee competitors, a prosthesis was specially constructed by the Disablement Services Centre, Withington Hospital, Manchester, to resemble as closely as possible the one worn. The prosthesis was weighed to obtain its mass and balanced on a knife edge to obtain the location of its segment centre of mass. The above-knee prostheses were disconnected at the knee and the mass and centre of mass of the prosthetic thigh and shank were obtained separately. From these measurements, and treating the residual stump as a cylinder of constant density, alternative segmental data were computed and used in subsequent calculations.

Data were smoothed using a Butterworth fourth order zero-lag filter with padded end points (Smith 1989) and a cut-off frequency of 7 Hz based on residual analysis (Winter 1990) and visual inspection of data. Derivatives were calculated by simple differentiation (Winter 1990) except for the horizontal and vertical speeds at touch-down and take-off, which were calculated using the method suggested by Yu and Hay (1996). This uses vertical displacement of the CM at touch-down and at the earliest frame available during the flight phase before touch-down and predicts vertical speed at touch-down based on the equations of motion. The same procedure is applied to the take-off using the vertical displacement of the CM after take-off. The equivalent points for the horizontal position of the CM were used to calculated the average horizontal speed at touch-down and take-off.

Several variables were computed from the kinematic data. These were related to the model of long jump performance for the key instants and phases as previously described. The point of touch-down was taken as the first frame in which the foot was clearly in contact with the ground. Similarly, take-off was taken as the first frame in which the foot was clearly off the ground. The point of MKF during the contact phase was established as the frame at which the measured angle at the knee of the touch-down leg reached a minimum. This point was assumed to represent the end of the compression phase and the start of the extension phase. It was also taken to represent the end of the pivot action as it has been shown that the CM is approximately over the ankle joint at the instant of MKF (Lees et al. 1993, 1994). Data are calculated at or between the key instants of touch-down, MKF and take-off and are presented as means and standard deviations for height (H) of the CM, horizontal (S_H) and vertical (S_V) speeds of the CM, and angles at the knee (A_{knee}) and hip (A_{hip}). In addition, for approach only, the peak approach speed ($S_{approach}$) as measured by the laser system was calculated. Also calculated, for touch-down only, were the leg angle (A_{leg}), which was defined by the angle made by the line joining the ankle joint to the CM with the vertical, and the speed of the touch-down foot (as measured by the speed of the ankle joint) relative to the CM ($S_{foot(rel)}$). The latter variable is sometimes referred to as 'sweepback speed' and is thought to aid the conversion of horizontal to vertical speed. For take-off only, the projection angle (A_{proj}) and angular extension velocity of the knee (AV_{knee}) were calculated. Official and actual (official + toe-to-board distance) distances were also presented.

The experimental tolerances in data were not explicitly determined, but they were assumed to be similar to errors using similar methods investigating long jump performances reported elsewhere (Lees et al. 1993). These authors reported percentage errors in displacement variables to be less than 2%, linear horizontal speed variables less than 4%, vertical speed variables less than 10%, angular displacement variables less than 9% and angular velocity variables less than 16%. It is unlikely that these are improved in this study owing to the lower sampling rate used (50 Hz compared to 100 Hz), although the error introduced by timing is not thought to be large. A more likely source of error is in the precise determination of the frame associated with each key instant and this is expected to lead to a greater variability in the data than reported elsewhere. A further source of error is the approximations made in re-computing the segmental data for below-knee and above-knee amputee athletes. The size of this error was tested by comparing the values of selected variables when using corrected segmental data with values obtained when these corrections were not made (i.e. treating the limb as intact).

Statistical procedures used were the Shapiro-Wilks test for establishing the non-normality of data sets, Pearson's product-moment correlation coefficient for establishing relationships and Student's t-test for establishing differences between the above- and below-knee amputees. A level of significance of $p<0.05$ was used to establish statistical significance. The variables selected for analysis were those thath define a performance model for the long jump. Consequently no adjustments were made to the alpha level to account for multiple comparisons.

3. Results

A typical graph of the vertical and horizontal speeds of a below-knee amputee is given in figure 1 for the touch-down to take-off phase. Above-knee amputees demonstrated similar velocity profiles. This graph shows the typical characteristics of long jump performance, with a vertical speed at touch-down close to zero, a reduction in horizontal speed from touch-down to MKF and a corresponding increase in vertical speed over this phase. The vertical speed reaches a positive peak just before take-off, with the horizontal speed remaining fairly constant from MKF onwards.

Descriptive data associated with long jump performance are given in table 1 for below-knee and above-knee amputees. Data for variables at the key instants of take-off, MKF and touch-down and also changes in these variables during the compression (touch-down – MKF), extension (MKF – take-off) and contact (touch-down – take-off) phases are included. These data were found to be normally distributed. Significant differences between below-knee and above-knee amputees in each variable are illustrated in table 1, together with comparative data reported on elite male able-bodied long jumpers obtained from the literature (Lees et al. 1994).

Significant differences ($p<0.05$) were found between the amputee groups for official distance jumped, actual distance jumped and approach speed, with the

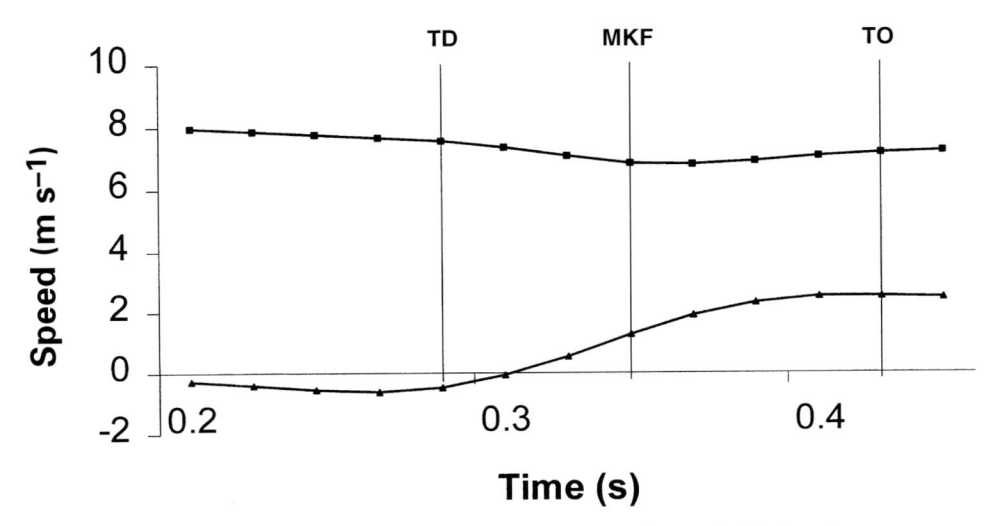

Figure 1. Vertical and horizontal speed of centre of mass of a typical below-knee amputee during the contact (TD – TO) phase of the long jump. (■) horizontal speed, (▲) vertical speed.

Table 1. Means and standard deviations (SD) for below and above-knee amputees at the key instants of touch-down (TD), maximum knee flexion (MKF) and take-off (TO) and between the key phases of compression (TD–MKF), extension (MKF–TO) and contact (TD–TO) phases. Also presented are data from able-bodied elite long jumpers from the literature. H = height of the centre of mass, S = speed, A = angle, AV = angular velocity.

	Below-knee amputees Mean	SD	Above-knee amputees Mean	SD	Elite able-bodied[†] Mean	SD
Official distance jumped (m)	5.94*	0.22	4.80	0.31	7.67	0.21
Actual distance jumped (m)	6.00*	0.21	4.87	0.32	7.79	0.24
Approach speed (ms^{-1})	9.00*	0.34	7.82	0.59	NA	NA
Key variables at TD						
H	1.08*	0.08	1.18	0.05	1.10	0.05
S_H	8.33*	0.45	7.64	0.70	9.96	0.52
S_H	−0.38	0.28	−0.39	0.18	−0.06	0.35
A_{knee}	158	7.1	154	6.4	167	4.7
A_{hip}	145*	8.9	136	12.1	149	8.2
A_{leg}	22.8*	2.4	19.0	3.7	24.8	3.7
$S_{foot(rel)}$	−5.90*	0.82	−4.57	0.47	NA	NA
$S_H(TD) \times A_{leg}(TD)$	189*	16.9	145	30.1	NA	NA
Key variables at MKF						
H	1.12*	0.06	1.19	0.06	1.15	0.06
S_H	7.51*	0.65	7.22	0.78	8.93	0.39
S_V	1.95*	0.29	1.00	0.24	1.96	0.42
A_{knee}	141	13.1	136.5	8.9	145	6.4
A_{hip}	160	10.8	152.8	10.4	160	8.7
Key variables at TO						
H	1.29*	0.06	1.35	0.07	1.34	0.06
S_H	7.50*	0.75	7.16	0.87	8.74	0.50
S_V	2.86*	0.25	2.34	0.29	3.02	0.32
A_{knee}	165	4.1	163	4.1	172	2.6
A_{hip}	201	5.9	199	8.3	194	6.9
A_{proj}	21.0*	2.9	18.4	3.8	19.1	2.1
AV_{knee} (peak)	8.57	1.69	8.3	2.40	10.4	1.90
Changes in key variables during compression						
ΔH (TD–MKF)	0.04	0.03	0.01	0.01	0.04	0.02
ΔS_H (TD–MKF)	−0.81*	0.39	−0.42	0.39	−1.02	0.25
ΔS_V (TD–MKF)	2.32*	0.51	1.39	0.27	2.02	0.52
ΔA_{knee} (TD–MKF)	−16.5	8.9	−17.4	6.4	−21.9	4.10
ΔA_{hip} (TD–MKF)	15.0	3.5	17.2	2.1	10.3	NA
Changes in key variables during extension						
ΔH (MKF–TO)	0.18	0.04	0.15	0.04	0.19	0.03
ΔS_H (MKF–TO)	−0.01	0.26	−0.06	0.10	−0.20	0.37
ΔS_V (MKF–TO)	0.91*	0.40	1.34	0.40	1.06	0.44
ΔA_{knee} (MKF–TO)	24.0	11.4	26.9	8.8	26.9	5.3
ΔA_{hip} (MKF–TO)	40.6	9.3	46.6	18.3	34.3	NA
Changes in key variables between TD–TO						
ΔH (TD–TO)	0.22*	0.05	0.17	0.04	0.24	0.04
S_H (TD–TO)	−0.83*	0.40	−0.48	0.38	−1.12	0.46
S_V (TD–TO)	3.24*	0.17	2.73	0.39	3.16	0.32

[†]Data taken from Lees et al. 1994.
*Significantly different from above-knee amputees at $p<0.05$.

below-knee amputees running faster and jumping further than the above-knee amputees. Other differences were noted for height of CM, horizontal speed, hip angle and leg angle at key instants. At touch-down the above-knee amputees had a greater CM height, lower horizontal speed, and a smaller hip and leg angle than the below-knee amputees. At MKF, the above-knee amputees had a greater CM height, and a lower vertical and horizontal speed than the below-knee amputees. These were also seen at take-off, where the above-knee amputees additionally had a smaller angle of projection.

The correlations defining expected relationships based on the model of performance are presented in table 2. The approach speed correlated significantly ($p<0.05$) with actual distance jumped for both amputee groups. None of the other projectile variables correlated with actual distance jumped with the exception of the horizontal speed at touch-down for the above-knee amputees. For both above- and below-knee amputees the leg angle at touch-down correlated with the gain in vertical speed during contact although this was not significant during the compression phase. The loss in horizontal speed (touch-down–take-off) correlated with gain in vertical speed (touch-down–take-off) for below-knee amputees only.

A graph of the relationship between approach speed and actual distance jumped is given in figures 2 and 3 for below- and above-knee amputees, respectively. A regression line was drawn through the data which yielded significant correlations as reported in table 2. The regression constants are given in table 3 for all successful jumps and also the best jump per competitor. There was no significant difference for either the gradient or intercept between the two groups although the gradients differed in a way which gave above-knee amputees a lower jump distance for a given speed, with this difference increasing as speed increased.

The error in selected variables at key points of touch-down and take-off due to the modified segmental data for missing or partial segments was less than 0.005 m (less than 0.5%) for CM vertical position, and negligible for speed, angle and angular velocity. This error was estimated by re-computing these variables using standard able-bodied segmental data. This finding means that errors in segmental data introduced by approximating stump lengths and prostheses' characteristics would have a minimal effect on the interpretations described below.

Table 2. Correlation coefficienty between selected long jump variables.

Variables			Below-knee amputees ($n=8$)	Above-knee amputees ($n=8$)
Actual distance jumped	vs.	approach speed	0.85*	0.83*
Actual distance jumped	vs.	A_{proj}	−0.33	−0.67
		S_H (TO)	0.55	0.83*
		S_V (TO)	0.13	−0.48
A_{leg} TD	vs.	ΔS_V (TD–TO)	0.95*	0.83
		ΔS_V (TD–MKF)	0.41	0.06
ΔS_V (TD–TO)	vs.	ΔS_H (TD–TO)	0.79*	0.63

*Significant at $p<0.05$.

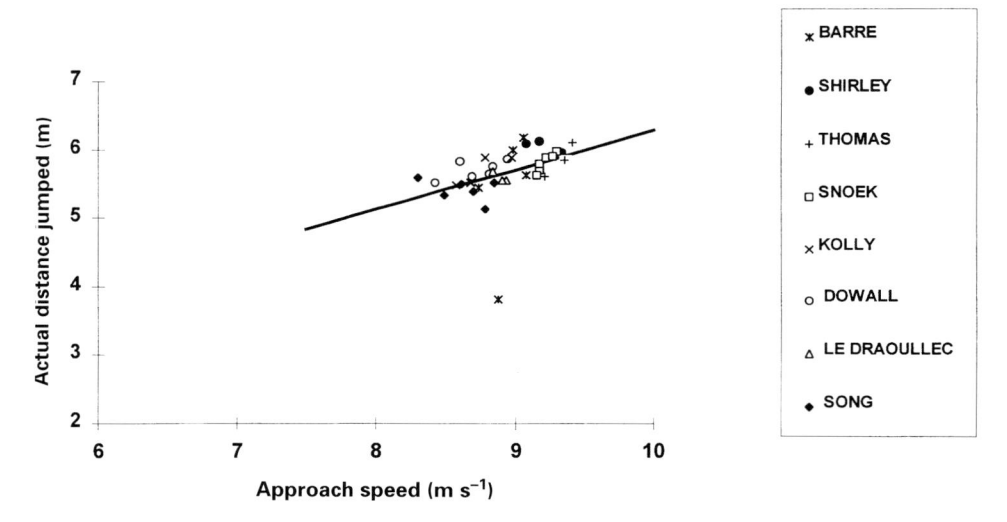

Figure 2. The relationship between approach speed and distance jumped for below-knee amputees.

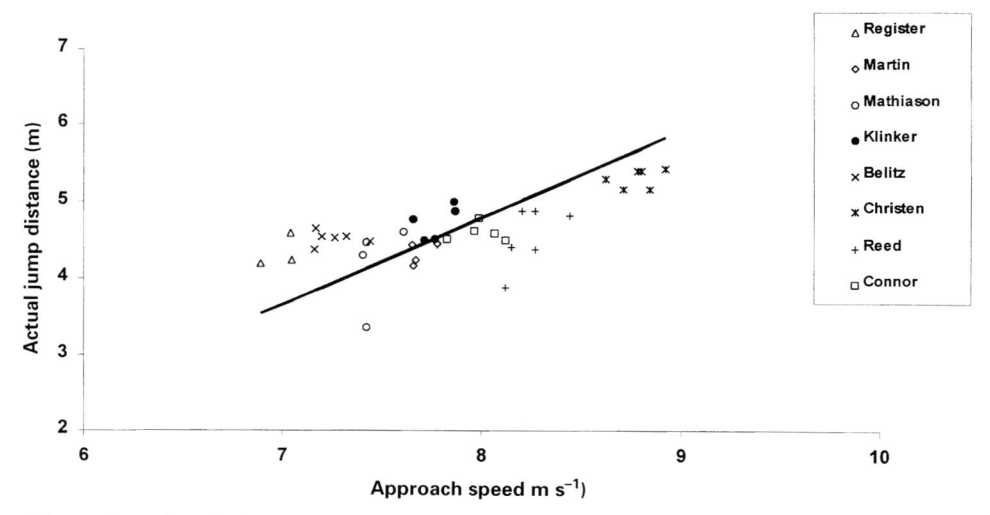

Figure 3. The relationship between approach speed and distance jumped for above-knee amputees.

4. Discussion

4.1. Overview

The first aim of this study was to establish the long jump characteristics of elite amputee athletes. The important technique variables were identified and compared between above- and below-knee amputees. In order to compare qualitatively the long jump characteristics of amputees with able-bodied performers, the data from elite male long jumpers reported by Lees et al. (1994) were used.

Table 3. Regression coefficients (with standard error) for approach speed (independent variable) and distance jumped (dependent variable) for above- and below-knee amputees for all successful jumps and for the best jump per competitor.

	Gradient	SE	Intercept	SE
All jumps				
Below-knee amputees	0.58	0.23	0.51	2.02
Above-knee amputees	0.50	0.01	0.66	0.73
Best jumps				
Below-knee amputees	0.52	0.13	1.28	1.20
Above-knee amputees	0.44	0.12	1.38	0.95

It was assumed that the error in the recorded variables would not exceed that previously reported for similar data although the lower frame rate used in this study may make the precise identification of key instants more difficult. It was suggested above that this might lead to a greater variability in data compared to that reported from studies in which a higher frame rate was used. The standard deviation data in table 1 indicates that the amputee groups covered a similar range of jump performance as the able-bodied athletes. Of the 29 performance variables reported where comparisons of variability can be made with able-bodied athletes, 17 (below-knee) and 19 (above-knee) variables had a variability that exceeded that reported for able-bodied athletes. However, only three (in the below-knee amputee group) were significantly different from able-bodied athletes and these were related to knee angle at MKF. It is not thought that this was the result of error, as the point of MKF corresponds to a zero velocity point and so will be less affected by the frame rate. It is likely that this variability is a true reflection of the performance variation existing within the amputee groups. Consequently, this and the other variables reported are assumed to be comparable, in terms of their accuracy, to other data reported in the literature.

4.2. *Performance and general characteristics*

The distance jumped by the above-knee amputees was, as expected, lower than that of below-knee amputees who in turn jumped lower distances than the able-bodied athletes. This was also found to be the case for the approach speeds. However, the relationship between approach speed and distance jumped was significant for both amputee groups even taking into account the few jumps that were poor, presumably due to some error made by the athlete during the approach. This finding has confirmed that the relationship between the approach speed and distance jumped, which has been widely reported for other able-bodied groups (Hay 1993) also holds for elite amputee long jumpers and should be of some value to amputee athletes and their coaches who may rely on this practical relationship to guide their training and performance. The regression constants for each group did not differ significantly, although across the range of approach speeds used, the below-knee amputees were able to generate an additional 0.5 m or so in jump distance. In long jumping this difference is important and suggests that there is a fundamental difference between the two groups in performance terms. Putting it another way, the above-knee amputee jumpers should not aspire to below-knee amputee levels of performance even if their approach speeds are comparable. The lower approach speed of the amputees is likely to be due to their prosthetic limb. A prosthetic limb causes asymmetry in movement and the difficulty of controlling this as well as an inability to tolerate high force levels acting on the residual stump will lead to a reduction of approach speed, and will be more evident in above-knee amputees.

The reduced performance of amputees may also be due to lower levels of physical capability. Although no tests of physical performance were made on the amputee athletes, it is likely that they are less well conditioned than their elite able-bodied counterparts.

4.3. Characteristics at touch-down

At touch-down the below-knee amputees are characterized by a greater horizontal speed, greater leg, hip, and knee angles, lower height of CM and higher foot relative speed (sweepback speed) when compared to above-knee amputees. The horizontal speed is closely related to approach speed and has been considered above. The greater leg angle is associated with a significantly larger angle of flexion of the hip and a slightly straighter leg as indicated by the larger knee angle. In order to achieve this the below-knee amputees must lean their trunk further backwards at touch-down than above-knee amputees whose trunk remains more upright. The backward lean of the trunk is seen in able-bodied jumpers and is thought to provide a stronger position to resist the large forces at touch-down.

It is not possible to assess directly the height of the CM at touch-down because the heights of the athletes were unknown. At take-off the height of the CM was higher in above-knee amputees compared to below-knee amputees suggesting that as a group they were taller. Dapena *et al.* (1990) suggested that the height of the CM of high jumpers at take-off is reliably related to their stature and it is assumed that this is applicable to long jumpers as they attain a similar take-off position. Thus, changes in height of the CM can be assessed in relation to the height of the CM at take-off and represented as a percentage of take-off height. The percentage gain in height from touch-down to take-off was 17.0% for below-knee amputees, 12.6% for above-knee amputees and 17.9% for able-bodied athletes, and can be used to support the notion that below-knee amputees achieved a lower position of the CM at touch-down relative to their take-off height than the above-knee amputees.

The slower approach speed of the above-knee amputees has the effect of reducing the severity of impact at touch-down. This is thought to increase as both the horizontal speed and the leg angle at touch-down. The 'severity' of touch-down is indicated by the product of these two variables and is significantly greater for below-knee amputees than for above-knee amputees. The leg angle at touch-down for below-knee amputees is only marginally smaller than that of able-bodied athletes but the horizontal speed is lower.

The backward speed of the foot relative to the CM (the sweepback speed) is greater in below-knee amputees. This supports the finding above that the below-knee amputees are better able to produce the required conditions at touch-down than are above-knee amputees. High sweepback speed is thought to help the extension of the hip and in so doing helps to bring the body over the support foot. This will enhance the effect of the pivot and improve the conversion of horizontal speed to vertical speed.

The vertical speed at touch-down does not differ significantly between the two amputee groups but it does have a higher negative value for each group compared to able-bodied jumpers. The amputee athletes appear to be less able to control their negative vertical velocity at touch-down and this will be a further factor limiting their performance. All except one below-knee athlete jumped from their intact limb, which means that for the majority of amputees the last stride was taken off their prosthetic limb. This will inevitably affect their capability to adopt the required positions at take-off last stride in order to prepare for touch-down and is likely to be more problematic for above-knee amputees who do not possess a fully functional knee joint. The ability to take off successfully at the last stride on the prosthetic limb may be a key factor in determining the performance of amputee athletes and can be responsible for lower horizontal speed and lower leg angle as well as higher negative speed at touch-down.

4.4. Characteristics at maximum knee flexion and the compression phase

The key instant of MKF is used to determine the end of the pivot, which is the mechanism used by able-bodied athletes to generate the majority of their vertical speed. The vertical speed achieved by below-knee amputees at MKF was significantly greater than for above-knee amputees but similar to able-bodied subjects suggesting that they were able to demonstrate a successful pivot. Considering the greater negative speed at touch-down for below-knee amputees compared to able-bodied athletes, the change in vertical speed during compression is impressive. The percentage of take-off vertical speed achieved at MKF is 68% for the below-knee amputees but only 43% for the above-knee amputees. These values show that above-knee amputees were less effective in their demonstration of the pivot and may indicate that they use a different technique.

The knee joint performs a shock-absorbing role at touch-down and the speed, leg angle and initial knee flexion angle all determine the load that the knee joint has to support. The above-knee amputees subjected themselves to a less severe touch-down than the below-knee amputees who in turn had a less severe touch-down than able-bodied athletes. This is reflected in the change in knee angle during the compression phase, which was smallest for the above-knee amputees and largest for the able-bodied athletes. It is interesting to note that the mean knee angle at MKF was similar for all groups even though there was a much larger variation between athletes in each amputee group.

At touch-down in long jumping the hip is flexed and in able-bodied jumpers it extends continually throughout the contact phase. The same hip motion was observed for both amputee groups in this study. Hip extension was quite small during the compression phase amounting to 14° for both groups but although this is slightly larger than for able-bodied jumpers, the difference cannot be considered to be meaningful due to the large variability associated with all groups. The smaller hip angle at touch-down demonstrated by the above-knee amputees may have some relevance to their performance as it produces a less strong position at touch-down to tolerate the forces of impact but does provide for a greater range of motion during the extension phase. A greater percentage of the final vertical speed may be achieved from hip extension and so compensate for the reduced pivot effectiveness as noted above. This compensation is in agreement with Nolan (1995) who found that both above- and below-knee amputees compensate for the partial loss of a lower limb by increasing range of motion at the intact limb hip joint.

4.5. Characteristics at take-off and the extension phase

At take-off the above-knee amputees are characterized by a lower horizontal and vertical speed than below-knee amputees, who in turn had lower speeds than those for able-bodied athletes. However, the gain in vertical speed over the contact phase by below-knee amputees exceeded that of able-bodied subjects, due to their ability convert horizontal speed to vertical speed through the pivot mechanism. Above-knee amputees were generally less able at converting their horizontal speed to vertical speed, which reflects their poor use of the pivot mechanism as noted earlier.

Above-knee amputees were able to generate a higher percentage of their vertical speed at take-off during the extension phase. This ability can be related to the greater range of motion used at the hip. The above-knee amputees had a smaller hip angle at touch-down and at MKF compared to below-knee amputees and able-bodied athletes but the hip angle at take-off did not differ between the three groups. Above-knee amputees therefore used a greater range of hip motion (49°) during the extension phase than below-knee amputees (41°) and able-bodied athletes (34°) supporting the notion that they use a compensatory strategy to overcome their inability to perform the pivot effectively.

4.6. *The long jump model*
The second aim of this study was to determine to what extent the long jump technique of above- and below-knee amputee athletes conformed to the model of performance exhibited by able-bodied athletes (Lees *et al.* 1994). The essential characteristics of the long jump model are defined by the position and motions of the body at touch-down and the ability of athletes to utilize a pivot for gaining vertical speed. These characteristics of able-bodied athletes are also found to a greater or lesser degree in each of the amputee groups. Both groups exhibit a lowering of the CM and an extended support leg at touch-down, creating a pivot which enables a conversion of horizontal to vertical speed.

The reduced levels of performance in both amputee groups can be related to their slower approach speed and this in turn can be related to the problems they encounter at the last stride prior to take-off due to their prosthetic limb. The above-knee amputees would be expected to have greater difficulty at this stage and as a result they appear to modify their technique to cope with demands of touch-down. This is seen in their more upright trunk, more flexed leg and a smaller leg angle at touch-down. As they are less able to use the pivot due to these constrained conditions at touch-down, they appear to adopt a compensation strategy which is to use a greater range of hip joint motion during the extension phase. This compensation is quite effective, as above-knee amputees are still able to demonstrate a strong relationship between approach speed and distance jumped, which is a noted feature of able-bodied performance. However, they are less capable of converting their horizontal speed to vertical speed than the below-knee amputees or the able-bodied athletes. These are clear indications of a departure from the model of performance as demonstrated by able-bodied long jumpers.

The below-knee amputees have a slower horizontal speed and a greater negative vertical speed at touch-down compared to able-bodied athletes, but many other variables which characterize the contact phase are similar. These include leg angle at touch-down and pivot effectiveness. They appear to be less able to generate vertical speed during the extension phase but they do appear to be better than able-bodied athletes at maintaining their horizontal speed through to take-off.

There are no specific criteria to judge whether a group of athletes conform to a model of performance based on several interacting variables. However, the data reviewed here suggest that below-knee amputees do conform to the model of performance defined by elite able-bodied long jumpers although their performance is at a slower speed. In other words, increased approach speed should lead to performances comparable with able-bodied subjects. It is therefore concluded that below-knee amputees demonstrate the same basic jumping technique as elite able-bodied long jumpers. On the other hand, although above-knee amputees demonstrate some aspects of the model, their inability to utilize the pivot mechanism effectively and the compensation made for this by using a larger hip range of motion during the extension phase are clear indicators that their technique has changed and departs substantially from the elite able-bodied model. For these athletes, increased approach speed will not lead to performances comparable with elite able-bodied athletes. As above-knee amputees are substantially constrained by their prosthesis and residual limb they have developed compensatory strategies to enhance their performance by substantially modifying their technique. Consequently the elite able-bodied model is not appropriate for improving the mechanics of above-knee amputees.

Acknowledgements

The authors would like to thank Ben Patritti and Judith Smith for their help with data collection, and Martin Twiste for reconstructing the prostheses.

References

BREAKEY, J. 1976, Gait of unilateral below-knee amputees, *Orthotics and Prosthetics*, **30**(3), 17–24.
DAPENA, J., MCDONALD, C. and CAPPAERT, J. 1990, A regression analysis of high jumping technique, *International Journal of Sport Biomechanics*, **6**, 246–261.
DEMPSTER, W. T. 1955, Space requirements of the seated operator. WADC Technical Report 55-159, Wright-Patterson Air Force Base, Dayton, OH.
ENOKA, R. M., MILLER, D. I. and BURGESS, E. M. 1982, Below-knee amputee running gait, *American Journal of Physical Medicine*, **61**(2), 66–84.
HAY, J. G. 1986, The biomechanics of the long jump, in K. B. Pandolf (ed.), *Exercise and Sports Science Reviews*, vol. 14 (New York: Macmillan), 401–446.
HAY, J. G. 1993, Citius, altius, longius (faster, higher, longer): the biomechanics of jumping for distance, *Journal of Biomechanics*, **26** (Suppl. 1), 7–22.
LEES, A., FOWLER, N. and DERBY, D. 1993, A biomechanical analysis of the last stride, touch-down and take-off characteristics of the women's long jump, *Journal of Sports Sciences*, **11**, 303–314.
LEES, A., GRAHAM-SMITH, P. and FOWLER, N. 1994, A biomechanical analysis of the last stride, touch-down and take-off characteristics of the men's long jump, *Journal of Applied Biomechanics*, **10**, 61–78.
MURRAY, M. P., MOLLINGER, L. A., SEPIC, S. B. and GARDNER, G. M. 1983, Gait patterns in above-knee amputee patients: hydraulic swing control vs constant-friction knee components. *Archives of Physical Medicine and Rehabilitation*, **64**, 339–345.
NOLAN, L. 1995, An investigation of gait characteristics during walking for above- and below-knee amputees. Unpublished Masters thesis, Liverpool John Moores University, Liverpool.
NOLAN, L., WIT, A., DUDZINSKI, K., LEES, A., LAKE, M. J. and WYCHOWANSKI, M. 1998, Adjustments in gait symmetry with walking speed in above- and below-knee amputees, *Proceedings of 3rd Annual Congress of the European College of Sports Science*, 15–18 July, Manchester, 160.
SIMPSON, K. J., WILLIAMS, S., CIAPPONI, T., HSIU-LING, W., NANCE, M. and VALLEALA, R. J. 1998, Regulation of locomotion by above-knee amputee performers during the approach phase of the long jump, *Proceedings of the 3rd North American Congress on Biomechanics*, 14–19 August, Waterloo, Ontario, 465–466.

Smith, G. 1989, Padding point extrapolation techniques for the Butterworth digital filter, *Journal of Biomechanics*, **22**, 967–971.

Williams, S., Simpson, K. J. and Del Rey, P. 1997, Visual targeting during the long jump approach of male below-knee amputee paralympic athletes, *Proceedings of the XVth International Symposium for Biomechanics in Sports*, 21–25 June, Denton, Texas, 49.

Winter, D. A. 1990, *Biomechanics and Motor Control of Human Movement* (New York: John Wiley).

Yu, B. and Hay, J. G. 1996, Optimum phase ratios in the triple jump, *Journal of Biomechanics*, **29**, 1283–1289.

17 Dynamic control and conventional strength ratios of the quadriceps and hamstrings in subjects with anterior cruciate ligament deficiency

C. D. Hole[†*], G. H. Smith[‡], J. Hammond[†], A. Kumar[§], J. Saxton[¶] and T. Cochrane[¶]

[†]Division of Sport Studies, University College Northampton, Northampton, UK

[‡]School of Medicine, University of Dundee, Dundee, UK

[§]Department of Orthopaedics, Northern General Hospital, Sheffield, UK

[¶]Sheffield Institute of Sports Medicine and Exercise Science, University of Sheffield, Sheffield, UK

Keywords: Quadriceps; Hamstrings; Isokinetic; Anterior cruciate ligament; Strength ratios.

The hamstrings:quadriceps muscle strength ratio has been used as an indicator of normal balance between the knee flexors and extensors. A more functional approach to this strength ratio would be to compare opposite muscle actions of antagonistic muscle groups. The dynamic strength control ratio (DSCR) should give a more appropriate measure relating to knee function. There is a lack of normative data relating to DSCR for anterior cruciate ligament (ACL) deficient subjects. Effects of ACL deficiency on isokinetic peak torque for eccentric and concentric muscle actions of the quadriceps and hamstrings, in conjunction with isometric peak torque, were examined in 10 patients awaiting reconstructive surgery (male = 8, female = 2; age = 32.8 ± 8.3 years; height = 1.77 ± 0.08 m; mass = 72.1 ± 12.5 kg). These variables were assessed using an isokinetic dynamometer. The results were considered in terms of the conventional ratio and DSCR. Anterior tibial drawer was measured using a knee ligament arthrometer to confirm clinical diagnosis of ACL rupture. The isokinetic peak torque data analysed were for angular velocities of 1.05 rad s^{-1} (60° s^{-1}). Significant strength deficits were apparent between normal and injured sides for: concentric isokinetic quadriceps action ($p<0.05$); isometric quadriceps action at 70° of knee flexion ($p<0.05$); isometric quadriceps action at 40° of knee flexion ($p<0.01$); eccentric isokinetic hamstrings action ($p<0.05$). With bilateral comparison, the conventional strength ratios showed no significant difference, as did the DSCR. The bilateral comparison of isometric strength ratios revealed significant losses in quadriceps strength for the injured side ($p<0.05$) but no significant losses in hamstring strength ($p>0.05$). Thus, differences can be seen in conventional ratios and DSCR for ACL-deficient subjects. This is an area of clinical interest with the increasing frequency of ACL reconstruction using hamstrings tendons.

1. Introduction

During extension of the knee, the quadriceps work concentrically, and/or the hamstrings work eccentrically. Dvir *et al.* (1989) and Aagaard *et al.* (1998) have

---

*Author for correspondence. e-mail: craig.hole@northampton.ac.uk

suggested that it is these two functions that should be analysed in relation to each other as the components of knee extension. Conversely, for knee flexion, the relationship of concentric hamstrings to eccentric quadriceps modes of muscle action should be looked at. Thus, these two functional ratios, described by Dvir *et al.* (1989: 88) as the dynamic control ratios, can be defined as:

$$\text{Hamstring}_{concentric} : \text{Quadriceps}_{eccentric} = \text{DSCR of knee flexion, and}$$
$$\text{Hamstring}_{eccentric} : \text{Quadriceps}_{concentric} = \text{DSCR of knee extension.}$$

Aagaard *et al.* (1998) suggested that it is a combination of these dynamic control ratios, conventional Hamstrings:Quadriceps (H:Q) strength ratios and measures of absolute strength that can serve to provide a more complete picture of the strength balances for dynamic and static muscle actions, thus giving a clear outline of the functional implications. It may also be useful when evaluating by bilateral comparison the strength differences in patients with ACL deficiencies, to compare directly like for like measures (e.g. Q_{con} injured leg:Q_{con} uninjured leg) as well as the dynamic control ratios. This may be especially useful when attempting to rehabilitate ACL-deficient knees in providing quantitative target torque by means of bilateral comparisons.

The anterior shear forces created by the resistance of the attachment site of the limb to the lever arm in relation to the rotatory force of the maximally contracting quadriceps (Kaufman *et al.* 1991) are thought to be counteracted by the eccentric contraction of the hamstrings. Thus, the functional DSCR of $H_{ecc}:Q_{con}$ may well be indicative of the joint stabilizing effect of the hamstring muscles during knee extension (Aagaard *et al.* 1998). There are important implications when the ACL-deficient knee is being assessed, as it is the ACL that is principally responsible for opposing anterior shear forces in the normal knee. The degree to which the hamstrings compensate (if at all) for ACL deficiency is not known, but it has been suggested that they induce an increased braking effect on motion as the knee becomes more extended (Aagaard *et al.* 1998). This braking has been reflected in the increasing values for the $H_{ecc}:Q_{con}$ as knee extension increases, with values of >1.1 being observed for high isokinetic angular velocities (4.19 rad s^{-1}, 240° s^{-1}). The $H_{con}:Q_{ecc}$ ratio has tended to be lower, for the same studies, with values of around 0.3–0.4 being recorded (Aagaard *et al.* 1995, 1998). These values would suggest that the hamstring muscles contribute to dynamic knee stability to a lesser extent during knee flexion. The values observed for knee extension should be treated with caution. The increasing value for this functional ratio with increasing degree of extension is affected by the muscle length–tension relationship. As the knee moves towards maximal extension, the hamstring muscles are much closer to their optimal length than are the quadriceps to theirs. Thus, if the functional ratio were calculated for a given angle close to full extension, then a high value would be expected. The present study aimed to address the issues of isokinetic dynamometry in relation to ACL deficiency, with special reference to peak torque of hamstrings and quadriceps for concentric, eccentric and isometric muscle actions.

2. Methods

Ten subjects with complete ACL ruptured knees awaiting reconstructive surgery (male = 8, female = 2; age = 32.8 ± 8.32 years; height = 1.77 ± 0.08 m; mass = 72.1 ± 12.5 kg) were randomly recruited from a National Health Service orthopaedic surgery waiting list. Each subject underwent a knee laxity assessment using the KT-

1000 knee ligament arthrometer (MEDmetric, San Diego, USA) to confirm clinical diagnosis of ACL rupture. Following confirmation of ACL rupture, subjects performed a series of seven sets of maximal intensity movements (table 1) using the REV 9000 isokinetic dynamometer (Technogym®, Gambettola, Italy). The speed of all isokinetic movements was fixed at 1.05 rad s^{-1} (60° s^{-1}). Each subject was seated upright so that the leg being assessed was in a straight line, with hip, knee and ankle all being aligned. The subject was then secured into the seat, with the cuff of the lever arm being appropriately attached to the lower limb. The dynamometer's procedure for gravity correction was then performed, and the outcome of this taken into account for all of the results attained. This process was repeated for both legs of each subject, with the positions of all of the various parts of the isokinetic dynamometer being kept the same across two assessments.

The importance of pain-free work was highlighted and each subject was reminded of the need for maximal effort throughout the whole series of assessed exercises. The range of motion limits were then set at 0° (knee extension of 180°) and 90° (90° of knee flexion). The assessment protocol is shown in table 1.

The subjects underwent repetition of the assessment after a rest period of 10 days to allow for recovery from muscle soreness. The values included for analysis were the highest peak torques achieved by an individual subject over the two assessment sessions. This method of selection was decided to be the best means of identifying the

Table 1. Protocol for peak torque assessment using the REV 9000 isokinetic dynamometer.

Series order	Description	Number of repetitions/ duration	Angular velocity/ angle of flexion
1	Maximal concentric continuous reciprocal extension-flexion	×8	60° s^{-1} (1.05 rad s^{-1})
	Rest period	60 s	
2	Maximal concentric-eccentric continuous reciprocal extension-flexion	×8	60° s^{-1} (1.05 rad s^{-1})
	Rest period	60 s	
3	Maximal concentric-eccentric-continuous reciprocal flexion-extension	×8	60° s^{-1} (1.05 rad s^{-1})
	Rest period	60 s	
4	Maximal isometric quadriceps contraction. Recovery of 30 s between repetitions	3 × 5 s	40°
	Rest period	60 s	
5	Maximal isometric hamstrings contraction. Recovery of 30 s between repetitions	3 × 5 s	40°
	Rest period	60 s	
6	Maximal isometric quadriceps contraction. Recovery of 30 s between repetitions	3 × 5 s	40°
	Rest period	60 s	
7	Maximal isometric hamstrings contraction. Recovery of 30 s between repetitions	3 × 5 s	40°

values that represented the maximal measure that the dynamometric assessment was seeking to record. This met Dvir's (1995) suggested criteria for analysing isokinetic peak torques. None of the data was normally distributed; therefore the Mann-Whitney U test was employed throughout the analyses.

3. Results

Table 2 shows the mean dynamic strength ratios with standard deviations for the injured subject group. The values represent the relationship between peak torque of the muscles (and their action) at an angular velocity of 1.05 rad s^{-1} (60° s^{-1}). The differences between peak torque for dominant and non-dominant legs of the subjects revealed no differences as being statistically significant for any of the peak torque values ($p > 0.05$). Table 3 shows the mean bilateral strength ratios for peak isometric torques of injured subjects.

4. Discussion

The main area of concern within the present study relates to the functioning of the thigh muscles in the ACL-deficient leg. The contribution of the thigh musculature towards knee joint stability, and thus knee functioning, has been well documented (Wilk *et al.* 1994, Chen *et al.* 1995, Snyder-Mackler *et al.* 1995, Li *et al.* 1996). The methods adopted in attempts to quantify this contribution in respect to dynamic functioning have incorporated the use of isokinetic dynamometry. This has provided quantitative data which can be analysed to identify trends in particular populations. Although there is a great deal of research in the field of isokinetics, there is as yet no published work that has used the REV 9000 isokinetic dynamometer. The significant difference ($p < 0.01$) in peak torques between the normal and injured sides, for concentric quadriceps muscle action, indicates that a strength deficit exists. This suggested deficit in quadriceps strength was apparent for both concentric and

Table 2. Mean strength ratios ($\pm$SD) for peak isokinetic torques of injured subjects.

Ratio	Injured leg	Normal leg
$Q_{con}:Q_{ecc}$	0.83 $\pm$ 0.13	0.84 $\pm$ 0.16
$H_{con}:H_{ecc}$	0.82 $\pm$ 0.13	0.73 $\pm$ 0.10
$H_{ecc}:Q_{con}$	0.70 $\pm$ 0.14	0.70 $\pm$ 0.09
$H_{con}:Q_{ecc}$	0.47 $\pm$ 0.11	0.42 $\pm$ 0.09
$H_{con}:Q_{con}$	0.47 $\pm$ 0.11	0.50 $\pm$ 0.05
$H_{ecc}:Q_{ecc}$	0.58 $\pm$ 0.15	0.59 $\pm$ 0.14

No significant differences present between ratios for normal and injured legs ($p > 0.05$).

Table 3. Mean bilateral strength ratios ($\pm$SD) for peak isometric torques of injured subjects.

Description	Angle of flexion	Ratio $\pm$ SD
$Q_{injured}:Q_{normal}$	70°	0.85 $\pm$ 0.12†
$H_{injured}:H_{normal}$	70°	0.99 $\pm$ 0.21
$Q_{injured}:Q_{normal}$	40°	0.90 $\pm$ 0.08†
$H_{injured}:H_{normal}$	40°	0.98 $\pm$ 0.19

†Denotes significant difference in bilateral isometric torque.

isometric muscle actions. The results for these measures of muscle strength in relation to ACL deficiency sit well with the current literature. Murray et al. (1980) identified deficits of up to 10% in concentric isokinetic peak torques for the quadriceps of the ACL-deficient leg, in comparison with the normal leg. Bonamo et al. (1990) reported quadriceps deficits of 11% at an angular velocity of 1.05 rad s^{-1}. Dvir (1995) suggested that significant quadriceps strength deficits of approximately 20% exist during at least the first year following tear of the ACL. Further to this, he proposed that the effect of ACL rupture on hamstrings strength is significantly less than that of the quadriceps. The results of the present study support that statement, with only the eccentric isokinetic hamstrings action showing a significant deficit ($p<0.05$) out of all of the peak torques relating to knee flexor strength.

All of the strength deficits observed as being significant for the bilateral comparison of the injured group were concerned with extension of the knee. Since the ACL is concerned with maintaining joint integrity during extension, and preventing hyper-extension, the injury to this tissue has been thought to induce reflex inhibition of the thigh musculature (Morrisey 1989). This means that the voluntary contraction of the quadriceps is impeded by afferent stimuli arising from the injury to the joint. This might account for some of the strength deficits identified. Other factors relating to muscle atrophy through a reduction in use may also have contributed significantly to this defecit. The effects on torque as a result of reduced innervation could only be clearly identified with employment of electromyography during the isokinetic assessment.

The loss of eccentric hamstring strength in the injured leg may be connected with the theory relating to the ACL-hamstring reflex. This refers to the firing of hamstring motor units in response to stretching of the ACL (Jennings 1994). This response is regarded as a protective role of the hamstrings during extension of the knee. It has been suggested by the same author that rupture of the ACL results in a loss of proprioception. In turn this loss may cause a reduction in activation of hamstrings during extension of the knee in the ACL-deficient subject, subsequently leading to a reduced eccentric strength in the hamstrings.

The lack of bilateral strength deficits observed for the other measures relating to peak hamstrings torque may be regarded as fitting expectations when the evidence from the literature is reviewed. Bonamo et al. (1990) showed that deficits of only 3% for hamstrings in bilateral comparisons existed. Dvir (1995) suggested that strength deficits associated with injury to the ACL subside in the longer term, and that full recovery of hamstring strength in the involved side may be expected. With the majority of injured subjects having sustained the ACL injury over 1 year prior to the assessment, this lack of deficits observed for peak torques of the hamstrings may have been as a result of this trend. The results for the conventional strength ratios ($H_{con}:Q_{con}$, $H_{ecc}:Q_{ecc}$) of the ACL-deficient subjects, did not show significant differences ($p>0.05$) between injured and normal sides.

The dynamic strength ratio for knee extension ($H_{ecc}:Q_{con}$) showed no difference between the injured and uninjured legs of the subjects. This might be accounted for by the strength deficit noted for concentric quadriceps action ($p<0.01$) being cancelled out by the strength deficit in the eccentric hamstrings action ($p<0.05$). This ratio was 9% greater than the value obtained from athletes by Aagaard et al. (1998), which would suggest a shift in the balance of strength for ACL-deficient subjects with the stabilizing effect of the hamstrings being relatively increased for both legs. The similarity between the injured and normal legs of the subjects is not accounted for by a stabilizing hamstring effect brought about by ACL deficiency, as this would only affect the injured leg. The low angular velocity used throughout the assessment sessions (based upon clinical medical advice) may have influenced the value attained for this functional ratio. Aagaard et al. (1998) showed that this ratio was affected by angular velocity, increasing as the angular velocity increased, suggesting that the contribution of the hamstrings to knee stabilization increased with angular velocity.

The differences between injured and normal legs for isometric quadriceps actions were well defined here, once again confirming the pattern of strength losses for the quadriceps of the injured leg. Deficits of 15% and 10% were revealed for angles of flexion of 70° and 40°, respectively. These deficits served to support the findings of the dynamic concentric action of the same muscle group, and once again illustrated the reduction in maximal functioning of the quadriceps muscles. The differences in the isometric hamstrings ratios were only 1% and 2% for the muscle actions at 70° and 40° of flexion respectively, and as such showed that no significant difference was present.

The findings of this study have generally fitted well with the evidence available in the current literature. The importance of increasing the pool of information relating to muscle function of the ACL-deficient knee cannot be underestimated for this increasingly common injury, as the mode of treatment ultimately depends upon acting on this information, and the related functional capabilities of the injured subject. Future research may seek to combine the assessment of the DSCR in conjunction with electromyography, providing a more detailed analysis of the activation of the individual muscles of the hamstrings and quadriceps throughout the range of motion of the knee for the ACL-deficient subject. More comprehensive research would give a clearer picture of the functional role of the hamstrings allowing a more discerning judgement as to the appropriateness of using hamstrings tendons for reconstruction of the ACL in all subjects.

References

AAGAARD, P., SIMONSEN, E. B., MAGNUSSON, S. P., LARSSON, P. T. B. and DYHRE-POULSEN, P. 1998, A new concept for isokinetic hamstring:quadriceps muscle strength ratio, *American Journal of Sports Medicine*, **26**, 231–237.

AAGAARD, P., SIMONSEN, E. B., TROLLE, M., BANGSBO, J. and KLAUSEN, K. 1995, Isokinetic hamstring/quadriceps strength ratio: influence from joint angular velocity, gravity correction and contraction mode, *Acta Physiologica Scandinavica*, **154**, 421–427.

BONAMO, J., COLLEEN, F. and FIRESTONE, T. 1990, The conservative treatment of the anterior cruciate deficient knee, *American Journal of Sports Medicine*, **18**, 618–623.

CHEN, C. Y., JIANG, C. C., JAN, M. H. and LAI, J. S. 1995, Role of flexors in knee stability, *Journal of the Formosan Medical Association*, **94**, 255–260.

DVIR, Z. 1995, *Isokinetics: Muscle Testing, Interpretation and Clinical Applications* (Edinburgh: Churchill Livingstone).

DVIR, Z., EGER, G., HALPERIN, N. and SHKLAR, A. 1989, Thigh muscle activity and anterior cruciate ligament insufficiency, *Clinical Biomechanics*, **4**, 87–91.

JENNINGS, A. G. 1994, A proprioceptive role for the anterior cruciate ligament: a review of the literature, *Journal of Orthopaedic Rheumatology*, **7**, 3–13.

KAUFMAN, R. K. R., AN, K. N., LITCHY, W. J., MORREY, B. F. and CHAO, E.Y. 1991, Dynamic joint forces during knee isokinetic exercise, *American Journal of Sports Medicine*, **19**, 305–316.

LI, R. C. T., MAFFULLI, N., HSU, Y. C. and CHAN, K. M. 1996, Isokinetic strength of the quadriceps and hamstrings and functional ability of anterior cruciate deficient knees in recreational athletes, *British Journal of Sports Medicine*, **30** 161–164.

MORRISSEY, M. C. 1989, Reflex inhibition of thigh muscles in knee injury: causes and treatment, *Sports Medicine*, **7**, 263–267.
MURRAY, P. M., GARDNER, G. M., MOLLINGER, L. A. and SEPIC, S. B. 1980, Strength of isometric and isokinetic contractions, *Physical Therapy*, **60**, 412–419.
SNYDER-MACKLER, L., DELLITO, A., BAILEY, S. L. and STRALKA, S. W. 1995, Strength of the quadriceps femoris muscle and functional recovery after reconstruction of the anterior cruciate ligament, *Journal of Bone and Joint Surgery*, **77-A**, 1166–1173.
WILK, K. E., ROMANIELLO, W. T., SOSCIA, S. M., ARRIGO, C. A. and ANDREWS, J. R. 1994, The relationship between subjective knee scores, isokinetic testing, and functional testing in the ACL reconstructed knee, *Journal of Orthopaedic and Sports Physical Therapy*, **20**, 60–73.

Part V
Human Factors and Psychology

18 Performance and human factors: considerations about cognition and attention for self-paced and externally-paced events

ROBERT N. SINGER

Department of Exercise and Sport Sciences,
University of Florida, Gainesville, Florida 32611, USA

Keywords: Motor skill learning; Attention; Cognition; Learning strategies.

The cognitive psychology school of thought has spawned models of sequential stages or phases of information processing associated with various tasks. It has encouraged the study of cognitions and attention as related to learning, performance and high levels of achievement in goal-directed complex activities in which movement is the medium of expression. Although more recently proposed dynamical systems models challenge the simplicity of this approach, there is little doubt that the ability to learn as well as to excel in performing movement skills depends to a great degree on the effective self-regulation of cognitive processes in a variety of situations. What to think about (or not think about) prior to, during and even after an event can have great consequences on present and subsequent performance. Relevant externally-provided and self-generated strategies should enable these processes to function at an optimal level, and are the subject of an increasing amount of research. For such purposes, it is convenient to categorize events as self-paced (closed) and externally-paced (open). Examples of both types of events exist in sport as well as in various occupations and recreational activities, with different information processing demands associated with each one. Any breakdown in a particular stage of processing will potentially lead to poorer performance. Special training techniques and strategies are evolving from the cognitive and psychophysiological research literature that might improve the level of functioning at each stage for either self-paced or externally-paced skills.

1. Introduction

The ability to acquire, master and demonstrate complex motor skills under varying environmental and stressful conditions is indeed a challenge for the performer. Understanding and explaining the mechanisms, cognitive processes and self-regulatory strategies that enable the acquisition of, and proficiency in, skills poses challenges for the scientist/scholar. Furthermore, the broker—a person in the middle who can translate research for the practical consumer—completes the process from research generation to applied meaningfulness; from the laboratory to interpretation, application and usability. That, too, is a challenge.

Cognitive psychologists advocating an information processing framework for understanding learning and performance (Barber 1988) have attempted to address this challenge. However, a challenge for those endorsing this approach in the study of skilled movement behaviours is to validate their value in the face of critics who represent the increasingly popular dynamical systems (ecological, self-regulation) approach (Turvey 1990, Kelso 1994). Dynamical systems models tend to neglect or minimize the role of cognitions and attention in human learning and the performance of complex movement patterns and skills. Obviously, cognitive models advocate quite the opposite.

Williams et al. (1992), in an overview article, and in an excellent in-depth analysis of scholarly perspectives in a recently published book (Williams et al. 1999), have summarized the leading disagreements between information processing and dynamical systems models. They examined sports-related tasks and discussed the artificiality and limitations of laboratory-oriented research paradigms associated with cognitive psychology. Proponents of perception-action approaches try to be more ecologically valid, but these approaches are not without their share of strengths and weaknesses. Whereas information processing models conveniently but perhaps artificially divide internal events into key states that act in sequence or overlap and stimulate actions, they do lead to somewhat concrete investigations. On the other hand, alternative models have a more dynamic and realistic view of real-world complex acts, but they do not easily lead to convenient research paradigms to test basic tenets.

This review represents an attempt to provide a rationale for the use of cognitive models, in spite of well-recognized shortcomings, primarily for the applied research that can be generated. Meaningful findings should lead to better learning and performance experiences for individuals ranging in level of demonstrated ability to execute complex movement skills, as represented, for example, in sports. Objectives will be to (1) overview briefly the cognitive psychology movement as related to scholarly and practical interests of sport psychologists, (2) provide a rationale for the use of a cognitive model, which emphasizes understanding the nature of information processing, cognitions, attention and strategies related to acquiring skill in complex motor skills as associated with sport, and (3) discuss the nature and cognitive demands of self-paced and externally-paced events, with implications for training and improving performance.

2. The cognitive psychology movement

Primarily through the efforts of Bruner et al. (1956) and Chomsky (1957), among other pioneers, the evolution of information processing perspectives and cognitive psychology was subsequently to have a tremendous impact on scholars and practitioners (Lachman et al. 1979). Earlier and present-day concerns are related to understanding more about the functional mental capabilities of the person (Glaser 1990). Cognitive operations on information processing, as related to attention, memory, decision-making and problem-solving, for example, have been subjected to much study (Bower and Hilgard 1981, Anderson 1990) and continue to be of great interest.

Furthermore, the expert-novice paradigm has provided insights into ways that mental operations differ as a function of skill, especially in regard to memory, attention and task-achieving strategies (Chase and Simon 1973, Chi et al. 1981, 1982). In addition, understanding expertise and automaticity goes hand in hand, especially in regard to attentional resources and their development. The insightful research of Shiffrin and Schneider (1977) described the nature of controlled (conscious and slow) processing of information associated with the beginner versus the automatic (effortless and rapid) processing displayed by the expert. More detailed analysis of the nature of automaticity has been offered by Logan (1988).

Determining cognitive functions related to the potential to accomplish contributes important perspectives to the body of knowledge associated with the psychology of learning in the classroom. In addition, there are practical implications for achieving mastery in other knowledge situations. These include performance of complex motor skills, as required in various occupations, sports and recreational activities (Abernethy 1988, 1990, 1993). Some obvious pragmatic questions arise. How can information processing operations be improved (Singer 1998)? What can a supervisor, instructor, teacher or coach do to assist a worker, student or athlete in the functioning of cognitive operations, thereby leading to quality learning and acceptable standards of performance (Singer and Chen 1994)? How can progression occur quickly from the beginning stages of learning (trying to know about something) to a high level stage (how to operationalize that knowledge, to use it effectively in various circumstances; Anderson 1982, 1983, 1990)?

Of even more important concern is what individuals can do for themselves to attain proficiency, considering self-direction and self-regulatory strategies. Indeed, significant self-control techniques (Carver and Scheier 1998) have been advanced in this regard. In addition, the nature of motivation was at one time considered to be very behaviourally-based (e.g. dependent upon reinforcement contingencies). Such thoughts have been modified with the increasing realization that personally-set expectancies, goals, attributions and self-perceptions in general are important determinants of achievement motivation in various domains (for examples in sport, see Biddle 1999, and relevant chapters in Singer *et al.* 1993b).

3. Implications and applications for sport psychology

Sport psychology research and practice has advanced considerably as a result of the theoretical orientation of cognitive psychology, issues raised and paradigms produced. Representative research contributions and books are associated with:

(1) the increased study and understanding of cognitive and attentional processes and their functions, as well as strategies associated with improved learning and performance in sports (Abernethy 1993, Moran 1996);
(2) a conceptual framework (the computer metaphor) for studying sequential stages that are internally activated, from selectively attending to pertinent sources of information (cues) to completing a movement skill (Singer 1980, Schmidt and Lee 1999);
(3) the formulation of practical guidelines to help athletes to use mental processes to their advantage (Lohasz and Leith 1997, Kirschenbaum *et al.* 1998, Williams 1998);
(4) the development of self-regulatory strategies to cope with stress, self-doubt and potential performance distractors, thereby enhancing achievement potential (Crews 1993, Hanton and Jones 1999); and
(5) the recognition of self-expectations, self-efficacy, attributions and other important self-perceptions associated with achievement motivation (Biddle 1999, Orbach *et al.* 1999).

When learning sports skills, the emphasis has traditionally been on the physical condition of the body and movement technique. Through the impact of cognitive psychology, skilful movement production is also considered to be the result of the effective activation of pertinent cognitive processes as required, on the one hand, or the ability to inhibit them, as in performance automaticity with advanced skill (McPherson and French 1991), on the other hand. Better athletes seem to manage their emotions and cognitions more effectively and appropriately to meet the challenges of achieving in performance contexts. They learn how to use procedural (application) processes for situational demands. Very skilled individuals are capable of managing internal control processes associated with information intake, initiating appropriate temporal and spatial movement plans, effectively controlling movement acts through the preprogramming of movements, and utilizing appropriate feedback for movement evaluation and subsequent corrections.

There is increasing acceptance of the doctrine of mind over matter, that a trained and disciplined mind can control biological functions and, in turn, performance. The first half of the Twentieth Century was dominated by the behaviouristic school of thought through which were studied ways for an outsider, e.g. the instructor, to modify environmental settings and orient the learner to be 'shaped' to acquire desirable knowledge and behaviours. Much of this information has proven to be valuable. The present popularity of cognitive psychology has indicated a completely different direction of focus: 'the structures and process of human competence and the nature of the performance system' (Glaser 1990: 29).

4. Sports and ergonomics

Many sports require the use of equipment, projectiles and other articles, and the athlete needs to function in harmony with them. The tennis player with a racquet, a baseball player with a bat, a hockey player with a stick and the golfer with clubs are but a few examples of the athlete-equipment-situation interface. Primary concerns in the ergonomics literature are with human-machine-environment interfaces. As Shephard (1988) put it, the goal of ergonomics is to determine the optimal matching of humans and machines. A reason for this goal would be to understand more about human potential and achievement. Athletes with their tools and musicians with their instruments also attempt to attain the ideal match and comfort zone that leads to optimal achievement. Concepts, research and applications associated with ergonomics (human factors, engineering psychology) in the 1940s have been expanded considerably (Schmidt and Lee 1999).

Indeed, much of the initial research associated with sports skills learning and performance as well as expertise stemmed from the ergonomics literature (Singer 1980). The information processing strategies used by fighter pilots and radar operators during World War II, as well as the influence of certain conditions on attention, vigilance and performance (Barber 1988), were later to be considered in sports contexts. Control, cybernetic and communications models that evolved following the war seemed to have greater explanatory power than the behaviouristic model for understanding skilled performance. Research programmes in sports psychology have matured considerably in recent years. Indeed, the study of sports expertise has contributed to theoretical and applied developments in ergonomics (Schmidt and Lee 1999).

A model for sports participation in integrating ergonomics into a sports framework has been proposed by Reilly and Shelton (1994), who considered the psychological demands of an activity. Hardy and Nelson (1988) discussed stress reduction and self-regulation within sport settings, with implications for worker productivity. Some recent developments in the study of the role of cognitive processes in motor learning have been described by Annett (1993). He also showed research and application connections between sports science and ergonomics.

From another perspective, sports scientists concerned with understanding relationships among attention, distraction, arousal and performance have used such tasks as simulated auto racing (Janelle *et al.* 1999). Implications could be drawn for dynamic sports settings in which attentional flexibility and decision-making are crucial under severe time constraints and competitive pressure. Applications could also be made to understanding car driving performance under situations of varying task demands. Finally, the study contributed to basic research on issues surrounding external distraction, emotions, the attentional narrowing concept and skilled performance. Themes and issues are of interest in ergonomics as well as in sports settings.

Indeed, there are many relevant connections. Emphasized in this review is the study of internal processes and mechanisms activated or suppressed with appropriate strategies as complex movement skills are being acquired and then when superior proficiency is demonstrated. One goal is to be able to identify cognitions and attentional processes that contribute to skill immediately prior to and during a brief act or during sustained activity. Indeed, the ability to regulate deliberately or subconsciously a number of internal processes related to mental activities and emotions is a big part of achieving. By attempting to identify how these operations contribute to successful movement execution for specific types of activities, remedial (i.e. training) programmes for deficits can be generated to improve the way in which a particular stage in information processing functions, as suggested elsewhere (Singer 1998). A basic premise in cognitive psychology is that achievement in any goal-oriented situation is possible with the activation of appropriate strategies.

5. Rationale for the use of an information processing model

The information processing approach, with the identification of mental operations that systematically become activated from selective attention to task-relevant cues to response production, has been conveniently associated with flow diagrams and schematics. The information processing model is largely based on the computer analogy and therefore it has no problem in explaining the functioning of operations of sequential stages, as computer programmes consist of symbolic representations defined by experts (Dinsmore 1992). Terms associated with the computer metaphor such as motor programmes, memory drum, encoding and storage conventionally suggest the characteristics of seriality, limited capacity and symbolic representation of brain functions. Also, a strong predictability in the value of strategy training is often implicitly acknowledged.

A major complaint is that this type of model is too simplistic. It presumably fails to account for a valid description of the internal-external dynamics that occur in real-world situations. Whereas sensations, perception, cognitions and action have been traditionally studied as separate operations, an alternative view (as mentioned earlier in this paper) is that they are not separable. Ecological and dynamic concepts (Turvey 1990, Cziko 1992) emphasize the interaction of action and perception, and offer alternative approaches for the study and understanding of motor behaviour. While information processing models depend to a great extent on computer analogies of human behaviour, the orientation of perception/action models seem to be closer to the biological sciences. Simplifying the concept, perception leads to action and action influences perception (Prinz 1990, 1997). Gibson's (1979) classic work on a comprehensive view of perceptual systems and Bernstein's (1967) intriguing analysis and concepts of the human system as containing self-control and self-regulating devices have had a great impact on the advancement of the ecological approach. There is no denial of the contributions of such perspectives, for example Turvey's (1990) work related to motor behaviour.

Of course, any attempt to isolate sequential mental stages involved in the execution of complex motor skills is somewhat artificial. This is especially true in time-constrained events that require rapid anticipation, decision-making and adaptive responses when there is uncertainty as to what may happen. Boxing and goal-tending in hockey are good examples of such situations. With a high level of proficiency, it appears that the athlete is performing in a state of automaticity. However, even the best of athletes may need to refine some aspect of their 'mental' performance, such as distributing attention to the most relevant minimal cues at the right time in order to predict and anticipate an opponent's intentions. Being in position early enough on the tennis court leads to a greater capability of making good decisions as to what to do in returning the opponent's shot and, in turn, attaining an offensive position.

In other words, the cognitive elements related to successful performance presumably can be identified in fast action situations or in those cases where there is plenty of time to prepare to act. We have contrasted externally-paced events (open skills; Singer 1998) with self-paced events (closed skills; Singer 1988) in order to examine relevant cognitive and attentional processes. The intention is to identify primary cognitive components associated with self versus externally-paced tasks in order to instruct beginners to initiate pertinent strategies to improve the way a component, or stage, contributes to the overall successful completion of a complex movement act. Even a more highly skilled individual may demonstrate some limitations in the way a cognitive stage is functioning in the context of the overall act. Specialized training techniques should improve deficits. The next section includes a discussion on self-paced and externally-paced events, with flow charts to highlight the hypothesized sequences of stages from preparation to execution.

6. Self-paced and externally-paced events

Preparatory and attentional factors (location, duration, distribution, intensity) primarily distinguish requirements associated with the execution of self-paced and externally-paced complex motor skills. In self-paced events, there is plenty of time to go through a routine, or ritual, before initiating the act. Externally-paced events typically require rapid anticipation, decision-making and reactions.

Sports such as golf, bowling and archery, or elements in sports, such as the serve in tennis, the foul shot in basketball or the serve in volleyball, allow time for the athlete to prepare and to execute when ready (within reasonable time limitations). Situations are stable and predictable. Repetitious practice is dedicated to developing consistency in the preparatory stage and in the act itself.

On the other hand, externally-paced acts involve anticipating the opponent's intentions, recognition of meaningful cues offered in the situation, rapid decision-making under severe time constraints and the initiation of an appropriate response in a timely fashion. Essentially, attending to the most meaningful yet minimal cues and information processing occur quickly. Examples of externally-paced skills are wrestling and boxing manoeuvres, returning a serve in volleyball or tennis, a hockey goalkeeper stopping a puck and a batter hitting a baseball. Meaningful practice includes a great variety of experiences in all kinds of related situations in order to acquire adaptive skills. Effective execution depends on adaptability, and top-level athletes demonstrate elements of automaticity and creativity in many situations.

7. Improving performance in self-paced acts

For self-paced activities, the 'Five-step Strategy' (Singer 1988), or some variation of it, seems to be particularly helpful in the learning process as well as for the production of high levels of skill. The Strategy contains five sub-strategies:

(1) *readying* by establishing a ritual that involves optimal positioning of the body, confidence, expectations and emotion;
(2) *imaging* a picture and the feeling of performing an act at one's best;
(3) *focusing attention* on a relevant external cue or thought;
(4) *executing* with a quiet mind; and
(5) *evaluating* if time permits, the quality of execution of the act and the outcome as well as the implementation of the previous four strategies.

For over a decade published research under laboratory and field conditions indicates the effectiveness of the Five-step Strategy and similar strategies in contributing to achievement (Singer and Suwanthada 1986, Singer *et al.* 1989, 1993a, Lidor *et al.* 1996). In the sequence of events (figure 1), the first three steps, separately or entwined together, involve activating and directing cognitions and attention for functional purposes related to the preparatory state. The ideal execution state involves inhibiting thought processes and performing with self-trust. Self-regulation techniques help to control and direct thought, emotional and physiological processes. In some sports such as golf, the preparatory routine also leads to a feeling of the rhythm needed to strike the ball the intended distance.

In the sequence of events associated with self-paced acts, attention is directed to focusing on the situation and challenge. In addition, an internal analysis addresses perceptions of self-efficacy, being under emotional control, and feeling good about the possibilities to do well. Attention is also oriented to creating an image and feelings of the intended act. Attention then has to be directed away from thoughts of self-doubt and the fear of failing, as well as awareness of such external distractors as noise, and the presence, movement and sounds of others. A useful technique is to

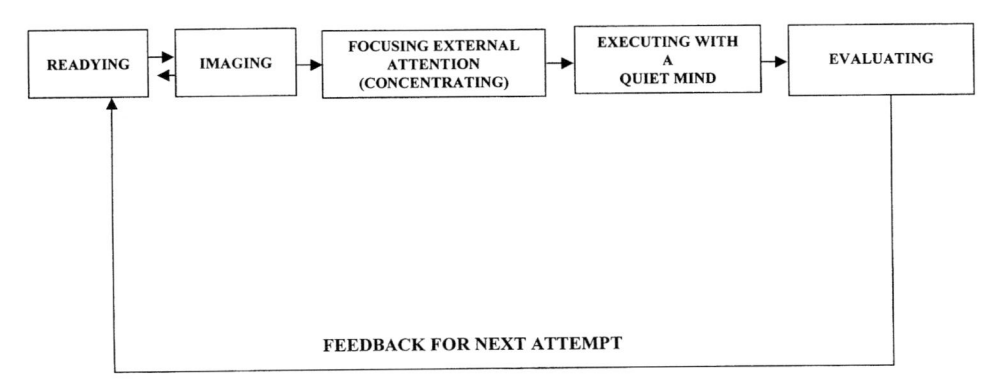

Figure 1. The Five-step Strategy for self-paced skills.

focus intently on the target, an object or a thought in order to block out internal and external distractors. The act is then performed with conscious attention suppressed, thereby allowing the body to execute what the image has created. Of course, performance potential is dependent on developed capabilities, and the image should reflect realistic expectations. The last stage, if time is available, suggests that attention should be generated toward analysing the performance—the outcome, the technique and the ability to implement appropriate strategies associated with the previous four stages.

Any breakdown at a particular stage can lead to poor performance. Figure 2 shows stages related to the 'Five-step Strategy'. It indicates the cognitive processes that may be activated in the different stages, as well as the suppression of them during the execution phase, for different purposes. Learners acquire self-paced tasks faster and skilled performers develop a high level of consistency with the finely-tuned use of strategies that can guide cognitions throughout the preparatory stage to the completion stage. The expert more than likely goes through the entire routine without deliberately thinking about the task.

Each of the five sub-strategies in the 'Five-step Strategy' have been researched and overviewed fairly exhaustively as independent contributors to the acquisition of skill. The research has included the examination of emotional readiness and perceptions of control over self (Hanin 1989, Jackson 1992, Hanton and Jones 1999), performance expectancies and self-efficacy (Biddle 1999), imagery (Ryan and Simmons 1982), concentration (Moran 1996), external focus on a target versus self-focus to the feelings of the movement (Hatfield *et al.* 1987, Singer *et al.* 1993a), blanking the mind while executing (Landers *et al.* 1991) and self-evaluation (Kitsantas and Zimmerman 1998).

The work of Wulf and her colleagues (Wulf *et al.* 1998, 1999, Wulf and Toole 1999) on focus of attention is an example of sub-strategy analysis with implications for instruction. This research supports the significance of an external focus of attention for learning. Their findings indicate that it is beneficial for learning to direct attention away from one's own movements (internal focus, a self-awareness strategy). Whereas the external focus of attention was on a target in the studies of Singer and his colleagues (Singer *et al.* 1993a), the research of Wulf and her colleagues has involved directing one's attention to the effect of the action. They have used the ski-simulator task, a stabilometer (dynamic balance task) and the pitch shot in golf each with similar results (Wulf *et al.* 1998, 1999, Wulf and Toole 1999).

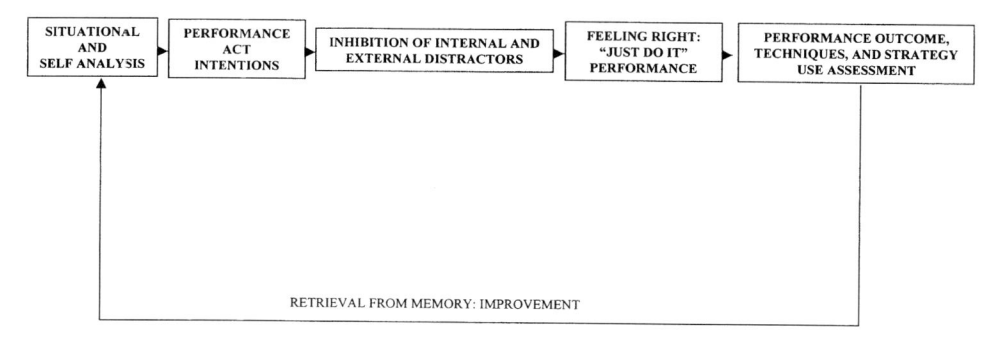

Figure 2. Cognitive processes associated with improved performance of self-paced skills.

In another study, Maddox *et al.* (1999) worked with beginning tennis players to study further the effect of the direction of attention on performance. The internal group was told to focus on the backswing and the contact point, while the external group was oriented to focus on the target area and the arc of the ball after contact. The external condition resulted in superior performance for retention and transfer.

Many guidelines and training programmes have appeared in the sport psychology literature (Singer *et al.* 1993b, Williams 1998) that might aid an athlete in improving the way each of these behaviours can be improved upon. If the Five-step Strategy or any global strategy was not particularly helpful, a critical analysis might indicate where one phase was dysfunctional. If, for example, a concentration problem was identified, attention focus would need to be trained (Schmidt and Peper 1998) so that thoughts do not wander to internal and external distractors. Skills in sport are typically associated with physical, physiological, biomechanical and technical capabilities and therefore the various ways in which cognitions and attention are involved in contributing to learning and accomplishing are rarely considered. Furthermore, an understanding of useful strategies, how to implement them and what is the appropriate timing, should help the person to produce a higher level of skill in a demanding self-paced activity.

In another direction, self-paced skills such as rifle shooting, archery and golf have been the subject of psychophysiological research to determine profiles of experts as they execute as well as when they perform at their best. Strategy implications have been derived from this type of research as to how to perform better by being in an ideal state immediately prior to and during the act. Of interest have been electrical activity in the brain, respiration rate, heart rate, EMG activity and visual search patterns.

Electroencephalogram (EEG) wave patterns, especially with regard to alpha level immediately prior to and during execution, presumably represent an important index reflecting attentional processing. Each hemisphere is associated with different processing activities, for example analytical-verbal (left) versus spatial (right). Conflicting research findings (Hatfield *et al.* 1987, Landers *et al.* 1994) have been reported in regard to alpha activity, in either or both hemispheres, associated with highly skilled athletes. More alpha activity in the left hemisphere, to a point, might be related to less active verbal-analytical processes. Thus, the athlete's state would be 'let it happen', to produce an act without deliberate awareness. For instance, Hatfield *et al.* (1984) studied elite marksmen and found an increase in alpha activity in the left hemisphere within 7.5 s of pulling the trigger. Alpha activity in the right hemisphere remained constant. In another study, Salazar *et al.* (1990) analysed the four best and worst shots of archers, and noted that an increase in alpha activity in the left hemisphere was related to the poorer performances. Perhaps too little or too much alpha activity contributes to undermining achievement potential.

Lacey (1967) proposed that heart rate (HR) decelerates and cortical activity lessens immediately before one initiates an act that requires an external attentional focus, as in target-aiming tasks. Golf-putting, archery and rifle shooting have been used to test Lacey's hypothesis. To provide further perspectives with regard to alpha activity, HR and performance, Radlo *et al.* (1999) also examined the effects of an internal focusing strategy versus an external one on dart-throwing skills. The external group (focusing on the centre of the target) performed better than the other group that was taught to direct attention toward self-movements when making a toss. Heart rate decelerated in the 'external' group immediately prior to dart release, while the opposite was true for the other group. Alpha EEG power was lower for the 'external' group. Implications are that an external strategy leads to more ideal psychophysiological functions that facilitate skilled performance.

Radlo (1999) compared the effects of the 'Five-step Strategy' on developing skill in an underhand dart-throwing task under non-competitive and competitive conditions in the author's laboratory. EEG activity and HR were assessed immediately before and during tosses. The Five-step Strategy contributed not only to significantly better performance scores as compared with the control group, but also to increased alpha activity and decreased beta activity in the left hemisphere during a 0–1.5 s period before the throw. The opposite occurred with the control group. The Strategy apparently led to the right side of the brain becoming the dominant processor, thereby minimizing verbal/analytical distracting information. Besides a decrease in left hemisphere activation, the Strategy led to progressive HR deceleration while the control group showed increased HR immediately prior to release of the dart. Emphasis on the strategy to focus attention externally to the target seemed to be related to the decelerated HR, consistent with other research (Lacey 1967, Boutcher and Zinsser 1990). In addition to comparisons made between the Strategy group and the control group, the four best and worst shots, according to Mean Radial Error (MRE) scores, were analysed. Best achievement was associated with an increase in alpha activity in the left hemisphere and a decrease in HR immediately before the toss. Benefits seem to be attributed to relaxing the left hemisphere, the right hemisphere becoming the dominant processor, and a decrease in HR.

In another investigation, Murray et al. (1999) attempted to implement visual extraction information from the environment concurrently with visual scanning instrumentation along with further processing of acquired information (as reflected by EEG activity). A shooting task was used with a laser shooting simulation. A significant association was observed between EEG spectral content, prolonged quiet eye periods and performance.

Visual search, or gaze, behaviours have been assessed in other research to determine differences between experts and novices in such activities as billiards, putting in golf and free-throw shooting in basketball. Fixation duration and location data imply degree of focus of attention (concentration) immediately prior to and during the act. For target skills, of interest is the extent of concentration on the critical point of aim. In billiards and putting in golf, gaze patterns shift back and forth until ready to perform the skill. Therefore, an ideal number of switches to appropriate locations probably exists, as well as an optimal duration of fixations at each location. Another consideration is the amount of time spent in the preparation phase prior to execution, considering task complexity, and comparing times of experts with those less skilled. These kinds of evidence imply that there are strategies that performers may want to use in order to improve upon their preparatory/attentional skills. In addition, consistency is a great contributor to level of performance. This problem was in part addressed by Ko and Singer (1999) with the use of visual gaze instrumentation. The 'Five-step Strategy' group performed better than the control group in an underhand dart-throwing task. The 'Five-step Strategy' group had better accuracy and consistency scores during practice and reflection sessions, with a longer quiet eye duration when compared with the control group.

Duration of fixation on a target preparatory to the movement act has been termed quiet eye duration by Vickers (1996). Presumably, longer quiet eye periods lead to better performance, and parameters, or movement requirements, are set during this time. This internal activity, also termed cognitive processing (response programming) would be longer for more complex tasks (Henry and Rogers 1960). In essence, visual control over movement skill in target-aiming tasks seems to be a crucial consideration (Vickers 1996). Critical parameters of the aiming movement to be performed are developed and refined.

As to the quiet eye duration, Vickers (1996 and 1997) found that the more highly-skilled free-throw shooters differed from the lesser-skilled among members of the Canadian women's national basketball team. Better shooters were recorded with quiet eye durations of almost 900 ms while the lesser skilled averaged less than 400 ms. Similar data were observed by Frehlich *et al.* (1999) with billiard players. Three shots standardized as to degree of increasing difficulty yielded quiet eye duration data of 500 ms versus 275 ms comparing two groups of players with different levels of skill. Vickers (1996) also noted that expert shooters spent more time in preparing for the shot. However, in the study conducted by Frehlich *et al.* (1999), the two groups of billiards players did not differ in time allotment.

Fixation patterns to different locations were also analysed in golf-putting (Vickers 1992) and in billiards play (Frehlich *et al.* 1999). In both studies, experts seemed to fixate on only a few important locations. Shifting gaze patterns in billiards such as on the cue ball, to the object ball, and back to the cue ball, was similar between the groups. However, the novice group repeated the sequence more frequently and spent less time at each location. The highly-skilled players demonstrated greater quiet eye durations in every condition. The higher skilled players were characterized as having more efficient scanpaths, as suggested by Abernethy (1988) of skilled performers in general.

In summary, it is becoming more obvious that the study of behavioural and psychophysiological aspects of expertise, as well as experimental manipulations in the adoption of particular useful strategies, provide important information for understanding how to perform self-paced tasks more capably. The refinement in the use of attentional and cognitive processes leads to more skilled performance. A similar analysis of externally-paced activities is provided in the next section.

8. Improving performance in externally-paced events

Turning to externally-paced events, performance improves when adaptive reactive behaviours function in an appropriate manner as changing circumstances dictate. With more time to judge what is happening and what might happen, the person is more likely to respond effectively. In the most demanding situations, the highly-skilled appear to function without the need to attend deliberately to all relevant sources of information and how to respond. Perception and action seem to be inseparable (Prinz 1990).

To reach that point, much practice and learning have to occur (Ericsson 1996). An information processing model suggests an approach to determining the primary cognitive and attentional components involved in externally-paced events that contribute to performance excellence. For the less skilled, realizing what strategies to use deliberately in somewhat sequential form should lead to better proficiency. With improvement, information processing becomes accelerated, anticipation and decision-making quicken, and behaviours seem to 'happen'. The transition in attention was explained well in the classic work of Schneider and Shiffrin (1977), in which automatic versus controlled processing were differentiated. Automatic processing, as compared with controlled processing, is rapid, effortless, autonomous and consistent, among other characteristic dissimilarities (Logan 1988). In very difficult tasks or those that are not well-learned, controlled processing (non-automatic) is usually required (Shiffrin 1988). Attention is allocated to various aspects of the situation and performance.

The question that naturally emerges has to do with what should be attended to so as to maximize performance. In order to demonstrate skill in an act that occurs in a situation in which there is uncertainty as to what will happen and very little time available to produce an acceptable, adaptive movement response, the performer must take advantage of relevant predictive information. The ability to anticipate events is of great value.

Traditional advice given to participants in externally-paced sports, such as tennis players and batters in baseball, is to keep their eye on the ball and to track it to the point of contact. Consider baseballs thrown at an exceptional velocity of 44.4 m s^{-1} (100 miles per hour) from 18.44 m away from the pitcher. Or consider a tennis ball served at 48.9 m s^{-1} (110 miles per hour) from 23.77 m away from the receiver. It is impossible to keep one's eye continuously on the ball as it is thrown by the pitcher to homeplate, or hit by the server to the receiver (Watts and Bahill 1990). A 44.4 m s^{-1} pitch takes 0.415 s to travel 18.44 m while a 37.8 m s^{-1} (85 miles per hour) pitch takes 0.48 s.

Bahill and LaRitz (1984) systematically varied ball velocities and trajectories, and studied gaze behaviours and head movements in a professional baseball player as compared with collegiate baseball players. The professional player was able to use minor head movements and smooth pursuit tracking until the ball was 1.68 m from the plate. The college players fell behind at the 2.74 m mark. Presumably, it appears that better hitters produce faster smooth pursuit eye movements, an ability to suppress the vestibulo-ocular reflex and the occasional use of an anticipatory saccade. The object is not necessarily to try to see the bat contact the ball but rather to estimate accurately the ball location relative to the batter.

Data derived from five expert table tennis players (Ripoll and Fleurance 1988) support the contention that the speed of the ball is generally superior to the oculomotor capacity of pursuit. Visual tracking occurred for part of the ball trajectory only. During contact, the ball was not well viewed. Apparently, placing the head, eyes, and racquet in the right place before making good contact with the ball is based on anticipatory information gained from ball velocity and location.

In isolated events in a number of reactive sports, it is important to be aware of (1) an opponent's capabilities, preferences and tendencies, (2) what most opponents might attempt to do under the particular circumstances, and (3) any meaningful sources of information revealed from the opponent's actions that give away his/her intentions. Attentional and cognitive processes would be devoted to:

(1) visual searching for the most relevant opponent and situational cues;
(2) anticipating what the opponent's intentions are on a probability basis;
(3) decision-making about what to do; and
(4) initiating the appropriate movement response considering temporal and spatial parameters.

Figure 3 illustrates a flow diagram of the possible sequence of events that transpire in an externally-paced situation. It suggests the need for a learner to activate a series of cognitive operations leading to execution of the desired movement. Selective attention, pattern recognition, anticipation, decision-making and response initiation lead to the actual completion of the act. To what degree the search for relevant cues precludes anticipation or anticipation occurs first is debatable. Indeed, they may be indicative of parallel processing. None the less, the beginner needs to master the use of such cognitions with appropriate strategies. A skilled movement act itself and variations of it are obviously necessary in order to advance in skill.

Comparisons of experts and novices in dynamic, reactive sports reveal a number of attentional, perceptual and decision-making differences. They are typically based on the study of visual search patterns (fixations, saccades) with the use of eye-tracking instrumentation. Most published research has been undertaken in laboratory settings with the subject responding in an artificial manner to a filmed opponent or opponents. Some examples are in tennis (Singer *et al.* 1996) and soccer (Williams *et al.* 1994). On less frequent occasions, such data have been recorded on-line, in somewhat real situations, such as in table tennis (Ripoll and Fleurance 1988), tennis (Singer *et al.* 1998) and volleyball (Vickers and Adolphe 1997). Returning to the laboratory setting, the film occlusion technique has also been used as a method of determining what elements in the sequence of routines in the opponent's act are most significant in predicting what he or she will do (Abernethy 1993).

Generalizations derived from research in which sports experts and novices have been compared, or only experts studied, have been made by Singer and Janelle (1999). The highly skilled in their specialization as compared to lesser skilled or novices:

(1) have more elaborate task-specific knowledge;
(2) make more meaning of available information;
(3) encode and retrieve relevant information more efficiently;

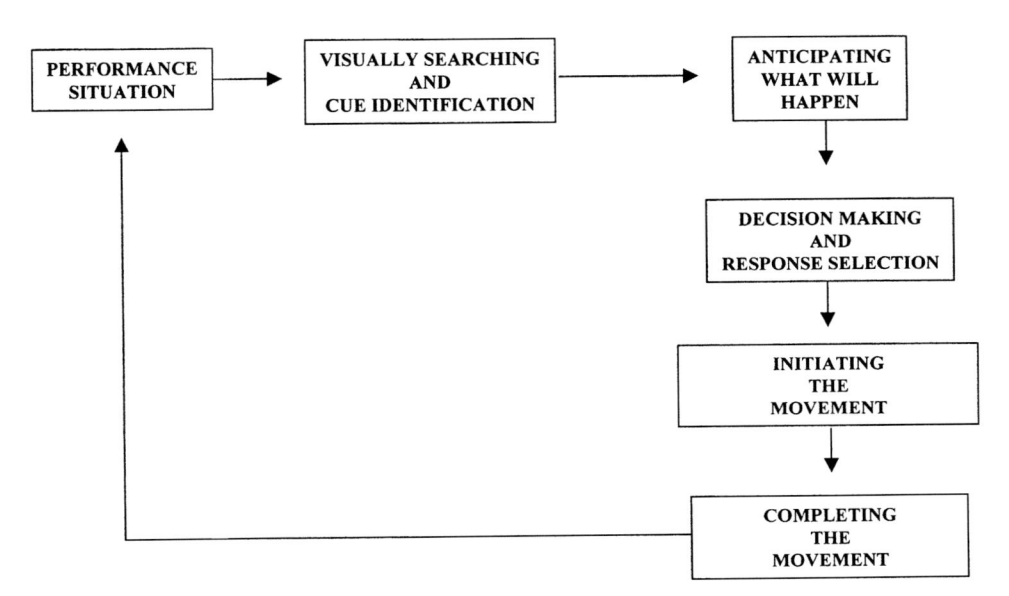

Figure 3. Cognitive and motoric elements involved in creating and generating an effective movement response in externally-paced situations.

(4) visually detect and locate objects and patterns in the visual field faster and more accurately;
(5) use situational probability information better; and
(6) make more rapid and appropriate decisions.

Based on many findings in the sport expert-novice research, attempts have been made to improve on the way cognitions and attention functions for the beginner and at the intermediate level. These efforts have primarily been initiated by sports scientists, and conducted in the laboratories in which elements of the real sport situation are simulated to a degree. Coaches and athletes seem to believe that such behaviours as attentional distribution, concentration, anticipation and decision-making are innate. There is little confidence in the sports world that these behaviours can be trained, but instead are due to heredity, and so very few serious attempts are made in this regard. Many coaches and athletes also believe that no meaningful knowledge exists as to how athletes can improve their perceptual/cognitive skills. Contemporary research challenges this idea, and results indicate that many of these attributes are indeed trainable.

Relatively few training studies have been reported. Nevertheless, positive effects have generally been observed, but under controlled laboratory conditions. The typical methodology involves an athlete, or would-be athlete, viewing a filmed opponent or opponents. Perceptual/attentional training techniques are used to guide early relevant cue pick-up, leading to better anticipated decision-making for the appropriate response. Dependent measures include a record of visual search patterns over trials. Fixation location and duration patterns presumably reflect attentional focus, ideally to important anticipatory cues in the sequence of the filmed opponent's actions. Choice reaction time measures, speed and accuracy in decision-making, suggest anticipatory skill.

Positive effects with video training have been reported by Christina *et al.* (1990) with a football linebacker, by Singer *et al.* (1994) with average tennis players, and by Burroughs (1984) with baseball batters. Such research is promising. Yet, the true pay-off of any training programme is the ability of the person to use knowledge and experiences gained and to demonstrate better performance in real competitive situations, such as on the football or baseball field, the tennis or squash court, and on the ice in an ice hockey arena. Thus far, a demonstration of transferability is lacking.

Researchers in recent years have become more creative in designing laboratory training and testing conditions with reasonable simulation of elements in the real sport condition. Measures used to demonstrate attentional focus, concentration, anticipation and decision-making are not easy to formulate and may be non-realistic to record in true competitive sport conditions. The trained coach can estimate athlete improvement based on keen observations in practices and contests, as well as on film. In addition, an athlete's self-perceptions and pronouncement of performance can be useful. From the hard-core researcher's perspective, these forms of 'data' are somewhat crude. Assessing transfer effectiveness is indeed a problem (Starkes and Lindley 1994).

Beyond measurement is the question raised earlier of whether laboratory-type experiences and learning are truly beneficial in the true sport setting. Knowing what to do has been termed semantic (or declarative) knowledge. Knowing how to do something, and to do it effectively in various situations, has been termed procedural knowledge. The first form of knowledge usually precludes the second. The utility of video training for transfer to the sports competitive environment has not been verified substantially (McPherson 1994), although it appears to be of value. Future researchers will no doubt be able to resolve the issue.

9. Concluding remarks

Cognitive psychology and an information processing approach has been explained and justified as a viable framework to study the learning and performance of complex movement skills, such as with sports skills. Indeed, a good deal of sport psychology research and practical applications having to do with improving learning conditions, learning processes and performance capabilities have emerged from cognitive psychology perspectives. Of interest has been how the beginner progresses to become highly proficient. Indeed, applications of this research can also be made to various types of occupational endeavours, recreational and leisure time pursuits, and such other performance areas as dance and music.

Intriguing insights and perspectives about human performance have been advanced by advocates of the ecological approach. They challenge many fundamental beliefs associated with an information processing framework used to describe cognitions and the sequence of internal events related to the acquisition and demonstration of skill. Each conceptual orientation still has unique value in contributing to the body of knowledge. The concern for understanding learning and conditions and special training programmes that promote it and high levels of performance stability favour the study of cognitional, attention and useful strategies.

At the present time, cognitive psychology and associated information processing models are indeed useful, for the many reasons explained earlier. Respected scholars have conjectured about their powerful influence on understanding the learning of movement skills. An example is Adams who had a great effect on the thinking of motor behaviourists for many years. It is of interest to note the final remarks in his historical publication about motor skills research through the decades. Adams (1987: 66) wrote that 'The temptation may arise, if cognitive factors turn out to be important determinants of motor behavior, to view movement as a slave to cognition and to decide that cognitive learning is about all there is to motor learning'. He added, 'That viewpoint would be a simplification, complex though it may be'.

His perceptions reflected the impact cognitive psychology was having at that time on research on all types of learning, including motor learning. He failed to realize the challenge that dynamical systems theorists were making as to the wisdom of endorsing computer symbolism and the notion of mental representations to explain skilled movement acts. However, the conjectures of Adams (1987) could be justified on the basis of the way motor skills were being researched; e.g. simple artificial tasks under contrived laboratory conditions.

The pendulum has swung considerably in the past decade or two. It is now acceptable to publish research about sports acts in even the most revered scholarly psychology journals. For the most part this was not the case in earlier decades. More and more well-trained motor learning/control and sport psychology researchers are generating exciting research. The themes of their research range considerably from the very technical basic orientation that contributes to understanding the nature of the development and refinement of skill, to practical applications for instruction and learning.

In the present paper, the author has attempted to remain under a cognitive psychology umbrella to describe advances in recent years to study real-world self-paced and externally-paced events. Such endeavours have generated more realistic laboratory simulations as well as the direct study of skills that are performed in their natural contexts. Although the emphasis has been directed to what a segment of sport psychologists study, their findings can be considered in the framework of interests of those scholars and practitioners who consider themselves to be human factors specialists.

The nature of cognitions and strategies, as well as the ability to suppress conscious activity at the appropriate time of preparation and execution of movement, constitutes a remarkably viable area of study. The challenge is to determine the cognitive demands imposed by any psychomotor task that requires technical skill to be performed at a level of acceptable proficiency. The most formidable goal is to determine ways that individuals can use cognitive processes wisely and efficiently. This is true throughout the learning process and when performing under stressful situations. With the acceptance of sports science as a legitimate scholarly enterprise, improved research methodology as well as more versatile and sophisticated instrumentation, many promising developments can be expected in the future.

Acknowledgements

Appreciation is extended to Dr Mark Williams and Dr Chris Janelle for their insightful and helpful comments on an earlier draft of this manuscript.

References

ABERNETHY, B. 1988, Visual search in sport and ergonomics: its relationship to selective attention and performer expertise, *Human Performance*, **1**, 205–235.

ABERNETHY, B. 1990, Expertise, visual search, and information pick-up in a racquet sport, *Perception*, **19**, 63–77.

ABERNETHY, B. 1993, Attention, in R. N. Singer, M. Murphey and L. K. Tennant (eds), *Handbook of Research on Sport Psychology* (New York: Macmillan), 127–170.

ADAMS, J. A. 1987, Historical review and appraisal of research on the learning, retention, and transfer of human skills, *Psychological Bulletin*, **101**, 41–74.

ANDERSON, J. R. 1982, Acquisition of cognitive skill, *Pychological Review*, **89**, 369–406.

ANDERSON, J. R. 1983, *The Architecture of Cognition* (Cambridge, MA: Harvard University Press).

ANDERSON, J. R. 1990, *Cognitive Psychology and its Implication*, 2nd ed. (New York: Freeman).

ANNETT, J. 1993, The learning of motor skills: sports science and ergonomics perspectives, *Ergonomics*, 37, 5–16.

BAHILL, L. T. and LARITZ, T. 1984, Why can't batters keep their eyes on the ball? *American Scientist*, **72**, 249–253.

BARBER, P. 1988, *Applied Cognitive Psychology: An Information-Processing Framework* (London: Methuen).

BERNSTEIN, N. A. 1967, *The Coordination and Regulation of Movement* (Oxford: Pergamon Press).

BIDDLE, S. J. H. 1999, Motivation and perceptions of control: tracing its development and plotting its future in exercise and sport psychology, *Journal of Sport and Exercise Psychology*, **21**, 1–23.

BOUTCHER, S. H. and ZINSSER, N. W. 1990, Cardiac deceleration of elite and beginning golfers during putting, *Journal of Sport and Exercise Psychology*, **12**, 37–47.

BOWER, G. H. and HILGARD, E. R. 1981, Theories of Learning, 5th ed. (Englewood Cliffs, NJ: Prentice-Hall).

BRUNER, J. S., GOODNOW, J. and AUSTIN, G. A. 1956, *A Study of Thinking* (New York: Wiley).
BURROUGHS, W. 1984, Visual simulation training of baseball batters, *International Journal of Sport Psychology*, **15**, 117–126.
CARVER, C. S. and SCHEIER, M. F. 1998, *On the Self-Regulation of Behavior* (Cambridge: Cambridge University Press).
CHASE, W. C. and SIMON, H. A. 1973, Perception in chess, *Cognitive Psychology*, **4**, 55–81.
CHI, M. T. H., FELTOVICH, P. and GLASER, R. 1981, Categorization and representation of physics problems by experts and novices, *Cognitive Science*, **5**, 121–152.
CHI, M. T. H., GLASER, R. and REES, E. 1982, Expertise in problem solving, in R. J. Sternberg (ed.), *Advances in the Psychology of Human Intelligence* (Hillsdale, NJ: Erlbaum), 7–75.
CHOMSKY, N. 1957, *Syntactic Structures* (The Hague: Mouton).
CHRISTINA, R. W., BARRESSI, J. V. and SHEFFNER, P. 1990, The development of response selection accuracy in a football linebacker using video training, *The Sport Psychologist*, **4**, 11–17.
CREWS, D. J. 1993, Self-regulation strategies in sport and exercise, in R. N. Singer, M. Murphey and L. K. Tennant (eds), *Handbook of Research on Sport Psychology* (New York: Macmillan), 557–568.
CZIKO, G. A. 1992, Purposeful behavior as the control of perception: implications for educational research, *Educational Researcher*, **21**, 10–18.
DINSMORE, J. 1992, *The Connectionist Paradigms: Closing the Gap* (Hillsdale, NJ: Erlbaum).
ERICSSON, K. A. 1996, *The Road to Excellence: The Acquisition of Expert Performance in the Arts and Sciences, and Sports and Games* (Mahwah, NJ: Erlbaum).
FREHLICH, S. G., SINGER, R. N. and WILLIAMS, A. M. 1999, Visual attention in experienced and inexperienced billiards players: is quiet eye duration the key to successful performance? *Journal of Sport & Exercise Psychology*, **21**, S46.
GIBSON, E. J. 1979, *An Ecological Approach to Visual Perception* (Boston, MA: Houghton-Mifflin).
GLASER, R. 1990, The reemergence of learning theory within instruction research, *American Psychologist*, **45**, 29–39.
HANIN, Y. L. 1989, Interpersonal and intragroup anxiety in sports, in D. Hackfort and C. D. Spielberger (eds), *Anxiety in Sports: An International Perspective* (New York: Hemisphere), 19–28.
HANTON, S. and JONES G. 1999, The effects of a multimodel intervention program on performers: II. Training the butterflies to fly in formation, *The Sport Psychologist*, **13**, 22–41.
HARDY, L. and NELSON, D. 1988, Self-regulation training in sport and work, *Ergonomics*, **31**, 1573–1583.
HATFIELD, B. D., LANDERS, D. L. and RAY, W. J. 1984, Cognitive processes during self-paced motor performance: an electroencephalographic profile of skilled marksmen, *Journal of Sport Psychology*, **6**, 42–59.
HATFIELD, B. D., LANDERS, D. L. and RAY, W. J. 1987, Cardiovascular-CNS interactions during a self-paced, intentional state: elite marksmanship performance, *Psychophysiology*, **24**, 542–549.
HENRY, F. M. and ROGERS, D. E. 1960, Increased response latency for complicated movements and a 'memory drum' theory of neuromotor reaction, *Research Quarterly*, **31**, 448–458.
JACKSON, S. A. 1992, Athletes in flow: a qualitative investigation of flow states in elite figure skaters, *Journal of Applied Sport Psychology*, **4**, 161–180.
JANELLE, C. M., Singer, R. N. and WILLIAMS, A. M. 1999, External distraction and attentional narrowing: visual search evidence, *Journal of Sport & Exercise Psychology*, **21**, 70–91.
KELSO, J. A. S. 1994, The informational character of self-organized coordination dynamics, *Human Movement Science*, **13**, 393–441.
KIRSCHENBAUM, D. S., OWENS, D. and O'CONNOR, E. A. 1998, Smart golf: preliminary evaluation of a simple, yet comprehensive, approach to improving and scoring the mental game, *The Sport Psychologist*, **12**, 271–282.
KITSANTAS, A. and ZIMMERMAN, B. J. 1998, Self-regulation of motoric learning: a strategic cycle view, *Journal of Applied Sport Psychology*, **10**, 220–239.
KO, W. and SINGER, R. N. 1999, Five-step Strategy and visual control in learning and performing a target aiming motor skill, *Journal of Sport & Exercise Psychology*, **21**, S69.

LACEY, J. I. 1967, Somatic response patterning and stress: some revisions of activation theory, in M. H. Appley and R. Trumbull (eds), *Psychological Stress* (New York: Appleton-Century-Crofts), 14–42.

LACHMAN, R., LACHMAN, J. L. and BUTTERFIELD, E. C. 1979, *Cognitive Psychology and Information Processing: An Introduction* (Hillsdale, NJ: Erlbaum).

LANDERS, D. M., HAN, M., SALAZAR, W., PETRUZZELLO, S. J., KUBITZ, K. A. and GANNON, T. L. 1994, Effects of learning on electroencephalographic and electrocardiographic patterns in novice archers, *International Journal of Sport Psychology*, **25**, 313–330.

LANDERS, D. M., PETRUZZELLO, S. J., SALAZAR, W., CREWS, D. J., KUBITZ, K. A., GANNON, T. L. and HAN, M. 1991, The influence of electrocortical biofeedback on performance in pre-elite archers, *Medicine and Science in Sports and Exercise*, **23**, 123–129.

LIDOR, R., TENNANT, K. L. and SINGER, R. N. 1996, The generalizability effect of three learning strategies across motor task performances, International Journal of Sport Psychology, 27, 23–36.

LOGAN, G. D. 1988, Automaticity, resources and memory: theoretical controversies and practical implications, *Human Factors*, **30**, 583–598.

LOHASZ, P. G. and LEITH, L. M. 1997, The effect of three mental preparation strategies on the performance of a complex response time task, *International Journal of Sport Psychology*, **28**, 25–34.

MADDOX, M. D., WULF, G. and WRIGHT, D. L. 1999, The effect of an internal vs. external focus of attention on the learning of a tennis stroke, *Journal of Sport & Exercise Psychology*, **21**, S78.

MCPHERSON, S. L. 1994, The development of sports expertise: mapping the tactical domain, *Quest*, **46**, 223–240.

MCPHERSON, S. L. and FRENCH, K. E. 1991, Changes in cognitive strategies and motor skill in tennis, *Journal of Sport and Exercise Psychology*, **13**, 26–41.

MORAN, A. P. 1996, *The Psychology of Concentration in Sport Performance: A Cognitive Analysis* (Hove, East Sussex, UK: Psychology Press).

MURRAY, N. P., JANELLE, B. M., HILLMAN, C. H. and HATFIELD, B. D. 1999, The eye as a window to the mind: concurrent recording of EEG and eye movement activity, Proceedings of the North American Society for the Psychology of Sport and Physical Activity Conference abstracts, *Journal of Sport & Exercise Psychology*, **21**, S84.

ORBACH, I., SINGER, R. and PRICE, S. 1999, An attribution training program and achievement in sport, *The Sport Psychologist*, **13**, 69–82.

PRINZ, W. 1990, A common coding approach to perception and action, in O. Newmann and W. Prinz (eds), *Relationships Between Perception and Action* (Berlin: Springer-Verlag), 167–195.

PRINZ, W. 1997, Perception and action planning, *European Journal of Cognitive Psychology*, **9**, 129–154.

RADLO, S. J. 1999, Effectiveness of Singer's Five-Step Strategy during competition: a psychophysiological investigation, *Journal of Sport & Exercise Psychology*, **21**, S88.

RADLO, S. J., STEINBERG, G. M., SINGER, R. N., BARBA, D. A. and MELNIKOV, A. in press, The influence of an attentional focus strategy on alpha brain wave activity, heart rate, and dart-throwing performance, *International Journal of Sport Psychology*.

REILLY, T. and SHELTON, T. 1994, Ergonomics in sport and leisure, Ergonomics, 37, 1–3.

RIPOLL, H. and FLEURANCE, P. 1988, What does keeping one's eye on the ball mean? *Ergonomics*, **31**, 1647–1654.

RYAN, E. D. and SIMMONS, J. 1982, Efficiency of mental imagery in enhancing mental rehearsal of motor skills, *Journal of Sport Psychology*, **4**, 41–51.

SALAZAR, W., LANDERS, D. M., PETRUZZELLO, S. J., CREWS, D. J., KUBITZ, K. A. and HAN, M. 1990, Hemispheric asymmetry, cardiac response, and performance in elite archers, *Research Quarterly for Exercise and Sport*, **61**, 351–359.

SCHMIDT, A. and PEPER, E. 1998, Strategies for training concentration, in J. M. Williams (ed.), *Applied Sport Psychology: Personal Growth to Peak Performance* (Mountain View, CA: Mayfield), 316–328.

SCHMIDT, R. A. and LEE, T. D. 1999, *Motor Control and Learning: A Behavioral Emphasis* (Champaign, IL: Human Kinetics).

SCHNEIDER, W. and SHIFFRIN, R. M. 1997, Controlled and automatic information processing: I. Detection, search, and attention, *Psychological Review*, **84**, 1–66.

SHEPHARD, R. J. 1988, Sport, leisure and well-beingv—an ergonomics perspective, *Ergonomics*, **31**, 1501–1517.

SHIFFRIN, R. M. 1988, Attention, in R. C. Atkinson, R. J. Herrnstein, G. Lindzey and R. D. Luce (eds), *Steven's Handbook of Experimental Psychology, Vol. 2: Learning and Cognition*, 2nd ed. (New York: Wiley), 739–811.

SHIFFRIN, R. M. and SCHNEIDER, W. 1977, Controlled and automatic human information processing: II. Perceptual learning, automatic attending, and a general theory, *Psychological Review*, **84**, 127–190.

SINGER, R. N. 1980, *Motor Learning and Human Performance: An Application to Motor Skills and Movement Behaviors*, 3rd ed. (New York: Macmillan).

SINGER, R. N. 1988, Strategies and metastrategies in learning and performing self-paced athletic skills, *The Sport Psychologist*, **2**, 49–68.

SINGER, R. N. 1998, From the laboratory to the courts: understanding and training anticipation and decision-making, in A. Lees, I. Maynard, M. Hughes and T. Reilly (eds), *Science and Racket Sports II* (London: E. & F. N. Spon), 109–120.

SINGER, R. N. and CHEN, D. 1994, A classification scheme for cognitive strategies: implications for learning and teaching psychomotor skills, *Research Quarterly for Exercise and Sport*, **65**, 143–151.

SINGER, R. N. and JANELLE, C. M. 1999, Determining sport expertise: from genes to supremes, *International Journal of Sport Psychology*, **30**, 117–150.

SINGER, R. N. and SUWANTHADA, S. 1986, The generalizability effectiveness of a learning strategy on achievement in related closed motor skills, *Research Quarterly*, **57**, 205–213.

SINGER, R. N., DEFRANCESCO, C. and RANDALL, L. E. 1989, Effectiveness of a global learning strategy practiced in different contexts on primary and transfer self-paced motor tasks, *Journal of Sport & Exercise Psychology*, **11**, 290–303.

SINGER, R. N., LIDOR, R. and CAURAUGH, J. H. 1993a, To be aware or not to be aware: what to think about while performing a motor skill, *The Sport Psychologist*, **7**, 19–30.

SINGER, R. N., MURPHEY, M. and TENNANT, L. K. 1993b, *Handbook of Research on Sport Pychology* (New York: Macmillan).

SINGER, R. N., CAURAUGH, J. H., CHEN, D., STEINBERG, G. M. and FREHLICH, S. G. 1996, Visual search, anticipation, and reactive comparisons between highly-skilled and beginning tennis players, *Journal of Applied Sport Psychology*, **8**, 9–26.

SINGER, R. N., CAURAUGH, J. H., CHEN, D., STEINBERG, G. M., FREHLICH, S. G. and WANG, L. 1994, Training mental quickness in beginning/intermediate tennis players, *The Sport Psychologist*, **8**, 305–318.

SINGER, R. N., WILLIAMS, A. M., FREHLICH, S. G., JANELLE, C. M., RADLO, S. J., BARBA, D. A. and BOUCHARD, L. J. 1998, New frontiers in visual search: an exploratory study in live tennis situations, *Research Quarterly for Exercise and Sport*, **69**, 290–296.

STARKES, J. L. and LINDLEY, S. 1994, Can we hasten expertise by video simulations? *Quest*, **46**, 211–222.

TURVEY, M. T. 1990, Coordination, *American Psychologist*, **45**, 938–953.

VICKERS, J. N. 1992, Gaze control in putting, *Perception*, **21**, 117–132.

VICKERS, J. N. 1996, Visual control when aiming at a far target, *Journal of Experimental Psychology: Human Perception and Performance*, **22**, 342–354.

VICKERS, J. N. 1997, Control of visual attention during the basketball free throw, *The American Journal of Sports Medicine*, **24**, 93–97.

VICKERS, J. N. and ADOLPHE, R. M. 1997, Gaze behavior during a ball tracking and aiming skill, *International Journal of Sport Vision*, **4**, 18–27.

WATTS, R. G. and BAHILL, A. T. 1990, *Keep Your Eye on the Ball: The Science and Folklore of Baseball* (New York: W. H. Freeman).

WICKENS, C. D. 1992, *Engineering Psychology and Human Performance*, 2nd ed. (New York: Harper-Collins).

WILLIAMS, A. M., DAVIDS, K. and WILLIAMS, J. G. 1999, *Visual Perception & Action in Sport* (London: E. & F. N. Spon).

WILLIAMS, A. M., DAVIDS, K., BURWITZ, L. and WILLIAMS, J. G. 1992, Perception and action in sport, *Journal of Human Movement Studies*, **22**, 147–204.

WILLIAMS, A. M., DAVIDS, K., BURWITZ, L. and WILLIAMS, J. G. 1994, Visual search strategies in experienced and inexperienced soccer players, *Research Quarterly for Exercise and Sport*, **65,** 127–135.
WILLIAMS, J. M. 1998, *Applied Sport Psychology: Personal Growth to Peak Performance*, 3rd ed. (Mountain View, CA: Mayfield).
WULF, G. and TOOLE, T. 1999, Physical guidance in complex motor skill learning: benefits of a self-controlled practice schedule, *Research Quarterly for Exercise and Sport*, **70,** 265–272.
WULF, G., HÖB, M. and PRINZ, W. 1998, Instructions for motor learning: differential effects of internal versus external focus of attention, *Journal of Motor Behavior*, **30,** 169–179.
WULF, G., LAUTERBACK, B. and TOOLE, T. 1999, The learning of an external focus of attention in golf, *Research Quarterly for Exercise and Sport*, **70,** 120–126.

19 Effects of cricket ball colour and illuminance levels on catching behaviour in professional cricketers

Karen Scott, Damian Kingsbury*, Simon Bennett, Keith Davids and Mark Langley

Psychology Research Group, Department of Exercise and Sport Science, Manchester Metropolitan University, Alsager, Cheshire ST7 2HL, UK

Keywords: Cricket; Slip-catching; Illuminance; Ball colour; Movement initiation time.

In recent years there have been many alterations to equipment and technology in professional cricket, including the introduction of white balls during day–night matches. In the present study simulated slip-catching performance and movement initiation time were examined in professional cricketers when ball colour and illuminance levels differed. Five male professional cricketers (mean age: 27.3 ± 1.4 years) volunteered to catch a total of 60 cricket balls, 20 (10 red and 10 white) under each of three illuminance levels (571, 1143 and 1714 lux). Balls were projected from a ball machine at 20 m s^{-1} (45 mph) over a distance of 8.4 m, to the subject's dominant side. Catching performance was measured using an established catching scale. Movement initiation times for each hand were also calculated for each trial using a motion-analysis system. Data were submitted to separate two-way (ball colour [2] × illuminance level [3]) repeated measures analysis of variance. No significant effects were obtained for ball colour or illuminance levels for either catching performance or movement initiation time. Neither ball colour nor light level (within the range tested) affected slip-catching performance and movement initiation times in professional cricketers. Therefore it was concluded that the changes made to ball colour and light conditions in professional cricket were not detrimental to catching performance.

1. Introduction

Professional cricket has undergone many changes in recent years in an attempt to promote the game to a larger audience, to increase participation at grass roots level and to attract more sponsorship and media coverage. The traditional features of the sport, such as white clothing, red ball and white sightscreens, have been replaced with coloured clothing, a white ball and black sightscreens for one-day matches, as observed in the 1999 World Cup competition. The same clothing and ball changes have also been made in the county one-day game with each county or country having its own coloured strip. At this time, little research has been conducted into the potential effects of these alterations on players' performance of relevant skills in professional cricket.

In cricket, catching is an integral part of fielding. Slip-catching is a perceptually-demanding task that requires the performer to stand behind the batsman and to

*Author for correspondence. e-mail: D.Kingsbury@mmu.ac.uk

respond to the ball being deflected by the edge of the bat. Several authors have highlighted the need to examine visual skills in skilled athletes, stressing that vision is particularly important in 'reactive' sports such as baseball, cricket and shooting. In these sports, objects to be intercepted can move at velocities above the tracking threshold for the human eye (Regan 1997). For example, Loran (1997: 3) has argued that cricket is an 'especially visually demanding sport' and he stressed the importance of skilled athletes being able to filter unwanted 'noise' and information from the surrounding environment (in a perceptual process known as contrast sensitivity). During slip-catching a professional cricketer must be able to pick up trajectory information from balls of different colours against textured backgrounds, typically under varying illuminance levels.

There has been only a limited amount of research conducted on the interactive relationship between ball colour and illuminance levels and catching behaviour. The lighting or illuminance level, that is the amount of light spread over the area receiving light, is measured in lux and is one of the most important parameters when specifying lighting requirements for performance of particular sports (Loran 1997). Luminance refers to the light being emitted or reflected from an important surface or object. The luminance of surfaces and objects in sport is dependent not only on the illuminance falling on them, but also on their reflective characteristics. Difficulties arise in fast ball games, such as cricket and baseball, when abrupt darkening occurs due to rapidly-moving heavy cloud causing a decrease in illuminance (or brightness) of the sky. The rules for professional cricket currently advise that the players can ask to leave the field of play when light levels fall to below 1000 lux.

These recommendations are broadly in line with data from Campbell and colleagues (Campbell et al. 1987, Perry et al. 1987, Rothwell and Campbell 1987) who suggested that conditions appear to be 'gloomy' at an illuminance of 800 lux. Campbell et al. (1987) examined visual reaction times (RT) using a computer-based pointer movement task and attempted to generalize findings to the game of cricket. They found that below this level, RT start to rise above baseline values with obvious implications for interceptive actions in cricket. They surmised from their results that a ball bowled at 40 m s^{-1} (90 mph) in good lighting conditions could elicit reaction 1.9 m (6 ft) from the batsman's face, yet when light levels diminished, RT was increased by approximately 33 ms per log factor of light reduction. It was argued that the net result was equivalent to the bat movement taking place 1.31 m (1.5 yds) nearer to the batsman. It is unclear to what extent these findings can actually be generalized from a computer-based visual RT task to the specific context of cricket and interceptive actions.

The other relevant factor to receive limited empirical attention is the effect of ball colour on catching performance. Morris (1976) investigated the performance of children catching different coloured balls (blue, yellow and white) against two different coloured backgrounds (white and black). The balls were projected directly towards subjects at a constant trajectory. The white balls produced the lowest catching performance scores while the blue and yellow balls produced significantly higher catching scores. The perception of the ball was, however, influenced by the ball colour and background colour combinations. Blue balls against a white background produced the highest catching scores, while white balls against the white background produced the lowest catching scores. Ball colour and background colour combinations of greater contrast seemed to enhance catching scores. An interesting feature of the data is that the influence of ball colour seemed to diminish with age and experience. Ball colour did not appear to influence performance outcomes for more highly skilled ball catchers (Morris 1976).

An alternative explanation for the putative effects of ball-background illuminance contrast was proposed by Koslow (1985). He found that RT was affected by chromatic cues, specifically the luminance contrasts present, i.e. the contrast created by viewing the ball on a particular background. The significant effect of colour was attributed to the luminance cues present. Koslow's (1985) findings suggested that the effect of ball colour was removed when the luminance effects were compensated for. The implication of these findings is that the luminance contrasts and not ball colour were responsible for the generation of actions. Koslow (1985) suggested that choice of ball colour for sports administrators should entail consideration of ball–background luminance contrast, as larger contrasts seem to induce faster reactions.

In previous research there have been few attempts to examine performance of natural interceptive actions in professional games players. Little effort has been made to capture the constraints of a specific sporting context (i.e. slip-catching in cricket). In sports such as cricket where players are required to co-ordinate a response to a moving ball, fielders need to utilize spatio-temporal information from ball flight to ensure that the hand is in the right place at the right time to catch the ball (Savelsbergh and Bootsma 1994). It is clear that the recent changes in professional cricket have expedited the need to investigate the effect of ball colour and illuminance level on the performance of interceptive actions. Therefore, the aim of this study was to examine the effects of illuminance levels and ball colour on two-handed slip-catching performance and movement initiation time in professional cricketers. For this purpose a range of values around the current advisory field-exit value of 1000 lux was investigated. A question of interest was whether there would be a decrement in catching performance and an increase in movement initiation time under the test conditions. Based on previous findings, it was not expected that there would be an effect of ball colour alone (Morris 1976, Koslow 1985).

2. Methods

2.1. Participants

Five male, professional cricketers, aged 24–33 years (mean: 27.3 ± 1.4 years), with previous experience of playing with both red and white cricket balls, volunteered to participate in the study. All subjects were tested for colour blindness and they completed an informed consent form and a short pre-test questionnaire. The subjects wore sports clothing of their choice in order to be comfortable when moving to catch the balls.

2.2. Experimental apparatus

The cricket balls (Duke Special Crown A, Duke, Worcester, UK) were the kind used in professional cricket, and were projected from a Bola Club Bowling Machine (Stuart and Williams, Bristol, UK). The balls were placed into the machine with the seam vertical to further ensure similar flight for each delivery.

To determine movement initiation time (defined as the time between ball projection and first movement of the hands of a subject) accurately, a pair of photoelectric timing gates were placed in front of the ball projection machine. Subjects started with hands placed on their thighs. Lever microswitches were placed on the thighs of the subjects, which the subjects held down prior to each trial and released when attempting to catch the balls. The switches were placed to be unobtrusive to catching performance. Both the photoelectric gate and lever microswitches were connected to an IBM-compatible PC with an analogue-to-digital converter to process the timing data.

To prevent anticipation of ball projection, background sheeting was randomly sprayed in patches (5–15 cm in diameter) with eight different colours (red, blue, green, yellow, brown, black, grey and white), and placed in front of the ball machine to mimic the texture appearance of a crowd. A hole was cut to allow ball projection, but the sheet still ensured that placement of the ball into the machine could not be seen. The remainder of the surrounding walls were covered by black curtains to reduce reflection from laboratory lighting. Light levels were adjusted using dimmer switches and were checked at two points in the room using an umpire's light meter (Megatron model ULM SN131; Megatron Ltd, London, UK). The light levels used were 571, 1143 and 1714 lux. The testing was filmed using a Panasonic F15 CCD Video Camera (Matsushita Inc., Tokyo, Japan) to obtain performance scores and also for checking order of trials against computer data.

2.3. Procedure

Six practice trials (3 with a red ball; 3 with a white ball) were administered under 1714 lux prior to testing to ensure that the subjects were familiar with the task and the equipment. The testing involved a total of 60 deliveries, consisting of 20 deliveries (10 white balls, 10 red balls) randomly projected under the three different lighting conditions in a random order. There was a 5-min rest period between each set of 20 trials. The stance adopted by the subjects was that which they were comfortable with, starting with their hands placed on the switches on the thighs. The chosen height of the bowling machine (0.8 m) was equivalent to the average waist height of a batsman playing a front foot shot, i.e. the height at which the ball would usually be contacted by the bat when the ball is deflected to the slip-catching area of the field. The balls were projected along a similar flightpath to the dominant side of the subjects over a distance of 8.4 m, at a speed of 20 m s^{-1} (45 mph). Subjects were required to catch the balls with both hands, and not to anticipate projection from the ball machine.

2.4. Data analysis

Outcome data, in the form of a catching performance score, were obtained using Wickstrom's (1983) established catching performance score scale (table 1) from the video footage collected. Movement initiation times for each trial were calculated using a motion analysis system (Elite, BTS, Milan, Italy). Owing to the bimanual nature of the task, the movement initiation times of the left and right hand were averaged for each trial to provide a single initiation time as in previous work (Kelso 1979, Tayler and Davids 1997).

3. Results

3.1. Outcome (performance) data

The scores achieved on the performance scale were averaged for each subject in each condition (table 2). The mean and standard deviation for each condition were then calculated. A 2 (ball) × 3 (light level) analysis of variance (ANOVA) was performed with repeated measures on both factors. No significant effects were noted for ball

Table 1. Catching performance score scale adapted from Wickstrom (1983).

Outcome score	Description
5 (Clean catch)	The ball is contacted and retained by the hands
4 (Assisted catch)	The ball is juggled and retained by the hands
3 (Hand contact)	The ball contacts the hand but is dropped
2 (Upper body contact)	Upper body (but no hand) contact
1 (Lower body contact)	Lower body (but no hand) contact
0 (No ball contact)	No cricket ball contact

Table 2. Mean catching scores and standard deviations by ball colour and light level.

	Condition					
Subject	Rb571	Wb571	Rb1143	Wb1143	Rb1714	Wb1714
1	3.6±2.1	4.2±1.8	4.8±0.7	4.6±0.9	4.1±1.7	4.9±0.4
2	4.4±1.0	4.4±1.0	4.8±0.6	5.0±0.0	5.0±0.0	4.4±1.0
3	4.6±0.7	5.0±0.0	5.0±0.0	5.0±0.0	5.0±0.0	5.0±0.0
4	4.2±1.1	4.6±0.9	4.1±1.7	3.8±1.0	3.8±1.0	4.2±1.1
5	4.8±0.7	5.0±0.0	5.0±0.0	4.8±0.8	4.8±0.0	5.0±0.0
Mean	4.3	4.6	4.7	4.6	4.5	4.7
SD	0.5	0.4	0.4	0.5	0.6	0.4

Rb = Red ball, Wb = white ball.

($F(1,4) = 0.81$, $p = 0.418$; effect size = 0.169), light level ($F(2,8) = 0.60$, $p = 0.571$; effect size = 0.131), or for ball by light level ($F(2,8) = 1.28$, $p = 0.329$; effect size = 0.243).

A 2 (ball) × 3 (light level) repeated measures ANOVA was performed on the standard deviations of performance scores (table 2). No effects were noted for ball ($F(1,4) = 2.19$, $p = 0.213$; effect size = 0.353), light level ($F(2,8) = 2.71$, $p = 0.127$; effect size = 0.403), or ball by light level ($F(2,8) = 0.37$, $p = 0.699$; effect size = 0.086).

3.2. Movement initiation time data

The mean movement initiation times were calculated for each subject in each condition (table 3). The mean and standard deviation for each condition were calculated. A 2 (ball) × 3 (light level) repeated measures ANOVA was performed on the ball and light level factors. No significant effects were noted for ball ($F(1,4) = 0.00$, $p = 0.947$; effect size = 0.001), light level ($F(2,8) = 0.10$, $p = 0.906$; effect size = 0.024), or for ball by light level ($F(2,8) = 1.32$, $p = 0.320$; effect size = 0.248).

A 2 (ball) × 3 (light level) repeated measures ANOVA was performed on the standard deviations of movement initiation times for each condition (table 3). No effects were noted for ball ($F(1,4) = 1.86$, $p = 0.244$; effect size = 0.318), or ball by light level ($F(2,8) = 1.67$, $p = 0.247$; effect size = 0.295). There was also no main effect for light level ($F(2,8) = 3.66$, $p = 0.075$; effect size = 0.477).

4. Discussion

The aim of the current study was to examine the effects of different coloured cricket balls and varying light levels on slip-catching performance and movement initiation time in professional cricketers. Slip-catching performance, under illuminance levels lower and higher than current field-exit values, was not affected by ball colour. A clear implication of this finding is that the changes made to ball

Table 3. Mean movement initiation times and standard deviations by ball colour and light level.

Subject	Rb571	Wb571	Rb1143	Wb1143	Rb1714	Wb1714
1	125 ± 26	121 ± 22	131 ± 20	124 ± 18	110 ± 11	120 ± 25
2	178 ± 22	183 ± 12	156 ± 20	175 ± 11	169 ± 14	168 ± 18
3	178 ± 17	167 ± 42	159 ± 18	157 ± 22	167 ± 23	166 ± 27
4	147 ± 13	126 ± 27	147 ± 16	146 ± 11	150 ± 5	136 ± 13
5	144 ± 19	153 ± 20	162 ± 21	171 ± 23	157 ± 15	163 ± 22
Mean	154	150	151	155	151	151
SD	23	26	12	214	21	21

Rb = Red ball, Wb = White ball. All values are in milliseconds.

colour in the professional one-day game are not detrimental to slip-catching performance.

The subjects in the present study were professional cricketers who were used to playing with both the red and white cricket balls. Since the introduction of the white ball, the players had some time in which to adjust to the experience of playing with it. It would be of interest to investigate the effect of manipulating ball colour with players of different skill levels, to determine if there are initial effects that diminish as players habituate to using the white ball. In this respect, there is clearly a need for work to be carried out with club level players. Morris (1976) noted that the effect of ball colour on catching performance diminished with age and experience. This finding could influence the timing of the introduction of the white ball into the youth ranks of the game of cricket. It may be beneficial to introduce the different ball colour at a very young age so that the players become accustomed to playing with either coloured ball.

An issue for future research is to examine the effect of ball colour and illuminance level on slip-catching performance with varying ball projection speeds. Only one speed of ball projection (20 m s^{-1}) was used in this study. This speed was chosen in pilot work because it was considered to be at the upper limit of those speeds safe for use under laboratory conditions. The selected speed was also considered to be representative of the task constraints of slip-catching for professional cricketers. If slower speeds had been used to vary the experimental task in the study, it is possible that the task would have been too easy for the subjects. Increasing projection speed could have resulted in the task becoming too difficult and potentially dangerous for the cricketers. Future research could utilize variable projection speeds with players of different skill levels (e.g. senior club level performers) as well as professional cricketers.

Although difficult to compare, owing to the differences between the catching tasks used, in some ways the findings differed from those of other researchers in the area. For example, Koslow (1985) found an effect of ball colour on both reaction time and depth perception in peripheral vision. Koslow (1985) determined that the luminance contrasts (the contrast created by viewing the ball on a particular background) was the predominant factor in perceiving and reacting to the ball when catching, and that ball colour itself was of secondary importance. An implication of this finding, which needs to be confirmed in relation to the data from the present study, is that the combination of both sightscreen and ball colour seems to outweigh the importance of ball colour alone. Future research needs to examine this luminance contrast effect in high-level performers of reactive ball sports such as cricket, baseball and tennis.

With reference to light level and movement initiation time, there were also no significant differences due to experimental manipulations. There was, however, a general trend towards greater variability in initiation time at the lower levels of illuminance. The level of variability is evident from the magnitude of the effect size calculated (0.477), which according to Cohen (1992) is approaching a moderately meaningful effect (0.5). This level of variation suggests that, although initiation times did differ at lower light levels, the subjects were of a level of expertise to prevent a significant deterioration in performance.

Although this study was specifically concerned with the effects of ball colour and light level on slip-catching performance, the luminance of the ball was important because of the laboratory conditions under which the experimental work took place. Illuminance (brightness) affects the luminance of objects. In normal environmental conditions (daylight), luminance is equally spread over an object and can be easily anticipated. Under artificial lighting conditions (such as those used in the laboratory), luminance becomes less predictable and may be less useful as a source of information for making judgements on movements (such as ball flight). The luminance of cricket balls needs to be measured in future work to facilitate comparison with previous work, such as that of Koslow (1985), where the luminance contrasts were held to be responsible for differences in reaction time and depth perception in peripheral vision.

The illuminance levels used in the study (571, 1143 and 1714 lux) were in a range above and below the level where professional players are instructed to leave the field of play for 'bad light' (1000 lux). As the present results show, catching performance did not appear to deteriorate under poorer lighting conditions. A wider range of light levels needs to be explored to examine whether there would be greater performance variability at lower levels of illuminance. The implication is that illuminance level for 'bad light' could be changed to ensure optimum performance from the cricketers. However, all aspects of the game would need to be considered before a change of this nature could be implemented. It should also be considered that it is the batsmen (and not the close catchers) who are asked about leaving the field for 'bad light'. Further empirical research also needs to examine the effects of lighting and ball colour on the batsmen's performance of interceptive actions.

The small effect sizes noted for both performance scores and movement initiation times for ball, light and ball by light interactions support the lack of significant effects in the data. Inspection of the data tables showed very little evidence of individual differences in the group data analysis. This consistency helped to strengthen the conclusions regarding the lack of differences in performance due to ball colour and illuminance level. An alternative explanation, which cannot be completely rejected at this stage, is that the lack of effects may have been due, in part, to the lack of statistical power because of the small sample size. The small sample size reflected the difficulty in recruiting professional cricketers to take part in this study, and there is a need to confirm the findings with data from a larger sample of professionals.

In conclusion, this study attempted to redress the lack of research into the change of ball colour from red to white in one-day cricket. There were no effects on either catching performance or initiation time during slip-catching in cricket as a function of illuminance levels. These findings apply only to the task of slip-catching, because the task parameters were chosen due to their relevance to the constraints of slip-catching in cricket. There is a need for more work on other interceptive actions in cricket, such as batting and fielding balls in the outfield.

Acknowledgements

The authors would like to thank Megatron Ltd, 165 Marlborough Road, London N19 4NE for the use of a Megatron ULM Light Meter during this study.

References

CAMPBELL, F. W., ROTHWELL, S. E. and PERRY, M. J. 1987, Bad light stops play, *Ophthalmic and Physiological Optics*, **7**, 165–167.
COHEN, J. 1992, A power primer, *Psychological Bulletin*, **112**, 155–159.
KELSO, J. A. S. 1979, On the co-ordination of two-handed movements, *Journal of Experimental Psychology: Human Perception and Performance*, **5**, 229–238.
KOSLOW, R. E. 1985, Peripheral reaction time and depth perception as related to ball colour, *Journal of Human Movement Studies*, **11**, 125–143.
LORAN, D. F. C. 1997, An overview of sport and vision, in D. F. C. Loran and C. J. MacEwen (eds), *Sports Vision* (Oxford: Butterworth-Heinemann), 1–21.
MORRIS, D. G. S. 1976, Effects ball and background colour have upon the catching performance of elementary school children, *The Research Quarterly*, **47**, 409–416.
PERRY, M. J., CAMPBELL, F. W. and ROTHWELL, S. E. 1987, A physiological phenomenon and its implications for lighting design, *Lighting Research and Technology*, **19**, 1–5.
REGAN, D. 1997, Visual factors in hitting and catching, *Journal of Sports Sciences*, **15**, 533–558.
ROTHWELL, S. E. and CAMPBELL, F. W. 1987, The physiological basis for the sensation of gloom: quantitative and qualitative aspects, *Ophthalmic and Physiological Optics*, **7**, 161–163.
SAVELSBERGH, G. J. P. and BOOTSMA, R. J. 1994, Perception-action coupling in hitting and catching, *International Journal of Sport Psychology*, **25**, 331–343.
TAYLER, M. A. and DAVIDS, K. 1997, Catching with both hands; an evaluation of neural crosstalk and co-ordinative structure models of bimanual co-ordination, *Journal of Motor Behaviour*, **29**, 254–262.
WICKSTROM, R. L. 1983, *Fundamental Motor Patterns* (Philadelphia, PA: Lea & Febiger).

20 Individual differences, exercise and leisure activity in predicting affective well-being in young adults

C. Sale†, A. Guppy‡* and M. El-Sayed†

†Research Institute for Sport and Exercise Sciences, Liverpool John Moores University, Liverpool, UK

‡Department of Psychology, Middlesex University, Enfield EN3 4SF, UK

Keywords: Well-being; Coping; Physical activity; Health behaviours; Individual differences.

This study focuses on the prevalence of exercise and health-related leisure activities (smoking, drinking), across groups of subjects, defined by personality and gender, in relation to subjective well-being. Results from a cross-sectional survey of 187 participants are reported. Males ($n=80$) reported more drinking ($p<0.001$) and smoking ($p<0.001$) than females, though they also reported higher habitual physical activity levels ($p<0.001$). Females ($n=107$) reported more frequent use of social support coping ($p<0.01$). There was a positive association between extraversion and self-reported habitual physical activity as well as alcohol consumption (even when controlling for gender). Neuroticism was not related to any of the exercise and leisure activity variables. Multiple regression analyses predicted 34% of variance for the depression-enthusiasm and 39% of the variance for the anxiety-contentment measures of affective well-being. Neuroticism ($p<0.001$) and avoidance coping ($p<0.05$) were the only significant predictors of both anxiety-contentment and depression-enthusiasm. It is concluded that the influence of individual differences such as personality and gender on coping behaviour and well-being is consistent with social learning theory research. Limitations of cross-sectional research designs necessitate caution with inferring causal paths. Recommendations for future research are presented concerning the use and value of repeated measures designs within research into exercise and well-being.

1. Introduction

There have been recent attempts to place exercise and health-related behaviours in the wider context of stress and coping processes. Rostad and Long (1996) reviewed a range of empirical studies, focusing on the role of exercise behaviours as coping strategies for stress. While the general conclusion suggested positive support for the efficacy of such strategies in coping with stress, none of the studies reviewed have incorporated measures of stress appraisal and coping processes in sufficient detail.

Rick and Guppy (1994) identified several exercise and health-related coping strategies regularly used by a sample of over 600 white collar employees. Taking regular exercise was reported as a frequent means of coping with work stress by 30% of employees. Other health behaviours, such as maintaining a healthy diet, were

*Author for correspondence. e-mail: a.guppy@mdx.ac.uk

frequently reported by nearly half of the sample. Factor analysis of the reported coping strategies identified a leisure activity oriented dimension encompassing 'taking regular exercise', 'turning to hobbies and pastimes', 'eating a healthy diet' and 'expending energy'. This factor was in addition to other more commonly reported coping dimensions such as problem-oriented coping, avoidance and social support seeking. The health and exercise coping dimension was significantly related to problem-focused coping, social support seeking and the use of formal relaxation methods. It was also significantly related to reported mental health, though use of formal relaxation methods was not significantly related to mental health.

Ingledew et al. (1996) considered health-related behaviours as well as more traditional measures of coping behaviour. Their main aim was to establish the relationship between the dimensions of coping that have been identified in research over the past 20 years and the health-related behaviours. They reported 'clear clusters' of exercise behaviours as well as the use of problem-focused, emotion-focused and avoidance coping. Their results indicated that exercise coping was significantly related to problem-focused coping, thus supporting the earlier findings of Rick and Guppy (1994), though no associations between exercise coping and well-being were reported by Ingledew et al. (1996).

Various researchers have suggested that exercise may be linked with dispositional characteristics of the individual, such as personality and perceived locus of control. For example, Courneya and Hellsten (1998) reported that exercise behaviour was positively linked with extraversion, and negatively correlated with neuroticism, within a student sample. Personality factors were related to the different types of exercise behaviours that participants adopted, with extraverts preferring to exercise with others rather than alone. An element to this research concerned the motives for exercising. Neuroticism was the only factor to correlate with the motivating factors of physical appearance and weight control. The personality dimension of 'openness' was solely related to exercise used as a mental health/stress relief coping strategy.

Some gender-based differences in general coping behaviours have been reported in previous research. Rick (1995) found that females were less likely to use regular exercise as a means of coping with stress, though they were significantly more likely to use social support coping strategies. Ingledew et al. (1996) reported that females were more likely to use eating and emotion-focused coping strategies, though they did not find significant gender differences in the adoption of exercise activity as a coping behaviour.

In predicting well-being from coping, a large number of studies have indicated the positive impact of problem-focused coping and the potentially negative impact of coping strategies such as avoidance (Parkes 1990, Rick and Guppy 1994). Previous research has also identified positive links between exercise activity and affective well-being. Rostad and Long (1996) discussed the positive impact of exercise-related programmes on measures such as state anxiety and depression. Yeung and Hemsley (1997) reported significant associations between reported physical activity and both positive and negative affect measures. However, in subsequent regression analyses, they found that trait measures such as neuroticism and extraversion were more important predictors of well-being.

Therefore, the objectives of the present study were to expand upon the previous research identifying a link between exercise and general coping behaviours; and to explore the direct effects of exercise and the use of general coping strategies on psychological well-being, while controlling for the effects of personality and individual differences.

2. Methods

2.1. *Measures*

Seven scales were used in this questionnaire-based study. Brief details of each of the scales employed are given below. As the questionnaire used here mainly incorporated well established, prevalidated scales, only a relatively small pilot study was employed. At three separate stages in the development of the questionnaire, following significant changes in format, the questionnaire was piloted to 30 respondents.

2.1.1. Demographic information: Information was obtained on age (year of birth) and gender (one male, two females).

2.1.2. Coping in general life: The 20-item version of the Cybernetic Coping Scale (Edwards and Baglioni 1993) with an additional four items from the Ways of Coping Check-List (Lazarus and Folkman 1984) were incorporated to yield general coping strategies but not specific coping responses. The items were selected to represent the coping dimensions of changing the situation, accommodation, devaluation, avoidance, and symptom reduction (from the CCS) and social support seeking (from the WCCL). Each item was graded on a five-point Likert scale, according to how often the respondent used a particular method to cope, with the scale ranging from 'never' ($=1$) to 'always' ($=5$).

2.1.3. Affective well-being: A 12-item scale incorporating adjectives designed to measure job-related well-being (Warr 1990) was included in the questionnaire. The scale is based on the two well-being axes reported by Warr (1990); namely job-related anxiety-contentment and job-related depression-enthusiasm. The scale was altered to make it applicable to university students by asking them to indicate how often the respondent had felt the way the items described over the past few weeks at university. Respondents were asked to rate each of the 12-items along a six-point frequency scale ranging from 'never' ($=1$) to 'all of the time' ($=6$). A high score thus indicates a greater perception of personal well-being.

2.1.4. Personality: Extraversion was assessed using the six-item improved short scale of extraversion from the Eysenck Personality Inventory (EPI) (Eysenck and Eysenck 1964). Neuroticism was assessed using the corresponding six-item measure of neuroticism from the EPI. Responses were graded along a four-point frequency scale ranging from 'almost never' ($=1$) to 'almost always' ($=4$).

2.1.5. Habitual alcohol consumption: This was designed as a means of gaining detailed information concerning the typical alcohol consumption patterns of respondents. Respondents were asked to report their typical weekly (7 days) alcohol consumption, recording both the type and amount of alcoholic beverage consumed. These data were then converted to represent the total number of units of alcohol consumed over the typical 7-day period. This methodology is similar to that used in a large number of investigations ranging from general population studies (Wilson 1980) to cross-national investigations of drink driving (Guppy and Adams-Guppy 1995). While it is acknowledged that there may be under-reporting of consumption through self-reported measures (Midanik 1992), it is felt less likely that correlational analyses are affected by such bias compared with estimates of population means.

2.1.6. *Physical activity*: Respondents were asked to rate their habitual activity levels along a four-point Likert type scale, with responses ranging from 'sedentary' (=1) to 'highly active' (=4).

2.1.7. *Smoking*: Smoking was assessed as a dichotomous variable with subjects reporting either 'yes' (=1) or 'no' (=0) to the question 'do you smoke?'.

2.2. *Procedures*

Cross-sectional data were collected over 3 weeks corresponding to weeks 4–6 of the second university academic semester. Data were collected by means of a self-completed questionnaire, administered in a classroom setting following lectures. The response to the questionnaire was voluntary and subjects were given written assurance that all individual data would be treated confidentially and anonymously. The local institution's Ethics Committee approved the study. All data were analysed using the Statistical Package for the Social Sciences (SPSS).

3. Results

3.1. *Sample*

Two-hundred and sixty-one questionnaires were administered, with 190 of these being completed and returned (response rate = 72%). Three participants were excluded from this sample because they provided incomplete data on many of the questions. Thus, statistical analysis was performed on 187 respondents (80 males, 107 females; mean age 24 ± 9 years). All respondents were students from the Psychology, Health Studies, Nursing, Engineering and Sports Science courses at Liverpool John Moores University.

3.2. *Scale-descriptive statistics*

The scale means $\pm$ SD, as well as measures of internal consistency (Cronbach's Alpha), are presented for each of the scales employed (table 1). As can be seen, the majority of the scales have reasonable levels of internal consistency, with accommodation coping (I = 0.57) being the lowest.

3.3. *Relationship between coping, exercise and health behaviours, and well-being*

Intercorrelations (table 1) identified some significant relationships between the coping and health-related behaviours and the measures of individual differences. Although males reported significantly more drinking ($r = -0.34$, $p < 0.001$) and smoking ($r = -0.27$, $p < 0.001$), they also reported higher habitual physical activity levels than females ($r = -0.29$, $p < 0.001$). Females reported more frequent use of social support coping ($r = 0.23$, $p < 0.01$), with these findings remaining statistically significant even after controlling for the effects of personality. There was a positive correlation between extraversion and self-reported habitual physical activity ($r = 0.19$, $p < 0.01$), as well as alcohol consumption ($r = 0.37$, $p < 0.001$), even when controlling for gender. Neuroticism did not significantly correlate with any of the exercise and leisure variables once gender had been partialled out.

Extraversion was significantly related to measures of anxiety-contentment ($r = 0.28$, $p < 0.001$) and depression-enthusiasm $r = 0.26$, $p < 0.001$), as was neuroticism (anxiety-contentment: $r = -0.56$, $p < 0.001$; depression-enthusiasm: $r = -0.51$, $p < 0.001$). Neuroticism shared nearly 30% of the variance in anxiety-contentment scores. Significant positive correlations with anxiety-contentment were also identified

Table 1. Correlation coefficients, scale means (±SD) and Cronbach's Alpha coefficients for each of the scales used.

Alpha	1	2	3	4	5	6	7	8	9	10	11	12	13	14
1. Anxiety-contentment	**0.90**													
2. Depression-enthusiasm	0.72^c	**0.85**												
3. Gender	−0.32^c	−0.22^b	—											
4. Extraversion	0.28^c	0.26^c	−0.19^a	**0.74**										
5. Neuroticism	−0.56^c	−0.51^c	0.33^c	−0.27^c	**0.65**									
6. Alcohol consumption	0.22^b	0.13	−0.34^c	0.37^c	−0.14	—								
7. Activity	0.20^b	0.11	−0.29^c	0.19^b	−0.10	0.04	—							
8. Smoking	−0.13	−0.13	−0.27^c	−0.03	0.12	0.07	−0.13	—						
9. Changing the situation	0.09	0.15^a	0.05	0.20^b	−0.09	0.02	0.00	0.01	**0.73**					
10. Accommodation	0.00	0.11	−0.09	0.09	0.05	−0.09	0.06	−0.10	0.13	**0.57**				
11. Devaluation	0.08	0.06	0.03	−0.02	−0.02	−0.05	0.07	0.13	−0.06	0.24^c	**0.87**			
12. Avoidance	−0.08	−0.10	0.02	0.05	−0.01	0.02	−0.09	0.03	−0.19^b	0.16^a	0.45^c	**0.82**		
13. Symptom reduction	0.08	0.13	0.03	0.16^a	0.02	0.02	0.17^a	0.03	0.25^c	0.29^c	0.09	0.03	**0.82**	
14. Seeking social support	−0.12	−0.08	0.23^b	0.11	0.18^a	−0.08	−0.01	0.00	0.22^b	0.17^a	−0.11	−0.11	0.42^c	**0.77**
Mean	3.81	4.43	1.57	3.09	2.23	18.15	2.66	1.76	3.43	3.06	2.94	2.81	3.56	3.51
SD	0.95	0.77	0.50	0.62	0.54	13.57	0.82	0.43	0.54	0.52	0.96	0.75	0.65	0.72

$^a p<0.05$; $^b p<0.01$; $^c p<0.001$.

for alcohol consumption ($r = 0.22$, $p < 0.01$), habitual physical activity ($r = 0.20$, $p < 0.01$). Problem-focused coping (changing the situation) was significantly correlated with depression-enthusiasm ($r = 0.15$, $p < 0.05$). Gender was significantly related to both anxiety-contentment ($r = -0.32$, $p < 0.001$) and depression-enthusiasm ($r = -0.22$, $p < 0.01$) with males reporting higher affective well-being.

Following gender-based interaction analyses in other areas of research on well-being (Laurent et al. 1997), interaction terms were calculated for gender and the two personality dimensions. These were correlated with the coping and health-related measures, with the main effects of gender and personality partialled out (reflecting procedures described by Cohen and Cohen 1983). The only significant interaction term revealed that habitual physical activity was associated with the gender × extraversion interaction term, although this was only a weak effect ($r = 0.18$, $p < 0.05$). Further examination of the simple effects revealed that, for females, extraversion was significantly related to physical activity ($r = 0.32$, $p < 0.001$).

Hierarchical multiple regression analyses (MRA) identified significant predictors of affective well-being scales. MRA predicted 39% of the variance for anxiety contentment and 34% of the variance for depression-enthusiasm. Neuroticism and avoidance coping significantly predicted levels of anxiety-contentment, once other variables had been statistically controlled (table 2). Neuroticism and avoidance coping also significantly predicted depression-enthusiasm (table 2).

4. Discussion

The present results extend the findings of Ingledew et al. (1996) on links between health-related behaviours (exercise, smoking, drinking) and more traditional dimensions of coping. However, evidence of significant correlations was limited, with only the frequency of habitual physical activity being significantly correlated

Table 2. Multiple regression analysis predicting anxiety-contentment and depression–enthusiasm.

Variable	Anxiety-contentment Standardized regression coefficient	p	Depression-enthusiasm Standardized regression coefficient	p
Gender	−0.04	0.61	0.01	0.95
Extraversion	0.07	0.31	0.11	0.15
Neuroticism	−0.47	0.00	−0.46	0.00
Alcohol consumption	0.11	0.12	0.04	0.56
Activity	0.08	0.24	0.01	0.87
Smoking	0.07	0.30	0.07	0.29
Changing the situation	0.01	0.92	0.05	0.52
Accommodation	−0.02	0.79	0.11	0.13
Devaluation	0.13	0.08	0.08	0.28
Avoidance	−0.15	0.04	−0.16	0.03
Sympton reduction	0.09	0.20	0.10	0.21
Seeking social support	−0.07	0.37	−0.08	0.28

$R = 0.62$; $R^2 = 0.39$; $R^2_{ADJ} = 0.34$; $F = 8.63$; d.f. = 12,163; $p < 0.00$
$R = 0.58$; $R^2 = 0.34$; $R^2_{ADJ} = 0.29$; $F = 6.91$; d.f. = 12,163; $p < 0.00$

with 'symptom reduction coping'. This lack of relationship between drinking and smoking behaviour and the coping dimensions such as avoidance could well reflect weaknesses in the measures used. Cooper *et al.* (1988, 1992) emphasized the distinction between alcohol consumption and the use of alcohol as a coping strategy. Furthermore, Polich and Orvis (1979) described measures of alcohol use as a coping strategy, which has qualitative advantages over simple measures of consumption. Thus, a more precise measure of how smoking and drinking are used to cope could improve the current methodology.

The results provided some support for the impact of individual differences on the use of coping behaviours, particularly health-related behaviours. The link between gender and social support coping has been reported previously (e.g. Rick and Guppy 1994), as has the finding that males have higher alcohol consumption patterns (e.g. Evans and Dunn 1995). These probably reflect wider processes of socialization and the development of expectations coping efficacy and normative beliefs (Abrams and Niaura 1988). The finding that males reported more frequent habitual physical activity contradicts research reported by Ingledew *et al.* (1996) and Ransford and Palisi (1996) where substantial gender differences were not observed, particularly with younger samples. It is possible, however, that the high proportion of sports science students in the present sample may be responsible for this anomalous finding.

The observation that extraversion was significantly related to drinking and habitual physical activity, as well as the reported frequency of the use of problem-focused coping (and to a lesser extent, symptom reduction coping) persisted after controlling for the effects of gender. The 'gender × extraversion' interaction term indicated that the strong association between extraversion and physical activity applied only to females. To some extent the link between extraversion and exercise supports the findings reported by Courneya and Hellsten (1998), though the lack of an association for neuroticism contradicts their results. While there is no sufficient explanation for such contradictory results within two similar student samples, it is clear from a number of studies (Daniels and Guppy 1994, Zapf *et al.* 1996) that the stability, as well as the interpretability, of relationships among measures in such studies would be better achieved with the use of repeated measures data sets.

From the multivariate analyses, in both cases the two significant contributors to the equations were neuroticism and avoidance coping. These findings support the previous literature, particularly Yeung and Hemsley (1997) that neuroticism was the strongest predictor of affective well-being. The negative contribution of avoidance coping is in line with Guppy and Weatherstone (1997), although the lack of a significant contribution from 'problem-focused coping' measures to the multiple regression analysis contradicts the findings of Rick and Guppy (1994) and Guppy and Weatherstone (1997). While to some extent this contradiction is felt to be related to differences in the coping measures used, there was a significant bivariate correlation between problem-focused coping ('changing the situation') and affective well-being (depression-enthusiasm). Further, from the bivariate analyses, both alcohol consumption and habitual physical activity were significantly related to affective well-being (anxiety-contentment). While the positive correlation between habitual physical activity and well-being is supported by, for example, Yeung and Hemsley (1997), the positive correlation between alcohol consumption and well-being conflicts with, for example, Graham and Schmidt (1999). The significant positive association between these two variables was rendered non-significant when the effects of gender were partialled out.

To some extent the significance of measures such as extraversion and neuroticism in predicting affective well-being in cross-sectional surveys is as anticipated. Various researchers have utilized extraversion and neuroticism as measures of dispositional positive and negative affectivity respectively (e.g. Parkes 1990), logically expecting high correlations with indicators of well-being. Consideration of the merits of repeated measures designs in well-being research (Zapf et al. 1996, Daniels and Guppy 1997) suggests that the influence of such traits may be reduced when time lagged well-being measures are used as covariates for current levels of well-being. Thus, with the development of more sophisticated methodologies, it may be less likely that trait measures have strong predictive power, once the effects of time-lagged outcome variables have been entered into prediction equations (Zapf et al. 1996).

The present research has extended the findings from several other recent reports in relation to exercise, coping and psychological well-being. The relationships between exercise behaviours, health behaviours (such as smoking and drinking) and other active and passive coping behaviours were examined. In a further extension of the work reported by Ingledew et al. (1996), such patterns of behaviour were related to outcome measures of psychological well-being. The role of personality in exercise, exemplified by Courneya and Hellsten (1998), was also expanded in the current study. While the value of the current empirical contribution can be seen in the light of such reports, and particularly where there may be an integration of research lines, there are limitations in the strength of conclusions that can be drawn from single phase, cross-sectional research designs. As Zapf et al. (1996) indicated, there may be confusion about causality in significant associations as well as difficulty in determining the influence of unmeasured variables. It is anticipated that the later stages of the current research programme, which will include repeated measures data, will link coping and exercise patterns more clearly, with subsequent perceptions of efficacy and well-being.

References

Abrams, D. and Niaura, R. 1987, Social learning theory, in H. Blane and K. Leonard (eds), *Psychological Theories of Drinking and Alcoholism* (New York: Guildford), 131–178.

Cohen, J. and Cohen, P. 1983, *Applied Multiple Regression/Correlation Analysis for the Behavioural Sciences* (New Jersey: Erlbaum).

Coper, M. L., Russell, M. and George, W. H. 1988, Coping expectancies and alcohol abuse: A test of social learning formulations, *Journal of Abnormal Psychology*, **97**, 218–230.

Cooper, M. L., Russell, M., Skinner, J. B., Frone, M. R. and Mudar, P. 1992, Stress and alcohol use: Moderating effects of gender, coping, and alcohol expectancies, *Journal of Abnormal Psychology*, **101**, 139–152.

Courneya, K. S. and Hellsten, L. M. 1998, Personality correlates of exercise behaviour, motives, barriers and preferences: An application of the five-factor model, *Personality and Individual Differences*, **24**, 625–633.

Daniels, K. and Guppy, A. 1994, Occupational stress, social support, job control and psychological well-being, *Human Relations*, **47**, 1523–1544.

Daniels, K. and Guppy, A. 1997, Reversing the occupational stress process: a note on sub-consequences of employee psychological well-being, *Journal of Occupational Health Psychology*, **2**, 156–174.

Edwards, J. R. and Baglioni, A. J. 1993, The measurement of coping with stress: construct validity of the Ways of Coping Checklist and the Cybernetic Coping Scale, *Work and Stress*, **7**, 17–32.

Evans, D. M. and Dunn, N. J. 1995, Alcohol expectancies, coping responses and self-efficacy judgements: a replication and extension of Cooper et al.'s 1988 study in a college sample, *Journal of Studies on Alcohol*, **56**, 186–193.

Eysenck, S. B. G. and Eysenck, H. J. 1964, An improved short questionnaire for the measurement of extroversion and neuroticism, *Life Sciences*, **3**, 1103–1109.

Graham, K. and Schmidt, G. 1999, Alcohol use and psychosocial well-being among older adults, *Journal of Studies on Alcohol*, **60**, 345–351.

Guppy, A. and Adams-Guppy, J. 1995, Behavior and perceptions related to drink driving among an international sample of company vehicle drivers, *Journal of Studies on Alcohol*, **56**, 348–355.

Guppy, A. and Weatherstone, L. 1997, Coping strategies, dysfunctional attitudes and psychological well-being in white collar public sector employees, *Work and Stress*, **11**, 58–67.

Ingledew, D. K., Hardy, L., Cooper, C. L. and Jemal, H. 1996, Health behaviours reported as coping strategies: A factor analytical study, *British Journal of Health Psychology*, **1**, 263–281.

Laurent, J., Salvatore, J. C. and Callan, M. K. 1997, Stress, alcohol-related expectancies and coping preferences: a replication with adolescents of the Cooper *et al.* (1992) model, *Journal of Studies on Alcohol*, **58**, 644–651.

Lazarus, R. S. and Folkman, S. 1984, *Stress, Appraisal and Coping* (New York: Springer).

Midanik, L. T. 1992, Reliability of self-reported alcohol consumption before and after December, *Addictive Behaviours*, **17**, 179–184.

Parkes, K. 1990, Coping, negative affectivity, and the work environment: additive and interactive predictors of mental health, *Journal of Applied Psychology*, **75**, 399–409.

Polich, J. M. and Orvis, B. R. 1979, *Alcohol Problems: Patterns and Prevalence in the U.S. Air Force*. A project AIR FORCE report prepared for the US Air Force.

Ransford, H. and Palisi, B. 1996, Aerobic exercise, subjective health and psychological well-being within age and gender subgroups, *Social Science and Medicine*, **42**, 1555–1559.

Rick, J. 1995, Perceptions of stress, coping behaviour and the impact of a low level intervention amongst white collar public sector employees. Unpublished PhD thesis, Cranfield University.

Rick, J. and Guppy, A. 1994, Coping strategies and mental health in white collar public sector employees, *European Work and Organisational Psychologist*, **4**, 121–137.

Rostad, F. G. and Long, B. C. 1996, Exercise as a coping strategy for stress: a review, *International Journal of Sport Psychology*, **27**, 197–222.

Warr, P. 1990, The measurement of well-being and other aspects of mental health, *Journal of Occupational Psychology*, **63**, 193–210.

Wilson, P. 1980, *Drinking in England and Wales* (London: HMSO for the Office of Population Census and Surveys).

Yeung, R. R. and Hemsley, D. R. 1997, Personality, exercise and psychological well-being: static relationships in the community, *Personality and Individual Differences*, **22**, 47–53.

Zapf, D., Dormann, C. and Frese, M. 1996, Longitudinal studies in organisational stress research: a review of the literature with reference to methodological issues, *Journal of Occupational Health Psychology*, **1**, 145–169.

21 Transfer and motor skill learning in association football

C. WEIGELT, A. M. WILLIAMS*, T. WINGROVE and M. A. SCOTT

Research Institute for Sport and Exercise Sciences, Liverpool John Moores University, The Henry Cotton Campus, 15–21 Webster Street, Liverpool L3 2ET, UK

Keywords: Skill acquisition; Ball control; Juggling; Performance.

Transfer of learning involves the influence of previous experiences on the performance or learning of new skills. It is defined as a gain (or loss) in the capability for performance on one task as a result of practice on another. The aim of the study was to examine the degree of transfer between various association football skills. Twenty intermediate male players participated in the study. During pre- and post-training tests, participants juggled a football as many times as possible within 30 s using feet or knees. Further tests required participants to control an approaching football inside a restricted area using the preferred and non-preferred kicking leg. Following performance on the pretest, two matched skill groups were obtained. One group participated in a 4-week training period in which feet-only ball juggling was practised for 10 min daily, while the remaining group acted as a control. Trained participants exhibited superior post-test performance on knee juggling and ball control with preferred and non-preferred leg tasks relative to the control group ($p < 0.05$). Findings indicate positive transfer of learning from juggling practice with the feet to juggling with the knees and a football control task. Implications for theory and practice are highlighted.

1. Introduction

The acquisition of movement skill is essential in both sport and ergonomic contexts. Typically, becoming skilled requires many years of effortful, goal-directed practice (Ericsson *et al.* 1993, Ericsson and Charness 1994). There has been a growing body of literature in sport and other domains concerned with the content and quality of practice sessions leading to efficient learning (for recent reviews, see Abernethy *et al.* 1997, Schmidt and Lee 1999, Williams *et al.* 1999). The role of coaches or practitioners in enhancing learning through appropriate intervention and instructional strategies has also been examined (e.g. Singer and Suwanthada 1986, Lidor *et al.* 1996). The ability to direct learners to transfer knowledge from one situation to another is seen as an important instructional goal. If this knowledge can be transferred across different movement tasks, then learners would be advantaged as they attempt to acquire new but related movement skills. Efficiency in practice and learning would be demonstrated. An abundance of skills is learned from childhood to old age and consequently, there is much opportunity for skills to be transferred from one learning situation to another. As such, improving understanding of the mechanisms underpinning transfer is an important issue in motor learning.

---

*Author for correspondence. e-mail: m.williams@livjm.ac.uk

Transfer of learning involves the influence of previous experiences on the performance or learning of new skills. It is defined as a gain (or loss) in the capability for performance on one task as a result of practice on another (Schmidt and Lee 1999). Although discussions about mechanisms of transfer were observed in the literature some time ago (e.g. Thorndike and Woodworth 1901), the phenomenon is still not well understood (Schmidt and Young 1987).

Typically, when investigating transfer from one practised skill to a new one, a between-participants design is employed. Two groups of participants are matched on their pretest performance. One group takes part in a training procedure while the other acts as a control. After the training period, both groups are tested again under identical conditions on the same task as performed in the pretest (to assess learning) and on a criterion task (to measure transfer). Any difference between the training group's performance on the transfer task compared to the control group is presumed to reflect either positive or negative transfer effects. If the trained group's performance on the post-training test is comparable with that of the control group, then it is deemed that no transfer has occurred. Transfer effects can also be general (affecting a wide range of processes and skills) or specific (affecting only particular processes and skills) in nature (Cormier and Hagman 1987).

A significant amount of research has been undertaken on optimum transfer conditions in sport (e.g. Singer and Chen 1994) and other settings (e.g. Annett and Sparrow 1985, Detterman and Sternberg 1993). This work has typically addressed the effectiveness of various instructional strategies in facilitating transfer as opposed to the transfer of movement skill from one performance or learning situation to another. In classroom settings accumulative cluster strategies (Cavanaugh and Borkowski 1979), organizational strategies (Borkowski et al. 1983) and verbalization strategies (Berry 1983, Brooks and Dansereau 1983) have been employed successfully to create positive transfer from one learning situation to another. Similarly, in sport settings, Singer et al. examined the effectiveness of various learning strategies on achievement in related motor skills (e.g. Singer and Suwanthada 1986, Lidor et al. 1996).

In the context of motor transfer, the relationship or similarity between the learned and transfer tasks is thought to be important. For example, Holding's (1976, 1987) 'transfer surface' model argues that positive transfer occurs when the two tasks have common elements or components. The concept of task commonality emphasized in the 'transfer surface' approach seems intuitively plausible and has its grounding in Thorndike's (1914) 'theory of identical elements'. According to Thorndike, elements could refer to general (e.g. purpose of the response) or specific (e.g. physical components or requirements) characteristics. Based on research using verbal learning tasks, Osgood (1949) modified Thorndike's view by proposing that the amount and direction of transfer are related to the degree of similarity between each task's stimulus and response characteristics. The 'transfer surface' approach was extended to motor skills by Holding (1976).

From a 'transfer surface' perspective, motor skills are similar when there are common characteristics or elements between tasks or performance situations. The expectation is that the higher the degree of similarity between component parts the greater the amount of positive transfer between skills or performance situations. According to Holding's (1976) model, the degree of transfer between related skills such as the tennis and volleyball serves is expected to be greater than skills which do not appear to share common elements such as the golf drive and free throw shot in basketball. The former tasks would appear to have common features (e.g. overhead ball toss and overarm throwing action to propel the ball over a net), while the latter tasks are likely to have less in common. Positive inter-task transfer would be predicted between the tennis and volleyball skills, while no transfer would be expected between the golf and basketball tasks.

The currently prevalent view, therefore, is that transfer results from the development of an abstract symbolic schema that can support performance across the two task situations (e.g. Gentner 1983, Anderson and Thompson 1989, Reed 1993). For transfer to occur, the learner has to have acquired a sufficiently general symbolic schema and has to interpret this as a representation of the transfer situation. From a motor control perspective, transfer may be deemed to occur when the generalized motor programme (schema) employed to carry out the task is either identical or has common subroutines or components (Schmidt and Lee 1999). Holyoak and Thagard (1989) proposed a similar argument based on current theorizing in the area of neural networks. They argued that transfer occurs when a pattern of properties and relations is recognized through the activation of nodes in a connectionist network. Although there is evidence to support this theoretical viewpoint, we still do not know much about what causes the transfer phenomenon. Consequently, despite the empirical evidence supporting transfer of learning, our theoretical understanding of these effects is rather weak (Cormier and Hagman 1987, Schmidt and Young 1987).

Although the common assumption in cognitive theories of transfer is that a degree of similarity between tasks is essential, positive transfer may occur between two apparently dissimilar tasks. For example, some coaches believe that training to juggle a ball is beneficial to a wide range of association football skills (Rahmatpanah 1994), particularly ball control (Hopper and Davis 1988). The implicit assumption is that training to juggle a ball enhances the performance of other football skills through both general (Rahmatpanah 1994) and specific transfer effects (Hopper and Davis 1988). A clear discrepancy between theory and practice is identified.

A typical transfer design is used in the present study to explore the mechanisms underpinning transfer of learning between various football skills that may be classified, on the basis of task analysis, as being similar or dissimilar. First, the transfer between two tasks, juggling with the feet and knees, which are presumed to be similar, was examined. Since the two tasks appear to have common stimulus and response characteristics, the 'transfer-surface' approach would predict positive transfer. In the juggling task the ball is under participant control (i.e. proprioceptive or efferent outflow information about ball flight is available) and it requires cyclical actions of the lower limbs to be coupled with the extraction of optical information from the ball's trajectory. Second, the degree of transfer between two essentially dissimilar tasks, ball juggling and ball control, is examined. According to the 'transfer-surface' approach, no transfer is expected because of the apparent dissimilarity between each task's stimulus and response characteristics. In contrast to juggling, the ball is not under participant control (i.e. there is no proprioceptive or efferent outflow information) and effective performance requires optical information from the ball's flight path to be coupled with a discrete action response.

2. Methods

2.1. Participants

Twenty intermediate male football players gave their informed consent to participate in the study (aged 19–40 years). All participants played regularly for the university's first and second football teams. Eighteen participants were right-leg dominant; the remaining two were left-leg dominant. Participants were naive to the purpose of the study.

2.2. Apparatus and tasks

During pre- and post-tests, participants juggled a football (size 5, 5 psi) as many times as possible within 30 s. Performance was assessed for feet (contact with dorsal parts) and knee juggling separately. The ball was not allowed to touch the floor between successive juggles. Participants were permitted to move around a designated 3 × 3-m juggling area. Juggling performance was assessed as the number of successful contacts within the allotted time. The best score from three trials was taken for further analysis.

Further tests required participants to control an approaching football with one touch only inside a 2 × 2-m target area. The ball was served using a ball projection machine (BMP; JUGS, Inc. Tualatin, OR, USA) over a distance of 9 m with a speed of 6 $m.s^{-1}$. The target area was marked on the floor and subdivided into a central zone and two surrounding zones. The central target area measured 1 × 0.5-m and was surrounded by a second zone measuring 1.5 × 1-m. The third zone encompassed the rest of the 2 × 2-m area. Higher scores were awarded when the ball stayed inside the centre zone (10 points), while fewer points (5 and 2 respectively) were awarded when the ball was stopped inside the two surrounding zones. Five practice trials preceded 12 test trials, providing a maximum score of 120 points. To control for possible order effects, half the participants attempted to control the ball with their preferred kicking leg first while the other half used their non-preferred leg.

2.3. Experimental design

Pretest performance was recorded under laboratory conditions for feet juggling, knee juggling and ball control with preferred and non-preferred legs. Two matched groups were created based on performance on the initial ball-juggling task. One group ($n = 10$) participated in a 4-week training programme (training group) in which feet-only ball juggling was practised for 10 min daily. The participants were not given any guidance as to potential practice or learning strategies. They were merely required to practice the task for the allotted period. Each participant reported daily practice times and scores in a diary. Diaries were collated as a record of juggling practice at the end of the training period to ensure that participants had followed the required protocol. Participants were also informed that they would be retested at the end of the training period. The remaining participants acted as a control group ($n = 10$), and, consequently, they did not undertake specific practice on the juggling task. These participants were not warned about future test requirements. During the course of the 4-week training period, both groups participated in regular practice and match play. Post-test assessment followed the same procedure as the pretest. For pre- and post-tests, the order of juggling and ball control tasks was counterbalanced. Furthermore, the ball control task was counterbalanced with regard to the use of preferred and non-preferred leg.

2.4. Data collection and analysis

Performance was recorded with a VHS camera (sampling rate 50 Hz) positioned next to the BPM resulting in a frontal view of the performer and the target area. In the juggling conditions, a flashing LED indicated the beginning and end of each 30-s trial. Performance was assessed from the video film using slow motion and frame-by-frame analysis. Results were analysed using 2×2 MANOVA in which Group (training, control) was the between-participants variable and Test (pre-, post-) the within-participants factor. Performance scores on each of the four ball skill tests (knee and feet juggling, ball control with preferred and non-preferred legs) were included as dependent variables. Percentage of transfer was calculated as the ratio of the performance score difference between the training and control groups divided by the sum of performance scores $\times 100$ (Magill 1998: 160).

3. Results

The MANOVA showed a significant main effect for Test ($F_{4,15} = 9.32$, Wilks' Lambda $= 0.28$, $p < 0.01$). Follow-up discriminant analysis indicated that performance on the ball control tasks with preferred (standard coefficient $= 0.89$) and non-preferred (0.39) legs more clearly distinguished pre- and post-test performance. Separate univariate ANOVAs on each of the dependent variables showed significant pre- to post-test differences for knee juggling ($F_{1,18} = 9.45$, $p < 0.01$) and ball control with preferred ($F_{1,18} = 35.26$, $p < 0.01$) and non-preferred ($F_{1,18} = 9.53$, $p < 0.05$) legs. A significant Group $\times$ Test interaction was observed ($F_{4,15} = 6.60$, Wilks' Lambda $= 0.36$, $p < 0.05$). Follow-up discriminant analysis revealed that performance on the feet juggling (standard coefficient $= 0.58$) and ball control with preferred leg (0.88) tasks provided the greatest contribution to this interaction effect. This was supported by separate univariate analysis of the dependent variables which showed a significant difference for feet juggling ($F_{1,18} = 5.30$, $p < 0.05$) and ball control with preferred leg ($F_{1,18} = 19.38$, $p < 0.01$) tasks only.

Post-hoc analysis using Tukey tests on the Group $\times$ Test interaction showed that there were no significant differences between groups on the pretest. The trained group improved significantly from pre- to post-test on the knee juggling and ball control with preferred and non-preferred leg tasks (all $p < 0.05$). Moreover, a trend towards better post-test performance for the trained participants on the feet juggling task was observed ($p = 0.10$, effect size $= 0.84$). No significant pre- to post-training differences were observed for the control group. The proportion of transfer from pre- to post-test for the trained group was 20% for knee juggling, 23% for ball control with the preferred leg and 9% for ball control with the non-preferred leg. The results are presented in figures 1–4.

4. Discussion

The objective of this study was to examine the extent of transfer between similar and dissimilar football skills. More specifically, a typical transfer design was employed to examine the direction and amount of transfer following 4 weeks of feet only juggling training on the performance of a knee juggling skill and ball control tasks with the preferred and non-preferred leg. The 'transfer surface' approach would predict positive transfer between knee and feet juggling and no transfer between feet juggling and ball control.

The results provided partial support for these hypotheses. The trained group demonstrated a significant improvement in performance on the knee-juggling test as a result of training on the feet-juggling task. In the pretest the trained group scored 40.0 juggles compared with 47.8 on the post-training test, a transfer improvement of 20%. This positive transfer was predicted since it was assumed that the knee and feet juggling tasks employed elements of the same abstract representation of movement. In both tasks, participants had access to proprioceptive or efferent outflow

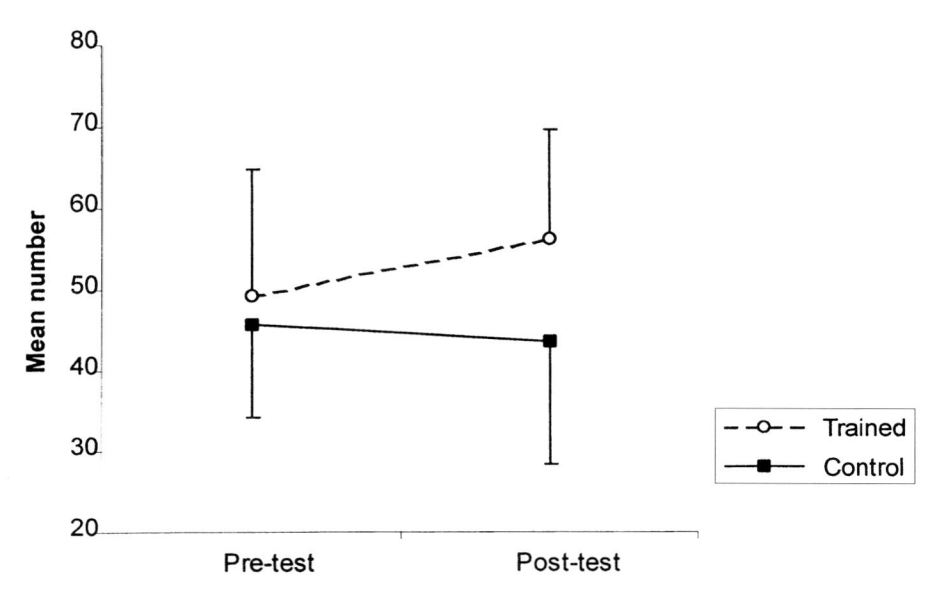

Figure 1. Mean number of successful juggles (feet) on pre- and post-tests across groups.

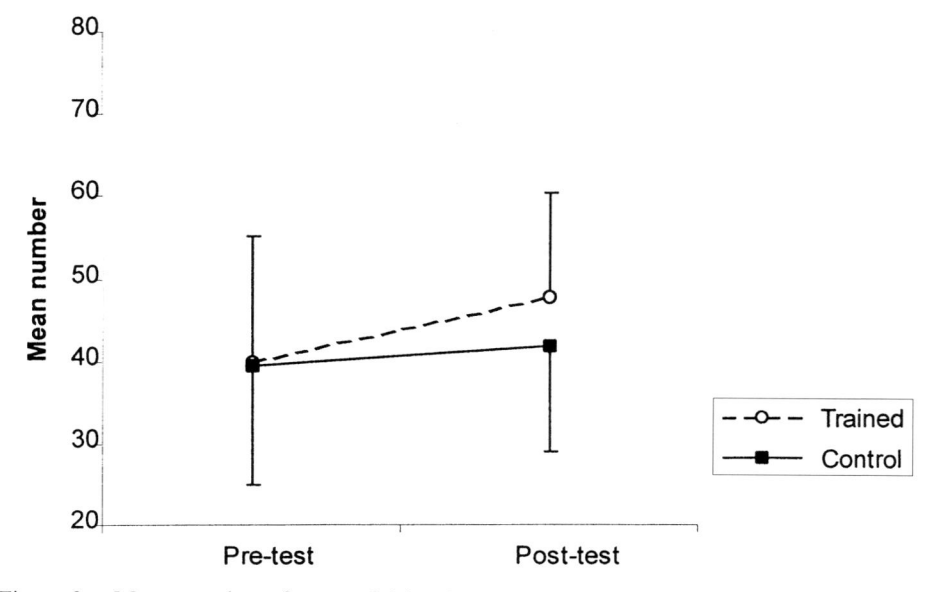

Figure 2. Mean number of successful juggles (knees) on pre- and post-tests across groups.

information about ball flight. Moreover, cyclical actions of the lower limbs need to be temporally synchronized and coupled with the extraction of optical information from the ball's trajectory.

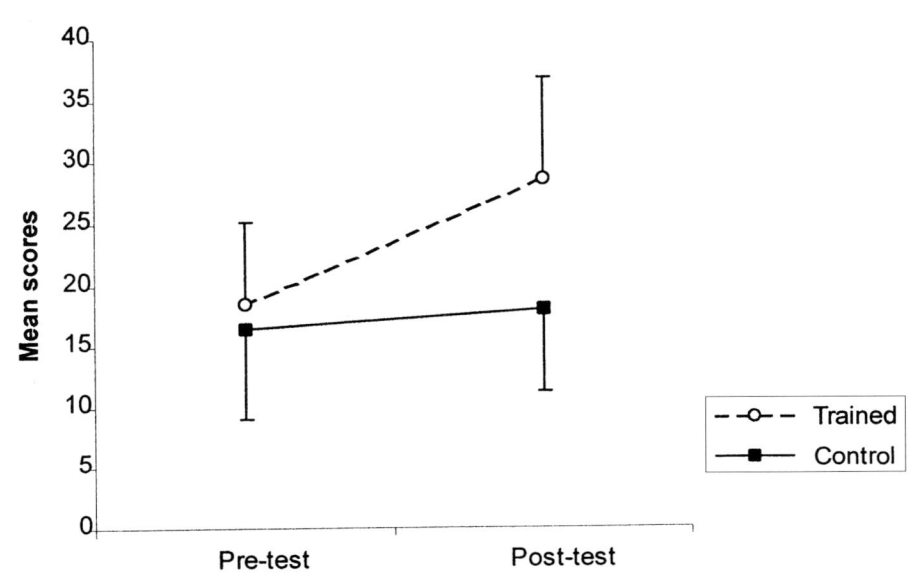

Figure 3. Mean absolute scores for ball control (preferred leg) on pre- and post-tests across groups.

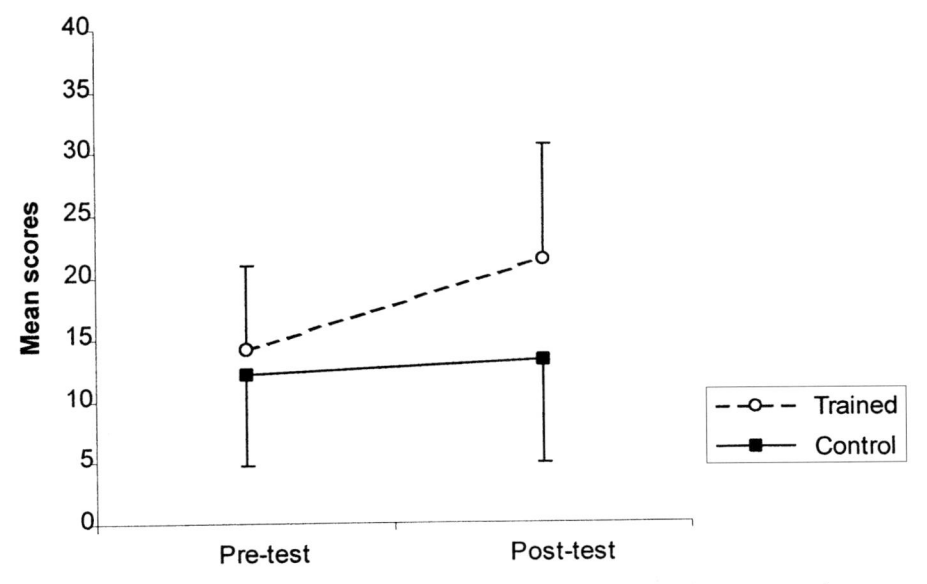

Figure 4. Mean absolute scores for ball control (non-preferred leg) on pre- and post-tests across groups.

Although many coaches believe that ball juggling training is beneficial to ball control (Hopper and Davis 1988), the positive transfer observed in this study between these two tasks was not predicted by the 'transfer surface' model. The trained group demonstrated significant improvements on the ball control with preferred and non-preferred leg tasks as a result of training. The trained group improved their performance from 18.5 to 28.6 using the preferred leg and from 14.2 to 21.4 with the non-preferred leg. These improvements resulted in 23 and 9% transfer effects respectively. Positive transfer was not expected since the two ball control tasks do not appear to have many common elements with juggling. For example, in the ball control task there is no proprioceptive or efferent outflow about ball flight since the learner does not control the trajectory. The task requires optical information from the ball's flight path to be coupled with a discrete rather than cyclical response. The timing characteristics differ markedly and consequently, it is unlikely that juggling and ball control employ the same abstract representation or generalized motor programme.

An alternative explanation for the transfer observed between the juggling and ball control tasks was proposed by Greeno et al. (1993) based on contemporary theorizing in ecological psychology. They oppose traditional views that cognitive representations play a mediating role in skill learning and that transfer depends on similarity between the representations involved in the initial learning and the transfer situations. Greeno et al. argued that although such mental representations may play an instrumental role, they are not fundamental to the transfer process. Transfer, in their view, occurs when the general properties and relations of the learner's interaction with the environment remain the same across the two tasks or situations. Their approach is concerned mainly with structures in the physical environment as opposed to structures inside the mind. Constraints on activity that result from the task structure and performance situation are more important than representations of structure. Positive transfer would be predicted between ball juggling and ball control because the two tasks require the pick up of similar invariant sources of optical and proprioceptive information to achieve a common affordance, namely to keep the ball under control.

Following on from the ideas of Gibson (1979) and Shaw et al. (1982), Greeno et al. (1993) conceptualized affordances as 'the support for particular activities created by relevant properties of the things and materials in the situations'. In a similar vein, Oudejans et al. (1996) described affordances as the behavioural possibilities that the situation offers the individual. The common affordance in juggling and ball trapping could be the behavioural potential for keeping the ball close to the feet (i.e. ball control). A source of perceptual information that is available in both tasks may therefore signify this common affordance. Initial support for this approach to understanding transfer would lie in identifying sources of invariant information that are common to both tasks. As both tasks are interceptive actions, a common invariant could be the optical information that regulates the action. For example, Bootsma and Oudejans (1993) described an optic variable that specifies the remaining time-to-contact between an approaching object (e.g. a ball) and the intercepting limb (e.g. the foot). In such situations the relative rate of dilation of an approaching object in combination with the relative rate of constriction of the optical gap between the object and the effector may be important. The affordance of close ball control common to the juggling and trapping tasks could be partly determined by this type of optical variable.

In summary, the present results suggest that practice on a juggling task leads to an improvement in performance on a ball control task. As the two tasks appear to lack common elements in relation to a stored symbolic code, an alternative explanation grounded in ecological theory has been offered. Greeno et al.'s (1993) ideas have yet to be fully examined using realistic sports-related tasks. Nevertheless, there would appear to be significant practical implications for structuring practice to ensure optimal transfer. In particular, future research should attempt to identify tasks that require the use of similar sources of information to achieve common affordances. For some time, ecological psychologists have been trying to identify key sources of optical information during interceptive actions. Current theorizing suggests that certain optical invariants may be employed across a range of related interceptive tasks (e.g. Michaels and Oudejans 1992, McLeod and Dienes 1996), implying potential avenues for exploring transfer. An explanation for transfer may therefore emerge as a by-product of the search for invariant perceptual information in sport-related tasks.

References

ABERNETHY, B., KIPPERS, V., MACKINNON, L. T., NEAL, R. J. and HANRAHAN, S. 1997, *The Biophysical Foundations of Human Movement* (Champaign: Human Kinetics).
ANDERSON, J. R. and THOMPSON, R. 1989, The use of analogy in a production system architecture, in S. Vosniadou and A. Ortony (eds), *Similarity and Analogical Reasoning* (New York: Cambridge University Press), 267–297.
ANNETT, J. and SPARROW, J. 1985, Transfer of training: a review of research and practical implications, *Programmed Learning and Educational Technology*, 22, 116–124.
BERRY, D. C. 1983, Metacognitive experience and transfer of logical reasoning, *Quarterly Journal of Experimental Psychology*, 35, 39–49.
BOOTSMA, R. J. and OUDEJANS, R. R. D. 1993, Visual information about time-to-collision between two objects, *Journal of Experimental Psychology: Human Perception and Performance*, 19, 1041–1052.
BORKOWSKI, J. G., PECK, V. A., REID, M. K. and KURTZ, B. E. 1983, Impulsivity and strategy transfers: metamemory as mediator, *Child Development*, 54, 459–473.
BROOKS, L. W. and DANSEREAU, D. F. 1983, Effects of structural schema training and text organisation on expositry prose processing, *Journal of Educational Psychology*, 75, 292–302.
CAVANAUGH, J. C. and BORKOWSKI, J. G. 1979, The meat–memory–memory 'connection': effects of strategy training and maintenance, *Journal of General Psychology*, 101, 161–174.
CORMIER, S. M. and HAGMAN, J. D. 1987, *Transfer of Learning: Contemporary Research Applications* (New York: Academic Press).
DETTERMAN, D. K. and STERNBERG, R. J. 1993, *Transfer on Trial: Intelligence, Cognition, and Instruction* (New Jersey: Alex).
ERICSSON, K. A. and CHARNESS, N. 1994, Expert performance: Its structure and acquisition, *American Psychologist*, 49, 725–747.
ERICSSON, K. A., KRAMPE, R. T. and TESCH-ROMER, C. 1993, The role of deliberate practice in the acquisition of expert performance, *Psychological Review*, 100, 63–406.
GENTNER, D. 1983, Structure-mapping: a theoretical framework for analogy, *Cognitive Science*, 7, 155–170.
GIBSON, J. J. 1979, *An Ecological Approach to Visual Perception* (Boston: Houghton-Mifflin).
GREENO, J. G., SMITH, D. R. and MOORE, J. L. 1993, Transfer of situated learning, in D. K. Detterman and R. J. Sternberg (eds), *Transfer on Trial: Intelligence, Cognition, and Instruction* (New Jersey: Alex), 99–167.
HOLDING, D. H. 1976, An approximate transfer surface, *Journal of Motor Behavior*, 8, 1–9.
HOLDING, D. H. 1987, Concepts of training, in G. Salvendy (ed.), *Handbook of Human Factors* (New York: Wiley), 939–962.
HOLYOAK, K. J. and THAGARD, P. 1989, Analogical mapping by constraint satisfaction, *Cognitive Science*, 13, 295–356.

HOPPER, C. A. and DAVIS, M. S. 1988, *Coaching Soccer Effectively* (Champaign: Human Kinetics Books).

LIDOR, R., TENNANT, K. L. and SINGER, R. N. 1996, The generalisability effect of three learning strategies across motor task performances, *International Journal of Sport Psychology*, **27**, 23–26.

MAGILL, R. A. 1998, *Motor Learning Concepts and Applications*, 5th edn (Singapore: McGraw-Hill).

MCLEOD, P. and DIENES, Z. 1996, Do fielders know where to go to catch the ball or only how to get there? *Journal of Experimental Psychology: Human Perception and Performance*, **22**, 531–543.

MICHAELS, C. and OUDEJANS, R. 1992, The optics and actions of catching fly balls: zeroing out optic acceleration, *Ecological Psychology*, **4**, 199–222.

OUDEJANS, R. R. D, MICHAELS, C. F., BAKKER, F. C. and DOLNÉ, M. 1996, The relevance of action in perceiving affordances: Perception of catchableness of fly balls. *Journal of Experimental Psychology: Human Perception and Performance*, **22**, 879–891.

OSGOOD, C. E. 1949, The similarity paradox in human learning: a resolution, *Psychological Review*, **56**, 132–143.

RAHMATPANAH, M. 1994, Why ball-juggling? *Scholastic Coach and Athletic Director*, **64**, 14–16.

REED, S. K. 1993, A schema-based theory of transfer, in D. K. Detterman and R. J. Sternberg (eds), *Transfer on Trial: Intelligence, Cognition, and Instruction* (New Jersey: Alex), 39–67.

SCHMIDT, R. A. and LEE, T. D. 1999, *Motor Control and Learning: A Behavioral Emphasis*, 3rd edn (Champaign: Human Kinetics).

SCHMIDT, R. A. and YOUNG, D. E. 1987, Transfer of movement control in motor learning, in S. M. Cormier and J. D. Hagman (eds), *Transfer of Learning* (Orlando: Academic Press), 47–79.

SHAW, R. E., TURVEY, M. T. and MACE, W. M. 1982, Ecological psychology: the consequence of a commitment to realism, in W. B. Weimer and D. S. Palermo (eds), *Cognition and the Symbolic Processes* (Hillsdale: Lawrence Erlbaum), vol. 2., 159–226.

SINGER, R. N. and CHEN D. 1994, A classification scheme for cognitive strategies: implications for learning and teaching psychomotor skills, *Research Quarterly for Exercise and Sport*, **65**, 143–151.

SINGER, R. N. and SUWANTHADA, S. 1986, The generalisability effectiveness of a learning strategy on achievement in related closed motor skills, *Research Quarterly for Exercise and Sport*, **57**, 205–214.

THORNDIKE, E. L. 1914, *Educational Psychology* (New York: Columbia University Press).

THORNDIKE, E. L. and WOODWORTH, R. S. 1901, The influence of improvement in one mental function upon the efficiency of other functions, *Psychological Review*, **8**, 247–261.

WILLIAMS, A. M., DAVIDS, K. and WILLIAMS, J. G. 1999, *Visual Perception and Action in Sport* (London: E & FN Spon).

Part VI
Methodological Studies in Sport and Ergonomics

22 Electromyography in sports and occupational settings: an update of its limits and possibilities

JAN PIETER CLARYS*

Experimental Anatomy, Faculty of Physical Education and Physiotherapy,
Vrije Universiteit Brussel, Brussels, Belgium

Keywords: History and bibliometry; Raw EMG; Rectified EMG; Surface-integrated electromyography; Normalization; Detection hazards.

The detection of the electrical signal from human and animal muscle dates from long before L. Galvani who took credit for it. J. Swammerdam had already shown the Duke of Tuscany in 1658 the mechanics of muscular contraction. Even if 'electrology or localised electrisation'—the original terminology for electromyography (EMG)—contained the oldest biological scientific detection and measuring techniques, EMG remained a 'supporting' measurement with limited discriminating use, except in conjunction with other methods. All this changed when EMG became a diagnostic tool for studies of muscle weakness, fatigue, pareses, paralysis, and nerve conduction velocities, lesions of the motor unit or for neurogenic and myogenic problems. In addition to the measurement qualities, the electrical signal could be induced as functional electrical stimulation (FES), which developed as a specific rehabilitation tool. Almost in parallel and within the expanding area of EMG, a speciality developed wherein the aim was to use EMG for the study of muscular function and coordination of muscles in different movements and postures. Kinesiological EMG and therewith surface EMG can be applied in studies of normal muscle function during selected movements and postures; muscle activity in complex sports; occupational and rehabilitation movements; isometric contraction with increasing tension up to the maximal voluntary contraction, evaluation of functional anatomical muscle activity (validation of classical anatomical functions); coordination and synchronization studies (kinematic chain); specificity and efficiency of training methods; fatigue; the relationship between EMG and force; the human–machine interaction; the influence of material on muscle activity, occupational loading in relation to lower back pain and joint kinematics. Within these various applications the recording system (e.g. the signal detection, the volume conduction, signal amplification, impedance and frequency responses, the signal characteristics) and the data-processing system (e.g. rectification, linear envelope and normalization methods) go hand in hand with a critical appraisal of choices, limits and possibilities.

1. Introduction

Recent developments in electromyographic signal processing, especially systems for analysing data, have upgraded electromyography (EMG), in particular surface electromyography (SEMG), into a data acquisition method for solving, detecting and discriminating ergonomic problems. The reason why it took so long for EMG to

*e-mail: jclarys@exan.vub.ac.be

reach this status is because EMG took three distinct different directions in the course of its development, each with various approaches and analytical techniques.

Clinical EMG is largely a diagnostic tool whereas kinesiological EMG is merely a means to study function and coordination. The fundamental EMG itself deals with single motor unit action potentials and the related time–frequency domain. Depending on the user, whether a physician, anatomist, ergonomist, physiologist, engineer, physiotherapist or neurologist, one encounters independent improvements in registration technology, different approaches to data acquisition and various but specific graphic representations, modelling techniques and software for treatment of the signal.

All these have facilitated a great number of applications in neurology, neurophysiology, neurosurgery, bioengineering, functional electrostimulation (FES), orthopaedics, zoology, ergonomics, occupational biomechanics and medicine, rehabilitation and physical therapy, sports medicine and sports science, and other areas. This dissemination of knowledge about the neuromuscular system, both in the normal and the disabled person, was already predicted (and in part described) by Duchene de Boulogne in 1855, 1862, 1867, 1872 and 1885. As movement is the prime sign of animal life, scientists have shown a perpetual curiosity about the origins of locomotion in human and other creatures. Among the oldest scientific experiments known are those concerned with the detection of electricity and function of muscle (Basmajian and de Luca 1985, Clarys and Lewillie 1992, Clarys 1994).

Von Humboldt (1797), especially in the correspondence supplements in their translation into French (Jadelot 1799), alluded on several occasions to the fact that he had been experimenting with the irritation of muscles and nerves for many years. He claimed he was unable to publish his results because of his other scientific interests or his long and frequent travels or because he felt the experiments needed more time, more repetition and more analysis under different circumstances. It is known, however, that he was working on this topic in 1792 (the year Galvani published *De Viribus Electricitatis in motu Musculari*). He handed a written manuscript on the matter to Professors Soemering and Blumenbach in 1795. The same year Pfaff (cited in von Humboldt 1797) published his book on the electricity and irritation of animal muscular tissue. He was already working in this field before Galvani's publication. Pierson and Sperling (1893) confirmed both statements. From the earlier works and especially from the correspondence published in them, it is clear that electricity generated by skeletal muscle had become an important research area for many scientists. Their experiments and discussions were to be found in *Le Journal de Physique* (France), *Journal der Physic* (Germany), *Journal Encyclopédique de Bologne* (Italy) and *Le Journal de Grenoble* (Switzerland). Several scientists studied the electrical phenomena of muscular and nervous tissue after Galvani's publications but we are led to believe that some of them studied these phenomena independently before Galvani in 1786 (cited by Trouvé 1893) and/or 1792 (Clarys 1994). This explains in part the early discussion of terminology such as: 'metallic irritation', 'animal electricity', 'Galvanic irritation', 'human electrology' and 'vital action'. It explains also the methodology used by those who stated that 'muscular irritation (or electricity) is not possible without metal or carbon excitators' and the findings of those scientists who confirmed that muscular electricity could be produced through irritation of humid animal tissue. According to von Humboldt (1797) and Jadelot (1799), it was Cotugno in 1786 and Vassali (1789) who investigated and suggested that soft animal tissue irritated muscle. It was their work that was the basis of the experiments of Galvani.

At the beginning of 1793, Volta wrote to Vassali that he believed Galvani's experiments were faulty and that they proved nothing (Jadelot 1799). Von Humboldt repeated and completed these experiments with success (figure 1) and convinced Volta that, in fact, the contrary was true (von Humboldt 1797 and letter of van Humboldt to Blumenbach 1795 cited by Jadelot 1799). Apparently nobody corresponded with Galvani informing him of these findings (Clarys 1994).

Looking at other sources, one encounters different names, periods and stories in addition to different experiments and observations. According to Basmajian and de Luca (1985), Francesco Redi in 1666 was the first scientist to make the logical deduction that muscles generate electricity since he suspected that the shock of an electric-ray fish was muscular in origin. From Trouvé (1893), it is known that Du Fay in 1698 stated that all living bodies, including the human body, have electrical properties. The most spectacular discovery was the giant sized book ($\sim 60 \times 40$ cm) on the 'Biological experiments of Jan Swammerdam' written by Boerhaave *et al.* (1737) in Old Dutch and Latin (and of which a very limited amount of copies were printed) (Clarys 1994) (figure 2). Here Swammerdam described various experiments on the irritation of nerves, on the mechanism of muscle contraction and on the relation between stimulation and contraction. He showed the results of these experiments to the Duke of Tuscany in 1658, about 130 years and more before the works of Galvani and von Humboldt who were arguing to get credited with the original findings. For a good understanding of the developments of EMG in the 20th

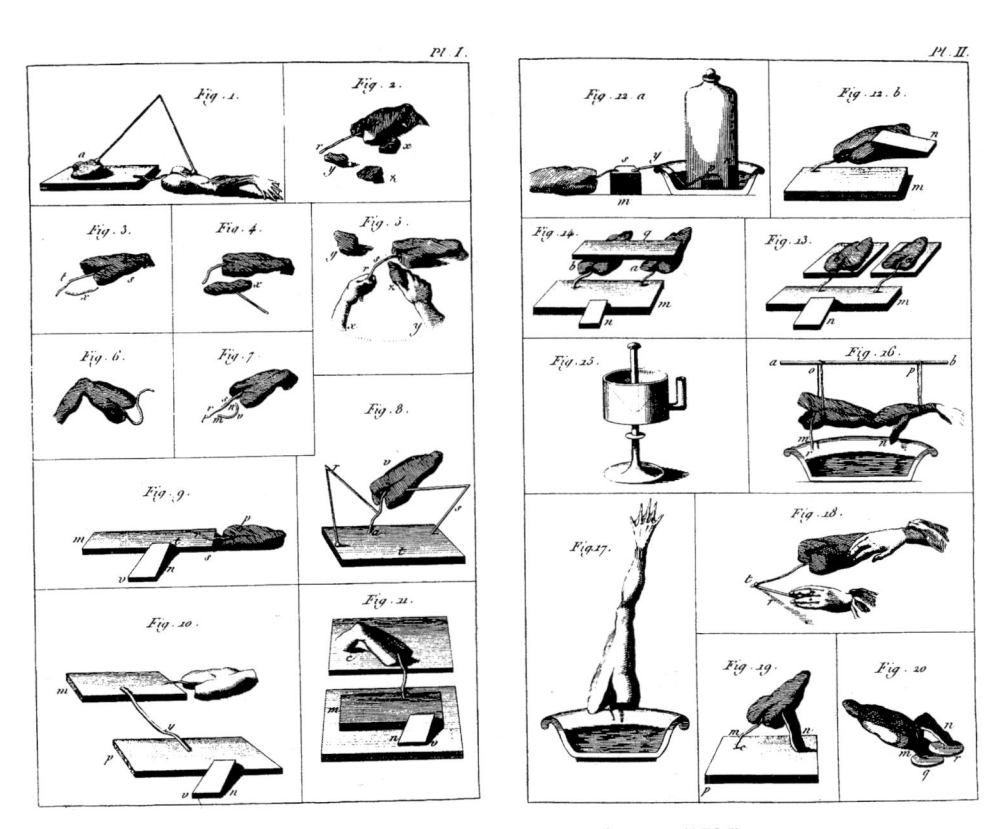

Figure 1. Von Humboldt's experiments (1795).

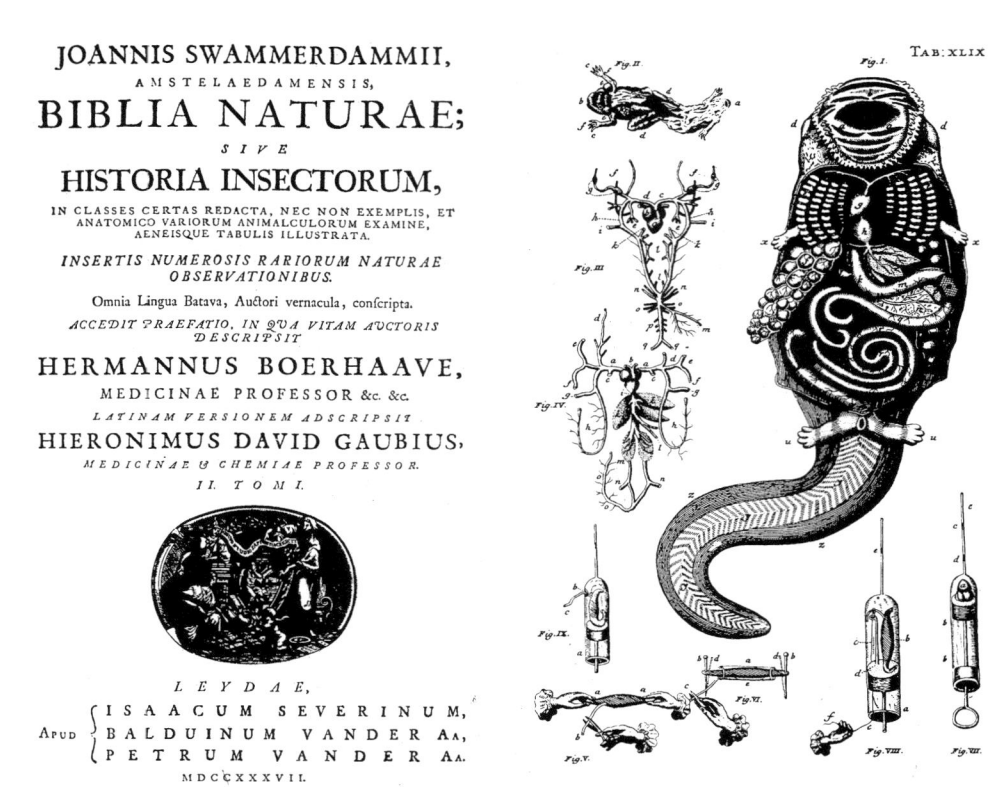

Figure 2. Front page of Boerhaave et al. (1737) and the experiments of Swammerdam (1658) (bottom right).

century, a series of historical landmarks in clinical, kinesiological and fundamental EMG are identified in table 1.

2. Technical and bibliometric considerations

Within EMG, in particular within the sports sciences and ergonomics, a speciality has been developed wherein EMG is used for studying muscular function and coordination. This area of research is usually called kinesiological EMG. The general aims are to analyse the function and coordination of muscles in different movements and postures, in healthy subjects as well as in the disabled, in skilled actions as well as during training, in humans as well as in animals, under laboratory conditions as well as during daily or vocational activities. Combining electromyographical, kinesiological, kinanthropometric, psychosocial and epidemiological data acquisition techniques often do this.

The research areas of kinesiological EMG can be summarized as follows. Areas incorporate studies of normal muscle function during selected movements and postures; studies of muscle activity in complex sports; occupational activity and rehabilitation movements; studies of isometric contraction with increasing tension up to the relative maximal voluntary contraction; evaluation of functional anatomical muscle activity (validation of classical anatomical functions); coordination and synchronization studies (kinematic chain); specificity and efficiency of training

Table 1. Historical landmarks in EMG before 1900†.

Author(s)	Contribution	Source
Swammerdam (1658)	Described different experiments on muscle and nerve irritation, depolarization and contraction	Boerhaave *et al.* (1737), Clarys (1994)
Redi (1666)	Made deduction that muscles generate electricity since he suspected the shock of a ray-fish was muscular in origin‡	Biederman (1898), cited in Basmajian and de Luca (1985)
Duverney (1679)	Described the function of muscles through the electricity (spirits) coming from the brain	Duverney (1761), Trouvé (1893)
Du Fay (1698)	Stated that all living bodies including humans have electric properties‡	Trouvé (1893)
Musschenbroek (1746)	Invented the Bottle of Leyden, the first electricity machine (improved by many since)	von Humboldt (1797), Trouvé (1893)
Jallabert (1750)	Described electrical stimulation of muscles for the purposes of medical re-education	Duchenne (de Boulogne) (1867)
Cotugno (1786) Vassal (1789) Galvani (1792) von Humboldt (1795) Jadelot (1799)	Experimented with muscular contraction without metal or carbon contact— the basis of Galvani's work Published *De Viribus Electricitatis in Motu Musculari* Made primitive electrodes and experimented with many forms of stimulation Translated the work of van Humboldt (1797) and commented on correspondence of 18th-century neurophysiologists	Pierson and Sperling (1893) von Humboldt (1797), Jadelot (1799) Galvani (1792) von Humboldt (1797) Jadelot (1799)
Matteucci (1838, 1844)	Stated that electrical currents originate in muscle and that they are related to voluntary contraction	Matteucci (1844)
Du Bois Reymond (1849)	Redesigned and improved the electricity machines and detected voluntary contraction *in vivo*	Du Bois Reymond (1849)
Duchenne (de Boulogne) (1855, 1867)	Described techniques and developed equipment for both registration and stimulation; described mapping of all motorpoints in segments and made the first myoelectric-powered orthosis. Author of the first EMG bestseller	Duchenne (de Boulgne) (1867, 1872)
Pierson and Sperling (1893) Trouvé (1893) Wedenski (1884)	Century literature up-date Century literature up-date Discovered that muscle becomes inhibited by high-frequency stimulation— the Wedenski block. Nobel prize for Medicine	Pierson and Sperling (1893) Trouvé (1893) Wedenski (1884)
Marey (1890)	Introduced the word 'electromyography'; made the first myograph and may be considered as the pioneer of biomechanics	Marey (1890)

†Revised and complete version of Clarys (1994).
‡No second supporting reference found and/or no original, nor copy.

methods; fatigue studies; the relationship between EMG and force; the human–machine interaction; the influence of equipment on muscle activity, and so on.

Since there are over 400 skeletal muscles in the human body and both irregular and complex involvement of the muscles may occur in neuromuscular diseases and in occupational or sports movements, it is impossible to sample all of the muscles of the entire body during the performance of complex motor skills. In addition, the EMG and the choice of integrated electromyography (iEMG) measurement, including the choice of the best normalization technique, depend on the specific demands on the type of subjects (e.g. athletes, manual handling workers, sedentary professions) and the specific demands of the field circumstances (e.g. swimming pool, office, alpine ski slope, hospital room, etc. whether in a vocational setting or in a simulation).

It is not within the scope of this review to discuss the instrumentation for recording electromyograms, nor to discuss functional anatomy. There is a wealth of literature concerned with the neuromuscular system, the recording system, the processing system and the information feedback system, which outlines the kinesiological and experimental data that should be reported (figure 3). Important dissemination roles have been fulfilled by:

1. The International Society of Electrophysiological Kinesiology (ISEK), in particular the ISEK conference proceedings (Wallinga *et al.* 1988, Anderson *et al.* 1991, Pedotti 1993, Shiavi and Wolf 1994, Hermens *et al.* 1996, Arsenault *et al.* 1998).
2. The European SENIAM's (Surface EMG for Non-Invasive Assessment of Muscles) concerted action of the Biomedical and Health Research Program, as part of Biomed II. The SENIAM project has two objectives. First,

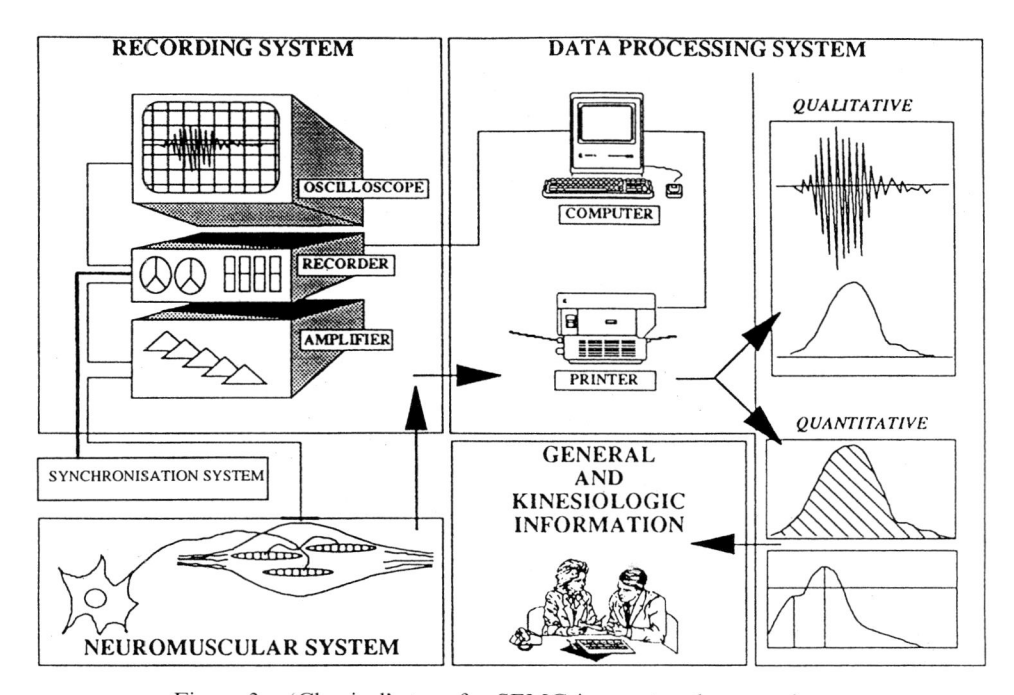

Figure 3. 'Classical' steps for SEMG in sport and occupations.

SENIAM enables scientists and clinicians working with surface EMG (SEMG) to exchange knowledge and experience on both basic and applied aspects of SEMG. The SENIAM project brings different disciplines together for the first time, enabling exchange and discussion and the creation of a common European body of knowledge on SEMG (e.g. by annual conferences 1996–99). The second objective of SENIAM is to develop recommendations for key items that presently prevent a useful exchange of clinical data and knowledge and a further maturation of SEMG. These items concern sensors, sensor placement, signal processing and modelling. The development of recommendations is essential to bring SEMG a major step forward as a general tool to be used in many applications focused on the assessment of the status and functioning of the neuromuscular system (Hermens *et al.* 1996a, 1997, 1998a,b, Hermens and Freriks 1997). Unfortunately, some of the contributions in these proceedings (1998a,b) report the mistakes/errors this author is cautioning against.
3. The *Journal of Electromyography and Kinesiology* (JEK) (edited by Solomonow, Maton and Moritani) spanning key topic areas such as muscle and nerve properties, motor units, physiological modelling, control of movement, motion analysis, posture, joint biomechanics, muscle fatigue, sports and exercize, measures of human performance, neuromuscular and musculoskeletal diseases, rehabilitation and functional electrostimulation. Important here is that the 'Units, Terms and Standards in Reporting EMG' (Winter *et al.* 1980) were rewritten and updated as the 'Standards and Guidelines of Reporting EMG Research' (Solomonow *et al.* 1996).
4. The journal *Ergonomics*, that was and remains the cradle of interdisciplinary applications in a range of vocational, simulation and work circumstances.

Whatever is not included in the above is to be found in the associated publications of biomechanics, neuroscience, sports science and orthopaedic societies respectively.

3. Critical appraisal of EMG studies, its limitations and hazards

The majority of activities in sport and occupational settings involve complex movement patterns often complicated by external forces, impacts and the equipment used during the movement. An electromyogram (or its derivatives) is the expression of the dynamic involvement of specific muscles within a determined range of that movement. The integrated EMG of that same pattern is the expression of its muscular intensity. However, intensity is not always related to force. For a review of EMG and force related to voluntary effort and isometric conditions, the reader is referred to the various literature sources. Mostly SEMG is used to investigate the activity of a series of muscles, seldom just one or two. The choice of these muscles is (based either on practical knowledge of the skill or on the basic anatomy literature. The functional EMG literature and early specific EMG work are rarely referred to.

The majority of scientists working in sport and occupational contexts measure EMG using surface electrodes. Skeletal muscles do not always stay in the same place during complex dynamic (sometimes ballistic) movements and the entire muscle belly may not be fully under the skin, but covered by parts of other bellies or tendons and subcutaneous adipose tissue (that is very variable both in composition and volume). It needs to be emphasized that the selection of muscles for EMG measurement requires careful consideration. Some of these choices can lead to erroneous registration, sometimes without being noticed by peer reviewers.

Despires (1974) placed his surface electrodes on M. sartorius, Broer and Houtz (1967) placed theirs on M. gracilis, and Strass (1990) and Toyoshima et al. (1971) on M. teres major. Measuring the EMG of these muscles under static conditions creates little or no problems, but under complex dynamic conditions the sartorius and gracilis muscles disappear from under the electrodes as does M. teres major, especially during arm motion above 90° abduction. It is therefore uncertain which muscles have contributed to the EMG patterns presented. Yoshizawa et al. (1978, 1987, 1989) and Oka et al. (1976, 1989) selected for their studies M. extensor carpi radialis brevis. This muscle has a very small superficial 'strip' accessible under the skin. The EMGs of this muscle are dubious and may give more information about M. extensor digitorum. The same problem arises when measuring M. semimembranosus (under M. semitendinosus) (Brandel 1973, Asang 1974, Elliot and Blanksby 1979, Gregor et al. 1981, 1985, Jorge and Hull 1983, 1986, Hull and Jorge 1985, Simonsen et al. 1985, de Proft et al. 1988), although the superficial muscle belly parts are greater in size than is the case with M. extensor carpi radialis brevis, the combination of displacement of the superficial M. semitendinosus with a lack of functional surface again gives different information from that which is expected (e.g. the cross-talk phenomenon).

Those authors who use wire electrodes do not necessarily have this problem, although measuring M. subscapularis in this way (Nuber et al. 1986, and others) — especially during front crawl in swimming and during golf movements — is questionable. This point of view of the anatomist who is confronted with these situations in the dissection room and palpation classes should not be discounted. That same anatomist will never measure M. sacrospinalis (Broer and Houtz 1967, Tokuyama et al. 1976, Yoshizawa et al. 1978, Oka et al. 1989), but instead chooses M. erector spinae. One group, however, reported measuring the EMG of M. tibialis posterior during skiing with unipolar active surface electrodes (Louie et al. 1984). It is assumed that this was a printing error.

In localizing the site of detection of the electrode on the skin, a variety of approaches has been applied: (1) over the motor point; (2) equidistant from the motor point; (3) near the motor point; (4) on the mid-point of the muscle belly; (5) on the visual part of the muscle belly; (6) at standard distances of osteological reference points (anthropometric landmarks); and (7) with no precision at all with respect to its placement.

The effects of electrode location on muscle fibre conduction velocity and median frequency estimates have been discussed in the SENIAM publications. The most reliable and most stable EMG values are to be obtained from the muscle belly area between the motor point and the most distal tendon. It follows that the position of the detection electrode must be chosen very carefully to minimize errors. Motor points are often located at the borders of muscles if projected to the skin, because other muscles or tendons cover part of those muscles.

As the motor point moves according to the level of contraction and the complexity of the movement, localizing the detection electrode over, near or equidistant from the motor point must be avoided. The motor point can in certain muscles disappear under another muscle. In other words, the region has to be large enough to accommodate the electrode. For complex skills in sport and occupational contexts, the muscle belly shortens in the proximal direction during concentric contractions and the electrode on the skin finds itself over the distal tendon. It is therefore proposed to place the electrodes over the visual midpoint of the 'contracted' muscle (Clarys et al. 1983, Clarys 1985).

In addition to localizing the electrode in its proper place on the skin over a muscle, it is also important to pay attention to the orientation of the electrode with respect to the muscle fibres. Bipolar surface electrodes have two detection surfaces. For optimal results, the two detection surfaces should be oriented so that the line between them is parallel to the muscle fibres. To accomplish this arrangement, it is assumed that the muscle fibres act along a line and that the muscles have a single arrangement of unipennate fibres. In some muscles, neither of these conditions is satisfied; in such cases it is advisable to place the electrode so that the line between the detection surfaces points to the origin and the insertion of the muscle. This orientation provides for consistent landmarks, so that the future placement of the electrode will have near-similar orientations and reduce the variation in EMG signal among the myoelectric measurements obtained from different contractions (de Luca and Knaflitz 1990).

Following the ISEK, SENIAM and JEK criteria, it is recommended to report the upper cut-off frequency, the lower cut-off frequency and the type of filter used in the amplifiers. If a DC-coupled amplifier is used, the input impedance and input current should also be reported. The type and material of the electrodes, the space between the contacts, the site and the preparation of the skin should also be documented (Solomonow et al. 1996). As to the processing of data, it is important to mention the use of raw EMG, iEMG, linear envelope, mean rectified EMG (MREMG), but also average EMG or ensemble average, together with the synchronization system and the normalization technique used, such as normalized to MVC or to 50% of the average of three MVCs, or to the highest peak (per movement or per subject) or to the mean of the subject ensemble average.

The linear envelope is the qualitative expression of the rectified and eventually averaged signal. Within a window choice this linear envelope can be smoothed, independent of its purpose. It should equally be clear once one starts smoothing, one cannot integrate anymore and it is unwise to use 'intensity' or 'activity level' in this case. Integration refers to the surface under the non-smoothed but rectified signal, to express the 'muscular intensity' phenomenon.

Because of the known variability of the EMG signal, not only between subjects, but also between different trials, these different normalization techniques have been developed to reduce variability. Generally, the EMG of maximum effort or the highest EMG value has been selected as the normalizing factor. In the main, the subject is asked to perform a maximal voluntary contraction (MVC) of the muscle (groups) being studied. This amplitude, either raw or rectified, is then used as a reference value (e.g. 100%). The use of the MVC reference is perfect in all static applications. For all dynamic activities the use of an isometric reference is debatable (Clarys et al. 1983, Yang and Winter 1984). Recently, the discussion about this normalization technique has been resurrected, because several investigators have found dynamic activities that exceeded the maximal isometric effort. Therefore, other normalization techniques have been developed in kinesiological EMG, e.g. normalization to the highest peak activity in dynamic conditions, to mean integrated EMG (ensemble average), to EMG per unit of measured force (net moment), and so on.

In an extensive review of sport specific and ergonomic studies using EMG, the missing information mostly concerned the normalization. In the majority of both sport and occupational studies in which a normalization technique is mentioned, the MVC technique has been used. This approach, however, is unreliable for several reasons:

> different maxima may be observed within the same subject repeating at different moments the same 'maximal' but isometric effort;
> different maxima will be observed at different angles of movement, both in the eccentric and concentric movement mode; and
> in addition, the question of linearity may arise when the values measured during isotonic dynamic–ballistic–complex sports movements exceed the 100% MVC. For example, Clarys *et al.* (1983) found dynamic percentages in swimming up to 160%, while Jobe *et al.* (1984) found up to 226% of MVC in baseball pitching.

Reason suggests that a statically obtained EMG cannot be a reference for dynamic EMG.

4. Closing remarks

To understand the nature and the development of EMG it is important, if not imperative, to have knowledge of its spectacular development in the previous millennium. Nevertheless it has taken about 300 years for EMG to emerge as an independent discriminating research methodology. Parallel to the technical hardware and signal-processing progression (not discussed in this review) is a requirement to be aware of problems with respect to the EMG acquisition and analysis methods. Users of EMG should possess a basic understanding of both fundamental electrophysiology and anatomical kinesiology before they start working with the tool. Valid conclusions have to be based on the assumption that the researcher knows exactly from which muscles the signals are being recorded.

To decide whether this condition has been met, different test procedures have been developed for collecting information, which will indicate temporal phases of the movement. These procedures assume knowledge of muscle function, and in the case of complex sports and occupational movements, a thorough knowledge of the sport and skill is concerned. ISEK and SENIAM have strongly suggested that some explicit standards should be met in reporting EMG research. A detailed description of the results of the test procedures and the circumstances under which the test procedures and the recording took place should be a requirement. This prerequisite has not always been met in the past. Another common mistake occurs when two independent EMGs from different muscles are available. It is not possible to state, as some investigators have implied, that because the amplitude, integral or some other measure of muscle A is greater than the corresponding measure from muscle B, muscle A is producing more force than muscle B. This comparison is complicated by many factors ranging from the size of the muscle fibres to the nature of the interface between the skin and electrodes (Cavanagh 1974, de Luca and Knaflitz 1990, Hermens *et al.* 1996, 1997, 1998).

Twenty-five years ago (Cavanagh 1974) the first warnings were given about use and misuse of EMG in physical education and the misuse still exists. These warnings stay valid today and apply to all movement studies, in both sport and occupational contexts.

References

ANDERSON, P. A., HOBART, D. J. and DANOFF, J. K. 1991, *Proceedings of the IIXth ISEK Congress* (Amsterdam: Excerpta Medica).
ARSENAULT, A. B., MCKINLEY, P. and MCFADYEN, B. 1998, *Proceedings of the XIIth ISEK Congress* (Quebec: Université de Montréal Press).
ASANG, E. 1974, Biomechanics of the human leg in alpine skiing, in R. C. Nelson and C. A. Morehouse (eds), *Biomechanics IV. International Series on Sport Sciences, vol. 1* (Baltimore: University Park Press), 236–242.
BASMAJIAN, J. V. and DE LUCA, C. J. 1985, *Muscles Alive, 5th edn* (Baltimore: Williams & Wilkins).
BOERHAAVE, H., GAUBIUS, H. D., SEVERINUS, I., VANDER, A. B. and VANDER, A. P. 1737, *Joannis Swammerdammii Biblia Naturae? Sive historia insectorum* (Leyden).
BRANDEL, B. R. 1973, An analysis of muscle coordination in walking and running gaits, in J. Vredenbregt and J. Wartenweiler (eds), *Medicine and Sport* (Basel: Karger), 278–287.
BROER, R. and HOUTZ, J. 1967, *Patterns of Muscle Activity in Selected Sports Skills: An Electromyographic Study* (Springfield: Charles C. Thomas).
CAVANAGH, P. 1974, Electromyography: its use and misuse in physical education, Journal of Health, Physical Education and Recreation, **23**, 61–64.
CLARYS, J. P. 1985, Hydrodynamics and electromyography: ergonomics aspects in aquatics. *Applied Ergonomics*, **16**, 11–24.
CLARYS, J. P. 1994, Electrology and localized electrization revisited, *Journal of Electromyography and Kinesiology*, **4**, 5–14.
CLARYS, J. P. and LEWILLIE, L. 1992, Clinical and kinesiological electromyography by 'Le Dr Duchenne (de Boulogne)', in A. Cappozzo, M. Marchetti and V. Tosi (eds), *Symposium on Biolocomotion: A Century of Research Using Moving Pictures-Formia. International Society of Biomechanics Series 1* (Rome: Promograph), 89–114.
CLARYS, J. P., MASSEZ, C., VAN DEN BROECK, M., PIETTE, G. and ROBEAUX, R. 1983, Total telemetric surface of the front crawl, in H. Matsui and K. Kobayashi (eds), *Biomechanics VIII-B. International Series on Biomechanics, vol. 4B* (Champaign: Human Kinetics), 951–958.
DE LUCA, C. J. and KNAFLITZ, M. 1990, *Surface Electromyography: What's New?* Monograph of the Neuromuscular Research Center (Boston: Boston University).
DE PROFT, E., CLARYS, J. P., BOLLENS, E., CABRI, J. and DUFOUR, W. 1988, Muscle activity in the soccer kick, in T. Reilly, A. Lees, K. Davids and W. J. Murphy (eds), *Science and Football* (London: E & FN Spon), 434–440.
DESPIRES, M. 1974, An electromyographic study of competitive road conditions simulated on a treadmill, in R. C. Nelson and C. A. Morehouse (eds), *Biomechanics IV* (Baltimore: University Park Press), 349–355.
DU BOIS REYMOND, E. 1849, *Untersuchungen uber thierische Electricitaet* (Berlin: G. Reimer).
DUCHENNE, G. B. (de Boulogne) 1855, 1862, 1867, *Physiologie des mouvements* (Paris: Librairie J. B. Baillières).
DUCHENNE, G. B. (de Boulogne) 1872, *De l'electrolisation localisée* (Paris: Librairie J. B. Baillières).
DUCHENNE, G. B. (de Boulogne) 1885, *Physiologie der Bewegungen* (Cassel: Theodor Fischer).
DUVERNEY, M. 1761, *Oeuvres anatomique*, tome I and II, C. A. Jombert (Paris: Librairie due Roi, pour le Genie et l'Artillerie, rue Dauphine).
ELLIOT, B. C. and BLANKSBY, B. A. 1979, The synchronisation of muscle activity and body segment movements during a running cycle, *Medicine and Science in Sports*, **11**, 322–327.
GALVANI, L. 1792, *De Viribus Electricitatis in Motu Musculari* (Bologna: Università di Bologna).
GREGOR, R. J., CAVANAGH, P. R. and LAFORTUNE, M. A. 1985, Knee flexor moments during propulsion in cycling — a creative solution to Lombard's paradox, *Journal of Biomechanics*, **18**, 307–316.
GREGOR, R. J., GREEN, D. and GARHAMMER, J. J. 1981, An electromyographic analysis of selected muscle activity in elite competitive cyclists, in A. Moercki, K. Fidelus, K. Kedzior and A. Wit (eds), *Biomechanics VII-B* (Baltimore: University Park Press), 537–541.

HERMENS, H. J. and FRERIKS, B. 1997, *The State of the Art on Sensors and Sensor Placement Procedures for Surface EMG*. The SENIAM EC Project (Enschede: Roessingh Research and Development).
HERMENS, H. J., HAGG, G. and FRERIKS, B. 1997, *European Applications on Surface EMG*. The SENIAM EC Project (Enschede: Roessingh Research and Development).
HERMENS, H. J., MERLETTI, R. and FRERIKS, B. 1996a, *European Activities on Surface EMG*. The SENIAM EC Project (Enschede: Roessingh Research and Development).
HERMENS, J. H., NENE, A. V. and ZILVOLD, G. 1996b, *Proceedings of the XIth ISEK Congress* (Enschede: Roessingh Research & Development).
HERMENS, H. J., RAU, G., DISSELHORST-KLUG, C. and FRERIKS, B. 1998a, *Surface EMG Application Areas and Parameters*. The SENIAM EC Project (Enschede: Roessingh Research & Development).
HERMENS, H. J., STEGEMAN, D., BLOK, J. and FRERIKS, B. 1998b, *The State of the Art on Modelling Methods for Surface EMG.*. The SENIAM EC Project (Enschede: Roessingh Research & Development).
HULL, M. J. and JORGE, M. 1985, A method for biomechanical analysis of bicycle pedalling, *Journal of Biomechanics*, **18**, 631–644.
JADELOT, J. F. N. 1799, *Experiences sur le Galvanisme et en général sur l'irritation des fibres musculaires et nerveuses* (Paris: Imprimerie Didot Jeune, Fuchs J. F.).
JOBE, W., TIBONE, E., PERRY, J. and MOYNES, D. 1984, An EMG analysis of the shoulder in throwing and pitching, *American Journal of Sports Medicine*, **11**, 3–5.
JORGE, M. and HULL, M. L. 1983, Preliminary results of EMG measurements during bicycle pedalling, in *Biomechanics Symposium*. ASME, vol. 56 (New York: American Society of Mechanical Engineers), 27–30.
JORGE, M. and HULL, M. L. 1986, Analysis of EMG measurements during bicycle pedalling, *Journal of Biomechanics*, **19**, 683–694.
LOUIE, J. K., KUO, C. Y., GUTIERREZ, M. D. and MOTE, C. D. 1984, Surface EMG and torsion measurements during snow laboratory and field tests, *Journal of Biomechanics*, **17**, 713–724.
MAREY, E. J. 1890, *Le vol des oizeaux* (Paris: G. Masson).
MATTEUCCI, C. 1844, *Traités des phénomènes électrophysiologiques* (Paris: Librairie des Sciences, rue de Médecine).
NUBER, W., JOBE, W., PERRY, J., MOYNES, R. and ANTONELLI, D. 1986, Fine wire electromyography analysis of muscles of the shoulder during swimming, *American Journal of Sports Medicine*, **14**, 63–65.
OKA, H., OKAMOTO, T. and KUMAMOTO, M. 1976, Electromyographic and cinematographic study of the volleyball spike, in P. V. Komi (ed.), *Biomechanics V-B*. International Series on Biomechanics, vol. 1B (Baltimore: University Park Press), 326–331.
OKA, H., OKAMOTO, T. and KUMAMOTO, M. 1989, Biarticular muscle activities during front handspring in tumbling, in *Proceedings of the XII International Congress of Biomechanics*, Los Angeles, 65.
PEDOTTI, A. 1993, *Proceedings of the IXth ISEK Congress* (Amsterdam: IOS Press).
PIERSON, E. and SPERLING, A. 1893, *Lehrbuch der Electrotherapie* (Leipzig: Ambr Abel).
SHIAVI, R. and WOLF, S. 1994, *Proceedings of the Xth ISEK Congress* (Nashville: B. E. Vanderbilt University).
SIMONSEN, B., THOMSEN, L. and KLAUSEN, K. 1985, Activity of mono- and biarticular leg muscles during sprint running, *European Journal of Applied Physiology*, **54**, 524–532.
SOLOMONOW, M., MATON, B. and MORITANI, Y. (eds) 1996, Standards & guidelines of reporting EMG research, *Journal of Electromyography and Kinesiology*, **6**, II–IV.
STRASS, D. 1990, Electromyographic evaluation of selected arm shoulder muscles at sprint swimmers in different technical positions, in D. McLaren, T. Reilly and A. Lees (eds), *Biomechanics and Medicine in Swimming* (London: E & FN Spon), 63–81.
TOKUYAMA, H., OKAMOTO, T. and KUMAMOTO, M. 1976, Electromyographic study of swimming in infants and children, in P. V. Komi (ed.), *Biomechanics V-B*. International Series on Biomechanics, vol. IB (Baltimore: University Park Press), 215–221.
TOYOSHIMA, S., MATSUI, H. and MIYASHITA, M. 1971, An electromyographic study of the upper arm muscles involved in throwing, *Research Journal of Physical Education*, **15**, 103–110.
TROUVÉ, G. 1893, *Manuel d'Electrologie Médicale*, ed. Octave Doin (Paris).

VASSALI, A. 1789, Dell'uso e dell'arrivita dell'areo conduttore nelle contrazioni dei musculi. Brugnatelli, *Giornale Phisico-Medico*, 94–99.
VON HUMBOLDT, F. A. 1797, *Versuche ÿber die gereizte Muskel- und Nerven Faser* (Berlin: H. A. Rottmann).
WALLINGA, W., BOOM, H. B. K. and DE VRIES, J. 1988, *Proceedings of the VIIth ISEK Congress* (Amsterdam: Excerpta Medica).
WEDENSKI, N. 1884, Wie rasch ermydet nerveder, *Zeischrift fur die Medicinische Wissenschaft*, 65–68.
WINTER, D. A., RAU, G., KADEFORS, R., BROMAN, H. and DE LUCA, C. J. 1980, *Units, Terms and Standards in Reporting of EMG Research* (ISEK).
YANG, J. and WINTER, D. 1984, Electromyographic amplitude normalization methods: improving their sensitivity as diagnostic tools on gait analysis, *Archives of Physical Medicine and Rehabilitation*, **65**, 517–521.
YOSHIZAWA, M., ITANI, T. and JONSSON, B. 1987, Muscular load in shoulder and forearm muscles in tennis players with different levels of skill, in B. Jonsson (ed.), *Biomechanics X-B*. International Series on Biomechanics, vol. 6B (Baltimore: University Park Press), 621–627.
YOSHIZAWA, M., KUMAMOTO, M., NEMOTO, Y., ITANI, T. and JONSSON, B. 1989, Effects of dynamic features of tennis rackets on muscular load of tennis players, in *Proceedings of the XII International Congress of Biomechanics* (Los Angeles), 64.
YOSHIZAWA, M., OKAMOTO, T., KUMAMOTO, M., TOKUYAMA, H. and OKA, H. 1978, Electromyographic study of two styles in the breaststroke as performed by top swimmers, in A. Asmussen and K. Jorgensen (eds), *Biomechanics VI-B*. International Series on Biomechanics, vol. 2B (Baltimore: University Park Press), 126–131.

23 The relationship between heart rate and oxygen uptake during non-steady state exercise

S. D. M. BOT* and A. P. HOLLANDER

Faculty of Human Movement Sciences, Vrije Universiteit, Van der Boechorststraat 9, 1081 BT Amsterdam, The Netherlands

Keywords: Oxygen consumption; Indirect measurement; Intermittent exercise.

In this study the validity of using heart rate (HR) responses to estimate oxygen uptake ($\dot{V}O_2$) during varying non-steady state activities was investigated. Dynamic and static exercise engaging large and small muscle masses were studied in four different experiments. In the first experiment, 16 subjects performed an interval test on a cycle ergometer, and 12 subjects performed a field test consisting of various dynamic leg exercises. Simultaneous HR and $\dot{V}O_2$ measurements were made. Linear regression analyses revealed high correlations between HR and $\dot{V}O_2$ during both the interval test ($r = 0.90 \pm 0.07$) and the field test ($r = 0.94 \pm 0.04$). In the second experiment, 14 non-wheelchair-bound subjects performed both an interval wheelchair test on a motor driven treadmill, and a wheelchair field test consisting of dynamic and static arm exercise. Significant relationships were found for all subjects during both the interval test ($r = 0.91 \pm 0.06$) and the field test ($r = 0.86 \pm 0.09$). During non-steady state exercise using both arms and legs in a third experiment, contradictory results were found. For 11 of the 15 subjects who performed a field test consisting of various nursing tasks no significant relationship between HR and $\dot{V}O_2$ was found ($r = 0.42 \pm 0.16$). All tasks required almost the same physiological strain, which induced a small range in data points. In a fourth experiment, the influence of a small data range on the HR-$\dot{V}O_2$ relationship was investigated: five subjects performed a field test that involved both low and high physiological strain, non-steady state arm and leg exercise. Significant relationships were found for all subjects ($r = 0.86 \pm 0.04$). Although the *r*-values found in this study were less than under steady state conditions, it can be concluded that $\dot{V}O_2$ may be estimated from individual HR-$\dot{V}O_2$ regression lines during non-steady state exercise.

1. Introduction

Continuous recording of the heart rate is commonly used to estimate the exercise intensity or physical strain during sports, work or daily activities. This indirect method is based on a linear relationship between heart rate (HR) and oxygen uptake ($\dot{V}O_2$) during steady state conditions (Åstrand and Rodahl 1986) and gives the opportunity of an easy, non-invasive and relatively inexpensive determination of the $\dot{V}O_2$. In most sports, occupational and daily activities the intensity varies constantly and no steady state is reached. Furthermore, numerous studies have shown that the $\dot{V}O_2$-HR relationship was modified when using different muscle masses or different modes of exercise (Vokac *et al.* 1975,

---

*Author for correspondence. e-mail: sbot@dds.nl

Maas et al. 1989, Rayson et al. 1995). Nevertheless, HR recordings are made in order to get an impression of the energy expenditure (Engels et al. 1994, Hollander et al. 1994, Montoliu et al. 1995, van der Beek and Frings-Dresen 1995). Whether and to what extent HR and $\dot{V}O_2$ are related to each other during activities with different muscle masses and unsystematically changing intensities is not known.

In several studies a linear relationship between HR and $\dot{V}O_2$ during non-steady state activities was found, however all tests were limited to progressive incremental exercise (Gilbert and Auchincloss 1971, Fardy and Hellerstein 1978, Fairshter et al. 1987, Matthys et al. 1996, Bernard et al. 1997). Statements about the presence or absence of a relationship between HR and $\dot{V}O_2$ have not been tested statistically (Edwards et al. 1973, Ballor and Volovsek 1992), or the specific nature of activities such as weight-lifting (Collins et al, 1991) and karate (Shaw and Deutsch 1982) may impede generalizations.

The indirect assessment of $\dot{V}O_2$ by measuring HR has mainly been limited to steady state exercise. Although not yet proven, a linear relationship between HR and $\dot{V}O_2$ during intermittent and non-steady state exercise is plausible. Bunc et al. (1988) concluded that the regulation of the HR at the onset of exercise might be similar to the regulation of $\dot{V}O_2$. Several studies have indicated that the time constant or mean response time for $\dot{V}O_2$ is similar to that for HR in the transition from rest or from unloaded cycling to a certain workload (Casaburi et al. 1977, Hughson and Morrisey 1983, Sietsema et al. 1989), which suggests a close relationship. Furthermore, heart rate gave a close estimate of $\dot{V}O_2$ during intermittent exercise in the study of Lothian and Farrally (1995).

The relationship between HR and $\dot{V}O_2$ may be different for exercise *engaging* a large muscle mass compared to exercise with smaller muscle masses (Vokac et al. 1975, Eston and Brodie 1986). Furthermore, the kind of exercise (dynamic or static) may influence the relationship (Kilbom and Persson 1981, Collins et al. 1991, Bhambhami et al. 1997.

In this study the relationship between HR and $\dot{V}O_2$ during dynamic and static non-steady state exercise with large muscle masses, small muscle masses, and with a combination of different muscle groups was examined in four different experiments. Interval tests were applied to examine the non-steady state character, whereas field tests were used as a general simulation of sports, work and daily activities. A simulation test was used to investigate the applicability of HR registration to estimate $\dot{V}O_2$ in an occupational setting.

The aim of this study was to investigate the validity of the use of HR-response in estimating the $\dot{V}O_2$ during non-steady state exercise. The applicability of HR measurement to predict $\dot{V}O_2$ will be extended if the relationship between HR and $\dot{V}O_2$ during non-steady state activities can be demonstrated.

2. Methods

2.1. *Experiment I: Dynamic exercise using a large muscle mass*

2.1.1. *Subjects and procedures*: Twenty-eight healthy subjects (12 females and 16 males) volunteered to participate in the first experiment after they had given written informed consent. Their physical characteristics are summarized in table 1. All subjects were familiar with the equipment and procedures of the tests. Each subject performed a maximal test on an electrically driven brake cycle ergometer (Lode Instruments, Groningen, The Netherlands) to determine maximum oxygen uptake

Table 1. Subject characteristics (mean ± SD) as obtained during the maximal test.

	Experiment I Interval test	Field test	Experiment II	Experiment III	Experiment IV
Number of subjects	16 (8m, 8f)	12 (8m, 4f)	14 (14m)	15 (5m, 10f)	5 (2m, 3f)
Age (years)	25 (±5)	33 (±10)	23 (±3)	32 (±8)	22 (±1)
Body mass (kg)	64 (±7)	76 (±12)	73 (±5)	68 (±9)	60 (±13)
Height (m)	1.76 (±0.08)	1.81 (±0.07)	1.86 (±0.05)	1.74 (±0.09)	1.70 (±0.08)
HR_{max} (beats min^{-1})	187 (±8)	186 (±9)	179 (±12)	186 (±12)	195 (±3)
$\dot{V}O_{2max}$ (1 min^{-1})	3.97 (±1.12)	3.98 (±0.92)	2.39 (±0.42)	2.58 (±1.04)	2.45 (±0.35)

($\dot{V}O_{2max}$) and maximum heart rate (HR_{max}). These maximal values were used to normalize the data. On a different day one of the two non-steady state tests, either an interval test on a cycle ergometer or a field test with various leg exercises, was performed. Both tests involved dynamic exercise with a large muscle mass. During all experiments, HR was monitored continuously using a short-range radio telemeter (Sport Tester PE4000, Polar Electro Inc., Kempele, Finland). During both the maximal test and the interval test, the $\dot{V}O_2$, minute ventilation ($\dot{V}E$) and respiratory exchange ratio (RER) were measured continuously using an on-line $\dot{V}O_2$ and CO_2 analyser (Oxycon Ox4, Mijnhardt, Bunnik, The Netherlands). Average values were calculated over 30-s periods. During the field test, 30-s samples of expired air were collected using portable Douglas bags (de Groot et al. 1983). Gas analysis took place using an electronic $\dot{V}O_2$ and CO_2 analyser (Oxylyser UG55 Capnolyser UG64, Mijnhardt, Bunnik, The Netherlands). The subject operated the two-way valve system that connected the Douglas bags with the mouthpiece. The two-way valve system contains a stopwatch to record actual air collection time. Both the gas analysing systems were calibrated against known gas mixtures before each test. For each 30-s air collection period, a simultaneous mean HR was calculated.

2.1.2. *Maximal test*: For this test an exercise protocol as described by Åstrand and Rodahl (1986: 364) was used. The test started with a warm-up period of 5–6 min at a low workload. At the end of this period, the $\dot{V}O_{2max}$ was predicted from HR and workload using the Åstrand and Åstrand nomogram (Åstrand and Rodahl 1986: 365) and the workload was increased. After 3 min, the $\dot{V}O_{2max}$ was predicted for the second time and the workload was changed in order to exhaust the subject in 3 min. The highest measured HR and $\dot{V}O_2$ during the maximal test were defined as HR_{max} and $\dot{V}O_{2max}$, respectively; the criterion for maximal exercise was the RER > 1.

2.1.3. *Interval cycling test*: Sixteen subjects performed an interval test on a cycle ergometer. The interval test consisted of 5-min warming up and a 30-min exercise protocol. The intensity varied between 0 Watt (0%) and a workload requiring a $\dot{V}O_2$ of 100% $\dot{V}O_{2max}$. The interval protocol consisted of exercise blocks with random duration (15, 30, 45, 60, 75, 90, 105 or 120 s) and random intensity (0, 10, 20, 30, 40, 50, 60, 70, 80, 90 or 100% $\dot{V}O_{2max}$). There was no rest between the exercise blocks. The most strenuous duration-intensity combinations (90-s: 90%; 105-s: 90%; 105-s: 100%; 120-s 90% and 120-s: 100%) were excluded to prevent early exhaustion of the subjects. An example of an exercise protocol is displayed in figure 1. Randomization of the protocol was performed for all subjects separately.

2.1.4. *Field test involving various leg exercises*: Twelve subjects performed a field test of three, 20-min periods. The exercise consisted of: (1) sitting; (2) standing; (3) walking; (4) jogging; (5) sideward movements I; (6) sideward movements II; (7) knee lifting; (8) 3 × 10 vertical jumps and (9) 6 × 10 m sprinting. For each subject the sequence of tasks was randomized. An example of the minute-to-minute exercise protocol is displayed in figure 2. In each period, 7 × 2-min exercise was alternated by 6 × 1 min standing. This spare minute was used to replace the Douglas bags. In the first and third 20-min period air was collected from 75 to 105 s of each 2 min of exercise; in the second 20-min period air was collected from 45 to 75 s of each 2 min of exercise.

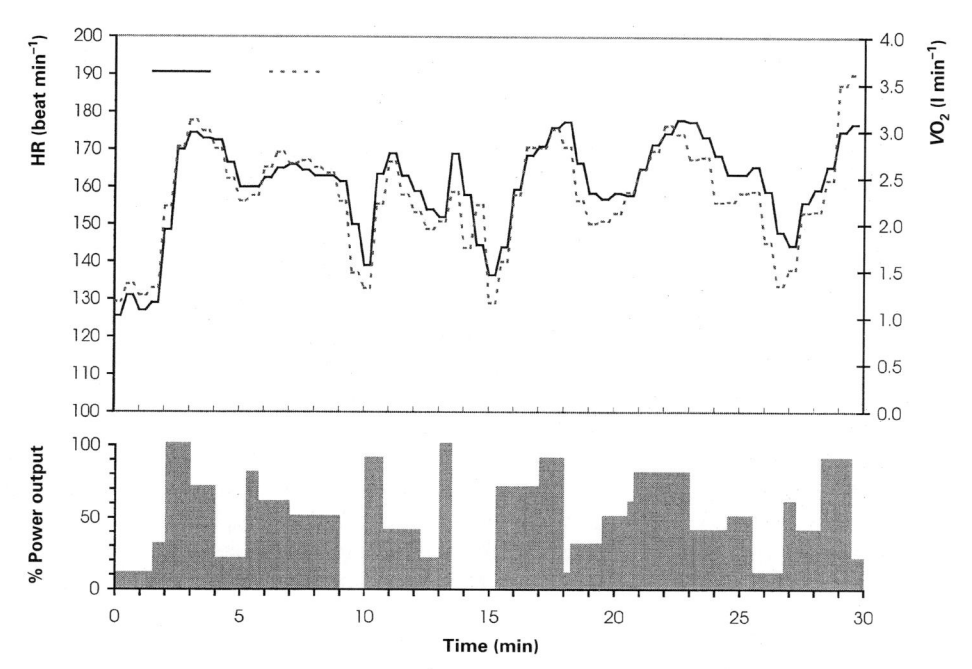

Figure 1. HR response (solid line) and $\dot{V}O_2$ (dashed line) for one subject during the interval test of Experiment I. The workload (shaded area) is given as the percentage of the workload at 100% $\dot{V}O_{2max}$.

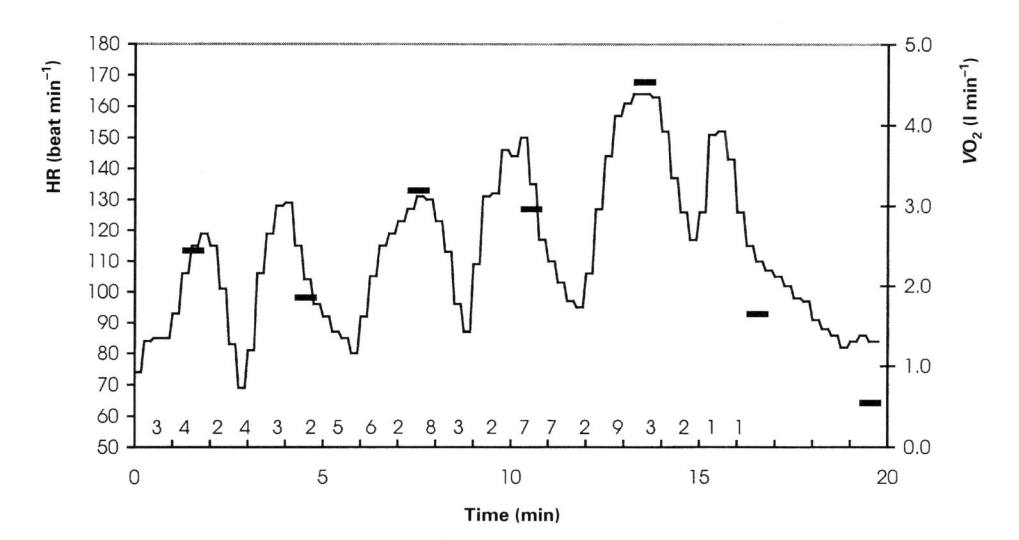

Figure 2. HR response (line) with the $\dot{V}O_2$ (bars) for one subject during the field exercise test of Experiment I. The numbers at the bottom of the chart indicate the task performed in the corresponding minute.

2.1.5. *Data analysis*: Linear regression analysis was used to determine the individual HR-$\dot{V}O_2$ relationships. Group regression lines were calculated for both tests using the non-dimensionalized data points of all subjects together. To obtain non-dimensionalized data the $\dot{V}O_2$ was expressed as a percentage of the $\dot{V}O_{2max}$ (%$\dot{V}O_{2max}$) and HR as a percentage of the HR_{max} (%HR_{max}).

The validity of the estimation of the $\dot{V}O_2$ from HR recordings during the non-steady state activities was evaluated by calculating the correlation coefficient (r) and the standard error of the estimate (SEE, determining the confidence intervals). The results were considered to be statistically significant when $p < 0.05$.

2.2. Experiment II: Dynamic and static exercise using a small muscle mass

2.2.1. *Subjects and procedures*: Fourteen healthy young men, not wheelchair bound, participated in this experiment. Their physical characteristics are summarized in table 1. Written informed consent was obtained from all the subjects. All subjects had experience with wheelchair propulsion and were familiar with the equipment and procedures. Each subject completed three different tests in a wheelchair (Ultra Light Premier, Everest and Jennings, Camarillo, California). On the first day all subjects performed a maximal wheelchair exercise test to determine the peak oxygen uptake ($\dot{V}O_{2peak}$) and peak heart rate (HR_{peak}) for maximal arm exercise. On another day, all subjects performed an interval test and a field test, both consisting of non-steady state exercise with small muscle masses. Maximal and interval tests were performed on a motor-driven-treadmill (Enraf Nonius, Delfi, The Netherlands; model 3446, belt width 1.25 m, length 3.0 m) with constant velocity (v) of 1.11 m s^{-1}. Behind the treadmill a pulley system was placed to regulate the power output (PO), described in the study of Rasche *et al.* (1993). An on-line electronic gas analyser (Oxycon Ox4, Mijnhardt, Bunnik, The Netherlands) continuously monitored $\dot{V}O_2$, $\dot{V}E$ and RER, which were averaged over 30 s. During the field test 30-s samples of expired air were collected in portable Douglas bags, which were fixed on to the back of the wheelchair. The experimenter operated the two-way valve system. The samples of expired air were analysed using an electronic $\dot{V}O_2$ and CO_2 analyser (Oxylyser UG55 Capnolyser UG64, Mijnhardt, Bunnik, The Netherlands). During all tests, continuous heart rate recordings were made and stored in memory every 15 s. For each air collection period a simultaneous mean HR was calculated.

2.2.2. *Maximal test*: For the maximal test, a continuous jump-max protocol as described by Rasche *et al.* (1993) was used. The test started with a warm-up period of 2–3 min at a low power output. The power output was increased every minute by raising the mass in the pulley system with a 0.5-kg load at a constant speed until the HR reached 150 beats min^{-1}; after that, 0.25 kg was added every minute to the pulley system. The test ended when the subject could no longer maintain the belt velocity. The highest measured HR and $\dot{V}O_2$ during the maximal test were defined as HR_{peak} and $\dot{V}O_{2peak}$, respectively; the criterion for maximal exercise was the RER > 1.

2.2.3. *Interval wheelchair test*: The protocol of the interval test was similar to that used in Experiment I (§2.1.3). The pulley weight varied between 10% and 90% PO_{max}. Randomization of the protocol was performed for all subjects separately.

2.2.4. *Field test involving various arm exercises*: The subjects performed eight different 1-min tasks in the wheelchair. For each subject the sequence of the tasks was randomized. The tasks were: (1) manoeuvring around two markers on the ground; (2)

push ups from the wheelchair; (3) sprinting, 10 m up and down; (4) wheelchair propulsion on a treadmill, 4 km h^{-1} and 4% incline; (5) sitting still, horizontally displacing light weights with the arms; (6) sitting still, vertically displacing light weights with the arms; (7) sitting still, horizontally displacing heavy weights with the arms; (8) sitting still, vertically displacing heavy weights with the arms.

The test was performed twice: the first time expired air was collected from 0 to 30 s of each task and the second time air was collected from 30 to 60 s of each task. Between each task, there was a rest period of 2 min. This time was used to change the Douglas bags and to prepare the subject for the next task.

2.2.5. *Data analysis*: The statistical analysis of experiment II was identical to that of experiment I.

2.3. *Experiment III: Dynamic and static exercises using both large and small muscle masses*

2.3.1. *Subjects and procedures*: Fifteen healthy subjects (10 females and 5 males), all experienced in nursing activities, volunteered to participate in a test that consisted of numerous nursing tasks in a hospital mock up. The physical characteristics of the subjects are shown in table 1. Written informed consent was obtained for each subject. Nine subjects completed a maximal test to determine $\dot{V}O_{2max}$ and HR_{max}. During the test 60-s samples of expired air were collected in portable Douglas bags. The experimenter operated the 2-way valve system. During all tests continuous HR recordings were made and data were stored in memory every 5 s. Heart rate and $\dot{V}O_2$ were measured simultaneously. All tasks were performed in a hospital simulation room.

2.3.2. *Maximal test*: Nine subjects completed maximal tests similar to those described in experiment I (§2.2.2.).

2.3.3. *Field test involving dynamic and static arm and leg exercises—nursing tasks in a hospital mock up*: The subjects were invited in couples, so that one could act as a patient, while the other subject performed the test. Afterwards they changed roles. The protocol was randomized for every subject. Each subject performed 14 nursing tasks: (1) washing upper-body and face of the patient; (2) brushing the teeth of the patient; (3) dressing the patient (t-shirt/sweater); (4) transferring the patient from a bed into a wheelchair; (5) transferring the patient from a wheelchair into a chair; (6) transferring the patient from a chair into a bed; (7) transporting the patient in a bed; (8) transporting the patient in a wheelchair; (9) turning the patient over from the back to his/her left side; (10) controlling the blood pressure; (11) supporting the patient while walking; (12) preparing lunch; (13) making the bed without the patient in it; and (14) reading and an administration task. Measurements were taken during the first minute of each task. Between the tasks a rest period of 1 min was taken to change the Douglas bags.

2.3.4. *Data analysis*: The statistical analysis of experiment III was identical to that of experiment I. No group regression lines were calculated.

2.4. *Experiment IV: Physiological low- and high-strain dynamic and static exercise using both large and small muscle masses*

2.4.1. *Subjects and procedures*: Five subjects performed a field test involving static and dynamic tasks with both arms and legs. Their physical characteristics are shown in table 1. Written informed consent was obtained for each subject. All were familiar with the equipment and procedures. During the test 60-s samples of expired air were collected in portable Douglas bags. The experimenter operated the 2-way valve system. During all tests continuous HR recordings were made and data were stored in memory every 5 s. Heart rate and $\dot{V}O_2$ were measured simultaneously.

2.4.2. *Field test involving both low- and high-strain arm and/or leg exercises*: The test consisted of two rounds of six low-strain activities: (1) washing the dishes; (2) keeping a tennis ball up with a beach-ball racket; (3) sweeping the floor; (4) cleaning a blackboard; (5) sitting; (6) carrying weights of 2×5 kg. These were followed by two rounds of high-strain activities: (7) jumping as high as possible; (8) inflating a tyre with a cycle pump; (9) stretching the arms above the head holding 2×3 kg weights; (10) mopping the floor; (11) lifting a crate (10 kg) from the ground onto a table and vice versa; (12) carrying weights of 2×10 kg. For each subject the sequence of activities was randomized. All tasks lasted 1 min and were alternated with a rest period of 1 min to change the Douglas bags.

2.4.3. *Data analysis*: The statistical analysis of experiment IV was identical to that of experiment I.

3. Results

3.1. Experiment I

For each subject on average, 59 simultaneous HR and $\dot{V}O_2$ data points were obtained during the interval cycling test and 20 data points were obtained during the field test. An example of heart rate and $\dot{V}O_2$ data of one subject obtained during the interval test is given in figure 1. An example of one period of the field test is presented in figure 2.

For each subject an individual linear regression line was determined and the correlation coefficient (r) and the SEE were calculated. The correlation coefficient of the individual linear regressions varied between 0.75 and 0.98 (mean = 0.90 ± 0.07) during the interval test; the mean SEE was 7.0 (± 1.9) % $\dot{V}O_{2max}$. During the field test the value of r varied between 0.88 and 0.98 (mean = 0.94 ± 0.04), with a mean SEE of 7.5 (± 2.7) % $\dot{V}O_{2max}$. The relationship between HR and $\dot{V}O_{2max}$ was significant for all subjects ($p < 0.05$). Using the non-dimensionalized data points for all subjects together resulted in the following group-wise linear regression lines ($y = \% \dot{V}O_{2max}$, and $x = \% HR_{max}$):

Interval test: $y = 1.42x - 51.55$ $r = 0.88$ SEE $= 8.7\% \dot{V}O_{2max}$

Field test: $y = 1.49x - 57.8$ $r = 0.89$ SEE $= 10.6\% \dot{V}O_{2max}$

3.2. Experiment II

On average 30 data points and 16 data points were generated during the interval test and the field test, respectively. Statistically significant relationships were found for all subjects during both the interval and the field test ($p < 0.05$). An example of an individual regression line is shown in figure 3. The individual value of r varied between 0.74 and 0.97 (mean = 0.91 ± 0.06) during the interval test. The mean SEE

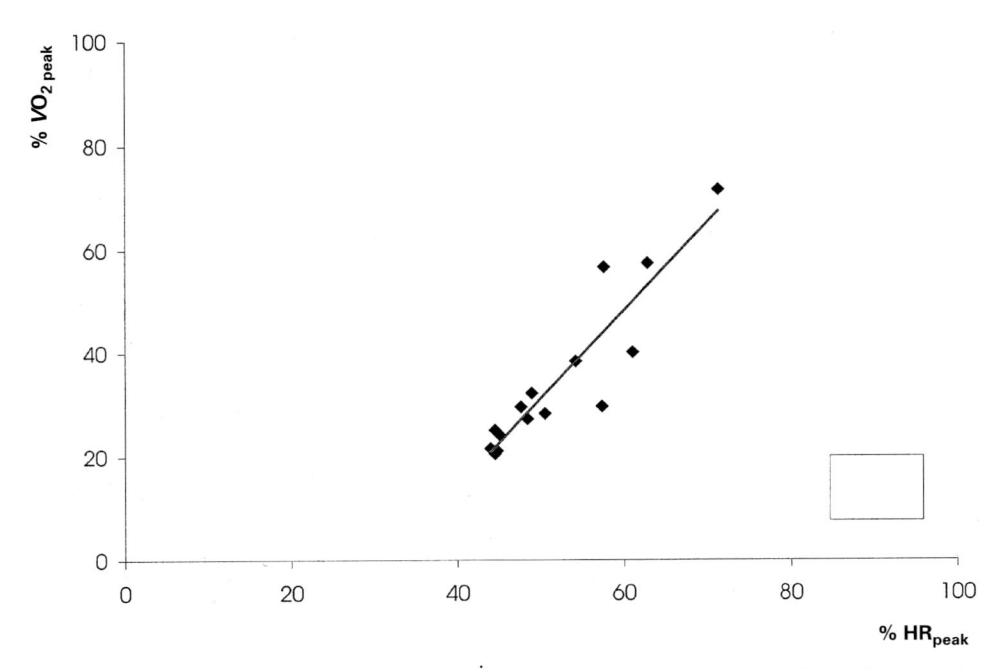

Figure 3. Scattergram of %HR$_{peak}$ and %$\dot{V}O_{2peak}$, obtained from the field exercise test of Experiment II for one subject.

was 6.7 (± 1.6) %$\dot{V}O_{2peak}$. During the field test the individual value of r varied between 0.63 and 0.94, (mean = 0.86 ± 0.09); the mean SEE was 9.9 (± 2.6) %$\dot{V}O_{2peak}$. Using the non-dimensionalized data points for all subjects together resulted in the following group-wise linear regression lines (y = %$\dot{V}O_{2peak}$, and x = %HR$_{peak}$):

Interval test: $y = 1.16x - 27.87$ $r = 0.79$ SEE = 10.8% $\dot{V}O_{2peak}$

Field test: $y = 1.31x - 43.02$ $r = 0.74$ SEE = 12.9% $\dot{V}O_{2peak}$

3.3. Experiment III

On average 14 simultaneous HR and $\dot{V}O_2$ data points were obtained for each subject. An example of heart rate and $\dot{V}O_2$ data of one subject obtained during the field test is given in figure 4. Taking all subjects and all tasks together the mean HR was 89 (± 6) beats min^{-1}. The individual r varied between 0.11 and 0.71 with a mean of 0.42 (± 0.16). For 11 of the 15 subjects no significant relationship between HR and $\dot{V}O_2$ was found. The mean SEE was 6.3 (± 2.9) %$\dot{V}O_{2max}$.

3.4. Experiment IV

During the test 24 simultaneous HR and $\dot{V}O_2$ data points were established for each subject. In figure 5, an example of data points and the regression lines of one subject is shown. Statistically significant relationships were found for all subjects for both the low-strain and high-strain activities and for all activities together ($p < 0.05$). For

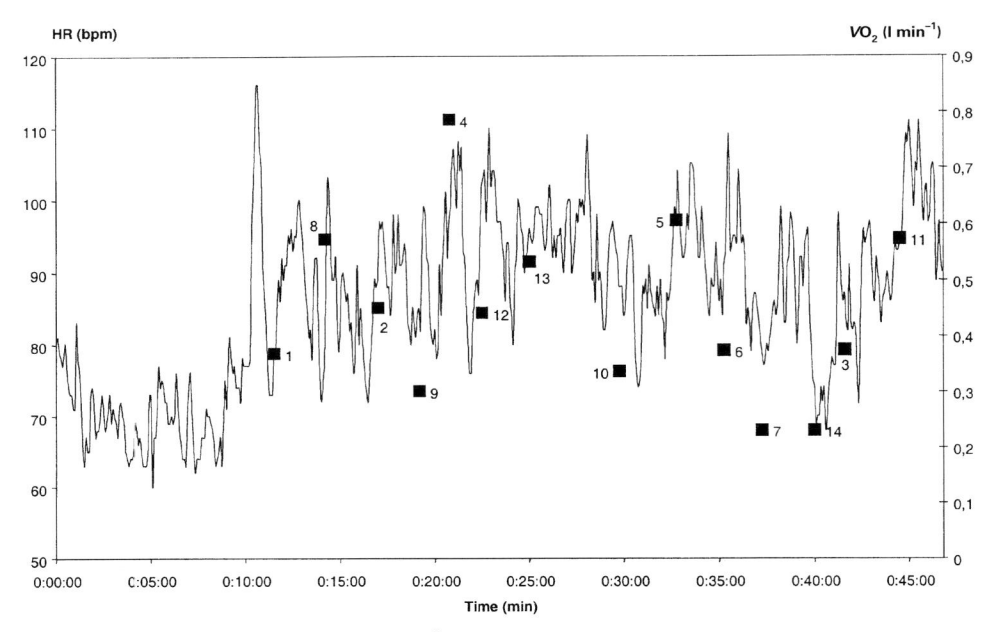

Figure 4. HR response (line) with the $\dot{V}O_2$ (squares) for one subject during the field exercise test of Experiment III. The numbers indicate the task performed in the corresponding minute.

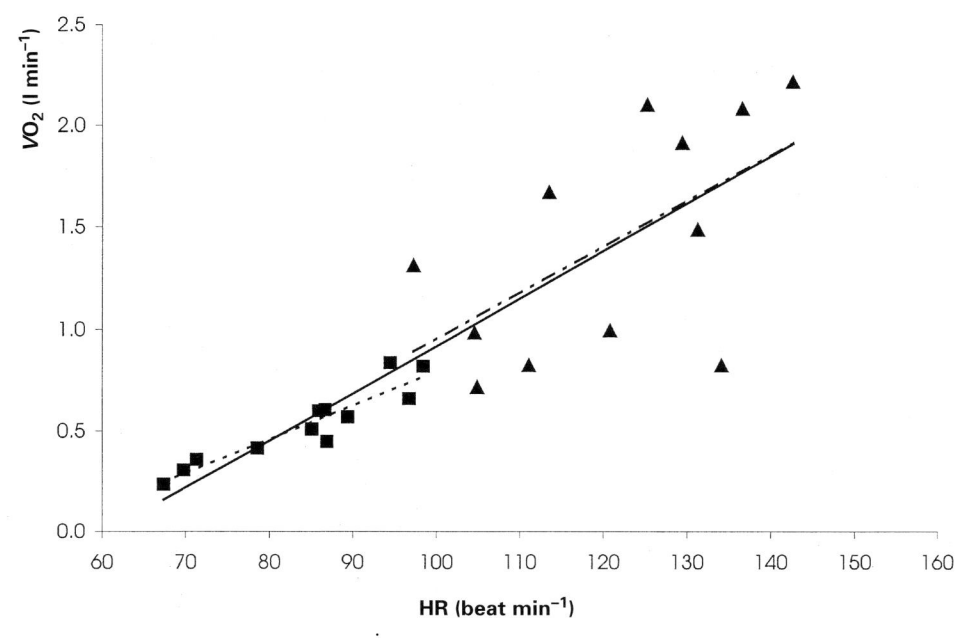

Figure 5. Scattergram of HR and $\dot{V}O_2$, during low strain activities (squares) and high strain activities (triangles), obtained from the field exercise test of Experiment IV for one subject. Regression lines for low strain (dashed line) and high strain activities (dashed and dotted line) as well as for all activities (solid line) are displayed.

the low-strain activities the values of r varied between 0.75 and 0.93 with a mean of 0.86 ($\pm$0.04) and an mean SEE of 3.6 ($\pm$0.3) % $\dot{V}O_{2max}$. The values of r for the high-strain activities varied between 0.55 and 0.76 with a mean of 0.64 ($\pm$0.08); the mean SEE was 9.9 ($\pm$1.0) % $\dot{V}O_{2max}$. Putting the data points of all activities together, the r varied between 0.81 and 0.89 (mean = 0.85$\pm$0.03) and a mean SEE of 7.4 ($\pm$0.4) % $\dot{V}O_{2max}$. There was no difference ($p<0.05$) between the r values of low-strain activities and those of all activities together.

4. Discussion

The reliability of the indirect assessment of $\dot{V}O_2$ by measuring HR has mainly been limited to steady state exercise. It is plausible that the relationship also exists during varying non-steady state activities. Knowledge of the regression of $\dot{V}O_2$ and HR permits the applicability of the indirect assessment of $\dot{V}O_2$ by measuring the HR. In this study the relationship between HR and $\dot{V}O_2$ during dynamic and static non-steady state exercise with large muscle masses (Experiment I), with small muscle masses (Experiment II), and with a combination of different muscle groups (Experiments III and IV), was investigated.

In order to make it possible to use a group regression equation for all the subjects, the $\dot{V}O_2$ was expressed as a percentage of $\dot{V}O_{2max}$ and the HR as percentage of HR_{max}. In Experiments I and II, all subjects performed a maximal test; in Experiment III only some of the subjects agreed to perform this test. Some subjects obtained a higher $\dot{V}O_2$ or HR during the interval or field test compared to the maximal test. For these subjects the highest observed $\dot{V}O_2$ and HR were used as $\dot{V}O_{2max}$ and HR_{max}. Repeated tests would possibly have led to a better assessment of the maximal values.

The individual regression equations resulted in significant linear relationships between HR and $\dot{V}O_2$ during dynamic, non-steady state leg exercise (Experiment I) and during non-steady state arm exercise (Experiment II). Also during non-steady state activities with different muscle groups (Experiment IV) significant correlation coefficients (r) were established for all subjects. However, during various nursing activities (Experiment III) a significant linear relationship between HR and $\dot{V}O_2$ was found for only four of the 15 subjects. The mean individual correlations found in the first experiment are higher than those found in the other experiments (table 2). This may be explained by the kind of exercise that was performed. The field tests of the other experiments (II–IV) involved static components. Static exercise increases the heart rate above the value expected from the metabolic rate (Åstrand and Rodahl 1986, Collins et al. 1991). Using Fischer's Z transformation, 95% confidence intervals were determined: only the mean r-value found in Experiment III was significantly different. The absence of a linear relation between HR and $\dot{V}O_2$ during this test may be explained by the low physical strain needed to perform all tasks, which led to a small range in data points. Divergences have more influence on the r-value when the range of data gets smaller. However, in Experiment IV high r-values were found within small ranges of data points. Furthermore during light and moderate exercise, emotional factors may affect the heart rate (Åstrand and Rodahl 1986). In contrast to the subjects in the other tests, the subjects who participated in Experiment III had no experience with the equipment and procedures of the test and had never been a subject in any study before. This could have induced fear or excitement during the test. Szabo et al. (1994) found that even a mild mental

Table 2. Mean correlation (r) between HR and $\dot{V}O_2$ and standard error of estimate (SEE) of the three experiments.

	Exercise	r	SEE
Experiment I			
Interval test	Leg	0.90	7.0
Field test	Leg	0.94	7.5
Experiment II			
Interval test	Arm	0.91	6.7
Field test	Arm	0.86	9.9
Experiment III			
Field test	Leg + arm	0.42*	6.3
Experiment IV			
Field test	Leg + arm	0.86	7.4

*Not significant ($p < 0.05$).

challenge was capable of inducing an increase in HR during low and moderate intensity cycling. Another explanation for the low correlation found in experiment III might be the lack of repeatability in the tasks. In Experiments I, II and IV the subjects repeated the task within 1 min, while in Experiment III different actions within one task were required.

The standard error of estimate (SEE) can express the validity of estimations based on a regression line. The SEEs found during the experiments are comparable: the individual SEEs in the present study varied between 3.9 and 11.7 % $\dot{V}O_{2max}$ during Experiment I; between 5.1 and 14.7 % $\dot{V}O_{2max}$ during Experiment II; and between 3.1 and 11.0 % $\dot{V}O_{2max}$ during Experiment III. The mean individual SEEs of the interval tests of Experiment I and II are slightly lower than the SEEs of the field tests in both experiments. This may be explained by the higher number of data points collected in the interval tests, that enables a better determination of the regression line.

Most studies determined r-values using the non-dimensionalized individual data points of all subjects together (Londeree and Ames 1976, Katch *et al.* 1978, Franklin *et al.* 1980, Vander *et al.* 1984, Hooker *et al.* 1993, Londeree *et al.* 1995). The reported r-values for steady state leg and arm exercise vary from 0.95 to 0.99. In the present study, lower values were found for non-steady state exercise: the group-wise r-values were 0.88 and 0.89 for leg exercise and 0.74 and 0.79 for arm exercise, for the interval test and the field test, respectively. The lower values found in the present study can be explained by a greater variability in HR during non-steady state exercise. During non-steady state exercise HR is not stabilized and hence shows a greater variability, as can be seen from the calculated SEEs: the group-wise SEE was remarkably higher during non-steady state exercise (table 3).

The high variability of data points in the group regression analyses compared to the individual regression lines indicate that estimated $\dot{V}O_2$ from group regression lines is inferior to estimated $\dot{V}O_2$ from individual regression lines. For a group-wise analysis, a careful evaluation of the required accuracy of the estimation of $\dot{V}O_2$ is needed.

5. Conclusions

The results of this study indicate a linear relation between HR and $\dot{V}O_2$ during both non-steady state leg exercise and non-steady state arm exercise. Although the r-

Table 3. Comparison of the group correlation coefficients (r) between HR and $\dot{V}O_2$, and standard error of estimate (SEE), during steady state exercise testing.

Authors	Mode*	Slope	Intercept	r	SEE
Franklin et al. (1980)	TM	1.33	−24.5	0.94	6.9
Katch et al. (1978)	CE	1.39	−44.8	0.97	7.9
Londeree and Ames (1976)	TM	1.37	−41.0	0.97	5.7
Londeree et al. (1995)	TM	1.30	−34.5	0.96	–
Londeree et al. (1995)	CE	1.41	−45.1	0.93	–
Londeree et al. (1995)	R	1.18	−21.0	0.94	–
Hooker et al. (1993)	ACE	1.41	−46.2	0.95	–
Vander et al. (1984)	ACE	1.43	−48.8	–	10.3
Present study					
Interval test, expt. I	TM	1.42	−51.6	0.88	8.7
Exercise test, expt. I	–	1.49	−57.8	0.89	10.6
Interval test, expt. II	WC	1.16	−27.9	0.79	10.8
Exercise test, expt. II	WC	1.31	−43.0	0.74	12.9

TM = treadmill; CE = cycle ergometer; R = Rower; ACE = arm crank ergometer; WC = wheelchair.

values were less strong than under steady state conditions, it can be concluded that the estimation of $\dot{V}O_2$ by measuring the HR is not limited to steady state exercise. The $\dot{V}O_2$ could be estimated from individual HR-$\dot{V}O_2$ regression lines during varying non-steady state activities. When the estimation of $\dot{V}O_2$ is based on group regression equations, a careful evaluation of the required reliability is needed.

Contradictory results were found during non-steady state activities with different muscle groups: only 4 out of 15 equations resulted in a significant relation ($r = 0.11 - 0.71$) during the hospital mock up ($p < 0.05$), while significant relations between HR and $\dot{V}O_2$ ($r = 0.81 - 0.89$) were found during similar tasks using both arms and legs in the last experiment. The absence of a linear relationship between HR and $\dot{V}O_2$ during the nursing tasks may possibly be explained by emotional factors that could have affected the HR. In further research concerning the relationship between HR and $\dot{V}O_2$ during non-steady state exercise with different muscle groups and/or static and dynamic components, emotional factors should be reduced as much as possible.

References

ÅSTRAND, P. and RODAHL, K. 1986, *Textbook of Work Physiology*, 3rd ed. (New York: McGraw-Hill).

BALLOR, D. L. and VOLOVSEK, A. J. 1992, Effect of exercise to rest ratio on plasma lactate concentration at work rates above and below maximum oxygen uptake, *European Journal of Applied Physiology*, **65**, 365–369.

BERNARD, T., GAVARRY, O., BERMON, S., GIACOMONI, M., MARCONNET, P. and FALGAIRETTE, G. 1997, Relationships between oxygen consumption and heart rate in transitory and steady states of exercise and during recovery: influence of type of exercise, *European Journal of Applied Physiology*, **75**, 170–176.

BHAMBHAMI, Y., BUCKLEY, S. and MAIKALA, R. 1997, Physiological and biomechanical responses during treadmill walking with graded loads, *European Journal of Applied Physiology*, **76**, 544–551.

BUNC, V., HELLER, J. and LESO, J. 1988, Kinetics of heart rate responses to exercise, *Journal of Sports Sciences*, **6**, 39–48.

CASABURI, R., WHIPP, B. J., WASSERMAN, K., BEAVER, W. L. and KOYAL, S. N. 1977, Ventilatory and gas exchange dynamics in response to sinusoidal work, *Journal of Applied Physiology*, **42**, 300–311.

COLLINS, M. A., CURETON, K. J., HILL, D. W. and RAY, C. A. 1991, Relationship of heart rate to oxygen uptake during weight lifting exercise, *Medicine and Science in Sports and Exercise*, **23**, 636–640.

DE GROOT, G., SCHREURS. A. W. and VAN INGEN SCHENAU, G. J. 1983, A portable lightweight Douglas bag instrument for use during various types of exercise, *International Journal of Sports Medicine*, **4**, 132–134.

EDWARDS, R. H. T., EKELUND, L-G., HARRIS, R. C., HESSER, C. M., HULTMAN, E., MELCHER, A. and WIGERTZ, O. 1973, Cardiorespiratory and metabolic costs of continuous and intermittent exercise in man, *Journal of Physiology*, **234**, 481–497.

ENGELS, J. A., GULDEN, J. W. J., VAN DER SENDEN, I. F., HERTOG, C. A. W. M., KOLK, J. J. and BINKHORST, R. A., 1994, Physical work load and its assessment among the nursing staff in nursing homes, *Journal of Occupational Medicine*, **36**, 338–345.

ESTON, R. G. and BRODIE, D. A. 1986, Responses to arm and leg ergometry, *British Journal of Sports Medicine*, **20**, 4–6.

FAIRSHTER, R. D., SALNESS, K., WALTERS, J., MINH, V.-D. and WILSON, A. 1987, Relationships between minute ventilation, oxygen uptake, and time during incremental exercise, *Respiration*, **51**, 223–231.

FARDY, P. S. and HELLERSTEIN, H. K. 1978, A comparison of continuous and intermittent progressive multistage exercise testing, *Medicine and Science in Sports*, **10**, 7–12.

FRANKLIN, B. A., HODGSON, J. and BUSKIRK, E. R. 1980, Relationship between percent maximal O_2 uptake and percent maximal heart rate in women, *Research Quarterly for Exercise and Sport*, **51**, 616–624.

GILBERT, R. and AUCHINCLOSS, J. H. 1971, Comparison of cardiovascular responses to steady- and unsteady-state exercise, *Journal of Applied Physiology*, **30**, 388–393.

HOLLANDER, A. P., DUPONT, S. H. J. and VOLKERIJK, S. M. 1994, Physiological strain during competitive water polo games and training, in M. Miyashita, Y. Mutoh and A. B. Richardson (eds), *Medicine and Science in Aquatic Sports* (Basel: Karger), 178–185.

HOOKER, S. P., GREENWOOD, J. D., HATAE, D. T., HUSSON, R. P., MATTHIESEN, T. L. and WATERS, A. R. 1993, Oxygen uptake and heart rate relationship in persons with spinal cord injury, *Medicine and Science in Sports and Exercise*, **25**, 1115–1119.

HUGHSON, R. L. and MORRISSEY, M. A. 1983, Delayed kinetics of $\dot{V}O_2$ in the transition from prior exercise. Evidence for O_2 transport limitation of $\dot{V}O_2$ kinetics: a review, *International Journal of Sports Medicine*, **4**, 31–39.

KATCH, V., WELTMAN, A., SADY, S. and FREEDSON, P. 1978, Validity of the relative percent concept for equating training intensity, *European Journal of Applied Physiology*, **39**, 219–227.

KILBOM, A. and PERSSON, J. 1981, Cardiovascular response to combined dynamic and static exercise, *Circulation Research*, **48**, Suppl. 1, 93–97.

LONDEREE, B. R. and AMES, S. A. 1976, Trend analysis of the %$\dot{V}O_2$ max–HR regression, *Medicine and Science in Sports*, **8**, 122–125.

LONDEREE, B. R., THOMAS, T. R., ZIOGAS, G., SMITH, T. D. and ZHANG, Q. 1995, %$\dot{V}O_{2MAX}$ versus %Hr_{MAX} regressions for six modes of exercise, *Medicine and Science in Sports and Exercise*, **27**, 458–461.

LOTHIAN, F. and FARRALLY, M. R. 1995, A comparison of methods for estimating oxygen uptake during intermittent exercise, *Journal of Sports Sciences*, **13**, 491–497.

MAAS, S., KOK, M. L. J., WESTRA, H. H. and KEMPER, H. C. G. 1989, The validity of the use of heart rate in estimating oxygen consumption in static and combined static/dynamic exercise, *Ergonomics*, **32**, 141–148.

MATTHYS, D., PANNIER, J. L., TAEYMANS, Y. and VERHAAREN, H. 1996, Cardiorespiratory variables during a continuous ramp exercise protocol in normal young adults, *Acta Cardiologica*, **51**, 451–459.

MONTOLIU, M. A., GONZALEZ, V. and PALENCIANO, L. 1995, Cardiac frequency throughout a working shift in coal miners, *Ergonomics*, **38**, 1250–1263.

Rasche, W., Janssen, T. W. J., Van Oers, C. A. J. M., Hollander, A. P. and van der Woude, L. H. V. 1993, Responses of subjects with spinal cord injuries to maximal wheelchair exercise: comparison of discontinuous and continuous protocols, *European Journal of Applied Physiology*, **66**, 328–331.

Rayson, M. P., Davies, A., Bell, D. G. and Rhodes-James, E. S. 1995, Heart rate and oxygen uptake relationship: a comparison of loaded marching and running in women, *European Journal of Applied Physiology*, **71**, 405–408.

Shaw, D. K. and Deutsch, D. T. 1982, Heart rate and oxygen uptake response to performance of Karate Kata, *Journal of Sports Medicine and Physical Fitness*, **22**, 461–468.

Sietsema, K. E., Daly, J. A. and Wasserman, K. 1989, Early dynamics of O_2 uptake and heart rate as affected by exercise work rate, *Journal of Applied Physiology*, **67**, 2535–2541.

Szabo, A., Péronnet, F., Gauvin, L. and Furedy, J. J. 1994, Mental challenge elicits 'additional' increases in heart rate during low and moderate intensity cycling, *International Journal of Psychophysiology*, **17**, 197–204.

Vander, L. B., Franklin, B. A., Wrisley, D. and Rubenfire, M. 1984, Cardiorespiratory responses to arm and leg ergometry in women, *The Physician and Sportsmedicine*, **12**, 101–106.

van der Beek, A. J. and Frings-Dresen, M. H. 1995, Physical workload of lorry drivers: a comparison of four methods of transport, *Ergonomics*, **38**, 1508–1520.

Vokac, Z., Bell, H., Bautz-Holter, E. and Rodahl, K. 1975, Oxygen uptake/heart rate relationship in leg and arm exercise, sitting and standing, *Journal of Applied Physiology*, **39**, 54–59.

24 Modelling handgrip strength in the presence of confounding variables: results from the Allied Dunbar National Fitness Survey

ALAN M. NEVILL†* and ROGER L. HOLDER‡

†School of Sport, Performing Arts and Leisure, University of Wolverhampton, Walsall Campus, Gorway Road, Walsall, WS1 3BD, UK

‡School of Mathematics and Statistics, University of Birmingham, Birmingham, UK

Keywords: Physical activity; Allometric models; Heteroscedastic errors.

Differences in handgrip strength, caused by risk factors such as physical inactivity, will be influenced by 'confounding' variables, e.g. age, body size. The aims of the study were to identify the confounding variables associated with handgrip strength and to assess the benefit that physical activity plays in maintaining grip strength within a population, having adjusted for differences in these confounding variables. The most appropriate linear body size dimension associated with grip strength was height rather than demispan. Non-linear associations with age and body mass were also identified. Handgrip strength peaked in the age group 25–34 years for male subjects and in the age group 35–44 years for female subjects. Similarly, handgrip strength increased with body mass until it peaked at a body mass of approximately 100 kg for male and 90 kg for female subjects; thereafter a rapid decline in grip strength was observed. Differences in handgrip strength were found to be significantly associated with levels of physical activity even having controlled for differences in age and body size (height, mass and percentage body fat), but the observed association was not linear. The level of physical activity necessary to maintain an optimal level of handgrip strength was found to be a balance of moderate *or* vigorous occasions of physical activity.

1. Introduction

Handgrip strength is an important indicator of physical capability in many aspects of daily life, especially among the elderly. The assessment of grip strength has also been used as a screening tool to help monitor diseases such as rheumatoid arthritis (Van Gestel *et al.* 1995). Clearly, there is a need to identify the effect of risk factors (e.g. physical inactivity, diet) on such indicators of physical capability within a population. For example, suppose one wishes to compare the grip strengths of two groups of subjects, those who do or do not take physical activity in their leisure time. It is well-known that in general strength increases proportionally with body size and declines with age. Since the subjects taking physical activity are likely to be a little younger, taller and have, on average, more body/muscle mass than the

*Author for correspondence. e-mail: a.m.nevill@wlv.ac.uk

group not taking activity, the grip strength measurements of the physical activity group are likely to be inflated for reasons other than just their participation in physical activity.

Variables such as age, height and body mass are known as confounding variables. Clearly, in order to compare the grip strength of mutually exclusive groups, the effect of body size, age and any other associated confounding variables must be removed before valid inference can be made about the benefits of physical activity.

In the past, the associations of various confounding variables with grip strength have been investigated either individually, using Pearson's product-moment correlations or simultaneously, using multiple linear regression or analysis of covariance (ANCOVA). Unfortunately, these linear regression or ANCOVA methods assume an additive association with the confounding variables and the error variance is assumed to be constant throughout the range of observations. Limitations of such a linear model are (1) that strength is likely to increase *proportionally*, but non-linearly with body mass (i.e. $m^{2/3}$), (2) the error variance is also likely to increase with larger measurements and conversely decrease with smaller measurements, and (3) the linear age term is not asymptotic, i.e. the model may predict *negative* handgrip strength estimates for older ages, and although beyond the range of observations, an equally absurd handgrip strength measurement at birth (age = 0). Clearly, a linear, additive regression model will only approximate handgrip strength for a limited group of individuals. For these reasons, researchers studying the physical capabilities of the elderly have suggested that any attempt to study the entire age span is inappropriate and, as such, recommend separate regression analyses for children, adults and the elderly. Since growing from childhood to adulthood and subsequently into old age is a continuous but non-linear process, the present authors have attempted to describe the population differences in handgrip strength using a continuous, biologically sound, non-linear regression model.

Hence the aims of the present study are as follows: (1) to confirm the known variables and to identify some less well-known confounding variables associated with handgrip strength measurements, recorded as part of the Allied Dunbar National Fitness Survey (ADNFS 1992); (2) to propose an appropriate proportional non-linear regression model for handgrip strength, and (3) to assess the benefit that physical activity plays in maintaining handgrip strength within the population, having adjusted for all the confounding variables.

2. Methods

2.1. Subjects

The Allied Dunbar National Fitness Survey (1992) randomly selected 4316 subjects, aged 16 years and over from the English population, each subject being interviewed by the Office of Population Censuses and Surveys (OPCS) trained employees about their health, lifestyle and physical activity. Of these subjects, 3024 agreed to take part in a physical appraisal that was conducted in one of three mobile laboratories in the 30 selected regional sites. Measurements of handgrip strength and complete records of other relevant information were available on 2632 (1270 male, 1362 female) subjects. This represented 87.0% of the subjects participating in the physical appraisal and 61.0% of the total subjects interviewed. Table 1 describes the number of men and women in the ADNFS

Table 1. The number of subjects (N) recording valid measurements of handgrip strength by gender and age group from the Allied Dunbar National Fitness Survey.

	Age (years)						
	16–24	25–34	35–44	45–54	55–64	65–74	All
Male (N)	176	253	265	251	182	143	1270
Female (N)	193	276	265	238	204	186	1362
All (N)	369	529	530	489	386	329	2632

sample that had valid measurements of handgrip strength by age groups (in 10-year intervals).

2.2. *Measurement of handgrip strength*

A maximum voluntary contraction of the subject's dominant hand was measured using a strain-gauge handgrip dynamometer developed at Nottingham University and supplied by Biomedical Engineering, Nottingham, UK (Bassey *et al*, 1986). The dynamometer consists of a gripping handle (with strain-gauge transducer) and amplifier with bar-graph and digital displays.

Details of the measurement protocol are described in full in the *Technical Report* (Fentem *et al*. 1994). Briefly, the subject was seated facing the dynamometer display. The test was explained to the subject after which an administrator gave a full demonstration. The three trials of maximum effort were separated by 30-s rest intervals, although one or two more additional trials were given if the last trial was more than 10% above the previous best.

2.3. *Measurements of body composition and size*

The maximum standing height of individual subjects was measured using a metal stadiometer (Holtain, Crymych, South Wales, UK), following a standard protocol (Weiner and Lourie 1981). Body mass was measured to the nearest 0.1 kg using a calibrated digital weighing machine. Among the other anthropometric measures, estimates of percentage body fat (% fat) were determined using the methods of Durnin and Womersley (1974), based on skinfold thickness at the biceps, triceps, sub-scapular and supra-iliac muscles. For subjects over 60 years of age skeletal size was measured as the demispan (the distance between the sternal notch and the second and third finger root) using a metal tape measure. Demispan was recorded in addition to the subject's height as a measure of skeletal size since, in this age group, there is evidence of varying levels of vertebral collapse and kyphosis (Bassey and Harries 1993).

2.4. *Physical activity questionnaire*

Each subject was asked a wide range of questions about his/her physical activity at work, in leisure time and in the household. For example, when asked about work activity, respondents were asked, 'So overall, would you say that in terms of physical effort your work is (1) very demanding, (2) fairly demanding, or (3) not very demanding'. Questions about physical activity were classified under three categories, (1) the intensity of activity (vigorous, moderate, light, none), (2) the frequency of activity lasting either greater or less than 20 min, and (3) past regular participation in sport and exercise since the age of 14 years. The level of intensity of activity was assessed according to the calorific expenditure of each activity (vigorous: ≥ 7.5 kcal min^{-1}; moderate: ≥ 5 and <7.5 kcal min^{-1}; light: ≥ 2 and <5 kcal min^{-1}; none <2 kcal min^{-1}). From this information a 'summary' variable of all vigorous and moderate physical activity occasions were created, whereby respondents were classified into one of six physical activity groups (levels).

> Level 1:12 or more occasions of vigorous activity.
> Level 2:12 or more occasions of moderate or vigorous activity.
> Level 3:12 or more occasions of moderate (no vigorous) activity.
> Level 4:5–11 occasions of at least moderate activity.
> Level 5:1–4 occasions of at least moderate activity.
> Level 6:No moderate or vigorous activity lasting 20 min.

2.5. Statistical methods

It is well-known that in general human muscle strength increases proportionally with body size (Nevill et al. 1998, Round et al. 1999) and declines in the elderly (Bassey and Harries 1993). Adapting a multiplicative model structure that has been used to describe a variety of human performance indices, e.g. leg power (Mockett et al. 1996, Nevill et al. 1996), strength and aerobic power (Nevill et al. 1998) and lung function (Nevill and Holder 1999), the proposed model for the ADNFS grip strength measurements (Y) is given as follows,

$$Y = \text{height}^{k_1} \text{ mass}^{k_2} \exp(b_0 + b_1 \text{mass} + b_2 \text{ age} + b_3 \text{ age}^2 + b_4 \text{ \%fat}) \quad (1)$$

where k_1, k_2, b_0, b_1, b_2, b_3 and b_4 are unknown parameters. Nevill and Holder (1999) recognized the advantage in incorporating body mass in the multiplicative model (Equation 1) as both a proportional allometric term as well as within the exponential function. By doing so, the terms provide a form of 'gamma function' within the model that is able to describe an initial increase in handgrip strength with heavier subjects, i.e. greater body mass (as a proportional or allometric term) that will eventually peak and then subsequently decline with excessive body mass (as a negative exponential term).

The model (Equation 1) can be linearised with a log transformation. A linear regression analysis on log(Y), can then be used to estimate the unknown parameters of the log-transformed model (Equation 2).

$$\begin{aligned}\log(Y) = &\, k_1 \log_e(\text{height}) + k_2 \log_e(\text{mass}) + \\ &\, b_0 + b_1 \text{ mass} + b_2 \text{ age} + b_3 \text{ age}^2 + b_4. \text{\%fat}\end{aligned} \quad (2)$$

Further categorical or group differences within the population (e.g. gender) can easily be explored by allowing some of the parameters in the log-transformed model to vary for each group (by introducing fixed indicator variables).

The proportional model (Equation 1) also assumes that the residual errors of Y are heteroscedastic (proportional), i.e. the error variance will increase with larger measurement means and conversely decrease with smaller means. Thus, by fitting the parameters using log-linear regression (Equation 2), it is assumed that the residual error variance of the log-transformed handgrip strength measurements remains constant throughout the range of observations.

If this constant error assumption is found questionable in the log-linear regression analyses, an alternative approach would be to model the error variance as described in Nevill and Holder (1999), adapting the methods originally proposed by Aitkin (1987). The recommended procedure is as follows: the parameters of the log-linear model (Equation 2), are fitted using backward elimination to obtain the parsimonious solution. Having saved the residual errors, these are squared, log transformed and then defined as the new dependent/response variable to identify any variance heterogeneity. The mean response model is then re-fitted but using *weighted* linear regression, weights taken as the reciprocal of the predicted error variance model. The process can then be repeated until a stable solution is obtained for both the mean response model (Equation 2) and the error variance model.

3. Results

The mean and standard deviation (SD) of handgrip strength measurements for the ADNFS male and female subjects by age groups are given in table 2. Note that the grip strength measurements peaked in the age group 25–34 years for male and in the age group 35–44 years for female ADNFS subjects, and declined thereafter, confirming the non-linear association with age (figure 1).

In the past, demispan has been chosen in preference to height as a measure of skeletal size in older subjects (Bassey 1987). For the ADNFS subjects over the age of 60 years ($N = 525$), the better predictor of handgrip strength was height ($r = 0.713$) rather than demispan ($r = 0.693$), confirming height in preference to demispan as the more appropriate linear component of skeletal body size when predicting handgrip strength (Equation 1). This was later confirmed when demispan was entered in addition to height in the proportional model (Equation 1) for handgrip strength. Height was retained as a significant predictor of handgrip strength whilst demispan became redundant (not significant).

When the proposed model (Equation 1) was fitted to the male and female handgrip strength measurements from the ADNFS using log-linear regression (Equation 2), the parsimonious solution is given in table 3.

Based on the fitted '$\log_e$ (mass)' and 'body mass' parameters reported in table 3 (both $p < 0.001$) and using elementary calculus, the model predicts that male handgrip strength will peak at body mass $= 1.4283/0.01367 = 104.5$ kg. Similarly, the model predicts that female handgrip strength will peak at body mass $= (1.4283 - 0.18715)/0.01367 = 90.8$ kg. These results can be clearly seen in figure 2 for both male and female ADNFS subjects. In order to assess whether this

Table 2. The mean and standard deviation (SD) of grip strength measurements by gender and age group from the Allied Dunbar National Fitness Survey.

	Age (years)					
	16–24	25–34	35–44	45–54	55–64	65–74
Male						
Mean (kg)	49.90	52.98	52.00	49.91	44.47	39.87
SD (kg)	9.66	10.11	10.15	9.74	9.25	8.74
Female						
Mean (kg)	30.11	31.35	31.80	30.07	27.21	23.23
SD (kg)	5.68	5.47	5.83	6.08	6.03	5.59

observed peak in handgrip strength was due to a relatively small number of overweight or obese subjects, the analysis was repeated sequentially removing the subjects with the greatest percentage of body fat. Even when 46.2% ($N = 1209$) of the subjects with percentage body fat greater than or equal to 30% were removed, the analysis on the remaining 'leaner' subjects still confirmed that the contributions of the $\log_e$ (mass)' and 'body mass' terms both remained highly significant ($p < 0.001$) but with *opposite* signs. This confirms a peak in handgrip strength with the heavy subjects even amongst this relatively lean group.

The number of male and female ADNFS subjects classified into one of the six 'summary' physical activity groups/levels are given in table 4. The mean handgrip

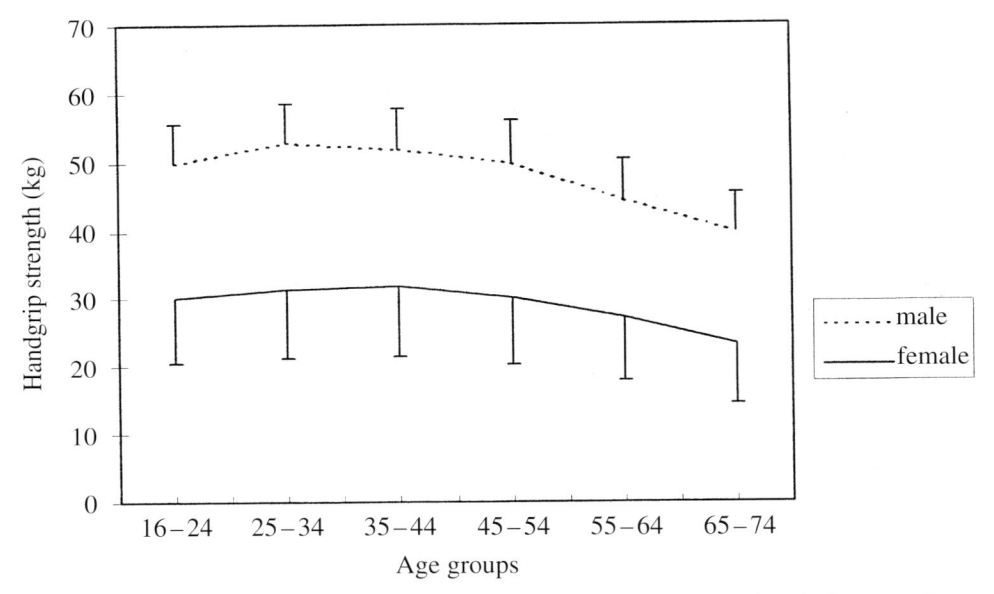

Figure 1. The relationship between the mean (standard deviation) handgrip strength measurements (kg) and age groups for the male and female ADNFS subjects.

Table 3. The fitted parsimonious log-linear regression model (Equation 2) for handgrip strength.

Fitted parameters	Coefficient	SD	t	p
Constant	−2.0004	0.6992	−2.86	0.004
Gender (female-male)	0.5203	0.2331	2.23	0.026
$\log_e$ (height)	1.1077	0.1258	8.80	<0.001
$\log_e$ (mass)	1.4283	0.2116	6.75	<0.001
$\log_e$ (mass) × female	−0.18715	0.05440	−3.44	0.001
Body mass	−0.013674	0.002598	−5.26	<0.001
Age	0.017118	0.001546	11.08	<0.001
Age^2	−0.000226	0.000017	−13.38	<0.001
Age^2 × female	−0.000011	0.000005	−2.07	0.038
%fat	−0.006167	0.001132	−5.45	<0.001

$S = 0.1971$; $R^2 = 68.3$.
S, estimated standard deviation of the residuals about the fitted model.

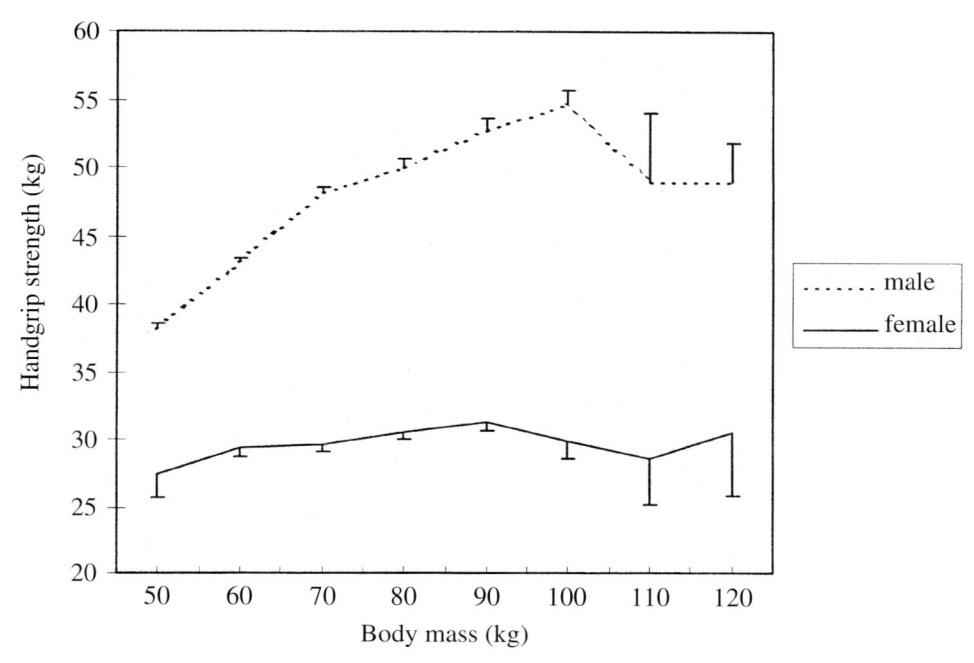

Figure 2. The relationship between the mean (standard error) handgrip strength measurements (kg) and body mass (kg) for the male and female ADNFS subjects.

Table 4. The number of male and female ADNFS subjects classified by the 'summary' physical activity groups (levels 1 to 6).

	Level 1	Level 2	Level 3	Level 4	Level 5	Level 6	All
Male	191	169	293	241	199	173	1266
Female	62	147	367	349	255	180	1360
All	253	316	660	590	454	353	2626

Level 1: 12 or more occasions of vigorous activity;
Level 2: 12 or more occasions of moderate or vigorous activity;
Level 3: 12 or more occasions of moderate (no vigorous) activity;
Level 4: 5–11 occasions of at least moderate activity;
Level 5: 1–4 occasions of at least moderate activity;
Level 6: No moderate or vigorous activity lasting 20 min.

strength values for the six physical activity groups/levels for the male and female ADNFS subjects are described in figure 3.

The unadjusted differences in grip strength (log transformed) due to the two main effects of 'gender' and 'physical activity' groups can be explored within the ADNFS population, using a two-way analysis of variance (ANOVA). The main effects of 'gender' (male versus females) and 'physical activity' groups (levels 1–6) were both highly significant ($F(1,2619) = 3159.06$; $p < 0.001$) and ($F(5,2619) = 32.66$; $p < 0.001$), respectively. There was no interaction between the two main effects ($p > 0.05$).

As described in §1, much of the unadjusted differences in handgrip strength can be, at least partially, explained by differences in the known associated/confounding variables described in table 3. For example, the more physically active groups/levels

(levels 1 and 2) are considerably younger than the less active groups (levels 5 and 6), as can be seen in figure 4.

Clearly, there is a need to adjust handgrip strength measurements within a population for differences in the confounding variables such as age, body mass and so on. This adjustment can be achieved by allowing the constant parameters in the

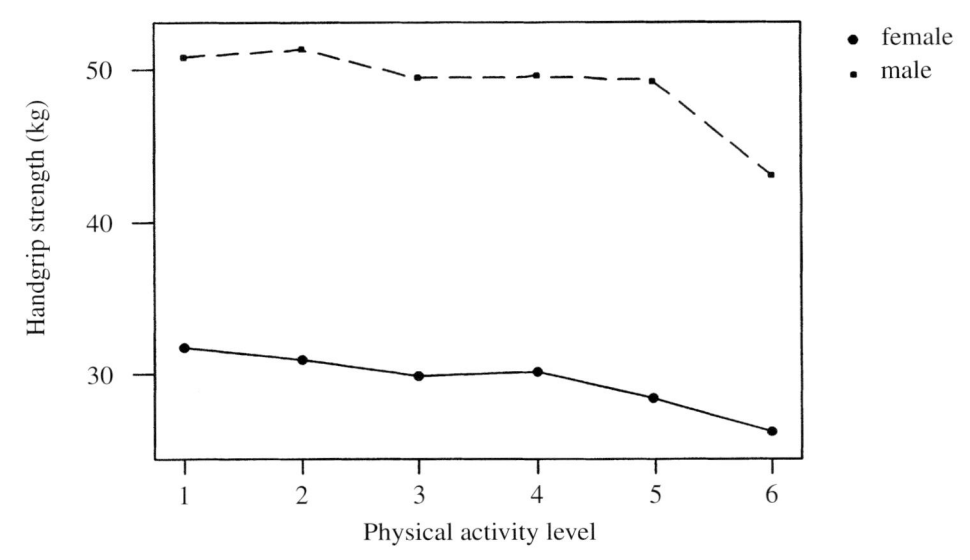

Figure 3. The relationship between the mean handgrip strength (kg) and the six physical activity groups/levels for the male and female ADNFS subjects.

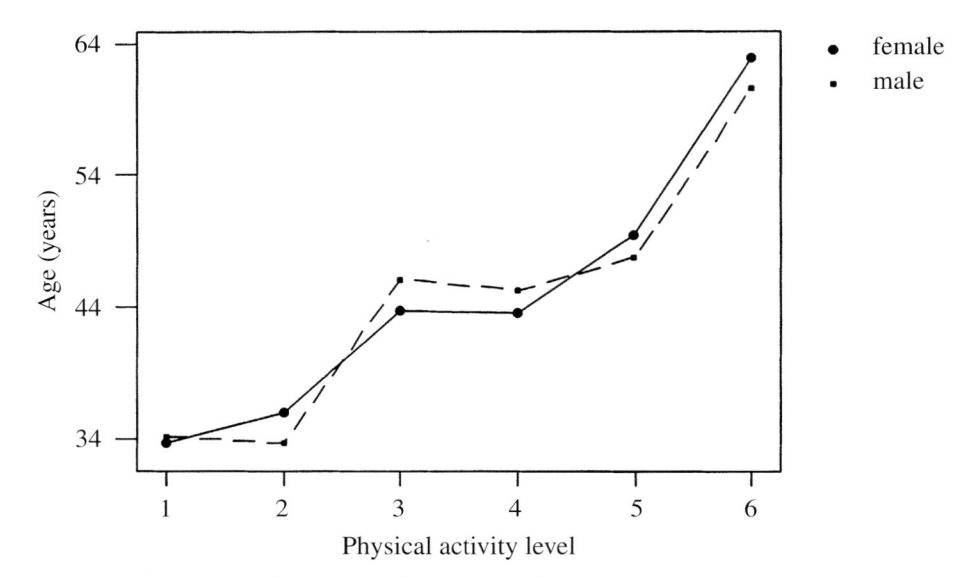

Figure 4. The relationship between the mean age (in years) of the male and female ADNFS subjects and the six physical activity groups/levels.

log-transformed model (Equation 2) to vary for each group (i.e. by introducing the main effect variable 'physical activity' in addition to 'gender' as [0,1] indicator variables). This regression analysis is equivalent to the analysis of covariance (ANCOVA), comparing the 'gender' and 'physical activity' group differences in log-transformed handgrip strength, adjusting for differences in body size (height, body mass and body fat) and age as the covariates. When the residuals from the ANCOVA were saved and re-defined as the new response variable, the error variance was found to increase linearly with body mass ($p<0.01$) and for female subjects, with age ($p<0.05$). The results from this analysis was saved as the fitted or predicted error variance (i.e. the 'fits' saved from the ANCOVA). The mean response model was then re-fitted but using *weighted* ANCOVA, weights taken as the reciprocal of the predicted error variance model.

This process was repeated three times after which a stable solution was obtained for both the mean response model and error variance model. The third re-fitted ANCOVA solution is described in table 5.

The significance of the 'gender' and 'physical activity' main effects ($F(1,2597)=7.30$; $p=0.007$) and ($F(5,2597)=5.93$; $p<0.001$), respectively, were greatly reduced, compared with the unadjusted F ratio statistics reported above. Having adjusted for all the confounding variables, the greatest handgrip strength (table 5) was found in the second 'vigorous *or* moderate' physical activity group (level 2) rather than the highest 'vigorous' physical activity group (level 1).

4. Discussion

The first most obvious findings from the regression model given in table 3 is that handgrip strength is approximately proportional to a linear dimension of skeletal size (height raised to the power 1.1077; SE = 0.126). Presumably, the larger the size of hand, the greater the handgrip strength. We can only speculate that height was selected in preference to demispan because the natural reduction in height in older subjects was better able to reflect the decline in grip strength observed in the older ADNFS subjects. However, in contrast to the model fitted by Bassey and Harries (1993) when investigating handgrip strength of men and women over 65 years of age, the decline with age in the present study was clearly not linear (table 2). This was confirmed by the significant positive age and negative age^2 terms in table 3 (both $p<0.001$). Using elementary calculus, the model predicts the rate of decline in handgrip strength for 60- and 70-year-old male subjects is approximately -1.0% and -1.5% per annum, respectively. Similarly, the model predicts the rate of decline is approximately -1.1% and -1.6% for 60- and 70-year-old female subjects, respectively. Interestingly, the model predicts rates of decline in handgrip strength of -1.9% and -2.1% per year for 80-year-old male and female subjects, although it is acknowledged that this age is beyond the range of ADNFS observations. These results concur with those rates reported by Bassey and Harries (1993), who estimated a 2% decline per annum in handgrip strength in both men and women aged over 65 years. Bassey and Harries (1993) suggested that this might be due to either a net loss in motor units that may accelerate with increasing years or may be caused by a temporary halt in physical activity caused by illness that results in a partial loss in grip strength, an event more likely to occur in the elderly.

The significant positive $\log_e$(mass) and negative body mass terms ($p<0.001$) predict that handgrip strength peaks at approximately 104 kg for male subjects and 91 kg for female subjects. This together with the negative percentage body fat terms,

Table 5. Physical activity and gender differences in handgrip strength (log transformed) using weighted ANCOVA, adjusting for differences in body size (height, mass and %body fat) and age.

Source of variation	df	Adjusted SS	Adjusted MS	F	p
Physical activity	5	128.5	25.7	5.93	<0.001
Gender	1	31.7	31.7	7.30	0.007
Covariates					
$\log_e$ (height)	1	300.2	300.2	69.22	<0.001
$\log_e$ (mass)	1	208.7	208.7	48.14	<0.001
$\log_e$ (mass) × female	1	64.5	64.5	14.87	<0.001
Body mass	1	123.1	123.1	28.40	<0.001
Age	1	511.8	511.8	118.02	<0.001
Age^2	1	704.3	704.3	162.43	<0.001
Age^2 × female	1	23.7	23.7	5.46	0.020
%fat	1	145.5	145.5	33.56	<0.001
Error	2597	11261.3	4.3		
Total	2611	35146.8			

Fitted parameters	Coefficient	SD	t	p
Constant	−2.3184	0.7256	−3.20	0.001
$\log_e$ (height)	1.0384	0.1248	8.32	<0.001
$\log_e$ (mass)	1.5297	0.2205	6.94	<0.001
$\log_e$ (mass) × female	−0.21437	0.05559	−3.86	<0.001
Body mass	−0.014715	0.002761	−5.33	<0.001
Age	0.016569	0.001525	10.86	<0.001
Age^2	−0.000215	0.000017	−12.74	<0.001
Age^2 × female	−0.000013	0.000006	−2.34	0.020
%fat	−0.006360	0.001098	−5.79	<0.001
Physical activity				
level 1	0.01581	0.01090	1.45	0.147
level 2	0.021794	0.009726	2.24	0.025
level 3	0.011256	0.007360	1.53	0.126
level 4	0.003109	0.007674	0.41	0.685
level 5	0.001161	0.008475	0.14	0.891
level 6	0.0†			
Gender (female-male)	0.6398	0.2367	2.70	0.007

†This parameter is set to zero because it is redundant.

confirms a muscular component associated with handgrip strength and that being excessively overweight is associated with a reduced handgrip strength.

The results from the unadjusted ANOVA to identify differences in grip strength between the 'physical activity' groups yielded a highly significant main effect ($F(5,2619) = 32.66$; $p < 0.001$). In absolute terms, the decline in handgrip strength was approximately linear with decreasing level of physical activity (figure 3). However, much of this trend was explained by differences in age, height, body mass and percentage body fat. When these confounding variables were included as covariates, the adjusted difference in handgrip strength was still significant ($F(5,2597) = 5.93$; $p < 0.001$), although much reduced. Furthermore, the linear decline in grip strength with decreasing levels of physical activity was no longer observed. Although an overall decline in handgrip strength was still identified with decreasing levels of physical activity, those undertaking 12 or more occasions of 'vigorous *or* moderate' physical activity (level 2) would appear to have the highest handgrip strength rather than those taking 12 or more occasions of 'vigorous' physical activity alone (level 1). This may be explained by the nature of vigorous activities, being predominately aerobic (running/jogging), that would not directly benefit the development of handgrip strength.

Finally, one may wish to consider why, having controlled for all the confounding variables, such as age, height, body mass and percentage body fat (as covariates), there still remains significant differences in handgrip strength associated with levels of physical activity. The most obvious explanation is that physical activity prevents or reduces the age-related loss of muscle and consequent decrease in strength. This could be the result of natural selection and elimination within an ageing population, with the survivors being the more active subjects. Alternatively there may be some preferential effect of activity on the muscles of the forearm or, conceivably, physical activity may influence the 'quality' of the muscle independently of the quantity. Nevertheless, whatever the cause, the results of the present study provide evidence to support the benefits of maintaining a physically active lifestyle.

Acknowledgements

The authors wish to thank The Sports Council and the Health Education Authority for access to and permission to publish the results from the Allied Dunbar National Fitness Survey. The authors would also like to thank Professor David Jones (Birmingham University) for his helpful advice when interpreting the results of their findings.

References

AITKIN, M. 1987, Modelling variance heterogeneity in normal regression using GLIM, *Journal of the Royal Statistical Society, Series C*, **436**, 332–339.

ALLIED DUNBAR NATIONAL FITNESS SURVEY 1992, *A Report on Activity Patterns and Fitness Levels: Main Findings* (London: The Sports Council and the Health Education Authority).

BASSEY, E. J. 1987, Demi-span as a measure of skeletal size, *Annals of Human Biology* **13**, 499–502.

BASSEY, E. J. and HARRIES, U. J. 1993, Normal values for handgrip strength in 920 men and women aged 65 years, and longitudinal changes over 4 years in 620 survivors, *Clinical Science*, **84**, 331–337.

BASSEY, E. J., DUDLEY, B. R. and HARRIES, U. J. 1986, A new portable strain-gauge hand-grip dynamometer, Journal of Physiology, **373**, 6P.

DURNIN, J. V. and WOMERSLEY, J. 1984, Body fat assessed from total body density and its estimation from skinfold thickness measurements on 481 men and women aged 16 to 72 years, *British Journal of Nutrition*, **32**, 77–85.

FENTEM, P. H., COLLINS, M. F., TUXWORTH, W., WALKER, A., HOINVILLE, E., COOKE, C., HARRIES, U. J. and RAYSON, M. 1994, *Allied Dunbar National Fitness Survey, Technical Report* (London: The Sports Council).

MOCKETT, S. P., FENTEM, P. H. and NEVILL, A. M. 1996, Leg extensor power and walking pace, *Medicine and Science in Sports and Exercise*, **28**, S188.

NEVILL, A. M. and HOLDER, R. L. 1999, Identifying population differences in lung function: results from the Allied Dunbar National Fitness Survey, *Annals of Human Biology*, **26**, 267–285.

NEVILL, A. M., MOCKET, S. P. and FENTEM, P. H. 1996, Scaling leg power for differences in body size, *Journal of Physiology*, **491P**, P81–82.

NEVILL, A. M., HOLDER, R. L., BAXTER-JONES, A., ROUND, J. and JONES, D. A. 1998, Modeling developmental changes in strength and aerobic power in children, *Journal of Applied Physiology*, **84**, 963–970.

ROUND, J. M., JONES, D. A., HONOUR, J. W. and NEVILL, A. M. 1999, Hormonal factors in the development of differences in strength between boys and girls during adolescence: a longitudinal study, *Annals of Human Biology*, **26**, 49–62.

VAN GESTEL, A. M., LAAN, R. F. J. M., HAAGSMA, C. J., VAN DE PUTTE, L. B. A. and VAN RIEL, P. L. C. M. 1995, Oral steroids as bridge therapy in rheumatoid arthritis patients starting with parenteral gold. A random double-blind placebo-controlled trial, *British Journal of Rheumatology*, **34**, 347–351.

WEINER, J. S. and LOURIE, J. A. 1981, *Practical Human Biology* (New York: Academic Press).

25 Prediction and validation of fat-free mass in the lower limbs of young adult male Rugby Union players using dual-energy X-ray absorptiometry as the criterion measure

W. BELL[†]*, D. M. COBNER[†] and W. D. EVANS[‡]

[†]University of Wales Institute Cardiff, Cyncoed, Cardiff CF23 6XD, UK

[‡]University Hospital of Wales, Heath Park, Cardiff CF14 4XW, UK

Keywords: FFM in lower limbs; Rugby players; DXA.

The aim was to derive and cross-validate prediction equations to estimate fat-free mass (FFM) in the lower limbs of young adult male Rugby Union players. Thirty players of mean ± SD age of 21.1 ± 2.1 years were recruited. Bone mineral mass, fat mass and lean tissue mass were measured with a Hologic QDR 1000/W whole-body scanner. Anthropometry included circumferences, segmental leg lengths and skinfold thicknesses. Players were divided randomly into prediction ($n = 15$) and cross-validation ($n = 15$) samples. Regression equations were derived from the prediction sample and validated on the cross-validation sample. Seven equations were formulated to predict leg FFM. The two equations providing the lowest standard errors of estimate were leg length with circumferences at the knee (0.7262) and calf (0.7382); the multiple correlation was 0.83 in both instances. Cross-validation statistics found no significant differences ($p > 0.05$) between measured (12.4 ± 1.5 kg) and predicted leg FFM (12.1–12.4 kg). The smallest mean difference was −0.05 kg, the largest 0.26 kg; these were equivalent to −0.4 and 2.1% of the measured leg FFM respectively. Correlations between measured and predicted leg FFM were reasonably high (0.79–0.90, $p < 0.001$). The ratio limits of agreement confirmed that there was little bias between measured and predicted leg FFM (1.00–1.02) and a good level of agreement (1.12–1.16). Because prediction equations tend to be age, gender and population specific, unless validated for other athletic groups, the present equations should be applied to male Rugby Union players with characteristics similar to those described.

1. Introduction

Generally speaking, the body composition of athletes has been described largely in terms of whole-body composition. This is used in a variety of ways; for example, to identify desirable weights for competition, monitor the effects of nutrition, and chart the progress of training and performance (Sinning 1996). Whole-body composition, however, furnishes little information about the regional distribution of tissue.

Variation in the anatomical distribution of major components of body mass (bone mass, fat mass, lean-tissue mass) is one of the distinguishing feature of individuals who participate in sport. In Rugby Union football, for instance, there are marked differences in size, shape, proportion and body composition between some of

*Author for correspondence. e-mail: wbell@uwic.ac.uk.

the playing positions (Bell 1979). This reflects, to some degree, the extent of player specialization, although in modern rugby there is a trend towards homogeneity of performance.

The quantity and nature of tissue in the lower limbs are influential in many sports. While the legs provide support for the upper-body mass, they also facilitate a variety of performance-related tasks. In Rugby Union these include fundamental skills such as running, kicking, jumping, scrummaging, rucking, mauling and tackling. Success or failure very often depends on how well these skills are performed and integrated in given circumstances.

Considering the extent to which the legs contribute to performance, there are surprisingly few procedures that estimate the quantity and nature of tissue in the lower limbs. A widely used strategy, which estimates the volume of muscle and bone in the legs, is the anthropometric model of Jones and Pearson (1969). Using segmental leg lengths, foot length, circumferences and skinfold thicknesses, the volume of the leg is partitioned into six truncated cones, the foot being regarded as a wedge. Validation of thigh and calf volumes was made by water displacement and fat volumes from X-rays. Similar anthropometric models have been constructed for both males (Katch and Katch 1974, Katch and Weltman 1975) and females (Katch et al. 1973).

Regression models, using body measurements to predict segmental mass and component tissue mass (skin, adipose tissue, muscle, bone), have been formulated from anatomical dissection of cadavers. About two-thirds of segments were predicted to within 10% of their true mass. It was suggested that the inability of the models to predict to within 5% of the measured segmental mass restricted their use to group comparisons only (Clarys and Marfell-Jones 1986a, b, 1994).

Dual-photon absorptiometry (DPA), a technique developed originally to assess skeletal mass *in vivo*, was used by Heymsfield et al. (1990) to quantify appendicular skeletal muscle mass in normal healthy adults. Strong significant correlations ($p<0.001$) were found between limb muscle mass estimated by DPA and anthropometric limb muscle areas, total body potassium (TBK), and total-body muscle mass based on TBK and total-body nitrogen models. It was concluded that this was a promising new approach to estimate skeletal muscle mass in the appendages.

Recent technological advances such as dual-energy X-ray absorptiometry (DXA) have greatly improved the resolution and precision with which bone and soft-tissue measurements can be made in segments of the body (Lohman 1996). This has allowed muscle mass, fat mass and bone mass, to be estimated in the upper and lower limbs (Fuller et al. 1992).

One of the weaknesses of most prediction equations is that they tend to be age, gender and population specific (Guo and Chumlea 1996). Many of the equations perform unsatisfactorily, largely because they have not been validated for specific athletic groups. Before an equation is used it is important to identify the degree to which it is valid for groups other than that from which it was derived. The more diverse the group the more unlikely it is that the equation will perform satisfactorily.

The aim of the present study, therefore, was to formulate and cross-validate prediction equations to estimate fat-free mass (FFM) in the lower limbs of young adult male Rugby Union players. Anthropometric dimensions were used as predictor variables and DXA fat-free mass (BMM + LTM) the criterion measure.

2. Methods

2.1. Participants

Thirty competitive young adult male Rugby Union players, representing all positions, were recruited to the study. Their mean age was 21.1 ± 2.1 years. One of the criteria for selection was that individuals had played representative rugby at a junior level of the game. At the time the measurements were taken, players were affiliated to the university and played in the Welsh National League. Since then a number of players have registered with first-class clubs. Informed consent was obtained from individuals and the appropriate Faculty Ethics Committee approved all procedures.

2.2. Anthropometry

Using the Harpenden range of equipment, anthropometry was conducted according to the recommendations of Weiner and Lourie (1981) and Cameron (1984). Stature was measured to the nearest 0.1 cm using a fixed stadiometer and body mass on a beam balance to the closest 0.1 kg. The measurements used by Jones and Pearson (1969) to determine lean leg volume (LLV) were taken on the left side of the body. Circumferences and segmental leg lengths were taken at the gluteal furrow, mid-thigh, above the knee, maximum knee, below the knee, maximum calf and minimum ankle. Skinfold thickness was measured at the anterior and posterior thigh and the medial and lateral calf. The mean technical error of measurement ($\sqrt{\Sigma d^2 / 2n}$) for sectional lengths was 1.5 mm, for circumferences 1.7 mm and for skinfold thickness 0.7 mm; these figures are comparable with those in Cameron (1984).

2.3. Body composition: dual-energy X-ray absorptiometry

Bone mineral mass (BMM), fat mass (FM) and lean-tissue mass (LTM) were measured by dual-energy X-ray absorptiometry (DXA) using a QDR 1000/W total body scanner with software version 5.47P (Hologic, Waltham, MA, USA). A detailed account of procedures and calibration can be found in Bell et al. (1995). Briefly, each subject was aligned with the long axis of the DXA couch and remained motionless during the scan (16 min). An X-ray tube mounted beneath the couch produces polenergetic spectra at two effective photon energies by switching between 70 and 140 kVP during alternate half-cycles of the mains supply. The X-ray beam was tightly collimated, the transmitted energy being measured by a detector positioned above the subject and rigidly coupled to the energy source. The X-ray beam was scanned in a rectilinear fashion to cover the entire area of the subject. During this process, transmitted energies were measured at a large number of points at both the low and high effective energies. The effective radiation dose of 5 μSv is very low (Lewis et al. 1994).

Dual-energy X-ray analysis was based on the differential attenuation of radiation at the two energies as it passes through bone and soft-tissue. The attenuation data were displayed as a digital image with each picture element (pixel) corresponding to individual measurement points through the subject. Pixels containing bone were identified by their value of differential attenuation and were decomposed into bone mineral and soft-tissue masses. Pixels not containing bone were decomposed into fat and lean soft-tissue components using a soft-tissue phantom, which is scanned simultaneously with the subject. When this process was complete, the total masses of BMM, FM and LTM compartments were obtained by summing the corresponding values for individual pixels. FFM mass was obtained from the sum of BMM and

LTM. The distinction between FFM and LTM was that by definition FFM is fat-free; LTM is sometimes described as bone-free lean tissue (Kohrt 1995, Lohman 1996). In the Hologic QDR 1000/W scanner the coefficient of variation for absolute whole-body BMM and FM was <1% (Herd *et al.* 1993); for LTM at the trunk it was <1% and for the upper and lower limbs 1.1% (Taaffe *et al.* 1994).

2.4. *Partitioning the digital image into regional segments*
FFM (BMM + LTM) in the lower limbs was determined by partitioning the digital image of each subject into regional anatomical segments comprising the head, right and left arms, trunk, and right and left legs (figure 1). Horizontal markers were placed inferior to the mandible, to delineate the head and at the level of the iliac crests. Markers midway between the head of the humerus and the glenoid fossa, and running vertically, connected upper and lower horizontal markers. This isolated the arms and at the same time defined the thorax and abdomen. The pelvic area was designated by a triangle comprising the horizontal marker at the iliac crests and oblique lines passing through the neck of each femur. The thorax, abdomen and pelvis comprised the trunk. The oblique markers running through the femoral neck separated the lower limbs from the trunk. Markers placed laterally and centrally separated the legs from the arms, and the legs from each other.

2.5. *Statistical analysis*
Players were randomly assigned to a prediction sample ($n=15$) and a cross-validation sample ($n=15$). There were no significant differences in stature, body mass, DXA mass, or absolute or relative amounts of body composition between the

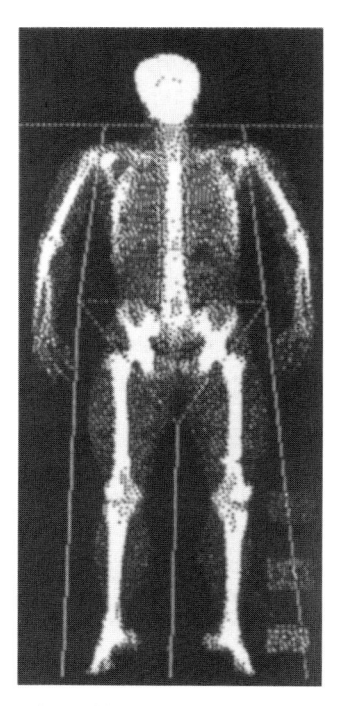

Figure 1. DXA image of a subject showing demarcation of regional segments.

two groups ($p > 0.05$). Multiple linear regression equations, using leg FFM as the dependent variable and anthropometric dimensions as independent variables, were formulated from the prediction sample and validated using the cross-validation sample. Derivation of equations was assisted by best subsets regression.

The statistics used for cross-validation included the significance of differences between measured (DXA) and predicted (derived equations) leg FFM and the correlations between them, a comparison of SD, standard errors of estimate (SEE = SD measured leg FFM $\times \sqrt{1-r^2}$), total error (TE = $\sqrt{\Sigma\text{(predicted-measured leg FFM)}^2/N}$) (Katch and Katch 1980, Lohman 1981), and 'ratio limits of agreement' (Nevill and Atkinson 1997). A paired Student's t-test assessed the significance of mean differences between measured and predicted leg FFM. The Pearson product-moment technique was used to compute correlations. All computations were carried out using the Minitab interactive statistical package (Minitab 1995). $p = 0.05$ was prescribed for all tests.

3. Results

The descriptive features of stature, body mass, DXA mass, and whole-body composition for prediction and cross-validation samples are displayed in table 1. There were no significant differences ($p > 0.05$) in any of the variables between validation and prediction samples.

Circumferences, leg length (floor–gluteal furrow), leg mass, tissue characteristics (BMM, FM, LTM and FFM) and LLV of the left leg are detailed in table 2. No significant differences ($p > 0.05$) were observed between the two groups.

3.1. Derivation of prediction equations

Seven equations for the prediction of leg FFM are listed in table 3, together with the SEE, adjusted R^2 and multiple correlation (R). The two equations providing the lowest SEE included leg length and circumferences at the maximal knee (0.7262, equation 4) and maximal calf (0.7382, equation 6). The multiple correlation was 0.83 in both cases. The percent of total variation of leg FFM accounted for by leg length and the respective circumferences was therefore 68–69%. The poorest fitting equation was leg length and the gluteal furrow circumference (SEE = 1.019, $R^2 = 39\%$, $R = 0.62$, equation 1). The remaining four equations had SEE ranging between 0.8576 and 0.9704 and multiple correlations between 0.67 and 0.75.

Table 1. Descriptive data (mean ± SD) for stature, body mass, DXA mass and whole-body composition for validation and prediction samples.

	Validation ($n = 15$)	Prediction ($n = 15$)
Stature (cm)	180.3 ± 5.4	183.6 ± 6.0
Body mass (kg)	88.1 ± 14.1	91.5 ± 11.7
DXA mass (kg)	87.3 ± 13.9	90.3 ± 11.8
FM (kg)	13.1 ± 7.2	14.2 ± 5.8
FFM (kg)	74.2 ± 7.9	76.1 ± 8.1
FM (%)	14.3 ± 5.7	15.4 ± 4.9
FFM (%)	85.7 ± 5.7	84.6 ± 4.9

FM, fat mass; FFM, fat-free mass.
Differences between groups not significant ($p < 0.05$).

Table 2. Anthropometric, tissue composition, and lean leg volume values (mean ± SD) of the left leg for validation and prediction samples.

	Validation ($n=15$)	Prediction ($n=15$)
Gluteal furrow circumference (cm)	62.4 ± 5.8	64.6 ± 4.2
Mid thigh circumference (cm)	56.0 ± 5.8	57.0 ± 3.0
Above knee circumference (cm)	42.1 ± 3.8	42.6 ± 2.3
Maximum knee circumference (cm)	39.5 ± 2.2	40.3 ± 1.9
Below knee circumference (cm)	35.5 ± 2.7	36.2 ± 1.6
Maximum calf circumference (cm)	39.3 ± 3.4	39.6 ± 1.6
Minimum ankle circumference (cm)	23.7 ± 1.8	23.6 ± 1.4
Leg length (cm)	83.2 ± 4.0	84.4 ± 4.5
Leg mass (kg)	14.9 ± 2.5	15.4 ± 1.9
Leg BMM (kg)	0.7 ± 0.1	0.7 ± 0.1
Leg FM (kg)	2.5 ± 1.4	2.8 ± 1.2
Leg LTM (kg)	11.7 ± 1.4	11.9 ± 1.3
Leg FFM (kg)	12.4 ± 1.5	12.6 ± 1.3
LLV (L)	11.6 ± 2.1	12.0 ± 1.3

BMM, bone mineral mass; FM, fat mass; LTM, lean tissue mass; FFM, fat-free mass; LLV, lean leg volume.
Differences between groups are not significant ($p > 0.05$).

Table 3. Regression equations for the prediction of FFM in the leg from leg length and leg circumferences.

Regression equation	SEE	Adj R^2 (%)	R
(1) Leg FFM = −9.4 + 0.0172 LL + 0.116 GFC	1.0190	39.0	0.62
(2) Leg FFM = −12.5 + 0.0168 LL + 0.191 MTC	0.9704	44.6	0.67
(3) Leg FFM = −12.8 + 0.0167 LL + 0.264 AKC	0.9453	47.5	0.69
(4) Leg FFM = −17.6 + 0.0144 LL + 0.449 MKC	0.7262	69.0	0.83
(5) Leg FFM = −18.3 + 0.0176 LL + 0.444 BKC	0.8576	56.7	0.75
(6) Leg FFM = −24.6 + 0.0201 LL + 0.511 MCC	0.7382	68.0	0.83
(7) Leg FFM = −14.7 + 0.0180 LL + 0.512 MAC	0.8744	55.0	0.74

GFC, gluteal furrow circumference; MTC, mid-thigh circumference; AKC, above knee circumference; MKC, max knee circumference; BKC, below knee circumference; MCC, max calf circumference; MAC, min ankle circumference; LL, leg length; R, multiple correlation; SEE, standard error of estimate.

3.2. Cross-validation of derived equations

Cross-validation statistics of the prediction equations are given in table 4. Mean differences between measured leg FFM (12.4 ± 1.5 kg) and predicted leg FFM from the equations (12.1 ± 1.2–12.4 ± 1.3 kg) were not significant ($p > 0.05$). The smallest mean difference was −0.05 kg (min ankle, equation 7), the largest 0.26 kg (max knee and below knee, equations 4 and 5); these were equivalent to −0.4 and 2.1% respectively of the measured leg FFM. The standard deviations of the predicted samples were generally smaller than that of the measured sample, but not significant ($p > 0.05$). The SEE ranged from 0.652 to 0.937 and the total error 0.714–0.965. Correlations between measured and predicted leg FFM were reasonably high, ranging from 0.79 to 0.90 ($p < 0.001$). The 'ratio limits of agreement' express a

Table 4. Cross-validation statistics of derived regression equations for the prediction of leg FFM.

Predictor variables	Predicted leg FFM (kg)	Mean diff (kg)†	SEE	Total error	Corr coeff‡	Ratio limits of agreement
1. LL+GFC	12.1±1.2	0.219	0.917	0.915	0.80	1.02 (*/÷1.15)
2. LL+MTC	12.2±1.4	0.192	0.713	0.714	0.88	1.02 (*/÷1.12)
3. LL+AKC	12.2±1.4	0.146	0.880	0.871	0.82	1.01 (*/÷1.15)
4. LL+MKC	12.1±1.3	0.260	0.937	0.948	0.79	1.02 (*/÷1.16)
5. LL+BKC	12.1±1.6	0.259	0.901	0.965	0.81	1.02 (*/÷1.16)
6. LL+MCC	12.2±2.1	0.173	0.652	0.931	0.90	1.01 (*/÷1.16)
7. LL+MAC	12.4±1.3	−0.049	0.831	0.804	0.84	1.00 (*/÷1.14)
Measured leg FFM	12.4±1.5					

GFC, gluteal furrow circumference; MTC, mid thigh circumference; AKC, above knee circumference; MKC, max knee circumference; BKC, below knee circumference; MCC, max calf circumference; MAC, min ankle circumference; LL, leg length; SEE, standard error of estimate. †mean differences between measured and predicted leg FFM not sig ($p>0.05$) ‡all correlations sig ($p<0.001$).

measure of bias together with the level of agreement. The results confirmed that there was little bias between measured and predicted leg FFM (1.0–1.02) and a good level of agreement (1.12–1.16).

4. Discussion

Using cross-sectional standards (Freeman et al. 1995) the mean stature (182.0±5.9 cm) for the rugby group as a whole ($n=30$) was located at the 75th centile and the mean body mass (89.8±12.8 kg) at the 98th centile; validation and prediction samples (table 1) were similarly placed. Since DXA measurements of BMM, FM and LTM are made independently, there is a direct source of measurement error; this is reflected in the difference between DXA and scale mass. In the validation sample the error amounted to 0.8 kg (0.9%), and in the prediction sample to 1.2 kg (1.3%) (table 1). Relative fat averaged ∼15% and FFM 85%; these are reasonably normal for this level and type of athlete, but much depends on the individual playing requirement (Bell 1979).

Anthropometric data and tissue composition of the left leg are presented in table 2. Most studies estimating LLV do not present their anthropometric data, only the derived LLV. As might be expected, rugby players have considerably larger LLV (11.6 and 12.0 litres) than normal males (7.4 litres) (Winter 1991). This difference highlights the importance of the lower limbs in Rugby Union, especially in those activities involving heavy use of the legs (e.g. scrummages).

There is no single definitive measure of leg length. Subischial length (stature less sitting height) is a frequently used measure, but others can also be employed. In the anthropometric method of Jones and Pearson (1969), leg length is taken as the height from the ground to the gluteal furrow; this is the measurement that has been used in the present study. In DXA, however, the leg is defined as the mass of tissue distal to an oblique line passing through the neck of the femur; it differs from the anthropometric model in that it includes a section of the gluteal muscles.

As far as the present authors are aware, there are no published studies that have described the directly measured tissue composition of the lower limbs in rugby players. Comparisons were therefore made with normal males. Leg mass in validation and prediction groups (table 2) were slightly larger (14.9 and 15.4 kg) than normal males (13.9 kg) (Fuller et al. 1992), although this may be a reflection simply of absolute size. As a proportion of leg mass, the validation group had a larger %FM, %LTM and %BMM than normal males, whereas the prediction sample had values similar to normal males.

Ideally, it would have been desirable to have larger sample sizes for both prediction and cross-validation groups. Vincent (1995) recommended that the ratio of subjects to independent variables in multiple regression should be >5:1 and preferably 20:1. The present ratio is at the lower end of this range. The SEE of the prediction equations ranged from 0.7262 for maximum knee circumference to 1.0190 for the gluteal fold, the corresponding R being 0.83–0.62 (table 3). The best fitting equations nominated circumferences from the lower rather than the upper segment of the leg. Clarys and Marfell-Jones (1994), likewise, found the best-fitting equation for the prediction of leg muscle mass to incorporate the tibiale–malleolare length and the corrected maximum calf circumference ($R^2 = 0.97$). Lower limb muscle mass has also been predicted using a theoretical model of body composition based on DXA (Fuller et al. 1992). Estimates of limb muscle mass were made from regression equations using total limb mass, limb skin mass, limb ash mass (taken as DXA bone mineral mass) and limb fat mass. In solving the equations some assumptions were made regarding the amount of fat in muscle, skin, adipose tissue and skeleton. Errors associated with the assumptions in predicting limb muscle mass ranged from 0.4% for fat to 4% for hydration and limb ash.

The accuracy of a prediction equation is naturally compromised when applied to other samples. Two of the factors not yet considered in this respect (Guo and Chumlea 1996) are the validity of the response variable and the relationships between predictor and response variables. Dual-energy X-ray absorptiometry is generally regarded as a precise method for the assessment of body composition, although little is known about the validity of the segmental tissue masses (Roubenoff et al. 1993, Lohman 1996). Leg length and leg circumferences are both suitable biological predictor variables; leg length is the size variable and limb circumferences indices of cross-sectional areas of FFM. The correlation between leg length and leg FFM was 0.56 ($p < 0.001$) and between leg length and leg circumferences 0.63–0.72 ($p < 0.001$).

The cross-validation statistics (table 4) showed good agreement between leg FFM estimated using the prediction equations and the measured leg FFM. The largest mean difference estimated mean leg FFM to within 2.1%. The total error of a prediction equation combines the SEE and the mean difference between predicted and measured leg FFM. Apart from equation (6), which has a small SEE (0.625), the SEE and total error agree reasonably well. Application of the equations to individuals in the cross-validation sample gave rise to mean errors ranging from 5 to 6.8% in two-thirds of cases.

Agreement between measured and predicted leg FFM can be expressed as the 'ratio limits of agreement' (Nevill and Atkinson 1997). The advantage of this procedure is that it facilitates a comparison of bias and agreement with other studies. Correlations between the mean difference of measured and predicted leg FFM, and the mean of both values, were positive ($r = 0.19$–0.42), the majority being larger than the equivalent correlations using log-transformed data; thus procedures were continued using log-transformed data. The first ratio of the 'limits of agreement' (table 4) expresses the relationship between measured and predicted leg FFM (bias), the second, the level of agreement between the two variables.

Equation (2) (LL and mid-thigh circumference) shows the best measure of agreement (1.12) with little bias (1.02), consequently 95% of ratios will be constrained within the range 0.91–1.14. In fact, there is very little to choose between the seven equations; all have good agreement ratios (1.12–1.16) with very little bias (1.00–1.02). External validation of the prediction equations using an independent sample of players of similar playing ability and constitution, would confirm their credibility.

In conclusion, seven regression equations were formulated to predict leg FFM in young adult male Rugby Union players. All equations were satisfactorily cross-validated. Because prediction equations tend to be age, gender and population specific, unless validated for use with other athletic groups, the present equations should be applied to Rugby Union players with characteristics similar to those described. It is prudent to remember that when a predictive equation is applied to individuals the associated errors will be larger than those for groups.

References

BELL, W. 1979, Body composition of rugby union football players, *British Journal of Sports Medicine*, **13**, 19–23.

BELL, W., DAVIES, J. S., EVANS, W. D. and SCANLON, M. F. 1995, The validity of estimating total body fat and fat-free mass from skinfold thickness in adults with growth hormone deficiency, *Journal of Clinical Endocrinology and Metabolism*, **80**, 630–636.

CAMERON, N. 1984, *The Measurement of Human Growth* (London: Croom Helm).

CLARYS, J. P. and MARFELL-JONES, M. J. 1986a, Anthropometric prediction of component tissue masses in the minor limb segments of the human body, *Human Biology*, **5**, 761–769.

CLARYS, J. P. and MARFELL-JONES, M. J. 1986b, Anatomical segmentation in humans and the prediction of segmental masses from intra-segmental anthropometry, *Human Biology*, **58**, 771–782.

CLARYS, J. P. and MARFELL-JONES, M. J. 1994, Soft tissue segmentation of the body and fractionation of the upper and lower limbs, *Ergonomics*, **37**, 217–229.

FREEMAN, J. V., COLE, T. J., CHINN, S., JONES, P. R. M., WHITE, E. M. and PREECE, M. A. 1995, Cross-sectional stature and weight reference curves for the UK, 1990, *Archives of Disease in Childhood*, **73**, 17–24.

FULLER, N. J., LASKEY, M. A. and ELIA, M. 1992, Assessment of the composition of the major body regions by dual-energy X-ray absorptiometry (DEXA) with special reference to limb muscle mass, *Clinical Physiology*, **12**, 253–266.

GUO, S. S. and CHUMLEA, W. C. 1996, Statistical methods for the development and testing of predictive equations, in A. F. Roche, S. B. Heymsfield and T. G. Lohman (eds), *Human Body Composition* (Champaign: Human Kinetics), 191–202.

HERD, R. J. M., BLAKE, G. M., PARKER, J. C., RYAN, P. J. and FOGLEMAN, I. 1993, Total body studies in normal British women using dual-energy X-ray absorptiometry, *British Journal of Radiology*, **66**, 303–308.

HEYMSFIELD, S. B., SMITH, R., AULET, M., BENSEN, B., LICHTMAN, S., WANG, J. and PIERSON, R. N. 1990, Appendicular skeletal muscle mass: measurement by dual-photon absorptiometry, *American Journal of Clinical Nutrition*, **52**, 214–218.

JONES, P. R. M. and PEARSON, J. 1969, Anthropometric determination of leg fat and muscle plus bone volumes in young male and female adults, *Proceedings of the Physiological Society*, **204**, 63–66P.

KATCH, V. L. and KATCH, F. I. 1974, A simple anthropometric method for calculating segmental leg limb volume, *Research Quarterly*, **45**, 211–214.

KATCH, F. I. and KATCH, V. L. 1980, Measurement and prediction errors in body composition assessment and the search for the perfect prediction equation, *Research Quarterly for Exercise and Sport*, **51**, 249–260.

KATCH, V. and WELTMAN, A. 1975, Predictability of body segment volumes in living subjects, *Human Biology*, **47**, 203–218.

KATCH, V., MICHAEL, E. D. and AMUCHIE, F. A. 1973, The use of body weight and girth measurements in predicting segmental leg volume of females, *Human Biology*, **45**, 293–303.

KOHRT, W. M. 1995, Body composition by DXA: tried and true? *Medicine and Science in Sports and Exercise*, **10**, 1349–1353.

LEWIS, M. K., BLAKE, G. M. and FOGELMAN, I. 1994, Patient dose in dual x-ray absorptiometry, *Osteoporosis International*, **4**, 11–15.

LOHMAN, T. G. 1981, Skinfolds and body density and their relation to body fatness: a review, *Human Biology*, **53**, 181–225.

LOHMAN, T. G. 1996, Dual-energy X-ray absorptiometry, in A. F. Roche, S. B. Heymsfield and T. G. Lohman (eds), *Human Body Composition* (Champaign: Human Kinetics), 63–78.

MINITAB, Inc. 1995, *MINITAB Reference Manual* (Philadelphia: State College).

NEVILL, A. M. and ATKINSON, G. 1997, Assessing agreement between measurements recorded on a ratio scale in sports medicine and sports science, *British Journal of Sports Medicine*, **31**, 314–318.

ROUBENOFF, R., KEHAYIAS, J. J., DAWSON-HUGHES, B. and HEYMSFIELD, S. B. 1993, Use of dual-energy X-ray absorptiometry in body composition studies: not yet a 'gold standard', *American Journal of Clinical Nutrition*, **58**, 589–591.

SINNING, W. E. 1996, Body composition in athletes, in A. F. Roche, S. B. Heymsfield and T. G. Lohman (eds), *Human Body Composition* (Champaign: Human Kinetics), 257–273.

TAAFFE, D. R., LEWIS, B. and MARCUS, R. 1994, Regional fat distribution by dual-energy X-ray absorptiometry: comparison with anthropometry and application in a clinical trial of growth hormone and exercise, *Clinical Science*, **87**, 581–586.

VINCENT, W. J. 1995, *Statistics in Kinesiology* (Champaign: Human Kinetics).

WEINER, J. S. and LOURIE, J. A. 1981, *Human Biology: A Guide to Field Methods* (London: Academic Press).

WINTER, E. M. 1991, Maximal exercise performance and lean leg volume in men and women, *Journal of Sports Sciences*, **9**, 3–13.

26 Variations of anatomical elements contributing to subtalar joint stability: intrinsic risk factors for post-traumatic lateral instability of the ankle?

E. BARBAIX*, P. VAN ROY and JAN PIETER CLARYS

Vrije Universiteit Brussel, Experimental Anatomy, Laarbeeklaan 103, B-1090 Brussels, Belgium

Keywords: Ankle; Instability; Sinus tarsi; Anatomical variation; Subtalar joints.

Ankle sprains are frequently followed by chronic lateral instability, often with talar hypermobility. This might be due to subtalar instability. Among intrinsic risk factors, anatomical variants are generally overlooked. In the subtalar region, anatomical variation is particularly frequent. On the talus as well as on the calcaneus, the anterior articular facets may be missing or fused with the medial facets, giving rise to three subtalar joint configurations: a three-joint configuration, a fused configuration with a relatively large anteromedial joint, and a two-joint configuration without anterior joint. Osteometry was performed on these joint facets (134 calcanei, 122 tali), demonstrating significant differences in the surface of these configurations and the existence of a supplementary supporting surface with grossly transverse orientation in the three-joint configuration. There are also several variants of stabilizing ligaments within the sinus tarsi. Some of these configurations might expose to increased risk of associated subtalar lesions, resulting in subtalar instability. A systematic look for these variants is recommended in order to evaluate the associated risk factors, eventually resulting in a better understanding, prevention and cure of sequellae.

1. Introduction

Residual complaints, mainly of lateral instability, appear in 11–33% of patients after lateral ankle sprain (Löfvenberg and Kärrholm 1993). Frelman (1965) drew attention to the fact that the frequency of these late complaints after severe ligamentous ruptures and after simple sprains is similar and that many patients complain of functional instability even in the absence of pathological talar tilt on stress radiographs. Complaints about instability are partly due to factors other than hypermobility of the talus within the tibiofibular mortise. A difficult to diagnose hypermobility of the subtalar joints due to lesions of the talocalcaneal ligaments is one of the possible causes. In an *in vitro* study, Kjaersgaard-Anderssen *et al.* (1988) demonstrated a moderate but significant increase of subtalar mobility after section of ligaments in the sinus tarsi.

According to Lysens *et al.* (1984), intrinsic risk factors for ankle sprain include age, sex, range of motion, general hypermobility among others. Anatomical variation is seldom considered as an intrinsic factor.

*Author for correspondence. e-mail: erik.barbaix@skynet.be

Considerable anatomical variation of the articular facets of the subtalar joints has been documented for both talus and calcaneus (Bunning and Barnett 1963, El-Eishi 1974, Gupta et al. 1977, Arora et al. 1979, Bruckner 1987, Forriol-Campos and Pellico 1988). The variation in number, shape, orientation and dimensions of the articular surfaces concerns not only accessory surfaces like the facies externae accessoriae whose frequency lies between 4.0 and 8.5% (Sarrafian 1993). Considerable variation also exists in the principal articular facets. Classically the subtalar joints are formed by three articular facets on the calcaneus supporting three corresponding facets on the plantar aspect of the talus. The posterior pair of facets gives rise to the posterior subtalar joint, the other pairs, called medial and anterior, participate in the talocalcaneonavicular joint. The sinus and canalis tarsi separate these two joints.

There are several anatomical variations from this classical model, which have been described and classified in different ways by the authors mentioned above. These variations concern mainly the medial and anterior facets; the latter is sometimes missing or it can be fused to varying extents with the medial facet.

A large project was initiated to study the possible role of anatomical variations of the configurations of the subtalar joints and ligaments on the outcome of ankle trauma. As a first step the relative frequencies of these variants were studied on dried bones and on cadaver material. Selected aspects with possible implications for stability like surface areas and orientation of facets and ligaments were quantified. Correlation between osseous and ligamentous variants were investigated.

As a next step the feasibility of identification of these variants through medical imaging techniques were evaluated. Hopefully this could lead to the validation of diagnostic procedures and to more adequate identification of structures at risk in trauma. In this communication variations of joint facets and the first observations on ligaments are addressed.

2. Material and methods

In an ongoing study on dried cadaver bones (134 calcanei, 122 tali) the configuration was identified using a classification:

- Type A: three facets:
    - A1: three distinct facets.
    - A2: confluence of medial and anterior facets, which remain clearly distinct through the presence of a ridge between them.
- Type B: two facets:
    - B1: fusion of medial and anterior facets with a narrowing subsisting at the site of fusion.
    - B2: complete fusion.
    - B3: anterior facet is missing.
- Type C: fusion of all three facets.

Type C has been described by Bunnings and Barnett (1965), Gupta et al. (1977), Arora et al. (1979) and by Forriol-Campos and Pellico (1988), but it has never been observed in Europeans. As it was not observed in this study either, it will not be mentioned further. Afterwards the A and B types were regrouped into three subgroups:

- classical three-facet (3F) configuration represented by type A1;
- fusion-configuration (Fus) represented by types A2, B1 and B2; and
- configuration with missing anterior facet (MAF) represented by type B3.

On all available calcanei, length and breadth were measured according to Martin and Saller (1957) as well as the surfaces of the different facets according to Bruckner (1987). The edges of the facets were marked with a charcoal pencil and then a non-elastic adhesive tape was firmly applied. On withdrawal of the tape the charcoal remaining on the tape reproduced the dimensions of the joint facets, which were then transposed on paper and scanned together with fragments of mm-paper. The scanned images were introduced in a graphic computer program (Image.93) where the surface areas and the accompanying mm-paper were measured in pixels. The results were transformed into cm^2 by dividing them by the number of pixels measured in 1 cm^2. Each measurement was performed three times and the mean of three results was used for statistical analysis. Two-tailed uncorrelated t-tests compared surface areas in different configurations. A significance level of $p=0.05$ was chosen.

On the scanned images the long axes of anterior (AF) and medial facets (MF) were determined by choosing a starting point (PS) near one extremity of the facet

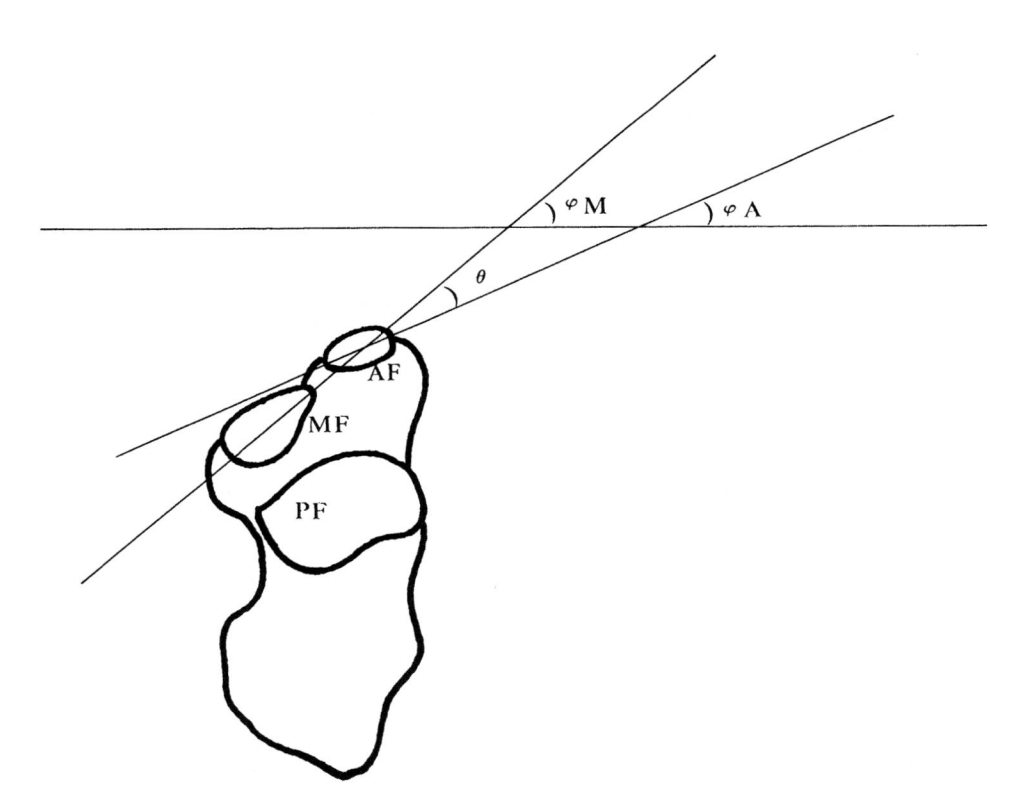

Figure 1. Orientation of the anterior and medial facets of calcaneus in a three-facet configuration. AF, anterior facet; MF, medial facet; PF, posterior facet; φA, orientation of long axis of AF; φM, orientation of long axis of anterior facet; θ, angle between MF and AF.

and looking for the most distant point (DP) on the opposite extremity. Starting from DP, the procedure was repeated towards the first extremity. After three-to-five repetitions starting from the successive DPs, the final DP coincided with the previous one. The long axis was defined as the segment joining the last two DPs. The angle between the long axis of a facet and the abscissa of the screen was measured in the same program using X- and Y- coordinates of both DPs: $\varphi = \text{ATAN}(\Delta Y \Delta X)$. The angle between two facets was calculated by subtraction $\theta = \varphi M - \varphi A$ (figure 1).

3. Results

3.1. Configurations

Table 1 provides data for the relative frequency of different configurations of subtalar facets on both ankle bones. Type 3F was dominant in the calcanei while fusion-configurations predominated in the tali.

3.2. Orientation of facets

On the available calcanei the long diameters of the roughly oval medial and anterior facets were not aligned. Between them an angle of $29° \pm 13°$ open in a lateral and distal direction was found, the long axis of the medial facet having a more longitudinal, the anterior one a more transverse direction. In the event of fusion of these facets, no such angle was found between the two components.

3.3. Surface areas (table 2)

Significant differences were found between the total articular surface of the MAF type calcanei (TMAF) and the total articular surfaces of the 3F- (T3F) or Fus-type (TFus) configurations ($p < 0.05$). The surface of fused anterior and medial facets (Fus) was significantly larger than the sum of the surfaces of the separated facets in type 3F (AF + MF). In this study, the difference between the total articular surfaces of 3F- and Fus-type calcanei (T3F versus Tfus) was not statistically significant ($p > 0.05$).

As the spread of individual values was very large (1 cm^2 for AF and 1.5 cm^2 for MF for mean surface areas of 0.78 and 1.56 cm^2 respectively) a 'surface/dimension index' (SDI) was calculated in an attempt to reduce the effect of the differences between small and large calcanei. This SDI can be considered an approximation of the articular surface relative to the dimension of the upper surface of the bone (table 3):

$$\text{SDI} = 100 \times \text{surface}(\text{mm}^2)/\text{length} \times \text{breadth}(\text{mm}). \tag{1}$$

Table 1. Relative frequency (%) of different configurations of subtalar facets on calcaneus and talus.

Calcanei ($n = 134$)	Tali ($n = 122$)
Type 3F = 61%	Type 3F = 39%
Fus type = 28%	Fus type = 49%
A2 = 3%	A2 = 22%
B1 = 11%	B1 = 6%
B2 = 14%	B2 = 21%
MAF type = 11%	MAF type = 11%

Table 2. Calcanei: surface area (cm^2) of anterior facets (AF), medial facets (MF), AF+MF in 3F configuration, fused AF and MF (Fus) in fused-facet configuration, posterior facets (PF) total surface of three facets in type 3F (T3F), in fusion types A2, B1 and B2 (TFus) and in configuration with mising anterior facet (TMAF). Number of calcanei (n), mean (M), standard deviation (SD), standard error of the mean (SEM).

	n	M	SD	SEM
AF	82	0.78	0.24	0.03
MF	96	1.56	0.38	0.04
AF+MF	82	2.39	0.48	0.05
Fus	38	2.7	0.54	0.09
PF	134	5.82	1.03	0.09
TA1	82	8.23	1.28	0.14
TF	38	8.42	1.40	0.24
TB3	14	7.04	1.25	0.36

Table 3. Calcanei: surface-dimension index for total articular surface of type 3F (T3F), for fusion types (TFus), for type MAF (TMAF), for fused AF and MF (Fus) and for AF+MF in type 3F.

	n	M	SD	SEM
T3F	82	23.30	2.41	0.27
TFus	38	25.23	2.71	0.45
TMAF	14	21.08	2.73	0.79
Fus	38	8.30	1.79	0.30
AF+MF	82	6.62	0.92	0.10

Using the SDI the significant differences of the total articular surface areas between the MAF-type and both other configurations were confirmed, as was the difference between the surface area of the fused facet and the sum of its separate counterparts in the 3F-configuration ($p<0.01$). Still using this SDI the difference in the total articular surface areas between 3F- and fused configurations was also significant ($p<0.05$). The fused surfaces occupied a larger fraction of the dimensions of the cranial surface of the calcaneus than did their separate counterparts.

4. Discussion

The frequencies of the different configurations found correspond well with the findings of Bunning and Barnett (1963) in their European subgroup. Forriol-Campos and Pellico (1988) and Bruckner (1987) found a lower frequency for type A and a higher frequency for type B, but part of their type B correspond to the present type A2. Bruckner studied 32 calcanei and 28 tali only; this lower number can induce considerable shifts in the frequency distribution.

On the calcanei MF and AF are 'normally' situated on two distinct apophyses, the sustentaculum tali and the anterior beak (rostrum) respectively. On the talus, both are always situated on the caput tali. As this caput tali is the convex partner of the joint situated within the concavity of the corresponding facets on the calcaneus, talar MF and AF are much closer to each other and hence are more frequently fused together. Among the available bones, there were 30 pairs of corresponding tali and calcanei. Whereas the correspondence was perfect between type B tali and calcanei, this was not the case for type A configuration: in seven pairs the calcaneus was type A1 but the corresponding talus was type A2 (fused with separating ridge).

It seems logical that articular facets forming an angle of $\pm 30°$ between them and with one of them transversely oriented, offer more stability, particularly against inversion than one single surface with only one more longitudinal long axis. This finding suggests the configuration with three facets to be more stable than those with two facets, but considering the great range of angles between the two facets (from 0° to 58°), even in the three-facet configuration some calcanei offer considerably more stability than others.

The total articular surface area was smaller when the anterior facet was missing (MAF type). The findings do not entirely confirm the conclusion of Bruckner (1987) who reported that configurations with fused anterior facets have a significantly larger total articular surface than those with three facets. Although the surface of the fused facet is smaller than the sum of the surfaces of its separate counterparts, the difference between the total surfaces of 3F- and Fus-configurations is only significant when the relative surface is considered. The larger the articular surface, the lower the pressure on articular cartilage from the same load. Whether this results in more degenerative damage in MAF and less in fused configurations is not known.

Anatomical variations of the ligaments of sinus and canalis tarsi were described by Smith (1958), Cahil (1965) and Clanton (1989). The classical description by Smith (1958) included ligaments arranged in three consecutive, approximately frontal layers. In the posterior layer Smith described the medially situated interosseous ligament of the tarsal canal. This ligament covers the medial two-third s of the sinus tarsi. It is arranged in grossly parallel fibres originating from the floor of the sinus and canalis tarsi, running in a cranial and medial direction to insert in the sulcus tali on the roof of the canal. In the intermediate layer Smith described a grossly triangular ligament with the top on deep layer of the extensor retinaculum on the dorsum of the foot, running round the talar neck to which it is often attached and spreading out towards the floor of the sinus tarsi to insert with three pillars. In the anterior layer he described the cervical ligament, originating from the floor of the sinus tarsi just medially to the origin of the M. extensor digitorum brevis, and inserting on the talar neck.

In an ongoing series of dissections of 26 embalmed feet a few major variations were observed, some already described by Smith (1958). In three feet a separate interosseous ligament was found, between the classical posterior and intermediate layers. It is a strong vertically oriented ligament situated at the transition from the large lateral and the canalicular medial parts of the sinus tarsi. In one case the medial insertion of the retinaculum was found to extend laterally into the canal, apparently replacing the absent interosseous ligament of the tarsal canal. In three feet the lateral pillar of the intermediate layer was largely separated from the rest of the ligament by a vertical hiatus. In one specimen the ligament of the canal was situated in the anterior layer and was continuous with the cervical ligament. In six of nine feet with a 3F configuration, a synovial fold was observed coming from between AF and MF and intruding into the talocalcaneonavicular joint. In some cases of the three-facet configuration, a fold of joint capsule separated the anterior joints. If it is strong enough, this capsular separation could add to the stability of the joint by limiting its range of motion. If it is just a thin synovial fold, it could be a structure at risk.

As the present study is still in progress, these descriptions should not be considered the final conclusion with regard to lateral instability of the ankle joint. They illustrate the occurrence of variations of the ligaments and capsules between talus and calcaneus. It is not clear at this moment whether or not these variations are related to specific configurations of the articular facets. The variations of joint capsule probably are. As the origin of the spring ligament is situated on the anterior aspect of the calcaneus exactly below and between the anterior and medial facets of the calcaneus, this structure should be studied in more detail in different calcaneal configurations.

Bruckner (1987) considered a larger articular surface to lead to greater mobility. Although this interpretation is questionable, in the particular case of fusion of the anterior and medial facets this hypothesis might be correct if one considers that in these configurations there is no capsule between the two components as there is in configurations with two distinct facets. At this time this should only be considered a hypothesis which needs confirmation by a detailed study of the relationship of the capsules and the surrounding structures.

5. Conclusions

The articular facets as well as the ligaments of the subtalar joint complex present several major anatomical variations. The anterior facet is missing in $\sim 10\%$ of the population. It is fused with the medial facet in $\sim 30\%$. A fold of joint capsule often separates the medial from the anterior facet in the three-facet configuration, possibly limiting the range of motion for inversion and eversion. In this configuration the more transverse orientation of the anterior facet also suggests more stability in inversion of the calcaneus. Some of these variants probably have also an impact on the position of the axis of movements between talus and calcaneus, resulting in different positions relative to load and ground reaction forces, and hence in more or less inversion momentum.

Anatomical variation might well be an overlooked intrinsic risk factor for chronic post-traumatic lateral instability of the ankle and of the subtalar joint in particular. They are not by themselves the primary explanation for complaints like pain or instability. These anatomical variants exist from the earliest days of life without causing any trouble. At the start of complaints a detailed analysis is needed to explain why the pre-existing variants have been traumatized.

It is suggested that the three-facet configuration offers more stability due to the more transverse orientation of its anterior facet and to the existence of a capsular fold within the talocancaneonavicular joint restricting the range of motion of inversion and eversion. As its total articular surface relative to the dimensions of the bone is larger, the pressure on the cartilage is probably lower than in other configurations. This fundamental anatomical research may provide insight into the understanding of ankle sprains in sports and certain occupational settings.

References

ARORA, A. K., GUPTA, S. C., GUPTA, C. D. and JEYASINGH, P. 1979, Variations in calcanean facets in Indian tali, *Anatomischer Anzeiger*, **146**, 377–380.

BRUCKNER, J. S. 1987, Variations in the human subtalar joint, *Journal of Orthopaedic and Sports Physical Therapy*, **8**, 489–494.

BUNNING, P. S. C. and BARNETT, C. H. 1963, Variations in the talocalcanean articulations, *Journal of Anatomy*, **97**, 643–648.

Bunning, P. S. C. and Barnett, C. H. 1965, A comparison of adult and foetal talocalcaneal articulations, *Journal of Anatomy*, **99**, 71–76.

Cahil, D. R. 1965, The anatomy and function of the contents of the human tarsal sinus and canal, *Anatomical Record*, **153**, 1–11.

Clanton, T. O. 1989, Instability of the subtalar joint, *Orthopaedic Clinics of North America*, **20**, 583–592.

El-Eishi, H. 1974, Variations in talar articular facets in Egyptian calcanei, *Acta Anatomica* (Basel), **89**, 134–138.

Frelman, M. A. R. 1965, Instability of the foot after injuries to the lateral ligament of the ankle, *Journal of Bone and Joint Surgery*, **47B**, 669–677.

Forriol-Campos, F. and Pellico, G. L. 1989, Talar articular facets in human calcanei, *Acta Anatomica* (Basel), **134**, 124–127.

Gupta, S. C., Gupta, C. D. and Arora, A. K. 1977, Pattern of talar articular facets in Indian calcanei, *Journal of Anatomy*, **124**, 651–655.

Kjaersgaard-Anderssen, P., Wethelund Journalo, Helmig, P. and Soballe, K. 1988, The stabilizing effect of the ligamentous structures in the sinus and canalis tarsi on movements in the hindfoot, *American Journal of Sports Medicine*, **16**, 512–516.

Löfvenberg, R. and Kärrholm, J. 1993, The influence of an ankle orthosis on the talar and calcaneal motions in chronic lateral instability of the ankle, *American Journal of Sports Medicine*, **21**, 224–230.

Lysens, R., Steverlynck, A. and van den Auweele, Y. 1984, The predictability of sports injuries, *Sports Medicine*, **1**, 6–10.

Martin, R. and Saller, K. 1957, *Lehrbuch der Anthropologie, 2 vols* (Stuttgart: Fischer).

Sarrafian, S. K. 1993, *Anatomy of the Foot and Ankle, 2nd edn* (Philadelphia: Lippincott).

Smith, W. 1958, The ligamentous structures in the canalis and sinus tarsi, *Journal of Anatomy*, **92**, 616–620.

27 Intra-articular kinematics of the normal glenohumeral joint in the late preparatory phase of throwing: Kaltenborn's rule revisited

J.-P. BAEYENS*, P. VAN ROY and JAN PIETER CLARYS

Vrije Universiteit Brussel (Belgium), Laarbeeklaan 103, 1090 Brussels, Belgium

Keywords: Glenohumeral joint; Intra-articular kinematics; Kaltenborn.

A new method to quantify intra-articular relationships between articular surfaces of the glenohumeral joint during discrete poses representing the late preparatory phase of throwing is presented. This method is based on 3D bone reconstructions from medical imaging data processed into finite helical axis parameters. With the shoulder moving in the anatomical planes from 90° abduction and 90° external rotation into the apprehension test pose, the centre of the humeral head posteriorly translated on the glenoid and rotated about a finite helical axis, which was positioned at the joint contact. The data are contrasted with Kaltenborn's convex–concave rule explaining intra-articular kinematics of the glenohumeral joint as a ball-and-socket joint. The data show at all conditions that the glenohumeral joint does not act as a ball-and-socket joint. Consequently, the mobilization techniques used in manual therapy, which are based on this convex–concave rule, should be adapted.

1. Introduction

MacConaill (1953) was one of the first to investigate intra-articular kinematics, based on osteokinematics and perceived movement of the articular surfaces. A specific arthrokinematic terminology was introduced based on the migration of contact of the articular surface of the distal (moving) segment on the articular surface of the proximal (reference) segment. Glide was defined as the tangential translation of the moving articular surface against the articular surface of the reference segment. Roll required the arc length of the articular surface of the moving segment to match the path on the articular surface of the reference segment so that the two surfaces have point to point contact. Spin occurred when the moving segment rotates but the contact point on the articular surface of the reference segment does not change.

Based on MacConaill's observations, Kaltenborn (1980) stated the convex–concave rule to describe intra-articular kinematics: 'when a concave articular surface moves on a convex articular surface, roll and glide occur in the same direction; when a convex surface moves on a concave surface, roll and glide occur in opposite directions'. This rule became a pivotal concept in manual therapy.

Trapped in a mechanistic approach of joint kinematics, the glenohumeral joint has always been considered as a ball-and-socket (or spinning) joint based on

*Author for correspondence. e-mail: j.baeyens@pi.be

morphological measurements which indicate that the radius of curvature of the glenoid cavity is only ~2 mm greater than the radius of the curvature of the humeral head, making the glenohumeral articular surfaces highly congruent (Soslowsky et al. 1992). With the idea of the glenohumeral joint as a ball-and-socket joint coupled to the concave–convex rule, mobilization techniques for the glenohumeral joint were introduced in which roll and glide occurred in the opposite direction. This concept has never been validated by intra-articular kinematics studies.

Study in vivo of the intra-articular kinematics of the glenohumeral joint encounters several methodological problems. First, 2D recordings of the glenohumeral motion are susceptible to systematic errors (de Groot 1999). Three-dimensional in vivo motion recordings of the shoulder girdle using markers attached to the skin are unreliable due to large bone to skin displacements (Cappozzo et al. 1996). Alternative, 3D methods have used Euler–Cardan rotations during shoulder elevation and have reliably recorded using roentgenstereophotogrammetry after implanting of radiopaque markers in the bone (Högfors et al. 1991), and with percutaneous pins (Harryman et al. 1992, Koh et al. 1998). Non-invasive methods have included 3D linkage systems to digitize the spatial coordinates of palpated skeletal landmarks (Pronk and van der Helm 1991, van der Helm and Pronk 1995, Ludewig et al. 1996), a goniometer facilitated by a scapular locating device (Pearl et al. 1992), the sensor of an electromagnetic tracking device mounted on a specially designed scapula locator (Johnson et al. 1993, MacQuade et al. 1995, MacQuade and Smidt 1998, Meskers et al. 1998), and a humeral inclinometer (MacQuade et al. 1995, MacQuade and Smidt 1998) or humeral receiver mounted on a circular cuff (Meskers et al. 1998). No translations were recorded by these methods but 3D reconstructions based on magnetic resonance imaging have been used in asymptomatic volunteers to quantify the translation of the centroid of the humeral head on the glenoid during internal and external rotation of the 0° abducted arm (Rhoad et al. 1998).

In these studies, the kinematic data were related to bone embedded coordinate systems. The considerable variability in the orientation of the glenoid surface with respect to the scapula (glenoid superior–inferior tilt and glenoid ante-/retroversion) (Matsen et al. 1998) necessitates the use of a coordinate system embedded on the articular surface of the glenoid to study the intra-articular kinematics of the glenohumeral joint. Coping with these methodological limitations, the intra-articular (dys)functions of the glenohumeral joint were studied in relation to the late preparatory phase of throwing, using helical CT data of discrete shoulder poses processed into finite helical axes. The finite helical axis parameters of rotation, shift and direction were related to an embedded coordinate system on the glenoid, whereas the position of the finite helical axis was related to the humeral head. For readers not familiar with the finite helical concept, see the appendix. The data of three normal glenohumeral joints are presented and their motion related to that predicted by Kaltenborn's rule.

2. Methods and materials

2.1. Step 1: Medical imaging data acquisition and 3D reconstruction

Three asymptomatic throwing shoulders of athletes were scanned by means of helical CT in three discrete shoulder poses. The display field of view was 25 cm, with a pixel size of 0.49 mm. Referred to the anatomical planes, the shoulder was set in 90°

abduction and variable external rotation (pose 1) (table 1, θ_{H12}). (Actually, the aim was to standardize pose 1 in 90° abduction and neutral rotation. However, the elbow had to be flexed in the gantry whereby the wrist became positioned over the shoulder joint. Subsequently, the shoulder had to be externally rotated to move the wrist out of the scanning beam, without the possibility to standardize this manoeuvre). Then, the shoulder was set in 90° abduction and 90° external rotation (pose 2). From a clinical point of view, the shoulder was put in the apprehension test pose with full external rotation, locked on an individual basis (pose 3). The medical imaging data on these three poses were used for 3D reconstruction of the bony configurations of the shoulder joint (figure 1).

Table 1. SD of the estimated helical axis parameters.

	Subject	σ_n (°)	θ_{H12} (°)	θ (°)	σ_θ (°)	t (mm)	σ_t (mm)	σ_s (mm)
GH_{12}	DB	1.44	35.92	30.32	0.53	3.75	0.22	0.41
	DH	0.56	72.42	67.39	0.44	−2.58	0.60	0.86
	JS	0.65	72.66	71.75	0.54	4.83	0.70	0.17
GH_{23}	DB	5.61		5.78	0.40	7.22	0.28	1.99
	DH	1.52		20.1	0.37	−12.86	0.40	0.53
	JS	3.67		8.1	0.36	5.38	0.33	1.23

GH_{ij}: glenohumeral, pose j related to pose i; θ_{H12}: amount of external rotation of the shoulder in the anatomical plane from pose 1 to pose 2; σ: SD of the estimated helical axis parameter **n** (direction vector), θ (rotation angle), t (shift) or **s** (position vector).

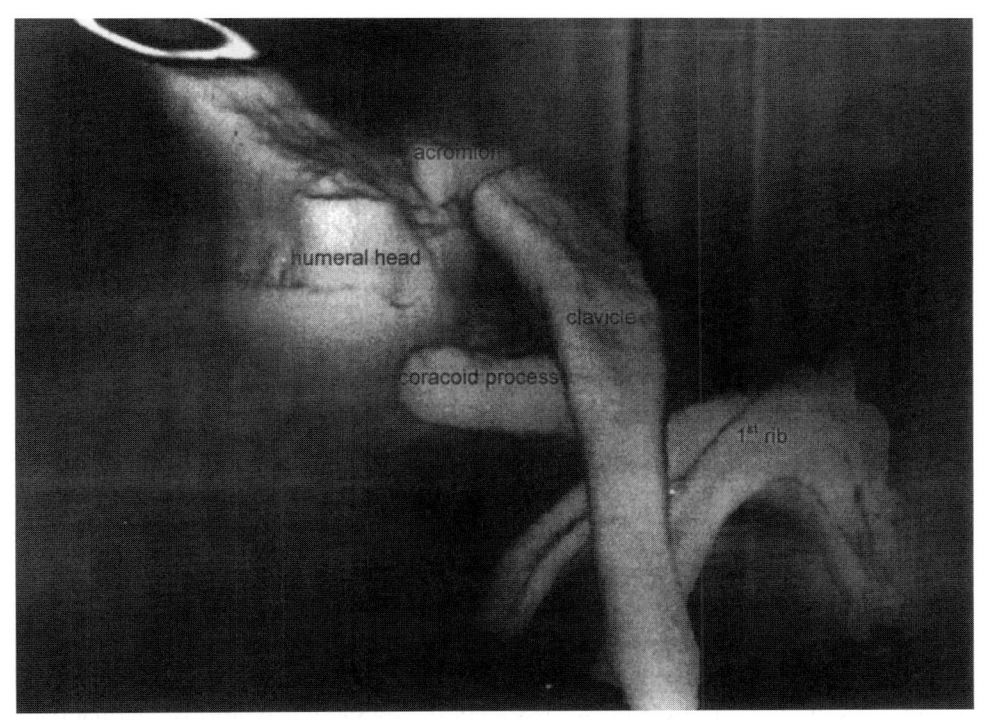

Figure 1. 3-D bone reconstruction of the shoulder based on helical CT data.

324   J.-P. Baeyens et al.

2.2. Step 2: Primary kinematic analysis

A series of bony landmarks was measured on the 3D reconstructions of these discrete shoulder poses. Rotation matrices and translation vectors were estimated using the approach of Veldpaus et al. (1988) and transformed into finite helical axis parameters (shift t, rotation angle θ, direction vector n, position vector s). Errors on the spatial parameters were estimated using Woltring and co-workers' (1985) model (table 1).

2.3. Step 3: Intra-articular kinematic analysis

After virtual disarticulation of the glenohumeral joint, a coordinate system was embedded on the glenoid cavity, built on three landmarks: i (superior), j (inferior) and k (anterior) at the edges of the glenoid (figure 2): with a superiorly directed unity vector $\mathbf{I_G}$, an anteriorly directed unity vector $\mathbf{J_G}$ and a laterally directed unity vector $\mathbf{K_G}$, and centred in the position vector $(\mathbf{i}+\mathbf{j})/2$ (figure 2).

Subsequently, the glenohumeral finite helical axis vectors $\boldsymbol{\theta}\ (=\theta.\mathbf{n})$ and $\mathbf{t}\ (=t.\mathbf{n})$ were decomposed on this coordinate system, with:

$$\mathbf{n} = n_{IG}\mathbf{I_G} + n_{JG}\mathbf{J_G} + n_{KG}\mathbf{K_G}$$
$$\boldsymbol{\theta} = \theta_{IG}\mathbf{I_G} + \theta_{JG}\mathbf{J_G} + \theta_{KG}\mathbf{K_G}$$
$$\mathbf{t} = T_{IG}\mathbf{I_G} + t_{JG}\mathbf{J_G} + t_{KG}\mathbf{K_G}.$$

Glenohumeral intra-articular horizontal abduction ($-$)/adduction ($+$) is researched by θ_{IG}, θ_{JG} indicating glenohumeral intra-articular abduction ($-$)/adduction ($+$), and θ_{KG} intra-articular glenohumeral external ($+$)/internal ($-$) rotation. The term t_{IG} refers to glenohumeral intra-articular superior ($+$)/inferior ($-$) translation, t_{JG} to glenohumeral intra-articular anterior ($+$)/posterior ($-$) translation, and t_{KG} to glenohumeral intra-articular lateral ($+$)/medial ($-$) translation. These glenoid-

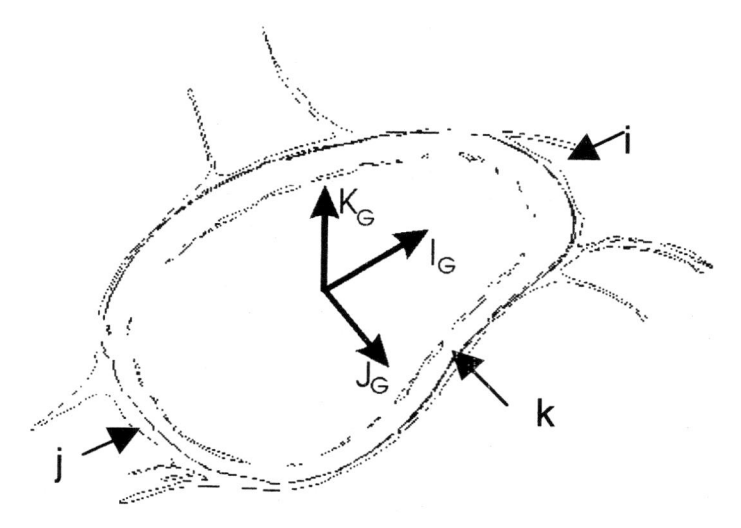

Figure 2.   Coordinate system embedded on the glenoid cavity. i, j, k, bony landmarks at the glenoid margin; $\mathbf{I_G}$, $\mathbf{J_G}$, $\mathbf{K_G}$, unit vectors of the coordinate system embedded on the glenoid with $\mathbf{I_G} = (\mathbf{m}-\mathbf{j})/|\mathbf{m}-\mathbf{j}|$, $\mathbf{J_G} = (\mathbf{k}-\mathbf{m})/|\mathbf{k}-\mathbf{m}|$ and $\mathbf{K_G} = \mathbf{J_G} \times \mathbf{I_G}$ in which $\mathbf{m}$ is the intersection of $\mathbf{i}-\mathbf{j}$ with its perpendicular through k.

referenced motions are not the same as motions that are referred to the traditional anatomical planes of the body.

The displacements of the articular surfaces at the contact area of the glenohumeral joint were explored graphically in the 'field of motion'. This field was defined as the set of planes, perpendicularly situated about the finite glenohumeral helical axis. Between the subsequent poses, corresponding profiles of the scapula at the contact area were first mapped (figure 3A). Then, the profile of the humeral head in pose i was related to its location in the subsequent pose i+1, taking care of the shift of the humeral head along the helical axis (figure 3B). The obtained centre of rotation was referred to a

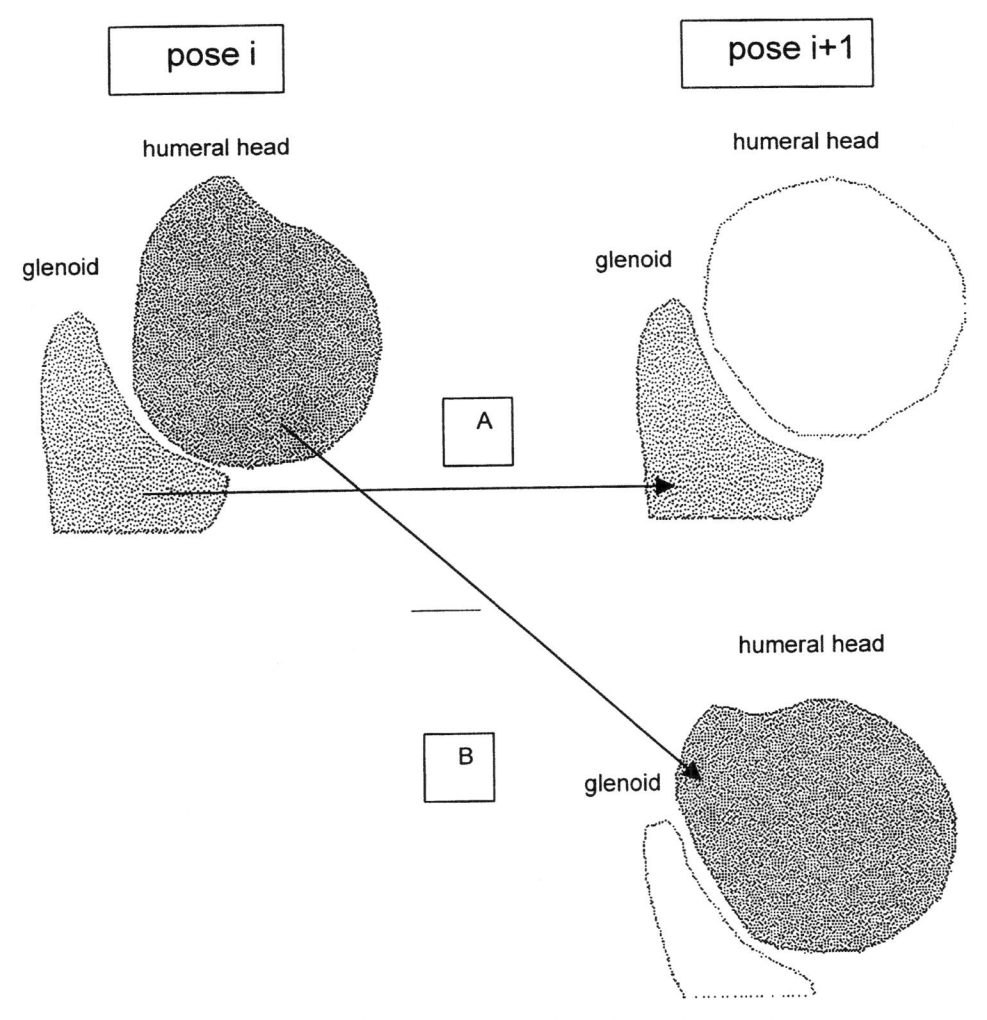

Figure 3. Corresponding profiles of the humeral head and the scapula between a pose i and the subsequent pose i+1. (A) Tracing and mapping of the corresponding profiles of the scapula. (B) Relating the orientation of the profile of the humeral head in pose i with its orientation in pose i+1, respecting the shift of the humeral head along the finite helical axis.

coordinate system, oriented with its *X*-axis parallel to the articular surface of the glenoid, and centred in the fitted curvature of the articular profile of the humeral head (figure 4).

3. Results

Although the shoulders differed in pose 1 with respect to the anatomical external rotation component (table 1), they all underwent spin from pose 1 to pose 2, with the finite helical axis centred in the humeral head. Intra-articularly, the glenohumeral motion from pose 1 to 2 was characterized by a comparable horizontal abduction component (θ_{IG}) with differing contributions of abduction (θ_{JG}) and external rotation (θ_{KG}) (table 2).

Intra-articularly, from pose 2 to 3, there was practically no external–internal rotation (n_{KG} $-0.06 \pm 0.11°$ with θ_{KG} having a magnitude of $0.71 \pm 0.68°$) (table 3). Intra-articular motion was merely restricted to the abduction–adduction plane (n_{JG}) with a smaller component of horizontal abduction (n_{IG}) (table 3). The finite helical axis was positioned on the $Y(-)$-axis against the glenoid cavity (figure 5).

All test cases were in pose 1 with an anteriorized position of the geometrical centre of the humeral head on the glenoid ($HH_{JG} = 6.91 \pm 0.40$ mm), translating into a posterior position on the glenoid ($HH_{JG} = -7.25 \pm 0.87$ mm) in pose 3 (table 4). HH_{JG} is the anterior–posterior position of the centre of the humeral head in the (I_G, J_G, K_G) coordinate system embedded on the glenoid cavity.

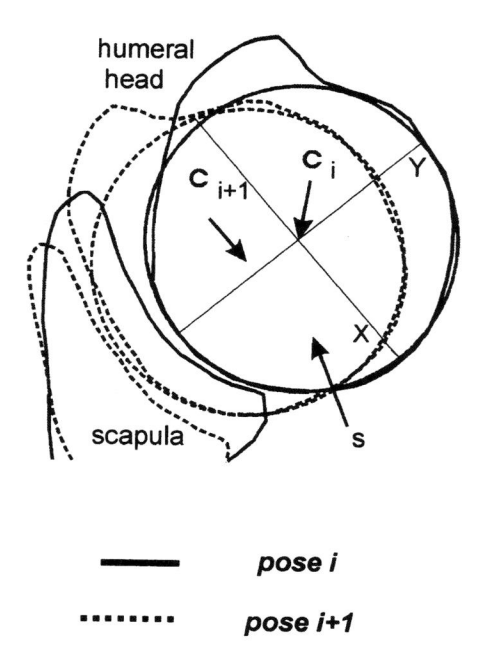

Figure 4. Position (s) of the glenohumeral finite helical axis from pose i to i+1. c, Centre of the fitted curvature of the articular profile of the humeral head. (*X*, *Y*) coordinate system with the *X*-axis oriented to the articular profile of the glenoid in pose i.

Table 2. Decomposed values of the direction vector **n** (n_{IG}, n_{JG}, n_{KG}) on the coordinate system [I_G, J_G, K_G] of the glenoid, and pitch t/θ of the glenohumeral finite helical axis from pose 1 to 2.

GH_{12}	n_{IG}	n_{JG}	n_{KG}	t/θ (mm/°)
DB	−0.51	−0.53	0.67	0.01
DH	−0.55	−0.61	0.58	−0.04
JS	−0.53	−0.83	0.13	0.07

GH_{12}, glenohumeral, pose 2 related to pose 1.

Table 3. Decomposed values of the direction vector **n** (n_{IG}, n_{JG}, n_{KG}) and the shift **t** (t_{IG}, t_{JG}, t_{KG}) on the coordinate system [I_G, J_G, K_G] of the glenoid, of the glenohumeral finite helical axis from pose 2 to 3, and the posterior pitch (t_{IG}/θ_{IG}) on the glenoid.

GH_{23}	n_{IG}	n_{JG}	n_{KG}	θ_{IG}	θ_{JG}	θ_{KG}	t_{IG} (mm)	t_{JG} (mm)	t_{KG} (mm)	t_{JG}/θ_{JG} (mm/°)
DB	−0.43	−0.90	0.05	−2.51	−5.2	0.27	−3.14	−6.50	0.34	−1.25
DH	−0.23	0.97	−0.01	−4.67	19.50	−0.18	3.00	−12.51	0.11	−0.64
JS	−0.13	−0.97	−0.21	−1.02	−7.82	−1.67	−0.68	−5.22	−1.12	−0.67

GH_{23}, glenohumeral, pose 3 related to pose 2.

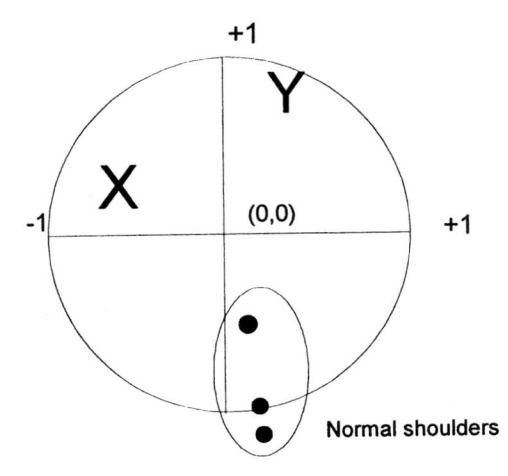

Figure 5. Position of the finite helical axis from pose 2 to 3, referred to the (X, Y) coordinate system related to the articular profile of the glenoid cavity and the centre (0,0) of the curvature of the articulating surface of the humeral head in pose 2.

4. Discussion

4.1. 3D intra-articular glenohumeral joint kinematics

In vitro research of glenohumeral joint kinematics assessed primarily on coupled motion, may be grouped into three categories: (1) when attempting to apply a pure glenohumeral rotation to a joint about one axis, rotations about the remaining two axes occurred automatically (Terry *et al.* 1991, Novotny *et al.* 1998); (2) when

Table 4. Anterior–posterior position of the centre of the humeral head (HH$_{JG}$) in the co-ordinate system (I$_G$, J$_G$, K$_G$) of the glenoid cavity.

	HH$_{JG}$	
Subject	Pose 1	Pose 3
DB	6.47	−7.62
DH	7.01	−8.51
JS	7.26	−6.42

generating a glenohumeral rotation about a single axis, additional torques were required in the remaining axes (Harryman et al. 1990); or (3) anteriorly induced translation produced spontaneous internal rotation, while posteriorly induced translation produced spontaneous external rotation (Gohlke et al. 1994). Rotations of the glenohumeral joint are accompanied by reproducible translations of the humeral head on the glenoid (Clark et al. 1990, Harryman et al. 1990, Novotny et al. 1998). Coupled translation has also been recorded in dynamic cadaveric shoulder models (Thompson et al. 1996, Wuelker et al. 1998), but may be dependent on input of muscle forces. Unconstrained coupled translations and rotations of the glenohumeral joint in the primary planes of motion have been investigated by Novotny et al. (1998), starting from a reference position of the humerus near 45° abduction, 0° flexion–extension and unconstrained internal–external rotation. For the lower load increments, the humeral head initially translated across the glenoid surface in the direction opposite the rotation of the humerus. The large standard deviations during the application of these small loads indicated a region of overall joint laxity that Panjabi et al. (1994) termed the 'neutral zone'. This may allow a wide range of possible paths of motion. Increasing the external moment load pushed the humeral head back along the glenoid surface in the direction of the rotation. The biomechanical studies show that during unconstrained motion, the glenohumeral joint does not display the expected behaviour of a ball-and-socket joint under all conditions.

The study of Novotny's group indicated that with an abduction moment load the humerus rotated externally while the humeral head translated medially, anteriorly and superiorly. The present data on the shoulder in pose 1 (90° abduction and variable external rotation) also demonstrated an anteriorized position of the geometrical centre of the humeral head on the glenoid. The present study demonstrated a constant horizontal abduction component from pose 1 to 2, with variable contributions of external rotation and abduction.

Using axillary radiographs, Howell et al. (1988) and Paletta et al. (1997) evaluated in vivo the active positioning of the arm in the cocked position of throwing. They noted a posterior translation of the centre of the humeral head to >4 mm related to the centre of the glenoid cavity. Using stereophotogrammetric measurements to evaluate glenohumeral contact patterns during elevation in the scapular plane, Soslowsky et al. (1992) demonstrated in vitro that with increasing elevation in the externally rotated starting position, the contact of the humeral head on the glenoid cavity shifted posteriorly. In a simulation of the cocked position of throwing, Novotny et al. (1998) located the position of the centre of the humeral head 5.7 ± 3.54 mm posteriorly. This accords well with the present in vivo data on the apprehension testpose ($J_G = -7.25 \pm 0.87$ mm). From pose 2 to 3, the finite helical axis was positioned eccentrically on the humeral head towards the glenoid cavity (figure 5). As such, the glenohumeral joint did not act as a ball-and-socket joint. Furthermore there was a remarkable shift along the helical axis, with a posterior component but variability concerning the lateral–medial or superior–inferior directions (table 3).

With the shoulder 90° abducted, the primary stabilizer limiting anterior or posterior translation is the inferior glenohumeral ligament complex (IGHLC), its anterior band becoming the primary constraint to anterior translation with additional external rotation (Matsen et al. 1998). Turkel et al. (1981) proposed a hammock-like supporting function for the IGHLC. With internal rotation and abduction, the anterior band of the IGHLC moves inferiorly to resist inferior translation as the posterior band fans out and shifts posterosuperiorly to prevent posterior motion. With external rotation and abduction, the posterior band moves inferior to the humeral head and the anterior band fans out and shifts anterosuperiorly to prevent anterior motion.

4.2. Kaltenborn's rule revisited

Based on the helical motion mechanism, the concave–convex rule and the appending terminology of roll and glide can be theoretically questioned:

1. With a discrete gliding motion from a convex articular profile on a concave articular profile, the finite centre of rotation is at the centre of the curvature of the stationary segment. The finite centre of rotation of a discrete rolling motion is located at the contact point. The spatial position of the finite helical axis could be referred to an axis connecting the centre of rotation for the roll with the centre of rotation for the glide. Such a description does not consider the decomposed values of the spatial position on the axis perpendicular to this roll–glide axis.
2. Furthermore, roll and glide are strictly 2D terms that do not include the shift along the helical axis. A moving segment not only rotates about the finite helical axis, but also shifts along it. Thus, in the field of motion, the related planar profiles of the articulating surfaces subsequently change in position.

Following the convex–concave rule, the roll and glide of a convex humeral head on a concave glenoid should be in opposite directions. With respect to glenohumeral external rotation, Matsen et al. (1998) demonstrated a posterior translation of the humeral head on the glenoid evolving out of specific capsuloligamentous tightening (i.e. in 90° abduction the anterior band of the inferior glenohumeral ligament and at lower angles the middle glenohumeral band, the anterior band of the inferior glenohumeral ligament and the subscapularis tendon). Consequently, the clinician must reconsider Kaltenborn's traditional concepts regarding intra-articular glenohumeral motion behaviour as well as restating its clinical impact. In the case of limited motion, mobilization of glenohumeral external rotation at 90° shoulder abduction with combined anterior gliding will stretch the tightening IGHLC. The effect is to enhance a pathokinematic behaviour as seen in anterior instability, which is characterized at the end of the apprehension test pose by a diminished posterior translation of the geometrical centre of the humeral head on the glenoid (Howell et al. 1988, Paletta et al. 1997). Then, glenohumeral mobilization should assess the roll behaviour characteristic of the late cocking motion. However, roll and glide are 2D arthrokinematic terms related to the plane of motion. From an individual point of view it is difficult to define therapeutically the plane of motion as well as the articular surface and thus the magnitude and direction of glide. Consequently, from a practical point of view it seems better to redefine manual therapeutic techniques for the glenohumeral joint in terms of rotation of the humerus and translation of the geometrical centre of the humeral head.

References

CAPPOZZO, A., CATANI, F., LEADINI, A., BENEDETTI, M. G. and DELLA CROCE, U. 1996, Position and orientation in space of bones during movement: experimental artefacts, *Clinical Biomechanics*, **11**, 90–100.

CLARK, J. M., HARRYMAN, D. T., SIDLES, J. A. and MATSEN, F. A. 1990, Range of motion and obligate translation in the shoulder: the role of the coracohumeral ligament, *Transactions of the Orthopaedic Research Society*, **15**, 273.

DE GROOT, J. H. 1999, The scapulo-humeral rhythm: effects of 2-D roentgen projection, *Clinical Biomechanics*, **14**, 63–68.

GOHLKE, F. E., BARTHEL, T. and DAUM, P. 1994, Influence of T-shift capsulorraphy on rotation and translation of the glenohumeral joint: an experimental study, *Journal of Shoulder Elbow Surgery*, **3**, 361–370.

HARRYMAN, D. T., SIDLES, J. A., HARRIS, S. L. and MATSEN, F. A. 1992, Laxity of the normal glenohumeral joint: a quantitative in vivo assessment, *Journal of Shoulder and Elbow Surgery*, **2**, 66–76.

HARRYMAN, D. T., SIDLES, J. A., CLARK J. M., MACQUADE, K. J. and GIBB, T. D. 1990, Translation of the humeral head on the glenoid with passive glenohumeral motion, *Journal of Bone and Joint Surgery*, **72A**, 1334–1343.

HÖGFORS, C., PETERSON, B., SIGHOLM, G. and HEBERTS, P. 1991, Biomechanical model of the shoulder joint: II. The shoulder rhythm, *Journal of Biomechanics*, **24**, 699–709.

HOWELL, S. M., GALINAT, B. J., RENZI, A. J. and MARONE, P. J. 1988, Normal and abnormal mechanics of the glenohumeral joint in the horizontal plane, *Journal of Bone and Joint Surgery*, **70A**, 227–232.

JOHNSON, G. R., STUART, P. R. and MITCHELL, S. 1993, A method for the measurement of three dimensional scapular movement, *Clinical Biomechanics*, **8**, 269–273.

KALTENBORN, F. M. 1980, *Manual Therapy for the Extremity Joints* (Oslo: Olaf Norlis Bokhandel).

KOH, T. J., GRABINER, M. D. and BREMS, J. J. 1998, Three dimensional in vivo kinematics of the shoulder during humeral elevation, *Journal of Applied Biomechanics*, **14**, 312–326.

LUDEWIG, P. M., COOK, T. M. and NAWOCZENSKI, D. A. 1996, Three-dimensional scapular orientation and muscle activity at selected positions of humeral elevation, *Journal of Orthopedic Sports Physical Therapy*, **24**, 57–65.

MACCONAILL, M. A. 1953, Movement of bones and joints: 5. The significance of shape, *Journal of Bone and Joint Surgery* **35B**, 290–297.

MACQUADE, K. J., WEI, S. H. and SMIDT, G. L. 1995, Effects of local fatigue on three-dimensional scapulohumeral rhythm, *Clinical Biomechanics*, **10**, 144–148.

MACQUADE, K. J. and SMIDT, G. L. 1998, Dynamic scapulohumeral rhythm: the effects of external resistance during elevation of the arm in the scapular plane, *Journal of Orthopedic Sports Physical Therapy*, **27**, 125–133.

MATSEN, F. A., THOMAS, S. C., ROCKWOOD, C. A. and WIRTH, M. A. 1998, Glenohumeral instability, in C. A. Rockwood and F. A. Matsen (eds), *The Shoulder*, 2nd edn (Philadelphia: W. B. Saunders), vol. **2**, 611–754.

MESKERS, C. G. M., VERMEULEN, H. M., DE GROOT, J. H., VAN DER HELM, F. C. T. and ROZING, P. M. 1998, 3-D shoulder position measurements using a six-degree-of-freedom electromagnetic tracking device, *Clinical Biomechanics*, **13**, 280–292.

NOVOTNY, J. E., CLAUDE, E. N. and BEYNNON, B. D. 1998, Normal kinematics of the unconstrained glenohumeral joint under coupled moment loads, *Journal of Shoulder and Elbow Surgery*, **7**, 629–639.

PALETTA, G. A., WARNER, J. J. P. and WARREN, R. F. 1997, Shoulder kinematics with two plane x-ray evaluation in patients with anterior instability or rotator cuff tearing, *Journal of Shoulder and Elbow Surgery*, **6**, 516–527.

PANJABI, M. M., LYDON, C., VASAVADA, A., GROB, D., CRISCO, J. J. and DVORAK, J. 1994, On the understanding of clinical instability, *Spine*, **19**, 2642–2650.

PEARL, M. L., JACKINS, S., LIPPITT, S., SIDLES, J. A. and MATSEN, F. A.. 1992, Humeroscapular positions in a shoulder range-of-motion-examination, *Journal of Shoulder and Elbow Surgery*, **1**, 296–305.

PRONK, G. M. and VAN DER HELM, F. C. T. 1991, The palpator, an instrument for measuring the three-dimensional positions of bony landmarks in a fast and easy way, *Journal of Medical Engineering Technology*, **15**, 15–20.

RHOAD, R. C., KLIMKIEWICZ, J. J., WILLIAMS, G. R., KESMODEL, S. B., UDUPA, J. K., KNEELAND, J. B. and IANNOTTI, J. P. 1998, A new in vivo technique for three-dimensional shoulder kinematics analysis, *Skeletal Radiology*, **27**, 92–97.

SOSLOWSKY, L. J., FLATOW, E. L., BIGLIANI, L. U., PAWLUK, R. J., ATHESIAN, G. A. and MOW, V. C. 1992, Quantitation of in situ contact areas at the glenohumeral joint: a biomechanical study, *Journal of Orthopedic Research*, **10**, 524–535.

TERRY, G. C., HAMMON, D., FRANCE, P. and NORWOOD, L. A. 1991, The stabilizing function of passive shoulder restraints, *American Journal of Sports Medicine*, **19**, 26–34.

THOMPSON, W. O., DEBSKI, R. E., BOARDMAN, D., TASKIRAN, E., WARNER, J. J. P., FU, F. H. and WOO, S. L.-Y. 1996, A biomechanical analysis of rotator cuff deficiency in a cadaveric model, *American Journal of Sports Medicine*, **24**, 286–292.

TURKEL, S. J., PANIO, M. W., MARSHALL, J. L. and GIRGIS, F. G. 1981, Stabilizing mechanisms preventing anterior dislocation of the glenohumeral joint, *Journal of Bone Joint Surgery*, **63A**, 1208–1217.

VAN DER HELM, F. C. T. and PRONK, G. M. 1995, Three dimensional recording and description of motions of the shoulder mechanism, *Journal of Biomechanical Engineering*, **117**, 27–40.

VELDPAUS, F. E., WOLTRING, H. J. and DORTMANS, L. M. 1988, A least squares algorithm for equiform transformation from spatial marker coordinates, *Journal of Biomechanics*, **21**, 45–54.

WOLTRING, H. J., HUISKES, R. and DE LANGE, A. 1985, Finite centroid and helical axis estimation from noisy landmark measurements in the study of human joint kinematics, *Journal of Biomechanics*, **18**, 379–389.

WUELKER, N., KORELL, M. and THREN, K. 1998, Dynamic glenohumeral joint stability, *Journal of Shoulder and Elbow Surgery*, **7**, 43–52.

Appendix: Finite helical axis

Arthrokinematics or joint kinematics compares the position and attitude of a coordinate system embedded on the distal segment (arbitrarily called the moving coordinate system) with the position and attitude of a coordinate system embedded on the proximal segment (arbitrarily called the stationary coordinate system). This relation can be quantified using a translation vector with a 3×3 rotation matrix. Starting from a series of at least three non-collinear bony landmarks, Veldpaus *et al.* (1988) presented an unweighted least-squares method to estimate these parameters.

Starting from the translation vector and the rotation matrix, two different interpretative models are used: the Euler–Cardan model and the helical axis model. In the Euler–Cardan model the 3×3 rotation matrix is represented in a set of three successive rotations (Euler–Cardan angles) required to align the attitude of the moving coordinate system with the attitude of the stationary coordinate system. In the Euler–Cardan approach, the value of the translation depends on the particular choice of the origin of the coordinate system.

In the finite helical axis approach, a finite displacement between subsequent poses is viewed in terms of a single helical displacement, with the moving segment having a shift (t) along and a rotation (θ) about a line with a specific position vector (**s**) and direction vector (**n**) in space (figure 6). The shift along the finite helical axis represents a real translation of the moving segment, invariant of the particular choice of the origin of the coordinate system embedded on the moving segment. The drawback of using the finite helical axis is the susceptibility to measurement errors of the bony landmarks, especially when the rotation is small. The error model of Woltring et al. (1985) predicts the standard deviation for the rotation angle (σ_θ), for the shift (σ_t), for the position vector (σ_s), and for the angular uncertainty of the direction vector of the finite helical axis (α_n).

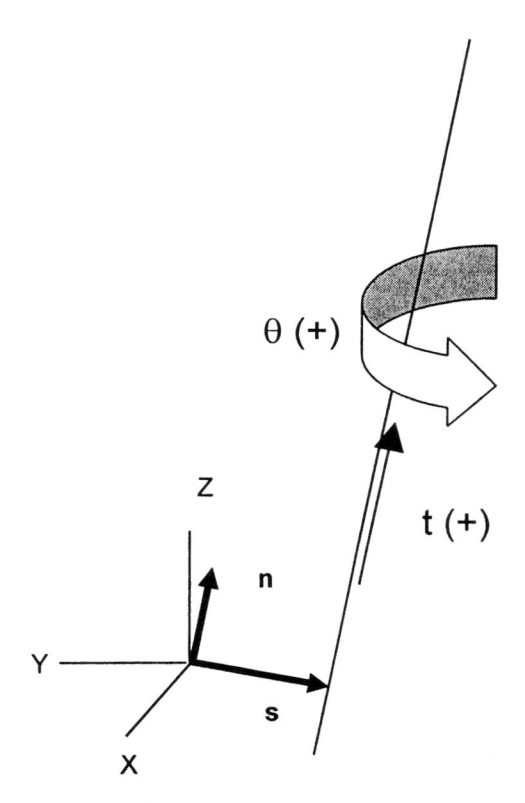

Figure 6. Helical axis parameters: direction vector (**n**), shift (t), rotation (θ) and position vector (**s**). (X, Y, Z), reference coordinate system.

Part VII
Ergonomics and Health in the Workplace

28 Developing a holistic understanding of workplace health: the case of bank workers

L. DUGDILL*

Research Institute for Sport and Exercise Sciences, Liverpool John Moores University, Henry Cotton Campus, 15–21 Webster Street, Liverpool, L3 2ET, UK

Keywords: Employees; Health; Lifestyle; Psychosocial.

Understanding health in the workplace from a holistic perspective requires appropriate research approaches. An exploratory, qualitative study with white-collar employees from a large banking organization in the North West of England is detailed here. The aim was to explore health from the perspective of the employee. Semi-structured interviews ($n=29$) elicited detailed responses, allowing the relationship between psychosocial and lifestyle factors to be examined. Psychosocial factors such as job design, ability to make decisions and control over work were all reported to be positive contributors to health in the work setting—and often were reported to be *more* relevant than individual lifestyle issues. A more holistic approach to developing workplace health programmes in the future is recommended, which allows the workplace setting to be considered as a complex system. Very few comprehensive workplace health programmes exist at present within UK organizations, and there is a tendency for programmes to be designed by the health professional 'expert' rather than developed in conjunction with, and by the workforce. It is argued that qualitative research methods can help to begin the dialogue needed for workplace health programme development.

1. Introduction

In the past decade, health promotion in the workplace within the UK has had a shift in emphasis, from changing individual lifestyles to changing organizational structures, in order to improve workers' health (Health Education Authority 1998). The assessment of workers' health still relies heavily on quantitative survey tools, such as questionnaires, which tend to be orientated towards the measurement of lifestyle behaviour (e.g. smoking, drinking, stress). Cassell and Symon (1994: 9) suggested that 'organisations are impressed by numbers, regarding these as "accurate" data', which may be one reason for an over-reliance on survey techniques in the past. It is likely that researchers in the organizational setting may have had to compromise their research design (tailoring the design to favour quantitative tools) in order to gain access to their research community. Limited methods of measurement have consequently led to development of workplace health programmes that have been narrow in focus and have lacked relevance to the target population. Such programmes have often struggled to sustain adherence and have failed to reflect the true complexity of the organizational setting and the needs of employees (Springett and Dugdill 1999).

*e-mail: l.dugdill@livjm.ac.uk

More positively, during the past decade there has been a slow shift in approach with researchers such as Karasek and Theorell (1990) taking a more holistic view of health in the workplace. Their work highlighted the importance of psychosocial factors, especially control and decision-making within the job context, in explaining well being.

Few studies, particularly within the health promotion arena, have attempted to explore these factors qualitatively, from the perspective of the worker. It is rare in workplace health studies for researchers to try and work inductively from a holistic theoretical framework of health (Dugdill and Springett 1994), which was attempted here. Bank workers were chosen as a sector that has experienced a rapidly changing organizational context over the last decade, with the introduction of information technology and consequent downsizing (O'Reilly 1994). It was expected that they would be able to articulate insights into the effect of individual and organizational issues on their health.

The primary aim of the study was to explore the relationships between psychosocial factors, lifestyle and health from the perspective of the bank worker. Hancock's Mandala of Health (figure 1) provided the researcher with a holistic, theoretical, systems model of health which encompassed relevant 'domains' of importance to health, e.g. work, lifestyle. It also provided the theoretical, deductive framework around which this study was designed.

2. Methods

2.1. Research design

The researcher adopted an interpretative stance (Rubin and Rubin 1995). This approach recognizes 'that meaning emerges through interaction and is not standardised from person to person ... it emphasises the importance of understanding the overall text of a conversation' (p. 31). Some limitations were imposed on the study from the onset. Only 45 min of work time was allowed to each participant to take part in the study, and this limited the research design. Complex action research designs and follow-up interviewing, although ideal, were not possible. As Bryman (1988: 17) stated, 'access to organisations is clearly very troublesome to researchers'. Therefore, given the time limitations, it was decided that the best approach was to interview a range of employees from different job types within the organization to get a broad view of their perspectives on workplace health. Rubin and Rubin (1995: 69) articulated the need to get a balanced argument when designing a qualitative study: 'you have to go for a balance in your choice of interviewees to represent all the divisions within the arena of study'. Here, the intention was to get a range of participants from different job types, of different ages and both sexes.

2.2. Sample

Male ($n=11$) and female employees ($n=18$) agreed to participate, from a range of different job types (managers, administrative assistants, cashiers, lending assistants) within the organization, and from different organizational sites within the city. The majority of participants ($n=24$) were from the two main office sites in the city, and who normally had telephone contact only with banking customers. The other four participants were from branch sites in different parts of the city and these individuals had daily contact with customers. Participants covered a wide age range, from 22 to 58 years; three were senior managers (all male). Four participants had been employed by the organization for >15 years; five had been employed for <3 years.

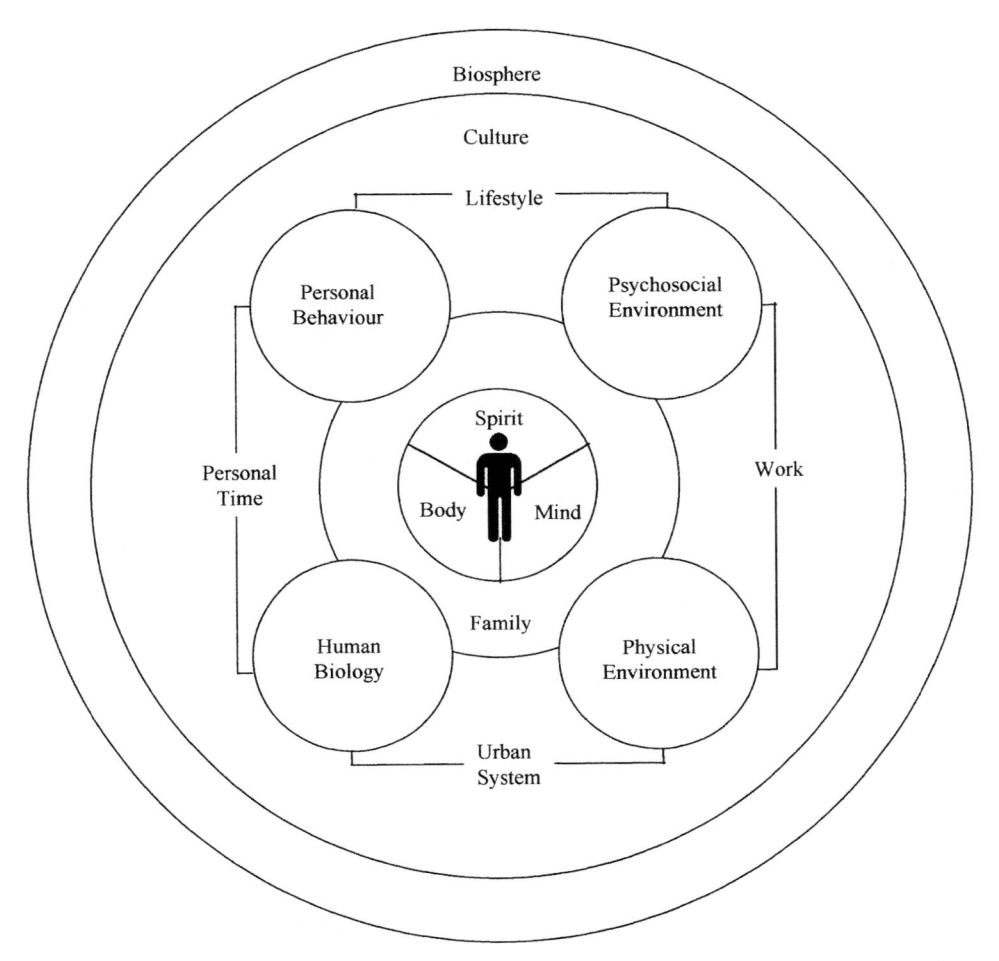

Figure 1. Hancock's Mandala of Health (1984).

2.3. Context

A semi-structured in-depth interview technique was employed (Rubin and Rubin 1995). The researcher, at the interviewees' place of work, carried out interviews over 1 month. Interviews were always conducted in a private office to ensure confidentiality and were recorded onto audiotape with the consent of the interviewee. Participants were encouraged to give honest responses, although interviewing within the place of work may have compromised some participants' ability to speak freely. At any point, the interview could be terminated or the tape recorder switched off at the request of the interviewee (although this was never requested).

2.4. Interviews

Guided by Hancock's Mandala of Health, an interview schedule (table 1) was designed by the researcher to cover most of the interrelated domains shown in the model. Twelve questions were asked covering issues regarding the employees' concept of health, contentment, psychosocial and lifestyle issues that affected their health both at home and work, and also the interrelationship between these two domains. The researcher was trying to gain understanding of the relative importance of these domains from the bank workers' perspective. The final few questions on the schedule were designed to get an idea of changes that could be made, and health at work priorities for the individual.

Interviews lasted on average 1 h; the shortest was 25 min, the longest 2.5 h. More senior members of staff (male) tended to give the longest interviews, possibly because they had more control over their work time and also because their experiences were fuller (due to length of employment) and therefore they had more to say. The researcher felt that all participants seemed at ease with the questions and answers were given quite readily. The questions were phrased as simply as possible. Prompts were used to elicit more detailed responses and to bring the interview focus back onto the topic of health. Prompts were especially important for questions 5–9, which were designed to elicit detailed responses to the issues of psychosocial and lifestyle influences on health at work: such prompts covered issues of decision-making, job content, control and support structures. Gaining subjective, multiple views of all these psychosocial variables was an important part of this study, as many previous reports have relied on psychological scales (with predefined variables) as a method of measurement (Karasek 1979).

All interviews were transcribed verbatim and analysed using a systematic content analysis approach (Burnard 1991, Bryman and Burgess 1996). A deductive framework for analysis was provided by the interview schedule but an inductive approach (Patton 1990) was used to allow themes to emerge from the data. Validation of the data through 'member checking' was not possible due to limited access to the study population. The credibility or trustworthiness of the research process was established in several ways, e.g. discussing the interview schedule with research partners before beginning the study, keeping notes after each interview to inform the analysis, and questioning the coherence of themes by follow-up prompts within each interview. The themes reported here relate particularly to the lifestyle, psychosocial, work and health domains of the model, not the entire data set generated by the interview schedule.

Table 1. Interview schedule for bank workers.

1. What does being healthy mean to you?
2. How would you describe your state of health at the moment?
3. Would you say you were happy at work?
4. Do you feel your physical work environment affects your health in any way?
5. Do you feel that the day-to-day content of your job impacts on your health in any way?
6. Do you have control within your job—and does that affect how you feel/your health?
7. How much support do you receive from colleagues—does this influence the way you feel/your health?
8. Are there any work factors which influence your lifestyle?
9. Do you feel there are any factors from your personal life (lifestyle) which influence both your work and your health?
10. Are there any changes which could be made (individual or organizational) to improve your health at work?
11. Do you feel the organization already has any strategies in place which are contributing to the health of its employees?
12. If you could pick one factor which you feel influences your health most at work what would it be?

3. Results and discussion

3.1. Employees' concepts of health

Very often, conceptual lay understanding of health, centres around an 'absence of disease' model, or 'functional' approach, reflecting the continued dominance of biomedicine within the health debate (Townsend and Davidson 1982). Participants' responses corroborated this view, as the following data exemplifies. For instance, interviewees often mentioned physiological constructs when describing health.

> Health is feeling well enough to get on with life in general ... not feeling you are full of cold and bedridden. (male, manager)

Sometimes this functional approach to defining health was also linked to state of mind:

> Feeling OK in yourself. (female, administrator)

> Being able to do things enthusiastically. (male, manager)

Or it was defined in a behavioural/lifestyle context:

> I think diet is one of the most important things. (female, supervisor)

> Trying to keep a balance of what you eat, trying to do some exercise ... and keeping your mind active. (male, personal assistant)

Also, participants reported awareness of the state of their health when they could not carry out tasks or functions they had previously been able to; or when health status changed from a usual, or optimum level, and some sort of medical intervention was perceived as necessary.

> I would say I'm a very healthy manager... visits to the doctor are very, very infrequent. (male, manager)

In summary, the responses to this question were very much focused around health being perceived as a functional and physical commodity and as such, only covered a few of the domains featured in Hancock's model. Only one male employee mentioned the importance of spiritual health. For the past century, Western cultures have been dominated by biomedical conceptualizations of health (McKeown 1979), so it was not surprising to find this ideology reflected in the responses from bank workers.

To get a broader perception of health, the researcher moved on to the topic of happiness at work, with the implicit understanding that health, subjective well-being and happiness were linked conceptually (Maddux 1997). Some participants were able to articulate this link as the following quote illustrates:

> I'm very interested in what I do therefore, it makes me want to work and I feel happy about it. So I suppose if you're happy about what you are doing you feel in a good state, don't you? (female, secretary)

Here the interviewee directly linked state of mind, feeling 'good' or positive, and happiness when asked about health.

3.2. Happiness at work

The mental health, subjective well-being and happiness of employees are increasingly being recognized as an important area for health promotion. Organizations have been slow to address these needs (Cooper and Williams 1994), possibly because these concepts are complex and will only be solved by equally complex and comprehensive workplace programmes. Consequently many organizations do not know how to begin to tackle these issues (Kavanagh et al. 1998).

This study revealed that the definitions and notions of health that were reported began to diverge when workers were asked about happiness and contentment at work, and psychosocial factors were mentioned. Day-to-day tasks and interactions, with colleagues and customers, were most prevalently commented on as important contributors to happiness and/or stress at work as illustrated below.

3.2.1. Job content:
The need for stimulation, responsibility, and task variability within the work setting were often reported as being important to happiness and state of mind.

> Throughout the time I've been in the bank I've been quite fortunate and done a variety of things... keeping my interest is quite important to me. If I do a repetitive job for a long period of time I become bored... I don't feel as good about it... and that affects my health in the long term. (female, supervisor)

Change within the job context was often viewed as important to maintaining interest, and hence a positive state of mind:

> I prefer customer contact, face-to-face rather than over the telephone, I don't really like that side of the job. However, our jobs are changing and becoming more sales oriented so we will be dealing with customers more directly in the future... I'm glad as it will not be so monotonous. (female, lending assistant)

A male member of staff, who indicated that lack of task variability within the job had affected his health, also picked up this theme:

> When I was on the lending part it was just monotonous, you knew what you were going to come in to face, 25 applications for loans,... it wasn't really challenging... it got me down in the end. (male, personal assistant)

Having challenging work, of a variable nature, and having the satisfaction of completing work-orientated goals were commonly cited as being the most important factors contributing to happiness at work, alongside supportive relationships.

3.2.2. Relationships:
Good relationships with colleagues were reported as a major source of happiness at work; however, poor or difficult relationships (either with colleagues or customers) were reported to be a major source of stress and

concomitant poor health. This concurs with Selye's (1946) early research, which highlighted the importance of good relationships to individual and organizational health. The following quote is a good illustration of this issue:

> I liked my other job before... it was a different set up, I think it was mainly the people... we used to get on really well together... (female, secretary)

Not all experiences reported were positive. Where stress was mentioned, the participants were asked how they felt their levels of stress influenced their health — overwhelmingly when stress was mentioned it was perceived as a negative influence on health (i.e. in 90% of all participants). For instance, managers who had to deal with staff reported this to be a source of stress:

> Dealing with people is the biggest cause of stress... you can have difficult staff to deal with, problem staff, I've got to think it over for a long time. Occasionally, I have to do a staff appraisal, and I have to think about it for a long time... I know it is going to be difficult and not an easy one to handle.... (male, manager)

Conversely, some managers found customers caused them most stress at work:

> My health is influenced mostly by stress... the stress of dealing with customers who are fighting a battle against the economic climate.... (male, manager)

Contact with customers varied between job type and job location. For instance, some workers in head office locations stated that high levels of customer contact over the telephone was unrewarding (see Section 3.2.1); however, staff located in branches also found face-to-face contact with customers stressful.

> I feel fairly relaxed and calm at the moment (branch manager who had recently moved to head office) ... the pressures here aren't the same as in the branches. (male, senior manager)

Several staff from branch offices commented on the potential threat of armed robbery. One of the interviewees who had been involved in such an incident several years earlier explained that she had to cope with the stress of remembering this incident every day when she came into work.

> The raid was several years ago but it is difficult to completely forget about it ... it is always in the back of your mind... and it makes work stressful. (female, branch cashier)

Stress counselling was offered to staff in the immediate aftermath of a raid, but it was not routinely available on a long-term basis. Another international banking corporation offered similar confidential counselling after bank-raids, including an immediate session to combat post-traumatic stress disorder (CBI 1993).

Some workers explained the linkage between job content, relationships and health. For example, one manager described the importance of job competence and satisfaction on self-esteem:

> I think the biggest influence on my health is probably job satisfaction... I'm not being big-headed when I say this but I feel very competent with the job I do.... (male, manager)

When asked why he felt so confident and positive, he went on to explain the importance of relationships and being in control.

> '... the way people react to me higher up the organisation... just the feeling that you are in control of your work, the ability to plan something and bring it to fruition and see the positive affects of it, that kind of thing makes me feel that I'm doing the job competently and confidently, and I can relax and think that I'm in charge of the job, the job is not in charge of me.... (male, manager)

Many of the interviewees ($n = 23$) mentioned psychosocial issues such as control, support and team-working without actually being directly asked about them. Further responses to more specific questions about psychosocial issues and work are explored below.

3.3. *Psychosocial factors and health at work*

Psychosocial factors such as job design, ability to make decisions and control over work situations, were perceived as the most important contributors to positive health status within the work setting, by the majority of workers interviewed ($n = 25$). Twenty-two interviewees suggested successful team working was essential for job satisfaction. These psychosocial variables were seen to be equally important to lifestyle behaviour in their contribution to health status, and *more* relevant within the organizational setting. The inability to take part in decision-making processes was frequently stated to be a source of stress among female workers, particularly those who had a job lower down the management hierarchy.

The following comment illustrates how issues of control and decision-making, within job design, were deemed important to health:

> Sometimes it's frustrating because you can't go in and change things, and that can make you feel tired and stressed. (female, personal banker)

> I think more control needs to be delegated downwards but I'm not sure where you would stop. (male, manager)

Site location also influenced levels of control reported within the job:

> It's more difficult when you are working in this office because you are working more closely with directors... so you haven't got as much control maybe as you have in branch offices. (male, manager)

Support and teamwork were also considered to be important:

> If you know you've got the support and back-up from the people you work with, then it makes you a lot happier. (female, systems trainer)

> Personalities within the teams are an important factor... if you are in a team that is pulling in the same direction, it makes a lot of difference.... (male, manager)

Asking for support was deemed to be more difficult by managers higher up the organizational hierarchy:

> Support for me is picking you up when you are down... but to do that, you have to tell them that you are down and that is more difficult for me... because it is an admission of vulnerability.... (male, manager)

It is important to state that the issues of support, control and teamwork were defined differently by participants. For example, to some control meant having control over day-to-day tasks and to others it meant being involved in organizational developments. Permitting a subjective definition was an important aspect of this study.

3.4. Interaction between work and lifestyle/personal behavioural domains

There were many examples of the influence of work on lifestyle, and the length of the working day often appeared to impinge on health behaviours outside of work:

> The long working day affects my social life and then I don't feel as well. (female, personal assistant)

> My job takes up evenings as well... so I don't have time to be active. (male, manager)

> Yeah, I do think work influences your lifestyle... you don't go to the gym and things like that... you have people who are fed up and tired all the time. (female, systems leader)

Other interviewees mentioned that most of their relaxation and leisure activity was curtailed within the weekend, for example:

> My average alcohol consumption is about 25 units a week but it is mainly at weekends... I think that has a negative impact on my health.... (male, manager)

It was much more common for work to be reported as influencing an individual's lifestyle than vice versa. It appears that workers are 'living to work' rather than 'working to live', and perhaps is reflective of the average working hours of employees within the UK (HEA 1997).

3.5. Interaction between work and home/lifestyle domains

Parker and Wall (1998: 35) stated: 'it is important to investigate how the nature of work affects life outside work... nevertheless, there is too little research to provide a coherent picture as to how work affects non-work life'. The distinction between these domains is becoming increasingly blurred in the banking sector, particularly as the demand for weekend branch opening and 24-h telephone banking services develops.

> The bank isn't a nine-to-five job anymore, it hasn't been for a long time....
> (male, manager)

An important part of this study was to gain understanding as to how the domains of work and home (non-work) interacted. There was very clear evidence that work greatly influenced the patterns of individuals' home lives:

> Work pushes all social activities into the weekends. (male, manager)

and also personal behaviour and lifestyle:

> Because I have so many accounts, your day becomes very, very full and it does affect you ... you go home and you are thinking about customers and accounts, suddenly you wake up in the middle of the night thinking I must do that tomorrow ... and you can't get back to sleep.... (male, manager)

Workers appeared to fall into two categories. The first included those who could easily 'switch off' from the pressures of home when they came to work (and vice versa) and the second, those who found it very difficult to make a clear separation between these two settings.

> Whatever problems I have had in the past I've tended to come into work as my escape route ... and just go for it, and think I'll cope with that when I get home. (female, systems manager)

> Very rarely do I let my personal life affect my work.... (female, personal assistant)

A positive home life greatly contributed to self-reported ability to cope with work:

> I'm the sort of person who talks myself up at work and that is because I feel very secure in my home life.... (male, manager)

Overall, bank workers perceived the interaction between home and work domains to be very important (figure 2). Some appeared to have developed better mechanisms for coping with the dual demands of work and home life, e.g. task prioritization, although these mechanisms were 'self-taught' rather than being developed via a staff development programme. Women employees in particular mentioned the multiple demands placed on them in work and home settings, which reflects similar findings from other research studies (Dean 1992).

4. Conclusions

The semi-structured in-depth interview was an excellent way of learning about health issues in the workplace, although the time and resource implications of implementation may limit the use of this methodology within organizations. The research process itself can facilitate organizational change if used in a participatory and open manner and if set within an action research framework (Springett and

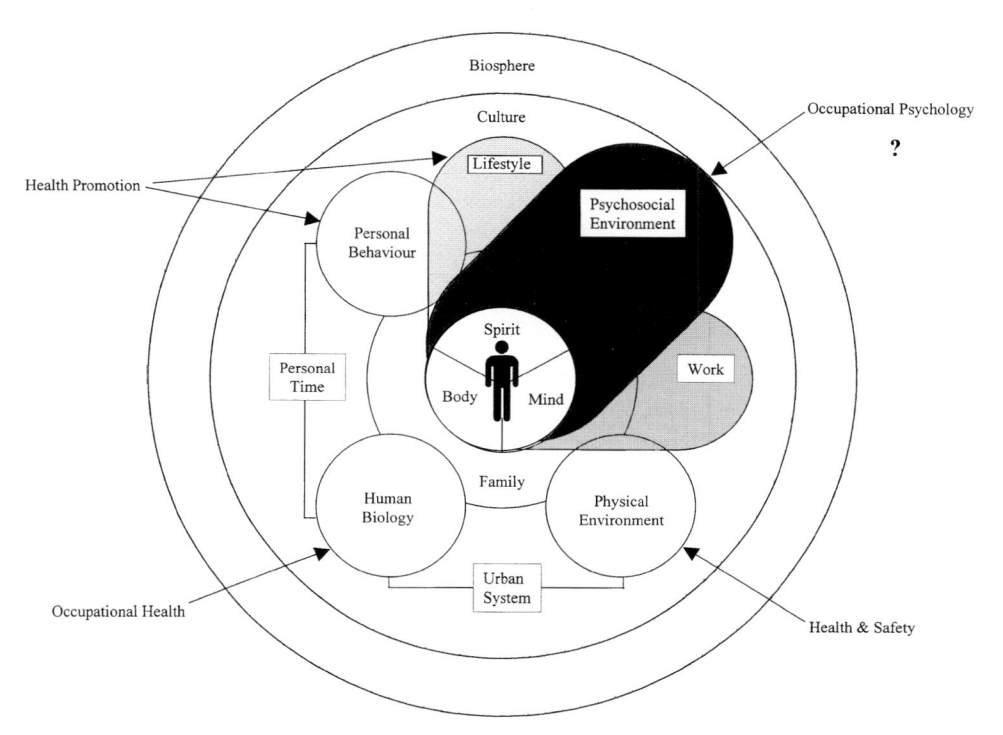

Figure 2. Reconfigured Mandala of Health showing relative importance of lifestyle and psychosocial domains.

Dugdill 1999), although participatory action research was beyond the scope of the present study.

The bank workers conceptualized health in a complex way but only when asked about health indirectly. Definitions of health primarily focused on 'absence of disease' or lifestyle. Primarily, psychosocial factors were cited as an influence on health within the organizational setting. These observations are not surprising and corroborate recent findings of a large study carried out within the National Health Service in the UK (HEA 1999). Employees believed home and work settings to be closely interconnected when influencing health.

Traditionally, UK workplace health activity has been focused around the disciplines of occupational health (represented by the human biology domain on figure 2), health and safety (represented by the physical environment domain) and health promotion (represented by the personal behaviour/lifestyle domains) (Kavanagh et al. 1998). This may explain why past workplace health programmes have failed to address the psychosocial dimension successfully—the tools of occupational psychology may help to redress this balance if this discipline can be incorporated within a health framework.

When the results of this study are transposed onto Hancock's Mandala of Health, it would appear that the interrelationship between the domains of lifestyle, psychosocial, work and individual health overlap. Also, the psychosocial domain appears to be a more influential factor in explaining health at work. The author (figure 2) suggests a reconfigured model of health. Finally, workplace health promotion strategies must start to address psychosocial, as well as individual lifestyle issues, and bridge the interface between home and work settings, if the health of the working population is to be improved.

References

BRYMAN, A. 1988, *Doing Research in Organisations* (London: Routledge).
BRYMAN, A. and BURGESS, R. G. 1996, *Analysing Qualitative Data* (London: Routledge).
BURNARD, P. 1991, A method for analysing interview transcripts in qualitative research, *Nurse Education Today*, 11, 461–466.
CASSELL, C. and SYMON, G. 1994, *Qualitative Methods in Organisational Research* (London: Sage).
CONFEDERATION OF BRITISH INDUSTRY 1993, *Working for Your Health: Practical Steps to Improve the Health of Your Business* (London: CBI).
COOPER, C. L. and WILLIAMS, S. 1994, *Creating Healthy Work Organisations* (Chichester: Wiley).
DEAN K. 1992, Double burdens of work: the female work and health paradox, *Health Promotion International*, 7, 17–25.
DUGDILL, L. and SPRINGETT, J. 1994, Evaluation of workplace health promotion: a review, *Health Education Journal*, 53, 337–347.
HANCOCK, T. 1984, *The Mandala of Health: A Model of the Human Ecosystem* (Toronto: Department of Public Health; University of Toronto).
HEALTH EDUCATION AUTHORITY 1997, *Health Update: Workplace Health* (London: HEA).
HEALTH EDUCATION AUTHORITY 1998, *Fit to Face the Future: Maintaining a Healthy Workforce for the NHS* (London: HEA).
HEALTH EDUCATION AUTHORITY 1999, *More to Work Than This: Developing and Sustaining Workplace Health in the NHS* (London: HEA).
KARASEK, R. A. 1979, Job demands, job decision latitude, and mental strain: implications for job redesign, *Administrative Science Quarterly*, 24, 285–308.
KARASEK, R. A. and THEORELL, T. 1990, *Healthy Work: Stress, Productivity and the Reconstruction of Working Life* (New York: Basic Books).
KAVANAGH, C., BARLOW, J. and DUGDILL, L. 1998, *Health at Work Strategy for Liverpool* (Liverpool: Liverpool Occupational Health Project).
MADDUX, J. E. 1997, Habit, health and happiness, *Journal of Sport and Exercise Psychology*, 19, 331–346.
McKEOWN, T. 1979, *The Role of Medicine: Dream, Mirage or Nemesis?* (Oxford: Blackwells).
O'REILLY, J. 1994, *Banking on Flexibility: A Comparison of Flexible Employment in Retail Banking in Britain and France* (Aldershot: Avebury).
PARKER, S. and WALL, T. 1998, *Job and Work Design: Organising Work to Promote Well-being and Effectiveness* (London: Sage).
PATTON, M. Q. 1990, *Qualitative Research and Evaluation methods* (Newbury Park: Sage).
RUBIN, H. J. and RUBIN, I. S. 1995, *Qualitative Interviewing: The Art of Hearing Data* (London: Sage).
SELYE, H. 1946, The general adaptation syndrome and the diseases of adaptation, *Journal of Clinical Endocrinology*, 6, 1–17.
SPRINGETT, J. and DUGDILL, L. 1999, *Health Promotion Programmes and Policies in the Workplace: A New Challenge for Evaluation*. Technical document (Geneva: World Health Organisation/Euro).
TOWNSEND, P. and DAVIDSON, N. 1982, *Inequalities in Health: The Black Report* (Harmondsworth: Penguin).

29 Effects of activity–rest schedules on physiological strain and spinal load in hospital-based porters

C. BEYNON*, J. BURKE, D. DORAN and A. NEVILL

Research Institute for Sport and Exercise Sciences, Liverpool John Moores University, Henry Cotton Campus, 1521 Webster Street, Liverpool, UK

Keywords: Activity–rest schedules; Spinal shrinkage; Physiological responses; Hospital porters.

Workers in physically demanding occupations require rest breaks to recover from physiological stress and biomechanical loading. Physiological stress can increase the risk of developing musculoskeletal disorders and repeated loading of the spine may increase the potential for incurring back pain. The aim of the study was to assess the impact of an altered activity–rest schedule on physiological and spinal loading in hospital-based porters. An existing 4-h activity–rest schedule was obtained from observations on eight male porters. This schedule formed the normal trial, which included two 5- and one 15-min breaks. An alternative 4-h schedule was proposed (experimental condition) that had two breaks each of 12.5 min. It was hypothesized that the experimental trial is more effective in promoting recovery from physiological strain and spinal shrinkage than the normal trial, due to the 5-min breaks being insufficient to allow physiological variables to return to resting levels or the intervertebral discs to reabsorb fluid. Ten males performed both test conditions and oxygen uptake VO_2, heart rate, minute ventilation VE, perceived exertion and spinal shrinkage were recorded. There were no significant differences in any of the measured variables between the two trials (p > 0.05). Median heart rates were 78 (range 71–93) and 82 (71–90) beats.min^{-1} for the normal trial and the experimental trial respectively, indicating that the activity was of low intensity. The light intensity was corroborated by the oxygen uptakes (0.75, range 0.65–0.94 l.min^{-1}). Spinal shrinkage occurred to the same extent in the two trials (2.12 ± 3.16 mm and 2.88 ± 2.92 mm in the normal trial and the experimental trial respectively). Varying the length and positioning of the rest breaks did not significantly affect the physiological responses or magnitude of spinal shrinkage between the two trials. More physically demanding work than the porters' schedule should induce greater physiological fatigue and spinal shrinkage. The ratio between activity and rest breaks would then become more important.

1. Introduction

Individuals engaged in sport or exercise that is intermittent in nature need adequate resting periods to allow the body sufficient time in between the high-intensity efforts to recover from the physiological consequences of exertion. If recovery is incomplete, performance in the subsequent bouts of activity is likely to suffer due to fatigue. This principle of allowing time for physiological systems to recover must also be observed in an occupational setting, as workers who perform physically demanding tasks are

*Author for correspondence. e-mail: humcbeyn@livjm.ac.uk; humjbur1@livjm.ac.uk

likewise subject to fatigue. Workers in physically demanding occupations can exhibit high levels of energy expenditure and associated elevations in heart rate. Furthermore, regular exposure to high levels of physiological strain can put workers at risk of developing musculoskeletal or cardiovascular disorders (Mathiassen 1993).

Frymoyer and Gordon (1989) suggested that since 1980 evidence has accumulated with respect to the importance of work and psychosocial factors in the onset of long-term disability rather than any anatomical pathology. This observation is particularly true of musculoskeletal disorders, which are the most commonly reported occupational disease in the workforces of the European Union, the majority of which affect the low back area. Biering-Sorensen (1983) reported that 52–60% of the general population group surveyed cited work as the primary cause of their back pain. Long-term loading of the spine is one of several factors potentially linked with the development of back pain (Althoff et al. 1992). A period of recovery allows reabsorption of fluid lost from the nucleus pulposus under compressive loading during physically demanding work. Prolonged periods of compressive loading and insufficient recovery may damage the endplates resulting in irreversible loss in disc height (Brinkmann et al. 1998), further disc degeneration and stiffness. Measurement of loss in stature, otherwise known as spinal shrinkage, has been used to determine the load on the spine (Eklund and Corlett 1984, Reilly et al. 1984).

One area of ergonomic intervention to reduce the physiological and biomechanical strain is the redesign of work–rest schedules, the rest periods allowing recovery from fatigue processes (Helander and Quance 1990). Excessive rest breaks decrease productivity because of reduced work time, but insufficient rest breaks result in cumulative fatigue, which will in turn lead to a fall in worker productivity and potentially cause ill-health. The optimum work–rest schedule, in terms of duration, frequency and time, must be established for each individual, depending on the occupational demands.

Studies of healthcare professionals such as nurses and physiotherapists have shown that they are under a considerable risk of developing musculoskeletal disorders (Hildebrandt 1995, Bork et al. 1996). Research considering the physiological and biomechanical stresses of hospital based porters is lacking. Porters' work is physically active, with long periods of standing and walking and some physically demanding elements such as pushing and pulling.

The aim of the present study was to assess the impact of an altered activity–rest schedule on physiological strain and spinal shrinkage in hospital-based porters. Total time of rest was the same in each trial. The exact time it takes for the spine to change from compression to expansion once the load has been removed is not certain. It was anticipated that two longer breaks of 12.5 min would promote more recovery as the 5-min breaks should not allow sufficient time to slow down the rate of discal compression or for the intervertebral discs to change from compression to expansion (Helander and Quance 1990). It was hypothesized therefore that a work–rest schedule incorporating two longer breaks is more effective than one of three breaks in alleviating spinal loading.

2. Methodology

2.1. Procedure development

Eight male hospital-based porters' were recruited from Southport and Formby District General Hospital to obtain work profiles. The mean age was 40 ± 8.6 years. Each porter was 'shadowed' by an observer for 2 h in which time the activities performed by the porter and the amount of time each action took were recorded. Actions included walking, standing, sitting and pushing or pulling while walking.

All activities were self-paced and the intensity of work was dependent on the requirements of the shift, reflecting the varying demands of different duties to different shifts. Heart rate was monitored every 15 s using a short-range radio telemetry system (Polar, Kempele, Finland). From the mean heart rates, the hardest work profile was ascertained. To simulate the hardest work programme observed, the percentage time spent walking, standing and so on was calculated from this work profile which then formed the basis of the laboratory protocol.

The formal work–rest schedule of the hospital porters was ascertained. Porters worked an 8-h shift with a 10-min break in both the morning and afternoon, and a 30-min break for lunch. A 4-h trial period was used to represent this work–rest schedule, with the three breaks constituting 25 min in total (figure 1). This schedule formed the first 4-h test protocol (normal trial). An alternate 4-h activity–rest schedule was proposed and constituted the second test session (experimental trial). The relative percentage of time that each action was performed was identical for each of the two test sessions. Rest breaks were distributed differently but the total time spent at rest was the same. A 4-h simulation was used due to the difficulty in obtaining subjects able to participate for a normal 8-h shift.

2.2. *Laboratory procedure*

Ten male subjects were recruited to participate in the study. The mean ($\pm$SD) age was 23.0 ($\pm$2.9) years; mean height was 1.80 ($\pm$0.05) m; mean body mass was 81.3 ($\pm$12.7) kg. Each subject was required to attend the laboratory on three separate occasions. A precision stadiometer measured changes in stature (spinal shrinkage) (Althoff *et al.* 1992), and on the first occasion subjects underwent training to familiarize themselves with this equipment. A cross was drawn on the spinous process of the first thoracic verterbrae (T1). The mark could be viewed by looking through a camera mounted behind the subject. The camera was connected to a linear

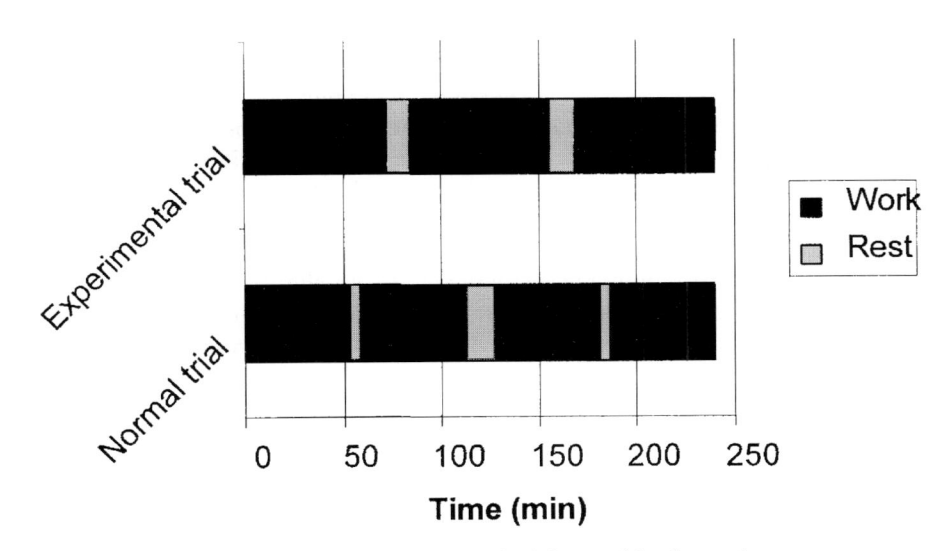

Figure 1.   Activity–rest schedules used in the study.

transducer. Relative stature was recorded by moving the camera so that the cross-hairs in the viewer lined up with the mark on the neck (Burton et al. 1994).

When the subject was familiarized with the stadiometer, he was asked to move away from and back on to the stadiometer and replicate the posture in quick succession. If the cross on the neck of the subject was exactly on the cross-hairs of the camera each time, the subject was deemed to be achieving the same position. This process was repeated several times and ensured that any changes during testing were attributable to changes in stature.

During this stage, subjects were also familiarized with a portable radio telemetry system for respiratory gas analysis (Cortex Biophysik GmbH, Borsdorf, Germany). Subjects were connected to the equipment, ensuring maximum comfort and then undertook 15 min of activity (3 min of pushing, pulling, standing, walking and sitting).

On each of the two subsequent trials, subjects were required to attend the laboratory having participated in no physical activity 24 h before testing. Subjects were asked to rest with trunk supine and legs raised, knees flexed and ankles supported (Fowler's position) for 20 min before each test session to constitute a controlled period of spinal unloading. Each subject performed the normal work–rest schedule and the experimental work–rest schedule, on two separate occasions. The order of testing was assigned randomly to the subjects. The research design adopted meant that the subjects effectively acted as their own control. Each subject was tested at the same time of the day to eliminate any effects of diurnal variation.

Heart rate and oxygen consumption $\dot{V}O_2$ were measured continuously throughout each trial using short-range radio telemetry for heart rate (Polar) and a Metamax gas analysis telemetry system (Cortex Biophysik) respectively. Spinal shrinkage was measured at set intervals throughout the work protocols and at the finish. Pretest data points were obtained to elicit the individual s natural shrinkage. These data points were extrapolated to determine the predicted shrinkage over the 4 h. Final shrinkage was the difference between the predicted and observed values. All subjects were standing before measurement of stature so heel compression would not affect shrinkage results (Foreman and Linge 1989). At the end of each test, subjects were asked to rate their perceived exertion (RPE) on a 6–20 scale (Borg 1970).

2.3. Analysis of data

Energy expenditure was calculated using the energy equivalents of oxygen for the non-protein respiratory quotient (McArdle et al. 1996). Differences in spinal shrinkage and RPE between the two trials were analysed using t-tests in Minitab (v.12). An average value for the physiological variables was obtained from each 4-h session. Owing to evidence of skewness in the physiological data for heart rate, $\dot{V}O_2$, minute ventilation $\dot{V}E$, and energy expenditure, a non-parametric one-sample Wilcoxon test was performed based on the median rather than the mean differences. In the Results, medians are quoted for these variables and means are quoted for spinal shrinkage and RPE.

3. Results

Table 1 shows values for mean spinal shrinkage and RPE and medians for heart rate, $\dot{V}O_2$, energy expenditure and $\dot{V}E$ for the normal and the experimental trial. Mean spinal shrinkage values for the normal and experimental trial did not differ significantly ($p > 0.05$). The large SDs for the spinal shrinkage results were a product

Table 1. Recorded variables (mean ± SD, or median and range) and level of probability.

	Normal trial	Experimental trial	p
Spinal shrinkage (mm)	2.10 (±3.16)	2.90 (±2.92)	0.47
Heart rate (beats.min^{-1})	78 (range 71–93)	82 (71–90)	0.353
$\dot{V}O_2$ (l.min^{-1})	0.75 (0.65–0.94)	0.81 (0.65–0.98)	0.155
Energy expenditure (kJ.h^{-1})	948.1	979.5	0.185
$\dot{V}E$ (l.min^{-1})	18.41 (15.50–22.80)	19.20 (15.00–24.00)	0.308
Perceived exertion	7.6 (±1.4)	7.8 (±1.5)	0.34

of the methodology in which the final value represented the difference between the observed and expected shrinkage for each subject. No significant differences were found for any of the physiological variables measured ($p > 0.05$). Median heart rates (HR) were low for both test sessions, being 78 and 82 beats.min^{-1} for the normal and experimental trials respectively.

The median $\dot{V}O_2$ for the normal and the experimental trial were 0.75 (range 0.65–0.94) and 0.81 (0.65–0.98) l. min^{-1} respectively. No significant differences were found in average energy expenditure between the two test sessions ($p > 0.05$). Subjects expended an average of 963.8 kJ.h^{-1} for both trials combined. There were no differences in $\dot{V}E$ between the two trials ($p > 0.05$); medians were 18.41 (15.5–22.8) and 19.20 (15.0–24.0) l. min^{-1}.

There was also no difference in the rating of perceived exertion between the two trials ($p > 0.05$). Mean RPE corresponded to a very light intensity for the normal and the experimental trial.

4. Discussion

Where occupational strain is concerned, it is not enough to make changes to the working environment and equipment with a view to increasing workers' comfort. It is also important to measure the workers physiological and biomechanical responses in view of the potential risk of injury (Mathiassen 1993). In the present study, a range of physiological variables was combined with physical (shrinkage) and subjective (perceived exertion) measures to establish whether an alteration of the work-cycle caused a reduced response on these criteria.

Beneficial effects of sitting on spinal shrinkage are not conclusive. Magnusson *et al.* (1990) reported decreased stature in a sitting position. In this present study the subjects laid down before testing and the shrinkage observed while sitting could be due to the shrinkage naturally observed when subjects move from a supine into a sitting posture (Leivseth and Drerup 1997). Eklund and Corlett (1984) considered stature loss under different seated conditions but a standing trial was not included in the research design so a comparison between sitting and standing could not be made. Only three subjects were used in their study.

Althoff *et al.* (1992) concluded that sitting reduced spinal stress compared with standing. Leivseth and Drerup (1997) also suggested that shrinkage was greater during standing work than seated work and a period of relaxed sitting caused significant stature gains. Sitting is therefore believed to allow for fluid reabsorption of the intervertebral discs and slow down or preclude spinal shrinkage. Spinal shrinkage may not cease totally during sitting but the rate of shrinkage may be slowed compared with the rate of shrinkage when standing.

There were no significant differences between the two trials in any of the measured variables. When an individual sits down following a period of standing while working, it is not known exactly how long it takes for recovery of stature to occur. Helander and Quance (1990) reported that 5 min was insufficient for the process of fluid reabsorption into the intervertebral discs to occur whereas a reversal in shrinkage occurred after 20 min. This present study aimed to quantify spinal shrinkage over 4 h comparing two 12.5-min breaks with one 15-min break, assuming that the 5-min breaks in the first trial were too short to facilitate stature gains.

No differences in spinal shrinkage were found between the two trials. Shrinkages were small as they represented the difference between the observed and expected shrinkage and not the absolute shrinkage occurring over the entire shift. The positioning and length of the rest breaks had no effect on spinal shrinkage. owing to time constraints, a 4-h test protocol was used. To simulate the porters activity–rest schedule correctly, a total rest of only 25 min was allowed. As a result of this limited break time, it was not possible to alter radically the existing activity–rest schedule in terms of modifying break lengths and so differences in break length between the two trials were small. No significant difference in spinal shrinkage was observed, possibly due to the relatively small modification made to the breaks.

An 8-h shift would have allowed for a greater rest period than was used in the current trials. The differences between the rest breaks in the trials would have been more pronounced and the chance of observing significant differences increased. In this study, the lack of difference between the trials may also be because the longest break of 15 min was still insufficient to allow for a stature gain. Having longer rest breaks in an 8-h trial could have overcome this problem by ensuring stature is at least partly regained in some of the longer rest breaks.

There were no differences between the two conditions in any of the physiological variables measured. Median heart rate, $\dot{V}O_2$ and energy expenditure all fell within the limits reported for low intensity work (Brouha 1960, Nag *et al.* 1980, McArdle *et al.* 1996). Mean RPE showed that the subjects perceived the workload to be relatively light. Owing to the low intensity of the work, it was unlikely that cumulative fatigue occurred in any subject. The observations suggest that physiological strain as indicated by aggregate metabolic, circulatory and subjective measures, is not a major occupational risk in porters. For this reason, the positioning and length of the rest breaks are likely to be more important in other occupations where the intensity of the work is high enough to cause cumulative fatigue. The low intensity of the work may be responsible for the lack of significant differences between the two trials.

The subjects used in this study were not experienced porters and are likely to have been fitter and younger than many porters in the real world. It should be noted that the hardest activity profile observed in porters (in terms of mean heart rate) was used for the laboratory simulation. It can be assumed that porters undertake work no more difficult overall than that which was simulated in the laboratory. Porters' work is self-paced and the intensity at which they work is dependent upon the needs of the hospital during their particular shift. If porters had undergone the testing it can be assumed that they would not have found the work any more strenuous than did the subjects who were used because the work was self-paced. The estimated energy cost of an 8-h shift was estimated as 7584.8 (1806 kcal) and 7836 (1865 kcal) kJ for the normal and experimental trial respectively. The metabolic load would constitute an elevation over resting levels of ~ 3.5-fold. These findings confirm that there was no great metabolic strain on the individuals and that periodic musculoskeletal efforts remain the most likely source of damage.

5. Conclusion

Heart rate, $\dot{V}O_2$, perceived exertion and spinal shrinkage were not affected by the positioning and length of the rest breaks examined during simulated porters' tasks. The rest breaks in the two trials were insufficiently long to demonstrate any differences in spinal shrinkage. Physiological variables and the RPE indicated the overall low intensity of the work schedule. Employing the experimental activity–rest schedule would provide no obvious advantage to hospital-based porters.

Acknowledgements

This work was supported by a financial contribution from the European Commission within the Biomedicine and Health research and technological development programme. Thanks to Dr Kim Burton of the University of Huddersfield for the loan of the stadiometer.

References

ALTHOFF, I., BRINCKMANN, P., FORBIN, W., SANDOVER, J. and BURTON, K. 1992, An improved method of stature measurement for quantitative determination of spinal loading: application to sitting postures and whole body vibration, *Spine*, **17**, 682–693.

BIERING-SORENSEN, F. 1983, A prospective study of low back pain in a general population. 1. Occurrence, recurrence and aetiology, *Scandinavian Journal of Rehabilitation Medicine*, **15**, 71–79.

BORG, G. 1970, Perceived exertion as an indicator of somatic stress, *Scandinavian Journal of Rehabilitation Medicine*, **2**, 92–98.

BORK, B. E., COOK, T. M., ROSECRANCE, J. C., ENGELHARDT, K. A., THOMASON, M. E. J., WAUFORD, I. J. and WORLEY, R. W. 1996, Work-related musculoskeletal disorders among physical therapists, *Physical Therapy*, **76**, 827–835.

BRINKMANN, P., FROBIN, W., BIGGEMANN, M., TILLOSTSON, M. and BURTON, K. 1998, Quantification of overload injuries to thoracolumbar vertebrae and discs in persons exposed to heavy physical exertions or vibration at the workplace, *Clinical Biomechanics*, **13**, 1–36.

BROUHA, L. 1960, *Physiology in Industry* (Oxford: Pergamon).

BURTON, A. K., TILLOTSON, K. M. and BOOCOCK, M. G. 1994, Estimation of spinal loads in overhead work, *Ergonomics*, **37**, 1311–1321.

EKLUND, J. A. E. and CORLETT, E. N. 1984, Shrinkage as a measure of the effect of load on the spine, *Spine*, **9**, 189–194.

FOREMAN, T. K. and LINGE, K. 1989, The importance of heel compression in the measurement of diurnal stature variation, *Applied Ergonomics*, **20**, 299–300.

FRYMOYER, J. W. and GORDON, S. L. 1989, Research perspectives in low-back pain. Report of a 1988 workshop, *Spine*, **14**, 1384–1390.

HELANDER, M. G. and QUANCE, L. A. 1990, Effect of work–rest schedules on spinal shrinkage in the sedentary worker, *Applied Ergonomics*, **21**, 279–284.

HILDEBRANDT, V. H. 1995, Back pain in the working population: prevalence rates in Dutch trades and professions, *Ergonomics*, **38**, 1283–1298.

LEIVSETH, G. and DRERUP, B. 1997, Spinal shrinkage during work in a sitting posture compared with work in a standing posture, *Clinical Biomechanics*, **12**, 409–418.

MAGNUSSON, M. L., ALMQUIST, M. and LINDSTROM, I. 1990, Measurements of time dependent height loss during sitting, *Clinical Biomechanics*, **5**, 137–142.

MATHIASSEN, S. E. 1993, The influence of exercise/rest schedule on the physiological and psychophysical response to isometric shoulder–neck exercise, *European Journal of Applied Physiology and Occupational Physiology*, **67**, 528–539.

MCARDLE, W. D., KATCH, F. I. and KATCH, V. L. 1996, *Exercise Physiology* (Baltimore: Williams & Wilkins).

NAG, P. K., SEBASTIAN, N. C. and MAVLANKAR, M. G. 1980, Occupational work-load of Indian agricultural workers, *Ergonomics*, **23**, 91–102.

REILLY, T., TYRRELL, A. and TROUP, J. D. G. 1984, Circadian variation in human stature, *Chronobiology International*, **1**, 121–126.

30 Implications of an adjustable bed height during standard nursing tasks on spinal motion, perceived exertion and muscular activity

D. E. Caboor[†*], M. O. Verlinden[†], E. Zinzen[†], P. Van Roy[†], M. P. van Riel[‡] and Jan Pieter Clarys[†]

[†]Department of Experimental Anatomy, Faculty of Physical Education and Physiotherapy, Vrije Universiteit Brussel, Laarbeeklaan 103, B-1090 Brussels, Belgium

[‡]Department of Biomedical Physics and Technology, Faculty of Medicine and Allied Health Sciences, Erasmus Universiteit Rotterdam, PO Box 1738, NL-3000 DR Rotterdam, The Netherlands

Keywords: Low back problems; Nurses; Posture; Task demands; Electromyography; Electrogoniometry; Back movements.

Manual handling is a source of occupational stress, particularly for nursing personnel. High levels of biomechanical strain are associated with lifting and transferring patients, especially when the tasks are performed in flexed and twisted positions that induce an increased risk of functional and musculoskeletal problems. The use of adjustable beds in nursing practice has been suggested as a means of influencing working postures and reducing the muscular demands on nurses. The purpose of this study was to investigate the effects on spinal motion, muscular activity and perceived exertion when nurses had the opportunity to adjust bed height. The measures recorded during the conduct of standardized patient handling tasks were the changes in posture (inclination) and in shape (sagittal bending, side bending, axial rotation). Muscular activity was measured using surface electromyography. Perceived exertion was rated using the 15-graded Borg scale. The range of motion was not influenced by the adjustment of bed height, but rather a shift of the time duration histogram was noticed in the direction of the erect, safer position. The time spent in the safe zone of spinal motion near the erect position was significantly increased and was significantly decreased in the potential health-hazardous zones of spinal motion in the extreme positions. No differences in muscular activity or in perceived exertion were found between the two bed height conditions for any of the muscle groups. It was concluded that the quality of spinal motion is enhanced when the opportunity of adjusting the bed height is offered.

1. Introduction

Low back problems (LBP) have a high frequency in the nursing profession and are often discussed with respect to their medical, psychosocial and economic consequences (Mayer *et al.* 1985, Spengler *et al.* 1986, Troup *et al.* 1987, Spengler and Szpalski 1990, Burton *et al.* 1997). Manual handling tasks in nurses are closely associated with the risk of suffering LBP. Stresses posed on the operators of manual

---

*Author for correspondence. e-mail: dcaboor@exan.vub.ac.be

handling tasks are manifested as strain on the musculoskeletal system, and when the task demands exceed the capacity of the musculoskeletal system the results can include discomfort and injuries (Dempsey 1998). Task demands are defined in terms of organization, material and devices, task performance and environmental characteristics. Worker capacity is defined in terms of personal characteristics, and biomechanical, physiological and psychological capacities. The ratio of task demands to worker capacity may be used to define the equilibrium between the task demands and the worker's capacity, being influenced by physiological, physical and psychosocial factors (Estryn-Behar et al. 1990, Parnianpour et al. 1990, Smedley et al. 1995, Burton et al. 1997).

In the occupational biomechanics' literature work intensity, static work postures (Gagnon et al. 1986, Estryn-Behar et al. 1990), frequent bending and twisting, lifting, pushing or pulling (Schultz and Andersson 1981, Adams and Hutton 1982, McGill and Norman 1987), velocity of performance (Buseck et al. 1988, Frigo 1990, Marras et al. 1995), and the distance to the object (Potvin et al. 1994) have been identified as risk factors. Task performance can be influenced by the design of the material and the devices available. De Looze et al. (1994) found that the adjustment of bed height led to lower values of time-integrated compression, peak shear force and time-integrated shear force at the L5–S1 level of nurses. The use of adjustable beds in nursing practice could influence the working postures of nursing personnel and lead to a reduction in task demands.

The rating of perceived exertion is commonly used in sports, training and occupational circumstances. Supplementary to physiological variables, Borg's scale is considered as a standardized tool for the description of the perception of physical effort (Noble and Robertson 1996).

Kinesiologal electromyography, in particular surface electromyography, is used extensively as a variable in specific ergonomic situations. Recent studies include investigations of static and dynamic lifting tasks (Yates and Karkowski 1992), fatigue during repetitive light work (Nakata et al. 1992), excessive drafts on shoulder muscles (Sundelin and Hagberg 1992), typewriting and keyboard use (Fernström et al. 1994), effects of precision and force demands in manual work (Milerad and Ericson 1994), symmetric and asymmetric lifting tasks in restricted postures (Gallagher et al. 1994), verifying spinal and abdominal muscle activity during garden raking (Kumar 1995), and during repetitive lifting tasks (Kim and Chung 1995), assessing load on upper M. trapezius in jet pilots (Harms-Ringdahl et al. 1996), monitoring the influence of the operating technique on muscular strain (Luttmann et al. 1996a) and muscular fatigue of surgeons in urology (Luttmann et al. 1996b). Marras and Granata (1997) developed an EMG-assisted model to assess spinal loading during free-dynamic lifting.

The purpose of this study was to investigate the implications on spinal motion, muscular activity and perceived exertion when nurses were given the opportunity to choose bed heights during the performance of typical nursing tasks.

2. Methods

2.1. Subjects

Eighteen right-handed nurses, 10 females (mean age 31 ± 7.6 years, mean height 1.664 ± 0.063 m, mean body mass 61.9 ± 5.3 kg) and eight males (32 ± 7.5 years, 1.752 ± 0.056 m, 75.8 ± 7.1 kg) employed in two Belgian hospitals, volunteered for the study. They were asked to perform four prescribed and standardized tasks both in standard bed height (SBHC) (0.515 m) and in individually adjusted bed height conditions (ABHC), in a simulated hospital room (Caboor et al. 1997). Each subject adjusted the bed manually and performed each task three times successively.

2.2. Task description

While the tasks were carried out, movement in three dimensions was recorded using an electrogoniometer and surface electromyography. The tasks to be performed were selected on the basis of a psychomotor task-analysis (Caboor et al. 1997), a method based on an expert survey to rank the task demands in the workplace. These highly classified loading tasks were (1) turning over and giving the bedpan to a quadriplegic patient starting in a resting position with the head of the bed inclined 24.5°, followed by removing the bed-pan and positioning the patient on the back, (2) washing the leg of a quadriplegic patient lying in bed, (3) transferring a right-sided hemiplegic patient in and out bed, starting in a resting position with the head of the bed inclined for 24.5°, followed by positioning upon a toilet seat and repositioning of the patient lying on the bed, and (4) helping a right-sided hemiplegic patient in and out bed, starting in a seated position, legs off the bed.

Transfer of a patient was the most prominent action during the conduct of the tasks. The standardization of the task performance was guaranteed by pretraining and a continuous coaching following the script.

2.3. Spinal motion: continuous movement registration

Spinal motion was measured using a continuously registering electrogoniometer (Snijders and van Riel 1987, van Riel et al. 1987). The spinal motion variables recorded were the changes in the position of the trunk (inclination) and the changes in the curvature (sagittal bending, side bending and axial rotation). The inclinometer was a resistive potentiometer whose wiper was attached to a pendulum that could move only in the sagittal plane with a range of 120°. The housing was attached to the skin at the L3 level and the position of the pendulum within it was dependent on the gravitational force. The two sensors for the bending motions were developed to measure changes in the length of the skin and were based on the principle of the extension of a small helical spring connected at one end to a strain gauge type transducer (Kulite load cell) and at the other to a thin thread. This tube was also a helical spring, being attached to the housing of the force meter at the end of the thread and taped to the skin. Measured between Th7 and L5 the length of the contour of the back increased if the subject bent forward and decreased when bending backwards. Lateroflexion gave rise to a difference between the distance on the left side of the spine and that of the right side. For measuring torsion between C7 and Th10 as a measure for the total rotation of the shoulders in relation to the pelvis, a single turn potentiometer with a flexible driving shaft of 0.15 m was used. The housing of the potentiometer and that of the end of the shaft were attached to the skin. The range of motion was divided in boundaries of 5° concerning inclination, 5 mm for the bending motions and 2° for axial rotation. Zone occupation frequency and time spent within each boundary band were calculated. Differences between bed height conditions were determined by the use of Student t-test statistics ($p<0.05$).

2.4. Electromyography

A classical surface electromyographic set-up was used (Clarys 1994). Surface-electrode recording was carried out with bar-like active bipolar Ag/AgCl electrodes with a diameter of 15 mm for each electrode (with 8 mm diameter detection area) at an interelectrode distance of 10 mm. Standard preparation of the skin (shaving, skin abrasion, alcohol to cleanse the skin) and the use of a conductive gel assured reliable measurements. The electrodes were oriented in the direction of muscle fibres, located on the bulkiest part of the contracted muscle. Activity was registered both at the left and at the right side of the body on the M. obliquus externus abdominis, the M. erector spinae and the M. biceps femoris. Those muscles were chosen for their functional relation to the lower back, the sacroiliac joints and the legs via the fascia thoracolumbalis and the ligamentum sacrotuberale (Vleeming 1990). The detected electromyographic signal was amplified by an AD 524 amplifier (gain 10) with an input impedance of 10^9 Ω and a CMRR of 90 dB at the level of the electrodes, followed by a preamplifier (gain 50, 10^{12} Ω, CMRR 85 dB). For digitalization (by a 12-bit analogue-to-digital converter) of the raw EMG signals, the authors sampled at a frequency of 2 kHz to store data on computer before further analysis. This may be considered as a low-pass filter of 1 kHz. In sequential order, visual inspection for considerable artefacts and baseline drifts was performed on the raw EMG signals. The synchronization signal indicated commencement of the different tasks and the different phases inherent to the task. Signals were full-wave rectified, smoothed by a time constant factor of 200 ms. The linear envelope was considered for qualitative analysis, the integrated EMG (iEMG) for a quantitative approach of the signals, iEMG indicating muscular intensity. In order to compare between males and females, between standard bed height (SBHC) and individually adjusted bed height conditions (ABHC), the highest dynamic EMG peak occurring during the task or movement under consideration was used for normalization purposes.

Using the RMS method, all task performances were averaged per subjects, per trials and per task performance for all subjects. Ensemble averages of three different muscle groups were calculated and compared in the different bed height conditions.

Pattern (linear envelope) inspection of the EMG showed considerable intra-individual variability between the trials. Therefore the global analysis was considered for quantitative purposes, while the detailed analysis was performed for qualitative purposes. Global analysis of the muscular activity as a result of all four tasks was carried out in order to determine the importance of the time factor in combination with the two ergonomic circumstances (SBHC and ABHC).

2.5. Rating of perceived exertion

At the end of the session, the nurses were asked to rate the perceived exertion on a 15-graded Borg scale (Noble and Robertson 1996). The rating runs from extremely light (score 6–7), fairly light (11) and fairly heavy (13) to extremely heavy (19–20).

3. Results

3.1. Task performance

Individuals could adjust the standard bed height in either direction, up or down. The adapted bed height differences were small. Sixteen nurses adjusted the bed height in a higher position and two lowered the bed. An average of 0.064 (SD 0.044, min −0.02, max +0.14) m only was employed by the nurses during adaptation of the standard bed height of 0.515 m. Experience of using and manipulating height-adjustable beds was mentioned by 16 nurses but only eight of them possessed any experience in using specific lifting skills.

3.2. Spinal motion

A variety of variables (torsion, inclination, side-bending, maximum amplitude) was considered, but the range of motion did not change significantly. Only a shift of the time duration histogram was observed in the direction of the erect and safer position. The occupation frequency was significantly changed within several boundaries of inclination (table 1), sagittal bending (table 2), lateroflexion (table 3) and torsion (table 4).

The occupation frequency within boundaries of movement was significantly increased in the safe zone of spinal motion near the erect position. It was significantly

Table 1. Inclination in adjusted bed height conditions versus standard bed height conditions: significant differences between the means of the zone occupation frequency.

Zone (°)	Difference	SD	p
0–5	86.67	134.06	0.000
5–10	160.39	151.70	0.003
10–15	268.06	331.03	0.002
15–20	344.22	391.35	0.001
20–25	391.06	404.14	0.001
25–30	387.78	417.20	0.007
30–35	328.22	457.49	0.045
35–40	–	–	ns
40–45	–	–	ns
45–50	–	–	ns
50–55	−191.83	456.6	0.026
55–60	−231.22	400.81	0.004
60–65	−334.28	430.61	0.000
65–70	−406.72	393.97	0.000
70–75	−353.06	341.46	0.000
75–80	−300.83	197.94	0.000
80–85	−230.50	180.20	0.002
85–90	−143.67	166.32	0.016
90–95	−85.17	135.25	0.046

Table 2. Sagittal bending in adjusted bed height conditions versus standard bed height conditions: significant differences between the means of the zone occupation frequency.

Zone (mm)	Difference	SD	p
−5–0	48.83	88.69	0.032
0–5	185.94	235.55	0.004
5–10	294.72	441.30	0.011
10–15	395.83	552.54	0.007
15–20	376.28	553.58	0.010
20–25	–	–	ns
25–30	−378.28	604.37	0.017
30–35	−476.94	427.55	0.000
35–40	−323.17	269.98	0.000
40–45	−185.78	277.27	0.011

ns, not significant.

Table 3. Lateroflexion in adjusted bed height conditions versus standard bed height conditions: significant differences between the means of the zone occupation frequency.

Zone (mm)	Lateroflexion occupation frequency		
	Difference	SD	p
20–15 left side	−250.11	440.90	0.028
15–10 left side	−294.50	476.62	0.018

Table 4. Torsion in adjusted bed height conditions versus standard bed height conditions: significant differences between the means of the zone occupation frequency.

Zone (°)	Lateroflexion occupation frequency		
	Difference	SD	p
Neutral position	229.50	377.77	0.020

decreased in the potentially health hazardous zones of spinal motion in the extreme positions ($p < 0.05$).

3.3. Electromyography

Analysis showed no significant changes in iEMG ($p > 0.05$). A comparison between left and right sides of the body indicated that male nurses in SBHC showed a higher iEMG in their left side, thus more muscular activity (intensity), intrinsic to the specificity of the task demands. Female nurses showed a higher iEMG activity right-sided during SBHC and made a shift in activity towards the left side of the body during ABHC, both during the flexion and extension movements. Male nurses did not show as consistently as the female nurses the differences in both bed height conditions. Besides, male nursing personnel showed the opposite pattern of muscular adaptation towards an ABHC.

Moreover, the EMG-patterns of the extension phases during performance of the actions were dominant and hence determined the whole of the activity of the total movement. The share of iEMG values as a result of the extension phases was probably more important in the overall movement than was the flexion phase. Comparing standard bed height conditions (SBHC) to adjusted bed height conditions (ABHC), for the whole set of four tasks, and taking into account the ensemble averages for 3 different muscle groups, the data suggest that there was no adaptation in terms of muscular activity towards an individually adapted bed height conditions (table 5).

3.4. Rating of perceived exertion

The measures on the 15-graded Borg scale were 12.85 ± 1.95 for SBHC and 12.93 ± 2.13 for ABHC. Perceived exertion did not differ significantly between the two bed height conditions ($p > 0.05$).

4. Discussion

The results of this study suggest that the range of motion was not influenced by the adjustment of the bed height, but rather by a shift of the time duration in the

Table 5. Mean integrated EMG for the comparison of different muscle groups in both standard and adjusted bed height conditions.

Combination	Standard bed height conditions SBHC	Adjusted bed height conditions ABHC	Significance (Mann-Whitney U)
1	40,04	40,62	ns
2a	39,18	40,51	ns
2b	40,89	40,74	ns
3a	39,44	39,91	ns
3b	39,88	41,38	ns

Combination 1 = all registered muscles.
Combination 2a = all left body-sided muscles.
Combination 2b = all right body-sided muscles.
Combination 3a = functionally coupled muscles (trunk right-sided muscles with the left-sided hamstring muscle).
Combination 3b = functionally coupled muscles (trunk left-sided muscles with the right-sided hamstring muscle).
ns, non-significant ($p \leqslant 0.05$).

direction of the erect, safer position. The definition of the zones of spinal motion was based on the studies on intradiscal pressure, the NIOSH guidelines and the OVAKO analysis. The zone nearest to the erect position was defined as the safe zone, the farthermost position as the hazardous zone and in between was the working zone. Nachemson (1975) and Wilke et al. (1999) found that the intradiscal pressure is dependent on posture, is lower in erect positions than in flexed positions, and increases with increasing forward bending and with increasing external load. The time spent in the safe zone of spinal motion near the erect position was significantly increased and was significantly decreased in the potential health-hazardous zones of spinal motion in the extreme positions.

The study showed an important implication for the quality of spinal motion when the opportunity of adjusting the bed height was offered, since the level of task demands was altered. The aim of interventions at this level is to influence the ratio of task demands to worker capacity towards a health promoting profile.

Adjusting the bed height did not show any consequence for muscular activity in any of the different muscle groups, at least within the conditions investigated. It is possible that if the bed height was altered more than the voluntary adjustment permitted in this study, a greater difference in muscular activity might be observed. It is therefore advised that nurses should be trained to obtain their individual optimal bed height conditions. According to de Looze et al. (1994), the major factor interfering negatively with possible favourable effects, is the restricted capacity of nurses to select an optimal bed position with respect to the specificity of the task.

The existence of a protective muscular system can never be eliminated consciously. Increased muscular activity is induced by a higher task demand, both at muscular and other soft tissue levels. Nachemson (1975), Andersson et al. (1977) and Schultz et al. (1982) found higher intradiscal pressures in response to increased muscular activity of the M. erector spinae. The observation of higher iEMG corresponds with higher compression forces in the spine, but de Looze et al. (1994) failed to confirm this relation. They observed lower time-integrated compression forces and peak and time-integrated shear forces, but could not discover any changes in the peak compression forces during the performance of nursing tasks in ABHC.

The perceived exertion in both bed height conditions did not reveal any significant difference. This might reflect that nurses did not have the capacity to adjust the bed height optimally.

5. Conclusions

Adjusting the bed height during standard tasks has significant implications for the quality of spinal motion. The use of adjustable beds in nursing practice can influence the working postures of personnel and cause a reduction of the task demands. The significant increase of time spent in the safe zone of spinal motion can influence the compression and shear forces in the lower back. In this way the lower values found by de Looze et al. (1994) in time-integrated compression, peak shear forces and time-integrated shear forces, are influenced by the increased quality of spinal motion. This is not the case for muscular activity, which did not alter with freely chosen but changeable bed heights. Perceived exertion indicated that changing the bed height had no significant psychological impact. The standard bed height was already close to optimal for the majority of the subjects.

Acknowledgements

The authors acknowledge the BioMed European Commission, Directorate General XII, contract MBH4-CT96-1057, for financial support.

References

ADAMS, M. A. and HUTTON, W. C. 1982, Prolapsed intervertebral disc, a hyperflexion injury?, Spine, **7**, 184–191.

ANDERSSON, G. B. J., ORTENGREN, R. and NACHEMSON, A. 1977, Intradiscal pressure, intraabdominal pressure and myoelecric back muscle activity related to posture loading, Clinical Orthopædics, **129**, 156–164.

BURTON, A. K., SYMONDS, T. L., ZINZEN, E., TILLOTSON, K. M., CABOOR, D., VAN ROY, P. and CLARYS, J. P. 1997, Is ergonomic intervention alone sufficient to limit musculoskeletal problems in nurses?, Occupational Medicine, **47**, 25–32.

BUSECK, M., SCHIPLEIN, O. D., ANDERSSON, G. B. J. and ILLMARINEN, J. 1988, Influence of dynamic factors and external loads on the moment at the lumbar spine in lifting, Spine, **13**, 918–921.

CABOOR, D., ZINZEN, E., VAN ROY, P. and CLARYS J. P. 1997, Job evaluation using a modified Delphi-survey, in Proceedings of the International Congress of The South African Society of Physiotherapy, University of Cape Town, 56–60.

CLARYS, J. P. 1994, Electrology and localised electrisation revisited (review), Journal of Electromyography and Kinesiology, **1**, 5–14.

DE LOOZE, M. P., ZINZEN, E., CABOOR, D., HEYBLOM, P., VAN BREE, E., VAN ROY, P., TOUSSAINT, H. M. and CLARYS, J. P. 1994, Effects of individually chosen bed height adjustments on the low-back stress of nurses, Scandinavian Journal of Work, Environment and Health, **20**, 427–434.

DEMPSEY, P. G. 1998, A critical review of biomechanical, epidemiological, physiological and psychological criteria for designing manual materials handling tasks, Ergonomics, **41**, 73–88.

ESTYRN-BEHAR, M., KAMINSKI, M., PEIGNE, E. and MAILLARD, M. F. 1990, Strenuous working conditions and musculoskeletal disorders among female hospital workers, International Archives of Occupational and Environmental Health, **62**, 47–57.

FERNSTRÖM, E., ERICSON, M. O. and MALKER, H. 1994, Electrographic activity during typewriter and keyboard use, Ergonomics, **37**, 477–484.

FRIGO, C. 1990, Three-dimensional model for studying the dynamic loads on the spine during lifting, *Clinical Biomechanics*, **5**, 143–152.

GAGNON, M., SICARD, C. and SIROIS, J. P. 1986, Evaluation of forces on the lumbo-sacral joint and assessment of work and energy transfers in nursing aides lifting patients, *Ergonomics*, **29**, 409–421.

GALLAGHER, S., HAMRICK, C. A., LOVE, A. C. and MARRAS W. S. 1994, Dynamic biomechanical modelling of symmetric and asymmetric lifting tasks in restricted postures, *Ergonomics*, **37**, 1289–1310.

HARMS-RINGDAHL, K., EKHOLM, J., SCHULDT, K., LINDER, J. and ERICSON, M. O. 1996, Assessment of jet pilots' upper Trapezius load calibrated to maximal voluntary contraction and a standardized load, *Journal of Electromyography and Kinesiology*, **6**, 67–72.

KIM, S. H. and CHUNG, M. K. 1995, Effects of posture, weight and frequency on trunk muscular activity and fatigue during repetitive lifting tasks, *Ergonomics*, **38**, 853–863.

KUMAR, S. 1995, Electromyography of spinal and abdominal muscles during garden raking with two rakes and rake handles, *Ergonomics*, **38**, 1793–1804.

LUTTMANN, A., SÖKELAND, J. and LAURIG, W. 1996a, Electromyographical study on surgeons in urology. I. Influence of the operating technique on muscular strain, *Ergonomics*, **39**, 285–297.

LUTTMANN, A., JÄGER, M., SÖKELAND, J. and LAURIG, W. 1996b, Electromyographical study on surgeons in urology. II. Determination of muscular fatigue, *Ergonomics*, **39**, 298–313.

MARRAS, W. S. and GRANATA, K. P. 1997, The development of an EMG-assisted model to assess spine loading during whole-body free-dynamic lifting, *International Journal of Electromyography and Kinesiology*, **7**, 259–268.

MARRAS, W. S., LAVENDER, S. A., LEURGANS, S. E., FATHALLAH, F. A., FERGUSON, S. A., ALLREAD, W. G. and RAJULU, S. L. 1995, Biomechanical risk factors for occupationally related low back disorders, *Ergonomics*, **38**, 377–410.

MAYER, T. G., SMITH, S. S., KEELEY, J. and MOONEY, V. 1985, Quantification of lumbar function. II. Sagittal plane trunk strength in chronic low back patients, *Spine*, **10**, 765–772.

McGILL, S. M. and NORMAN, R. W. 1987, Dynamically and statically determined low back moments during lifting, *Journal of Biomechanics*, **18**, 877–885.

MILERAD, E. and ERICSON, M. O. 1994, Effects of precision and force demands, grip diameter and arm support during manual work: an electromyographic study, *Ergonomics*, **37**, 255–264.

NACHEMSON, A. 1975, Towards a better understanding of back pain, a review of the mechanics of the lumbar disc, *Rheumatology and Rehabilitation*, **14**, 129–135.

NAKATA, M., HAGNER, I. and JONSSON, B. 1992, Perceived musculoskeletal discomfort and electreomyography during repetitive light work, *Journal of Electromyography and Kinesiology*, **2**, 103–111.

NOBLE, B. J. and ROBERTSON R. J. 1996, *Perceived Exertion* (Champaign: Human Kinetics).

PARNIANPOUR, M., NORDIN, M., SKOVRON, M. L. and FRANKEL, V. H. 1990, Environmentally induced disorders of the musculoskeletal system, *Medical Clinics of North America*, **74**, 347–358.

POTVIN, J. R., McGILL, S. M. and NORMAN, R. W. 1994, Trunk muscle and lumbar ligament contributions to dynamic lifts with varying degrees of trunk flexion, *Spine*, **16**, 1099–1108.

SCHULTZ, A. B. and ANDERSSON, G. B. J. 1981, Analysing loads on the lumbar spine, *Spine*, **6**, 76–82.

SCHULTZ, A. B., ANDERSSON, G. B. J., ORTENGREN, R., HADERSPECK, K. and NACHEMSON, A. 1982, Loads on the lumbar spine, validation of a biomechanical analysis by measurements of intradiscal pressures and myoelectric signals, *Journal of Bone and Joint Surgery*, **64**, 713–720.

SMEDLEY, J., EGGER, P., COOPER, C. and COGGON, D. 1995, Manual handling activities and risk of low back pain in nurses, *Occupational Environmental Medicine*, **52**, 160–163.

SNIJDERS, C. J. and VAN RIEL, M. P. 1987, Continuous measurement of spine movements in normal working situations over periods of 8 hours or more, *Ergonomics*, **30**, 639–653.

SPENGLER, D. M., BIGOS, S. and MARTIN, N. 1986, Back injuries in industry, *Spine*, **11**, 241–245.
SPENGLER, D. M. and SZPALSKI, M. 1990, Newer assessment approaches for the patient with low back pain, *Contemporary Orthopædics*, **21**, 371–378.
SUNDELIN, G. and HAGBERG, M. 1992, Effects of exposure to excessive drafts on myoelectric activity in shoulder muscles, *Journal of Electromyography and Kinesiology*, **2**, 36–41.
TROUP, J. D. G., FOREMAN, T. K., BAXTER, C. E. and BROWN, D. 1987, The perception of back pain and the role of psychological tests of lifting capacity, *Spine*, **12**, 645–657.
VAN RIEL, M. P., GROENEVELD, W. H. and SNIJDERS, C. J. 1987, Continue houdings-en bewegingsregistratie op de werkplek, *Klinische Fysica*, **4**, 184–190.
VLEEMING, A. 1990, The sacro-iliac joint: a clinical–anatomical, biomechanical and radiological study. PhD thesis, Erasmus University, Rotterdam.
WILKE, H. J., NEEF, P., CAIMI, M., HOOGLAND, T. and CLAES, L. E. 1999, New *in vivo* measurements of pressures in the intervertebral disc in daily life, *Spine*, **24**, 755–762.
YATES, J. W. and KARWOWSKI, W. 1992, An electromyographic analysis of seated and standing lifting tasks, *Ergonomics*, **35**, 889–989.

31 Work-based musculoskeletal problems: initiatives to improve health

DIANA LEIGHTON*

Health and Safety Executive, Human Factors Unit, Magdalen House, Trinity Road, Bootle, Merseyside L20 3QZ, UK†

Musculoskeletal disorders (MSD) are the leading cause of occupational ill-health in the UK and the level of suffering and costs associated with the current prevalence of MSD is of concern. In partnership with others, the Health and Safety Executive (HSE) is committed to tackling MSD. This will be achieved in part through the implementation of new wide-ranging, integrated policy initiatives and strategies, and through the HSE's core activities of publication of guidance, publicity campaigns and workplace visits by inspectors. A summary of these activities is presented, demonstrating that with new initiatives and partnerships, the HSE will continue to work towards greater awareness of occupational health issues and more effective management of health risks in the workplace.

1. Introduction

Musculoskeletal disorders (MSD) constitute the leading cause of occupational ill-health in the UK. Figures from a household survey indicate that MSD affect an estimated 1.2 million people (60%) out of an estimated 2 million total reported cases (Jones *et al*. 1998).

The level of suffering and costs associated with the current prevalence of MSD is of concern. It is considered a high priority issue and further action is needed to prevent or reduce the risks associated with the onset of MSD. In partnership with others, the HSE is committed to tackling MSD in the UK. This will be achieved in part through the implementation of new wide-ranging, integrated policy initiatives and strategies and through the core activities of publication of guidance, publicity campaigns and workplace visits by inspectors.

There is still a need to understand more fully the risk factors for MSD, the nature of these disorders and what can be done to prevent and treat them. Research programmes continue to play an important part in the HSE's approach, ensuring that sound advice is given to industry concerning risk factors and control measures. Research is commissioned in line with developing strategy and needs to be highly specific, of good quality and linked to operational and policy priorities.

The Health and Safety Commission (HSC) and the HSE are statutory bodies whose overall purpose is to ensure that risks to people's health and safety from work activities are properly controlled. The HSC advises and responds to the Department of Environment, Transport and the Regions and the Government Minister with

---

*e-mail: len.morris@hse.gsi.gov.uk
†Former address: Now at: Dept. of Public Health, Liverpool John Moores University, Gt. Crosshall Street, Liverpool

overall responsibility for health and safety. The HSE is a distinct statutory body that advises and assists the HSC. The HSE has day-to-day responsibility for enforcing health and safety legislation in over 650,000 establishments under general guidance from HSC. The HSC is a tripartite body—its members, appointed by the Secretary of State, are either independent or represent industry, local authorities and trade unions. It appoints the three members of the 'Executive' that in turn employs over 4000 staff, referred to as the HSE.

The combined aims of the two bodies are to protect the health, safety and welfare of workers, and to safeguard others, principally the public, who may be exposed to risks from work activities. To achieve these aims, the HSE provides assistance to those people who have responsibility for control of the work environment, that is the duty holder. The duty holder may be the employer, who has a duty to look after the health and safety of the employees, but employees and the self-employed also have a duty to look after their own health and safety. All have to safeguard others who may be exposed to risks from work activities.

The HSE operates in a number of ways to assist the duty holder. It has a regulatory role (policy), to develop the health and safety laws, codes and standards that cover safe working right across industry. The HSE and local authority inspectors promote and enforce health and safety legislation, usually by advising people how to comply with the law but may also require employers to make improvements by way of enforcement notices or prosecution where necessary. The HSE sets standards and also publishes guidance on good practice in health and safety issues. To support this guidance, it both carries out and funds technical research.

Since the early 1980s the HSE has pursued a number of activities with the aim of reducing the occurrence of work-related MSD. These include funding over 40 extramural research projects; the introduction of new regulations and associated guidance for employers, for example the Manual Handing Operations Regulations 1992, the Health and Safety (Display Screen Equipment) Regulations 1992; and health risk management awareness campaigns 'Good Health is Good Business' and 'Lighten the Load'.

The HSC published its first Strategic Plan in May 1999 to reflect its commitment to improve standards of health and safety (HSC 1999). The 3-year plan incorporates strategic themes, major issues that need to be addressed over the longer term. Each theme is underpinned by a number of key programmes with clear targets and measurable outcomes.

This review provides a summary of the HSE's main activities in relation to MSD covering each of its operations: policy, technical research, information provision and workplace inspection. Where appropriate, reference will be made to the HSC Strategic Plan, its themes and key programmes. A message is that there is a need to work in partnership with other stakeholders to engender a holistic approach to the prevention and management of MSD.

2. Policy initiatives

Following a review of occupational health policy (the Health Risk Reviews) in the early 1990s, the HSE has been working towards an integrated strategy for the prevention of MSD, bringing together existing initiatives on manual handling, display screen equipment and upper limb disorders. The priority status the government has now given to occupational health issues and MSD has been welcomed within the HSE as it provides a valuable opportunity to make real progress towards a reduction in ill-health.

2.1. *Healthy workplace initiative*
In the Green Paper on Public Health, the government identified the workplace as a major setting for targeting in relation to improving the health of the working population. To show the way forward, the HSE and the Department of Health combined in launching the Healthy Workplace Initiative with a joint Statement of Intent to work together to improve the health in the workplace, thereby underlining that improving health is everybody's business. The main aims of the initiative are to get employers working in partnership with occupational health professionals on health problems. The theme for the first 3 years is to be 'Back Pain'.

2.2. *'Back in Work' initiative*
Back pain is the leading cause of sickness absence from work. The HSE has estimated that 11 million working days were lost in the UK in 1995 from MSD including back pain. Around 650,000 people reported a workplace-linked back pain disorder (Jones *et al.* 1998). The workplace sectors with the highest incidence of back pain are healthcare, construction, agriculture, water, food and retail sectors.

The 'Back in Work' initiative supports the HSC's strategic theme 'to raise the profile of occupational health' and is part of the key programme of preventing and managing MSD (HSC 1999). The overall aim of the initiative is to prevent work-related back pain and to provide early access to assessment, treatment, rehabilitation and return to work. Successful outcome will be a reduction in the number of days off work due to back pain.

'Back in Work' is a campaign that will encourage employers and employees to work together to reduce the misery and cost of back pain to those in the workplace. It will seek to do this by increasing awareness of the issue and encouraging employers to take action in partnership with others.

The initiative, launched in May 1999, will begin by supporting and evaluating a series of pilot projects. It should be possible through these projects to demonstrate and promote good practice and show that this initiative is effective in reducing the number of days lost due to back pain. All selected pilot projects will be supported to a maximum of £50,000 per project and 12 months. Each pilot that has been supported will be evaluated and the full results will form the basis of a good practice guide to be published at the end of the year 2000.

2.3. *Occupational health strategy*
The HSE is at present facilitating the development of a long-term occupational health strategy for the UK. The strategy will complement the public health strategies for England, Scotland and Wales. On the 12 August 1998, the HSE published a discussion document to widen the debate on some of the important issues that had to be discussed before a new strategy could be developed, for example 'What is occupational health?' (HSE 1998a).

The strategy is likely to introduce initiatives that will be relevant to MSD. Examples are improving compliance with the law, collecting essential knowledge and ensuring that suitable frameworks are in place to make advice, information and support available.

The HSE's officials considered all the comments received during 1999. Recommendations were then presented to the HSC to consider in the summer of 1999. The development of this strategy also supports the HSC Strategic Plan (HSC 1999).

2.4. *HSE's strategy on MSD*
The components of a new strategy specifically on MSD are still under development but it is already apparent that the strategy will need to be based on an understanding that MSD may be due to a combination of work and other activities. Factors such as sport and leisure, past practices such as carrying a heavy school bag as a child, and lack of knowledge (e.g. poor posture) may increase individual susceptibility to musculoskeletal injury at work. A strategy will therefore be ineffective without the involvement of other stakeholders in order to promote good lifestyle practices and this in turn emphasizes the need to take a broad-based approach. Prevention should be the main target of the strategy, but other important issues include prompt diagnosis, correct treatment and rehabilitation in order to prevent chronicity.

The MSD strategy should be seen as a flexible framework to integrate the components for control of MSD. It is also essential that a strategy reflects the changes in work practices and organization and the potential for musculoskeletal problems that specific work may precipitate. For example, there is the rapid growth in the service sector, the large rise in the number of people working with display screen equipment, the ageing workforce, and so on.

Recent developments must also be integrated into the MSD strategy and the outcomes of the broader government and HSE initiatives discussed above will therefore need to be incorporated within it. In particular the overall occupational health strategy will heavily influence the development and eventual content of the MSD strategy.

3. Information provision

3.1. *Guidance*
The HSE publishes a large number of leaflets and books on a wide range of topics. Guidance may be work-sector specific or concerned with a particular ill-health condition or risk factors. Guidance is available on the practical applications of regulations made under the Health and Safety at Work (HSW) Act 1974 and to assist in the implementation of regulations resulting from European Directives, e.g. 90/269/EEC on the manual handling of loads.

Many guidance documents contain information relevant to MSD and related risk factors. For example, 'Manual handling in the drinks industry' and 'A pain in your workplace? Ergonomic problems and solutions'. The latter describes the potential work-related causes of back pain, sore shoulders or numb fingers and shows cost effective ways of reducing the risk of experiencing these symptoms in the form of illustrated case-studies.

Guidance is regularly updated to incorporate the findings of research. The revised guidance on the Manual Handling Operations Regulations (HSE 1998b) provides more detail concerning the increased risk to pregnant workers, the risks associated with lifting loads from floor level and how the risk may be affected if frequency of handling is increased when the weight of the load is reduced. In addition, regulations and accompanying guidance implementing a European Directive must undergo evaluation. Evaluation enables investigation of employers' and employees' awareness of, interpretation and response to the regulations. Evaluation is particularly useful and enables the HSE to determine the level of compliance with statutory requirements and to structure revised documents according to the needs of the readership.

3.2. *Good health is good business (GHGB)*
The aim of this public initiative is to raise awareness of occupational health and improve employers' competence in the management of health risks in the workplace. Each phase of the campaign since 1995 has focused on the management of a range of risks, e.g. MSD, noise, vibration. Practical advice and guidance including case study material has been produced to show employers how they can take action to manage risks effectively.

Phase IV of the GHGB campaign began in October 1999, starting with advertising and publicity for the first 6 months. Inspection activities then start in April 2000. Phase IV is built around the theme 'Making it Happen'. There is a shift in emphasis from giving advice and raising awareness towards more enforcement action, where appropriate to secure compliance with current legislation.

The campaign itself is a key programme in the HSC Strategic Plan (HSC 1999). The success of the campaign will be measured against the target of a 10% reduction in the number of firms taking no action to control health risks by 2003. Overall, GHGB supports the strategic theme 'to raise the profile of occupational health'.

4. Technical research
Historically, the HSE has commissioned research on an incremental basis in association with ongoing policy development or operational needs, an approach which has largely been successful. The HSE has obtained accurate information from its research programme about workplace risk factors, ill-health and control measures in order to develop guidance, to support publicity campaigns and workplace visits by inspectors.

4.1. *Research activities*
Research is an instrumental tool in the HSE's approach for tackling MSD. Epidemiological investigations will enable the scale and cost of problems and the effectiveness of interventions to be assessed. Prevention initiatives of the HSE are heavily dependent on sound scientific knowledge derived from the scientific literature and the HSE's own research programmes. The research commissioned ranges from studies of disease mechanics to compilation of practical case studies.

The HSE recently commissioned a research project to identify case-studies where practical solutions to manual handling problems have been identified by home carers. Home carers have far less control over their work environment than hospital nurses for example, and hence have greater constraints imposed by their work activities. They comprise a group of workers for whom detailed guidance on the management of risks associated with manual handling activities is not available. The case studies collected in this project will relate to real problems that have been solved and will be published by the HSE as a guidance document.

A programme of research spanning a 10-year period was conducted as a consequence of concern about the physical demands imposed on supermarket cashiers by some checkout systems. The research included an epidemiological survey of almost 2000 cashiers, an ergonomics investigation of the workplace, and a laboratory-based biomechanics study of checkout designs. The HSE has published three documents concerned with the research, each aimed at a different group of individuals: store employees; Local Authority Environmental Health Officers (who have responsibility for enforcing health and safety legislation in supermarkets/retail outlets); the third document is a technical report of the research and will be of interest to a wider audience but including managers and safety representatives (Mackay et al. 1998).

This study of supermarket cashiers is an example of how research can lead to the provision of information and increased awareness of ergonomic risks and control measures amongst employers, employees, inspectors and health and safety professionals/practitioners. The information obtained from the research has fed back into standards of design, training material and purchasing policies of companies. In addition, the research tools developed for this purpose have subsequently been applied in a wide range of industries and settings (Dickinson 1998) and have prompted employers to take an active stance over the management of musculoskeletal disorders.

4.2. *Musculoskeletal research strategy*
The review of occupational health policy by the HSE identified the need for a more strategic approach to the development of the musculoskeletal research portfolio (Morris et al. 1998). A review of existing and completed research, research workshops, seminars and issues emerging from recent scientific literature, all helped to develop a forward strategy. It became apparent from the review of extramural research projects that significant gaps existed in the research portfolio, in particular with regard to risk factors for the development of upper limb disorders and the interactions between them. Amongst others, a specific need was identified for basic research leading to quantitative guidelines for physical parameters such as force, frequency, duration in order to supplement the information in current guidelines for manual handling risk factors (HSE 1998b).

The workshops succeeded in bringing together key experts in order to discuss research needs relating to occupational low-back pain, upper limb disorders and diagnostic criteria for work-related MSD. The final seminar held at Chester in 1997 built upon the conclusions of the earlier workshops. It focused on developing trends in musculoskeletal research with sessions on health outcomes, psychosocial factors, exposure assessment and control.

The combined outcome of these activities was the identification and general agreement of research needs by a small number of external experts including academics and occupational health practitioners. Further opportunities for discussing the strategy were provided at scientific meetings, including the Annual Conference of the Ergonomics Society. The HSE is now in the process of finalizing a national musculoskeletal research agenda, with the hope that it will stimulate collaborative research partnerships both within the UK and internationally and thus maximize the benefits to be gained from available research resources.

The research strategy constitutes a number of research themes and priorities that can be used to plan research programmes in the short and medium term (5–10 years). The research topics/themes identified through the consultative process include work to identify causal risk factors, assessment of exposure to risk factors, health surveillance, workplace interventions, health outcomes and the provision of ergonomic guidance for prevention. Within each theme a number of research issues are proposed which form the basis for specific research programmes.

As research findings are evaluated, the direction and content of the HSE sponsored musculoskeletal research are likely to change. The MSD research strategy will obviously provide a framework for the development and evaluation of proposals, but it will also retain the facility to accommodate unforeseen scientific developments.

Following wider consultation with researchers, technical experts and other stakeholders, the published strategy will help ensure that limited research resources are appropriately targeted and that research findings are evaluated against agreed national objectives. Opportunities for collaboration will be greatly enhanced, enabling governmental, academic, charitable and industrial research resources to be shared.

5. Inspection

The main object of workplace inspection by the HSE and Local Authority inspectors is to secure compliance with health and safety legislation and to ensure that a good standard of protection is maintained. Inspectors have important statutory powers and can, among other things: enter premises where work is undertaken without giving notice; give informed or written advice; issue notices to improve or prohibit work practices (improvement and prohibition notices); and prosecute individuals and companies for breaches of the law.

Inspectors are trained in health and safety issues relating to health and safety legislation, particular work hazards (e.g. pesticides), industries (e.g. mining, nuclear) and means of securing compliance. Inspectors now also receive training in ergonomics and this includes risk assessment and enforcement issues related to manual handling and repetitive work activities.

The GHGB campaign has until now aimed to increase awareness of occupational health risks and improve employers' competence in health risk management. As already stated, the next phase of the initiative will move more towards enforcement activity, where appropriate, to secure compliance. The impact of this initiative on a reduction in the prevalence of MSD may take time to become apparent. The HSE will use a range of indicators to assess the effectiveness of the campaign, e.g. health data, inspection reports. Likewise it is not easy to measure the effects of previous enforcement and prosecution activities although the benefits may also be evident indirectly. In 1996/97 the HSE alone issued 81 improvement notices (HSE 1997) relating to the Manual Handling Operations Regulations 1992. Enforcement action may also be taken under the HSW Act 1974, the Management of Health and Safety at Work Regulations 1992, or the Provision and Use of Work Equipment Regulations 1992, so there is sufficient scope to ensure compliance and promote better work practices.

Medical Inspectors and Occupational Health Inspectors working in the HSE's Employment Medical Advisory Services (EMAS) provide an important contribution to the HSE's work on occupational health. The EMAS inspectors now have the same enforcement powers as other HSE inspectors. They are able to take action where health risks in the workplace are not properly controlled or where occupational health standards are poor.

6. Concluding remarks

This review has described how the HSE is working to reduce the prevalence of work-based musculoskeletal disorders. Tackling MSD is a priority within each of its operations: policy; technical research; information provision; and workplace inspection. It is appropriate that just as MSD have multifactorial aetiology, the HSE's activities are diverse and innovative, encompassing an holistic approach.

With new initiatives and partnerships, e.g. Healthy Workplace, the HSE will continue to work towards greater understanding and awareness of occupational health issues and more effective management of health risks in the workplace

Acknowledgements

The author acknowledges the following HSE personnel in recognition of their assistance in the production of this paper: Dr Claire Dickinson, Len Morris, Nigel Watson, Dr Ron McCaig, Dr Jim Neilson, John McElwaine, Malcolm Darvill and Margaret Kellett.

References

DICKINSON, C. E. 1998, Interpreting the extent of musculoskeletal complaints, in M. A. Hanson (ed.), *Contemporary Ergonomics* (London: Taylor & Francis), 36–40.

Health and Safety Commission 1999, *Health and Safety Commission Strategic Plan for 1999/2002* (Sudbury: HSE Books).

Health and Safety Executive 1997, *Health and Safety Statistics 1996/97* (Sudbury: HSE Books).

HSE 1998a, *Developing an Occupational Health Strategy for Great Britain*. HSE Discussion Document DDE8 (Sudbury: HSE Books).

HSE 1998b. *Manual Handling Operations Regulations 1992*. Guidance on Regulations (L23) 2nd edn (Sudbury: HSE Books).

JONES, J. R., HODGSON, J. T., CLEGG, T. A. and ELLIOTT, R. C. 1998, *Self-reported Work-related Illness in 1995: Results from a Household Survey* (Sudbury: HSE Books).

MACKAY, C., BURTON, K., BOOCOCK, M., TILLOTSON, M. and DICKINSON, C. 1998, *Musculoskeletal Disorders in Supermarket Cashiers* (Sudbury: HSE Books).

MORRIS, L., MCCAIG, R., GRAY, M., MACKAY C., DICKINSON, C., SHAW, T. and WATSON, N. 1998, Prevention of musculoskeletal disorders in the workplace—a strategy for UK research, in M. A. Hanson (ed.) *Contemporary Ergonomics* (London: Taylor & Francis), 46–50.

32 Will the use of different prevalence rates influence the development of a primary prevention programme for low-back problems?

E. Zinzen[†]*, D. Caboor[†], M. Verlinden[†], E. Cattrysse[†], W. Duquet[‡], P. Van Roy[†] and Jan Pieter Clarys[†]

[†]Department of Experimental Anatomy, Faculty of Physical Education, Vrije Universiteit Brussel, Laarbeeklaan 103, B-1090 Brussels, Belgium

[‡]Department of Human Biometry, Faculty of Physical Education, Vrije Universiteit Brussel, Pleinlaan 2, B-1050 Brussels, Belgium

Keywords: Back pain; Epidemiology; Factor analysis; Nurses; Statistics.

To determine relations to low-back problems (LBP), different prevalence rates are used. The disadvantage of using different selection criteria is that studies are not comparable, except where they provide the same results. The present aim was to establish whether different prevalence selection criteria lead to different answers on a newly formed set of questionnaires. Since this set is new, reliability tests were performed (test–retest and calculations of Cronbach's Alpha, Cohen's Kappa and the intraclass correlation). Results of the questionnaire should form the cornerstones of a primary prevention programme. Altogether 1783 nurses in four Flemish (Belgian) hospitals were questioned. Information was gathered on work circumstances, education, general health, psychosocial factors, leisure activities, family situation and musculoskeletal problems. Four different datasets with variables related to lifetime prevalence LBP, annual prevalence LBP, point prevalence LBP and a set with all related variables were constructed. The variables demonstrating a relation with LBP differed slightly depending on the kind of prevalence used (lifetime, annual, point). A factor analysis on each set of prevalence related data failed due to the lack of homogeneity of the variables. Fear avoidance, coping aspects and musculoskeletal problems in other regions then the lower back were, in all circumstances, the most discriminating variables. Their discriminating power, however, differed depending on the kind of prevalence used. The differences were too small to influence the construction of the prevention programme. It is concluded that in developing a primary prevention programme any of the prevalence rates can be used. The combination of the three types of prevalence rates studied provides the most complete and reliable image.

1. Introduction

Low-back pain interferes with leisure and occupational activity and so research into low-back problems (LBP) has increased over the last decade. Comparison between different studies, however, is not always possible due to the use of different definitions of LBP (on the basis of area, chronic/acute, idiopathic, symptoms, etc.) and different

*Author for correspondence. e-mail: emzinzen@vub.ac.be

'time' selection criteria (prevalence/incidence: annual, point or lifetime). The present study is focused on which 'time' selection criterion is best to use. Prevalence rates are often used as a selection criterion in order to determine variables related to or causing LBP. A lifetime prevalence indicates the number of individuals who had experienced an episode of LBP at least once in their lifetime. Statistics based on comparisons between 'lifetime prevalent' subjects and subjects whom never experienced LBP can reveal links to the occurrence of LBP. Variables determined by lifetime prevalence, however, may be influenced by the long-term memory of the subjects. Subjects have to describe the situation they were in when they experienced their painful episode. This event can be far in the past. On the other hand subjects who experienced LBP in the previous year (annual prevalence) or who are experiencing LBP at the time of the study (point prevalence) are included in the lifetime prevalence results.

Table 1 provides an overview of research concerned with LBP prevalence rates and is sorted by occupation to ease its interpretation. Table 1 provides a motivation for the development of a primary prevention programme due to the rather high prevalence rates of LBP, especially in the nursing profession. The purpose of such a programme is to prevent healthy people (who had never experienced LBP) from getting LBP and to prevent relapses in those who already experienced LBP.

It is clear from table 1 that the three different prevalence rates are rarely used together. The studies were based mostly on the lifetime prevalence, which is the weakest selection criterion as most LBP episodes heal spontaneously within 8 weeks. This results in a poor recollection of the variables involved (Anderson 1986, Andersson et al. 1991, Frymoyer and Andersson 1991).

The use of each type of prevalence has been supported (and argued) by different research groups in the various studies (table 1). Constructing a large enough sample is one of the major problems researchers have to face. Therefore, it becomes evident why most of the LBP studies have used lifetime prevalence rates, but it can be questioned if these studies really generate the information required. The aim in this study therefore is to find out if different prevalence selection criteria lead to different results in the development of a primary prevention programme.

2. Methods

The results presented here are a part of a larger research project with the aim to construct a programme for prevention of LBP in nurses (Zinzen 1998). The project started in 1992 with the construction of a questionnaire in Dutch. The results of this questionnaire were meant to provide the necessary information to form the corner stones of the prevention programme. The questionnaire used in the 'Study of musculoskeletal pain in nursing personnel' (Skovron et al. 1987a) was adopted as a base. In order to get a total overview of all possible LBP-related variables, this questionnaire was supplemented with psychosocial items such as the 'Pain Locus of Control (PLC)' questionnaire (Main and Waddell 1991), the 'Coping Strategies Questionnaire (CSQ)' (Rosenstiel and Keefe 1983, Main and Waddell 1991), the 'Modified Zung Depression Inventory (MZDI)' (Main and Waddell 1984) and the 'Fear-Avoidance Beliefs Quest (FABQ)' (Waddell et al. 1993). All these questionnaires were carefully translated in the presence of Dr Skovron (author of the basic questionnaire) and Dr Burton (Spinal Research Unit, Huddersfield, UK). Some questions were altered to fit the Flemish context (education degrees, nursing level, etc.). A Dutch 'general health' questionnaire, 'VOEG' (Hartgers et al. 1987) was added, being complemented with questions concerning sleeping habits and working clothes.

Table 1. Overview of different LBP prevalence rates (%).

Study	Lifetime prevalence (incidence) (%)	Point prevalence (%)	Annual prevalence (%)	n	Age (years)	Sex	Occupation (country)
In general occupations							
Bergenudd and Nilsson (1988)	–	28	–	323	55	M	all (S)
Bergenudd and Nilsson (1988)	–	30	–	252	55	F	all (S)
Biering-Sorenson (1982)	62.6	12.0	–	449	30–60	M	all (S)
Biering-Sorenson (1982)	61.4	15.2	–	479	30–60	F	all (S)
Burton et al. (1989)	51	14	–	545	16–65	MF	all (UK)
Frymoyer et al. (1983)	69.9	–	–	1221	28–55	M	all (USA)
Hirsch et al. (1969)	48.8	–	–	692	15–72	F	all (S)
Leighton and Reilly (1995)	58.9	25.1	57.8	315	±32	MF	all (UK)
Magora and Taustein (1969)	–	12.9	–	3316	–	MF	all (Israel)
Magora (1970)	12.9	–	–	3316	–	MF	all (Israel)
Nagi et al. (1973)	–	18.0	–	1135	18–64	MF	all (?)
Svensson and Andersson (1982), Svensson et al. (1988)	61	31	–	716	40–47	M	all (S)
Svensson and Andersson (1982), Svensson et al. (1988)	67	–	–	1640	38–64	F	all (S)
Valkenburg and Haanen (1982)	51.4	22.2	–	3091	20–	M	all (NL)
Valkenburg and Haanen (1982)	57.8	30.2	–	3493	20–	F	all (NL)
Videman et al. (1989)	47	–	–	112	–	MF	all (S)
In specific occupations							
Bongers et al. (1990)	55.0	–	–	133	21–56	M	Helicopter pilots (?)
Bongers et al. (1990)	11.0	–	–	228	20–61	M	airforce officers; ground personnel (?)
Brown (1975)	35	–	–	–	–	M	industry worker (UK)

(Continued)

Table 1. *Continued*

Study	Lifetime prevalence (incidence) (%)	Point prevalence (%)	Annual prevalence (%)	n	Age (years)	Sex	Occupation (country)
Brown (1975)	46	–	–	–	–	F	industry worker (UK)
Burdorf and Zondervan (1990)	–	–	61	–	±42	M	crane operators (NL)
Burdorf et al. (1990)	–	–	59	–	–	M	concrete workers (NL)
Burdorf et al. (1991)	–	–	31	–	–	M	maintenance engineers (NL)
Hildebrandt (1995)	–	–	51	–	–	M	farmers (NL)
Hult (1954)	60.0	–	–	1193	25–59	M	industry worker (S)
Magora (1970)	21.6	–	–	–	–	MF	heavy industry workers (Israel)
Riihimäki et al. (1989)	90	–	–	852	25–49	M	machine operators (Sf)
Riihimäki et al. (1989)	75	–	–	1013	25–49	M	office workers (Sf)
Serratoz and Mendiola-Anda (1993)	18.2	–	–	132	±28.2	M	shoemakers (sewing) (It)
Törner et al. (1991)	81	–	–	47	31–50	M	welders (?)
Törner et al. (1991)	24	–	–	33	31–50	M	office clerks (?)
Nurses							
Bosser and Lehman (1982)	75	–	–	143	–	MF	nurses (F)
Caillard et al. (1987)	29.6	–	–	723	–	MF	nurses (F)
Cassou and Gueguen (1985)	32	21	–	135	–	MF	hospital personnel (F)
Cato et al. (1897)	72	–	–	52	23–61	MF	nurses (USA)
Cedercreutz et al. (1987)	38	–	–	125	±22.1	F	candidates for studying nursing (Sf)
Dehlin et al. (1976)	46.8	18.0	–	267	±28.7	F	nursing aids (S)
Knibbe and Friele (1996)	87.0	–	–	355	±34.2	F	nurses at home (NL)

(Continued)

Table 1. Continued

Study	Lifetime prevalence (incidence) (%)	Point prevalence (%)	Annual prevalence (%)	N	Age (years)	Sex	Occupation (country)
Larese and Fiorio (1994)	48	–	–	425	±37.2	MF	general hospital nurses (It)
Larese and Fiorio (1994)	32	–	–	198	±34.2	MF	nurses in oncological centre (It)
Lecomte-Devooght (1982), cited in Maroy (1989)	22	–	–	–	–	MF	nurses (B)
Leighton and Reilly (1995)	61.4	24.4	58.8	1134	±36	MF	nurses (UK)
Magora (1970)	16.8	–	–	–	–	MF	nurses (Israel)
Moens et al. (1993)	–	18.0	63.0	4256	–	F	nurses at home (B)
Skovron et al. (1987a,b)	43	–	–	787	36.2	MF	nurses (USA)
Stappaerts (1988, 1989a,b)	65	–	–	930	±30	F	nurses (B)
Stappaerts (1988, 1989a,b)	73.0	–	–	227	±30	M	nurses (B)
Stubbs and Buckle (1984)	–	17	–	3912	–	MF	nurses (UK)
Turnbull et al. (1992)	60	–	–	1363	–	MF	nurses (UK)
Videman et al. (1984)	85	–	–	–	20–65	MF	nursing aids (S)
Videman et al. (1984)	79	–	–	–	20–65	MF	nurses (S)
Videman et al. (1989)	64	–	–	98	–	MF	nurses (S)

S, Sweden; UK, United Kingdom; USA, United States of America; NL, the Netherlands; Sf, Denmark; It, Italy; F, France; B, Belgium.

Before distributing the set of questionnaires, several nurses and head nurses checked the questions for their clarity. In order to perform a test–retest study, 435 question booklets were distributed twice at a 6-month interval.

For the present study, 1783 questionnaires were distributed. Of them 1216 were fit to be analysed. To check reliability, Cronbach's Alpha and differences between test–retest (Cohen's Kappa, Anova, Friedman, intraclass correlation) were calculated.

Six different experimental groups were constructed on the basis of the three prevalence rates (lifetime, annual and point) and compared by means of ANOVA, Kruskall–Wallis and Chi-square where appropriate in the following way (all significance levels $p<0.05$):

- (A) Lifetime prevalence LBP subjects (1: $n = 644$) versus subjects who never had a spell of LBP (2: $n = 572$);
- (B) Annual prevalence LBP subjects (3: $n = 499$) versus subjects who did not experience LBP in the past year (4: $n = 717$);
- (C) Point prevalence LBP subjects (5: $n = 340$) versus subjects who did not suffer LBP at the time of the study (6: $n = 876$).

This resulted in three sets (A–C) of variables related respectively to each type of LBP prevalence. 'Related' has to be interpreted in the sense that a significant difference was noticed between the experimental groups and their respective controls. Some variables showed a significant difference between experimental and control group only in one of the three prevalence conditions. Therefore, a fourth set (D) was created consisting of all the variables out of the three experimental sets (A–C) related to LBP. At first it was intended to reduce the variables in each set by means of a factor analysis. On the basis of the identified factors, a discriminant function would be constructed. This function consisting of different LBP-related factors would form the basis for proposing a primary prevention programme. The four developed programmes (out of the four discriminant functions: A–D) would be compared to evaluate differences between the usage of the three types of prevalence rates and the combination of all three as selection criteria. Since the factor analysis failed, a slightly modified methodology was used (see Results).

3. Results

3.1. *Reliability*

Cohen's Kappa figures for the dichotomous variables are displayed in table 2 and indicate in general an average to substantial concordance between the variables. For the continuous variables, the differences between the test–retest procedures were calculated by paired ANOVA (parametric distribution) and paired Friedman (non-parametric distribution) tests. Where possible, the intraclass correlation coefficient was calculated. All differences observed between the questions were due to the 6-month interval, e.g. the answer on the question 'What is your age (years + months)?' was significantly younger the first time than the second time. In other words these differences had no relevance to the problem posed. All intraclass correlations ranged from 0.74 to 0.99, which can be considered as very good.

In addition to the test–retest procedure, Cronbach's Alpha gives an indication of internal consistency of the questionnaires. The calculation of Cronbach's Alpha is only possible for questions measuring the same kind of variables; therefore it was calculated only for the subquestionnaires listed in table 3.

The internal consistency of the subquestionnaires was very high.

Table 2. Cohen's Kappa for the dichotomous variables compared with results of a similar study by Main and Waddell (1991).

Kappa	Percentage of questions this study	Percentage of questions Main and Waddell (1991)
>0.60	41.2	36
0.41–0.59	29.4	50
0.21–0.40	23.5	5
<0.21	5.9	9

Table 3. Cronbach's Alpha of selected sub questionnaires.

Questionnaire ($n=1216$)	Cronbach's Alpha
General Health (VOEG)	0.92
Modified Zung Depression Index	0.81
Pain Locus of Control	0.96
Fear Avoidance Beliefs	0.98
Coping Strategies	0.95
WORK.APGAR	0.82

3.2. Prevalence related variable

Lifetime prevalence of LBP in Flemish nurses was 53%. The annual prevalence was 41% and point prevalence 28%.

Table 4 shows all variables of the questionnaire where a significant difference ($p<0.05$) was found between the respective prevalence groups and their controls. The positive relation indicates that the prevalence group had significantly 'more' of this variable whereas the negative relation indicates a significantly 'less' value for that variable. For example, the three prevalence groups indicated they were working longer in the same position (staff nurse, head nurse, etc.) whereas they considered their work circumstances as less optimal in comparison to the respective control groups. Many variables investigated are not mentioned in table 4, as for those specific variables differences between subjects with LBP (all prevalence rates) and without could not be confirmed.

3.3. Factor analysis

Four groups of variables were created, one for each prevalence rate and one containing all variables in table 4. To reduce the amount of variables, four different factor analyses were performed, although all with the same result. The Kaizer–Meyer–Olkin (KMO) value, which gives an indication of the homogeneity of the variables, varies best at around 1 indicating a high homogeneity of the variables and meaning that the factors formed are reliable. In the present study and in the four conditions, KMO varied between 0.06 and 0.08, indicating that the factors created are not reliable due to the lack of homogeneity between the variables and therefore can not be used in further analysis.

3.4. Discriminant analysis

Since it was not possible to create reliable factors, a discriminant analysis was performed on the four datasets separately. Results of the four calculated discriminant functions were virtually the same. The functions consisted of ± 40 variables (range 38–42) with a discriminating power of $\pm 83.5\%$ (range 82.3–84.7%). Each time after the first step in forming the discriminant function, the maximum discriminating power was almost reached. The addition of the other variables raised the discriminating power of the function only slightly. This result together with the lack of homogeneity of the variables (see factor analysis) gives the impression that either the first variable selected in each function is the one and only discriminating variable or each variable on its own possesses a strong discriminating power. To be conclusive about this, a series of discriminant analyses were performed limited to one step and each time the variable found in the previous discriminant analysis was deleted. In this way it was possible to determine the discriminating

Table 4. LBP prevalence related variables.

Variable (question)	LP LBP	AP LBP	PP LBP
Positive related (subjects with LBP have 'more' of this variable compared with subjects without LBP)			
Time in work position	×	×	×
General Health (VOEG)	×	×	×
Modified Zung Depression Index	×	×	×
Cause of LBP on the job	×	×	×
Experienced LBP with cause on the job the past year	×	×	×
Cause of LBP outside the job	×	×	×
Experienced LBP with cause outside the job in the past year	×	×	×
Changed of department due to LBP	×	×	×
Sick leave	×	×	×
Musculoskeletal disorders (in all different body regions)	×	×	×
Pain level (in general) in worst circumstances	×	×	×
Pain level (in general) now	×	×	×
Experiencing always pain	×	×	×
Resting	×	×	×
Pain during and after work and at bedtime	×	×	×
Pain in all statures	×	×	×
Pain Locus of Control (PLCB15: pain responsibility)	×	×	×
Fear Avoidance of physical activity	×	×	×
Fear Avoidance of work	×	×	×
Coping Strategies: Pain behaviours	×	×	×
Lifting patients alone	×	×	×
Age when started nursing profession	×		×
History of smoking	×		×
Amount of pregnancies	×		×
Amount of children	×		×
Menopause	×		×
Pain level in the best circumstances	×		×
Coping Strategies: catastrophising	×		×
Date of birth	×		×
Menstrual cramps	×	×	×
Risk on cut and needle wounds	×	×	×
Risk for LBP while supporting patients	×	×	×
Weight	×	×	×
Sometimes no pain	×		×
Responsible for children < 5 year	×		×
Amount of persons demanding attention in the family	×		×
Responsibility for other persons in the family	×		×
No. of days worked/7 days		×	
Frequency of alcohol consumption		×	
Risk of contamination		×	
Wearing trousers and blouse		×	
Sex (more men)		×	
Height		×	
Being a nursing assistant			×
Being a nursing aid			×
Secondary level of education			×
Amount of deliveries			×
Amount of days sick leave			×

(Continued)

Table 4. *Continued*

Variable (question)	LP LBP	AP LBP	PP LBP
Alternating pain			×
Same pain levels			×
Pain before going to work			×
Diagnosis by a GP			×
Diagnosis by a company GP			×
Treatment of musculoskeletal disorders			×
Coping strategies:			
Diverting attention			×
Ignoring pain sensation			×
Self statements			×
Increasing physical activity			×
Praying and hoping			×
Sleeping on slats			×
Negative related (subjects with LBP have 'less of this variable compared with subjects without LBP)			
Being highest level nurse	×		×
Having more pain in the legs than in the back	×	×	
Lifting with help	×	×	
Space in the nursing room	×	×	
PAW: job satisfaction		×	×
Enough lifting education	×		
Space in the patient room		×	
Wearing an apron		×	
No. of hours work/week			×
Time spent on evaluating patients			×
Having the highest nursing education			×
Alcohol consumption			×
Frequency of alcohol consumption			×
Sometimes no pain			×
Sleeping on a board			×

LP, Lifetime prevalence; AP, annual prevalence; PP, point prevalence; LBP, low-back problems).

power of each variable on its own. Table 5 gives an overview of the discriminating values of each variable in the four test conditions limited to the 15 variables with the highest discriminating power.

4. Discussion

4.1. *Reliability*

According to Main and Waddell (1991), a Cohen's Kappa >0.60 means a substantial concordance between the variables, and 0.41–0.59 can be considered as an average concordance while 0.21–0.40 are an indication of a moderate concordance. Kappa <0.21 indicates no concordance between the test–retest values. Referring to the results of the questionnaire used in the present study, most of the questions showed an average to substantial concordance.

Test–retest results (Cohen's Kappa) after 6 months agree with the results of Main and Waddell (1991) after 24 h. No differences could be found either with ANOVA or Friedman tests, except for purely time-related questions. The calculated

Table 5. Overview of variables with their discriminating power (%) for each LBP prevalence condition (in descending order and limited to the 15 variables with the highest discriminating power).

Lifetime prevalence variable	%	Annual prevalence variable	%	Point prevalence variable	%	Combined prevalence variable	%
FABWORK	85.8	FABWORK	77.3	pain level now	88.9	FABWORK	80.8
FABPHYS	84.1	FABPHYS	76.6	pain in right hip	73.3	FABPHYS	79.8
Coping Strat: Catastrophy	83.2	Coping Strat: pain behaviour	76.5	pain in left knee	73.3	Coping Strat: praying/hoping	79.5
Coping Strat: Pain behaviour	83.2	pain while standing	75.2	pain in right knee	72.8	Coping Strat: diverting attention	79.5
Sometimes no pain	82.8	having pain is not my choice	75.2	pain in left elbow	72.7	Coping Strat: ignoring pain sensation	79.5
Pain level in best circumstance	82.8	pain while pushing	75.2	pain in left ankle/foot	72.6	Coping Strat: self statements	79.5
Pain level in worst circumstance	82.8	pain while trunc flexion	75.2	pain in right elbow	72.4	Coping Strat: increasing physical act.	79.5
Always pain	82.6	pain while trunc rotation	75.0	pain in right wrist/hand	72.4	Coping Strat: pain behaviours	79.5
Pain while lifting patients	82.4	pain while trunc sitting	74.9	Pain in the higher back	72.2	Coping Strat: catastrophysing	79.5
Pain was never my choice	82.3	pain while lifting patients	74.9	pain in left hip	72.2	diagnosis by GP	79.2
More pain in the legs than in the back	82.1	pain after work	74.5	pain in left ankle/foot	71.5	diagnosis by company GP	78.9
Pain level now	82.1	pain while going to sleep	74.4	pain in right calf	71.4	treatment for musculoskeletal disorders	78.4
Pain while trunk flexion	82.0	pain during work	74.2	pain in right shoulder	71.3	PLCB15:	78.0
Pain while standing	80.9	more pain in the legs than in the back	73.8	pain in left calf	71.3	pain while lifting patients	77.9
Pain while sitting	80.8	pain now	73.8	pain in left wrist/hand	71.3	pain while trunc flexion	77.8

FABWORK, fear avoidance believes related to work.
FABPHYS, fear avoidance believes related to physical activity.
Coping Strat, coping strategies.
GP, general practitioner.
PLCB15, pain locus of control related to level of pain and medication.

intraclass correlations were very high, as would be expected. Also, the internal consistency of the subquestionnaires is very high, in line with the results of Skovron *et al.* (1987a, b) who found Cronbach's Alpha >0.80 for the subquestionnaires. These findings suggest that the translation into the Dutch language was done thoroughly and that this translation had no effect on the already high reliability of the English questionnaires.

4.2. *Prevalence related variables*

For a detailed discussion of the results of the questionnaire, see Zinzen (1998). Here, the focus lies on the differences and/or equalities between the different selection procedures on the basis of prevalence rates.

It is clear (table 4) that many of the positively related variables with LBP are found within the three prevalence rates. However, the amount of variables related to LBP was much higher in the point prevalence group. Memory effects may explain this. In other prevalence rates, the questionnaire engages the long-term memory of the subjects while the point prevalence subjects describe what they are feeling now. Notable is the equality between lifetime prevalence and point prevalence for the positively related variables and the equality between lifetime prevalence and annual prevalence for the negatively related variables. These seem to be population-related findings since otherwise they are hard to explain.

Only one variable had contradictory results depending on the prevalence rate used: 'frequency of alcohol consumption'. In the annual prevalence group a greater alcohol consumption was related to the occurrence of LBP while in the point prevalence group less alcohol consumption seems to be significant. No explanation for this conflict is apparent. Except for this single variable, the overall use of the three different prevalence rates as a selection criterion reveals approximately the same related variables.

4.3. *Factor analysis and discriminant analysis*

The failure to extract factors and the fact that a lot of variables on their own possess a high discriminating power lead to the suggestion that LBP (independent of prevalence used) is related to a wide range of variables. Some of the variables were high in discriminating power, whereas others were low. Which variable eventually causes LBP seems unpredictable. The combination, however, of a lot of high contributors causes a much greater likelihood of sprain and it may be clear that prevention should be directed in the first instance towards these variables without neglecting the smaller contributors. The four different selection methods, used in this study, all show clearly that LBP is a multidisciplinary problem.

The weight (discriminating power) of the variables differed depending on the selection method used. With the exception of the point prevalence, 'fear avoidance beliefs of work and of physical activity' was the strongest discriminating variable in relation to LBP. This study, not being longitudinal, cannot determine a cause. However, subjects (in this study nurses) with fear and avoidance behaviour were the most likely to have (or develop) low-back problems. Coping strategies took second place in the rank of strongly LBP-related variables while stature, movements and musculoskeletal problems in other body regions were somewhat less important. It is remarkable that typical work-related items were not strongly discriminating for LBP in this nursing population. These findings are supported by Burton *et al.* (1989), Burton and Tillotson (1991) and Haldeman (1990) who also noticed that physical load items alone could not explain the high prevalence of LBP and suggested that psychosocial items could be of great importance.

Comparing the lifetime prevalence group, the annual prevalence group and the combined group, it is concluded that the weight given to the different variables differed but that in general the same variables emerged as being strongly discriminating. For the point prevalence LBP subjects, the musculoskeletal problems in other body regions were the most discriminating variables. This may be explained by the fact that subjects tended to describe what they felt at the time of the study. It appears that LBP is related to accidents since other body regions are affected at the same time. Or, are these results confounded by psychological variables? Are the results being influenced by a tendency of LBP sufferers to complain? This study can not be conclusive about these questions but perhaps the following gives a direction. In the related variables (table 4) one can detect that most of the coping items show a relation with point prevalence LBP only. The coping aspects in the discriminant analysis for point prevalence are not powerful. However, after combining the three prevalence rates, all coping aspects became the second most powerful discriminating variables while in lifetime and annual prevalence, they hardly showed a relation with LBP. The coping sub-items showing a relation were strongly discriminating. It is therefore considered that the high occurrence of musculoskeletal pain in other body regions is merely the result of poor coping strategies which could not be noticed if the combined prevalence group was not constructed.

From tables 4 and 5 similar conclusions may be derived for each prevalence rate in respect to the construction of a primary prevention programme. Despite the fact that the factor analysis failed to group the variables, it became possible on the basis of the discriminating power to distinguish four main groups: psychosocial variables, health variables, ergonomic factors and variables related to hospital policy. All these variables are more or less detectable by each of the prevalence rates used. The weight given to them differed slightly depending on the different prevalences. In primary prevention the purpose is to prevent healthy people (who had never experienced LBP) from getting LBP and to prevent relapses in those who already experienced LBP. The weight of the variable becomes less important when preventive actions in this field are undertaken.

The primary prevention model resulting from this investigation is given in figure 1. The effectiveness of this model is still to be investigated.

The data from this study clearly suggest that the three types of prevalence rates for LBP reveal more or less the same variables as related. These variables possessed a slightly different discriminating weight. The combination of the three prevalence rates revealed the importance of coping strategies which was not clear from using the point prevalence selection criterion alone. It is suggested therefore that future studies in the field of LBP should use the combination of the three prevalence rates. This combination will avoid possible confusion in comparing various studies. In the construction of a primary prevention programme however, the use of one out of three prevalence rates will provide sufficient information.

5. Conclusions

- The newly formed Dutch questionnaire proved a reliable tool when examining LBP.
- Variables related to LBP differed slightly depending on which prevalence rate was used to select the population.
- Variables related to LBP lack homogeneity and find their origin in multiple disciplines.

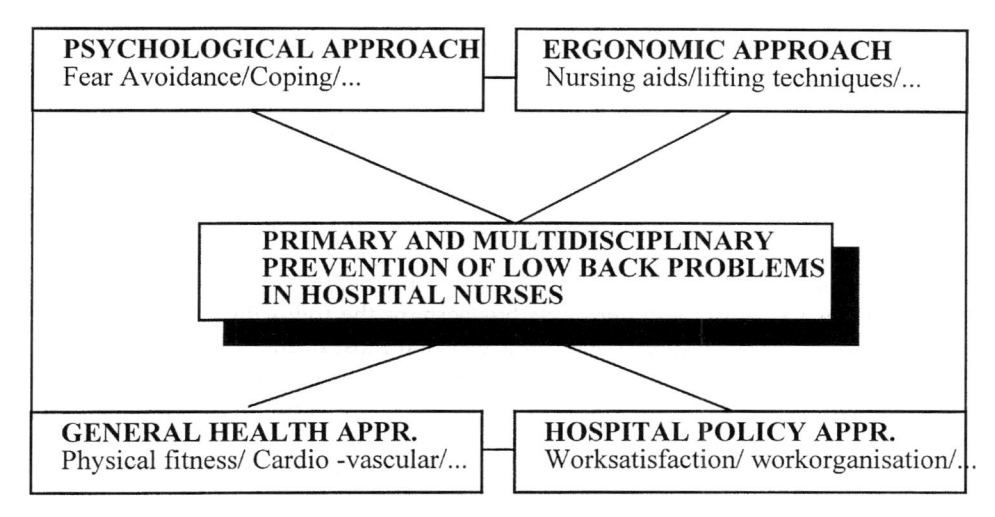

```
┌─────────────────────────────────┬─────────────────────────────────┐
│ PSYCHOLOGICAL APPROACH          │ ERGONOMIC APPROACH              │
│ Fear Avoidance/Coping/...       │ Nursing aids/lifting techniques/...│
└─────────────────────────────────┴─────────────────────────────────┘
              ┌──────────────────────────────────────┐
              │ PRIMARY AND MULTIDISCIPLINARY        │
              │ PREVENTION OF LOW BACK PROBLEMS      │
              │ IN HOSPITAL NURSES                   │
              └──────────────────────────────────────┘
┌─────────────────────────────────┬─────────────────────────────────┐
│ GENERAL HEALTH APPR.            │ HOSPITAL POLICY APPR.           │
│ Physical fitness/ Cardio-vascular/...│ Worksatisfaction/ workorganisation/...│
└─────────────────────────────────┴─────────────────────────────────┘
```

Figure 1. The primary prevention programme.

- The discriminating power of variables related to LBP differ slightly depending on the prevalence rate used. This difference will not significantly influence the construction of a primary prevention programme. Nevertheless, it is believed that taken in account variables related to the three prevalence rates will provide the most accurate and reliable information in the construction of a primary prevention programme.

Acknowledgements

The authors acknowledge the BioMed European Commission, Directorate General XII, for financial support.

References

ANDERSON, J. A. D. 1986, Epidemiological aspects of back pain, *Journal of the Society of Occupational Medicine*, **36**, 90–94.

ANDERSSON, G. B. J., POPE, M. H., FRYMOYER, J. W. and SNOOK, S. 1991, Epidemiology and cost, in M. H. Pope, G. B. J. Andersson, J. W. Frymoyer and D. B. Chaffin (eds), *Occupational Low-back Pain: Assessment, Treatment and Prevention* (St Louis: Mosby-Yearbook), 95–113.

BERGENUDD, H. and NILSSON, B. 1988, Back pain in middle age: occupational workload and psychologic factors: an epidemiologic survey, *Spine*, **13**, 58–60.

BIERING-SORENSON, F. 1982, Low-back trouble in a general population of 30-, 40-, 50-, and 60-year-old men and women: study design, representativeness, and basic results, *Danish Medical Bulletin*, **29**, 289.

BONGERS, P. M., HULSHOF, C. T. J., DIJKSTRA, L., BOSHUIZEN, H. C., GROENHOUT, H. J. and VALKEN, E. 1990, Back pain and exposure to whole body vibration in helicopter pilots, *Ergonomics*, **33**, 1007–1026.

BOSSER, J. and LEHMANN, R. 1982, Une expérience de prévention des lombalgies en milieu hospitalier, *Archives des Maladies Professionnelles*, **6**, 511–512.

BROWN, J. R. 1975, Factors contributing to development of low-back pain in industrial workers, *American Industrial Hygiene Association Journal*, **35**, 26–31.

BURDORF, A., GOVAERT, G. and ELDERS, L. 1991, Postural load and back pain of workers in the manufacturing of prefabricated concrete elements, *Ergonomics*, **34**, 909–918.

BURDORF, A. and ZONDERVAN, H. 1990, An epidemiological study of low-back pain in crane operators, *Ergonomics*, **33**, 981–987.

BURTON, A. K., TILLOTSON, K. M. and TROUP, J. G. D. 1989, Prediction of low-back trouble frequency in a working population, *Spine*, **4**, 939–946.
BURTON, A. K. and TILLOTSON, K. M., 1991, Prediction of the clinical course of low-back trouble using multivariable models, *Spine*, **16**, 7–14.
CAILLARD, J. F., CZERNICHOW, P., DOUCET, E., JAMOUSSI, S., REBAI, D., JULIEN, F. and PROUST, B. 1987, Le risque lombalgique professionnel à l'hôpital: étude au centre hospitalier régional de Rouen, *Archives des Maladies Professionnelles*, **48**, 623–627.
CASSOU, B. and GUEGUEN, S. 1985, Prévalence et facteurs de risque de la lombalgie: une enquête épidémiologique et rétrospective parmi le personnel d'un hôpital parisien, *Archives des Maladies Professionnelles*, **1**, 23–29.
CATO, C., OLSON, D. K. and STUDER, M. 1987, Incidence, prevalence, and variables associated with low-back pain in staff nurses, *American Association of Occupational Health Nurses (AAOHN) Journal*, **37**, 321–327.
CEDERCREUTZ, G., VIDEMAN, T., TOLA, S. and ASP, S. 1987, Individual risk factors of the back among applicants to a nursing school, *Ergonomics*, **30**, 269–272.
DEHLIN, O., HEDENRUD, B. and HORAL, J. 1976, Back symptoms in nursing aides in a geriatric hospital: an interview study with special reference to the incidence of low-back symptoms, *Scandinavian Journal of Rehabilitation Medicine*, **8**, 47–53.
FRYMOYER, J. W. and ANDERSSON, G. B. J. 1991, Clinical classification., in M. H. Pope, G. B. J. Andersson, J. W. Frymoyer and D. B. Chaffin (eds), *Occupational Low-back Pain: Assessment, Treatment and Prevention* (St Louis: Mosby-Yearbook), 44–70.
FRYMOYER, J. W., POPE, M. H., CLEMENTS, J. H., WILDER, D. G., MACPHERSON, B. and ASHIKAGA, T. 1983, Risk factors in low-back pain: an epidemiological survey, *Journal of Bone and Joint Surgery*, **65-A**, 213–218.
HALDEMAN, S. 1990, Failure of the pathology model to predict back pain, *Spine*, **15**, 718–724.
HARTGERS, C., VAN DEN HOECK, J. A., COUTINHO, R. A. and VAN DER PLIGT, J. 1987, Psychopathology, stress and HIV-risk injecting behaviour among drug users, *British Journal of Addiction*, **87**, 857–865.
HILDEBRANDT, M. V. 1995, Musculoskeletal symptoms and workload in 12 branches of Dutch agriculture, *Ergonomics*, **38**, 2576–2587.
HIRSCH, C., JONSSON, B. and LEWIN, T. 1969, Low-back symptoms in a Swedish female population, *Clinical Orthopaedics*, **63**, 171.
HULT, L. 1954, The Munkfors Investigation, *Acta Orthopaedica Scandinavica: Supplementum*, **1**, 16.
KNIBBE, J. J. and FRIELE, R. D. 1996, Prevalence of back pain and characteristics of the physical workload of community nurses, *Ergonomics*, **39**, 186–198.
LARESE, F. and FIORITO, A. 1994, Musculoskeletal disorders in hospital nurses: a comparison between two hospitals, *Ergonomics*, **37**, 1205–1211.
LEIGHTON, D. J. and REILLY, T. 1995, Epidemiological aspects of back pain: the incidence and prevalence of back pain in nurses compared to the general population, *Journal of Occupational Medicine*, **45**, 263–267.
MAGORA, A. 1970, Investigation of the relation between low-back pain and occupation, *Industrial Medicine*, **39**, 465–471.
MAGORA, A. and TAUSTEIN, I. 1969, An investigation of the problem of sick-leave in the patient suffering from low-back pain, *Industrial Medicine*, **38**, 398.
MAIN, C. and WADDELL, G. 1991, A comparison of cognitive measures in low-back pain: statistical structure and clinical validity at initial assessment, *Pain*, **46**, 287–298.
MAIN, C. J. and WADELL, G. 1984, The detection of psychological abnormality in chronic low-back pain using four simple scales, *Current Concepts in Pain*, **2**, 10–15.
MAROY, L. 1989, Arbeidsongeschiktheid wegens ruglijden bij verplegenden: een schets van het sociale zekerheidsrecht, *Acta Hospitalia*, **2**, 65–75.
MOENS, G. F., DOHOGNE, T., JACQUES, P. and VAN HELSHOECHT, P. 1993, Back pain and its correlates among workers in family care, *Occupational Medicine*, **43**, 78–84.
NAGI, S. Z., RILEY, L. E. and NEWBY L. G. 1973, A social epidemiology of back pain in a general population, *Journal of Chronic Diseases*, **26**, 769.
RIIHIMÄKI, H., TOLA, S., VIDEMAN, T. and HÄNNINEN, K. 1989, Low-back pain and occupation: a cross-sectional questionnaire study of men in machine operating, dynamic physical work, and sedentary work, *Spine*, **14**, 204–209.

Rosenstiel, A. K. and Keefe, F. J. 1983, The use of coping strategies in chronic low-back pain patients: relationship to patient characteristics and current adjustment, *Pain*, **17**, 33–44.

Serratoz-Perez, J. N. and Mendiola-Anda, C. 1993, Musculoskeletal disorders among male sewing machine operators in shoemaking, *Ergonomics*, **36**, 793–800.

Skovron, M. L., Mulvihill, M. N., Sterling, R. C., Nordin, M., Thougas, G., Gallacher, M. and Speedling, J. 1987a, Work organization and low-back pain in nursing personnel, *Ergonomics*, **30**, 359–366.

Skovron, M. L., Nordin, M., Sterling, R. C. and Mulvihill, M. N. 1987b, Patient care and low-back injury in nursing personnel, in S. S. Asfour (ed.), *Trends in Ergonomics/Human Factors 4* (Amsterdam: Elsevier), 855–862.

Stappaerts, K. 1989a, Lage rugpijn: een onderzoek bij verplegenden en richtlijnen voor preventie, in *Hermes* (Leuven: KUL), 7–34.

Stappaerts, K. H. 1988, Lage rugpijn bij verplegenden, *Tijdschrift voor ziekenverpleging*, **42**, 651–655.

Stappaerts, K. H. 1989b, Low-back pain and related sick leave in nurses, *Physiotherapy Practice*, **5**, 193–200.

Stubbs, D. and Buckle, P. 1984, The epidemiology of back pain in nurses, *Medical Education*, **32**, 935–938.

Svensson, H. O. and Andersson, G. B. J. 1982, Low-back pain in forty to forty-seven year old men. 1. Frequency of occurrence and impact on medical services, *Scandinavian Journal of Rehabilitation Medicine*, **14**, 47–53.

Svensson, H. O., Andersson, G. B. J., Johansson, S., Wilhelmsson, C. and Vedin, A. 1988, A retrospective study of low-back pain in 38- to 64-year-old women: frequency of occurrence and impact on medical services, *Spine*, **13**, 548–552.

Törner, M., Zetterberg, C., Andén, U., Hansson, T. and Lindell, V. 1991, Workload and musculo-skeletal problems: a comparison between welders and office clerks (with reference also to fishermen), *Ergonomics*, **34**, 1179–1196.

Turnbull, N., Dornan, J., Fletcher, B. and Wilson, S. 1992, Prevalence of spinal pain among the staff of a district health authority, *Occupational Medicine*, **42**, 143–148.

Valkenburg, H. A. and Haanen, H. C. M. 1982, The epidemiology of low-back pain, in A. A. White and S. L. Gordon (eds), *Symposium on Idiopathic Low-back Pain* (St Louis: Mosby), 9–22.

Videman, T., Nurminen, T., Tola, S., Kuorinka, I., Vanharanta, H. and Troup, J. D. G. 1984, Low-back pain in nurses and some loading factors of work, *Spine*, **9**, 400–404.

Videman, T., Rauhala, H., Asp, S., Lindström, K., Cedercreutz, G., Kämppi, M., Tola, S. and Troup, J. D. G. 1989, Patient-handling skill, back injuries, and back pain: an intervention study in nursing, *Spine*, **14**, 148–156.

Waddell, G., Newton, M., Henderson, I., Somerville, D. and Main, C. J. 1993, A Fear-Avoidance Beliefs Questionnaire (FABQ) and the role of fear-avoidance beliefs in chronic low-back pain and disability, *Pain*, **52**, 157–168.

Zinzen E. 1998, An epidemiological, anthropometrical and body composition study to the prevalence of musculoskeletal inconveniences of the cervical and lumbar spine in hospital nurses. Unpublished PhD thesis (in Dutch), Vrije Universiteit Brussel.